Nonlinear Stochastic Problems

NATO ASI Series

Advanced Science Institutes Series

A series presenting the results of activities sponsored by the NATO Science Committee, which aims at the dissemation of advanced scientific and technological knowledge, with a view to strengthening links between scientific communities.

The series is published by an international board of publishers in conjunction with the NATO Scientific Affairs Division

A	Life Sciences	Plenum Publishing Corporation
B	Physics	London and New York
C	Mathematical and Physical Sciences	D. Reidel Publishing Company Dordrecht, Boston and Lancaster
D	Behavioural and Social Sciences	Martinus Nijhoff Publishers
E	Engineering and Materials Sciences	The Hague, Boston and Lancaster
F	Computer and Systems Sciences	Springer Verlag
G	Ecological Sciences	Heidelberg

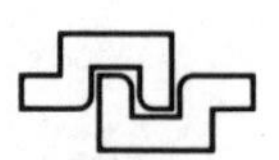

Series C: Mathematical and Physical Sciences No. 104

Nonlinear Stochastic Problems

edited by

Richard S. Bucy

Department of Aerospace Engineering,
University of Southern California, Los Angeles, CA, U.S.A.

and

José M. F. Moura

Department of Electrical Engineering,
Instituto Superior Técnico, Lisbon, Portugal

D. Reidel Publishing Company

Dordrecht / Boston / Lancaster

Published in cooperation with NATO Scientific Affairs Division

Proceedings of the NATO Advanced Study Institute on
Nonlinear Stochastic Problems
Armaçao de Pera, Algarve, Portugal
May 16-28, 1982

Library of Congress Cataloging in Publication Data

NATO Advanced Study Institute on Nonlinear Stochastic Problems With Emphasis on Identification, Signal Processing, Control and Nonlinear Filtering (1982 : Algarve, Portugal)
Nonlinear stochastic problems.

(NATO ASI series. Series C, Mathematical and physical sciences ; v. 104)
"Published in cooperation with NATO Scientific Affairs Division."
Includes index.
1. Stochastic systems—Congresses. 2. Nonlinear theories—Congresses.
I. Bucy, Richard S., 1935- . II. Moura, José M. F., 1946- .
III. North Atlantic Treaty Organization. Division of Scientific Affairs.
IV. Title. V. Series.
QA402.N335 1982 003 83-8688
ISBN 90-277-1590-4

Published by D. Reidel Publishing Company
P.O. Box 17, 3300 AA Dordrecht, Holland

Sold and distributed in the U.S.A. and Canada
by Kluwer Boston Inc.,
190 Old Derby Street, Hingham, MA 02043, U.S.A.

In all other countries, sold and distributed
by Kluwer Academic Publishers Group,
P.O. Box 322, 3300 AH Dordrecht, Holland

D. Reidel Publishing Company is a member of the Kluwer Academic Publishers Group

Printed in The Netherlands

TABLE OF CONTENTS

PREFACE

This volume corresponds to the invited lectures and advanced research papers presented at the NATO Advanced Study Institute on **Nonlinear Stochastic Problems with emphasis on Identification, Signal Processing, Control and Nonlinear Filtering** held in Algarve (Portugal), on May 1982. The book is a blend of theoretical issues, algorithmic implementation aspects, and application examples.

In many areas of science and engineering, there are problems which are intrinsically nonlinear and stochastic in nature. Clear examples arise in identification and modeling, signal processing, nonlinear filtering, stochastic and adaptive control. The meeting was organized because it was felt that there is a need for discussion of the methods and philosophy underlying these different areas, and in order to communicate those approaches that have proven to be effective.

As the computational technology progresses, more general approaches to a number of problems which have been treated previously by linearization and perturbation methods become feasible and rewarding. Because nonlinearity is a characterization by default, there is no unifying framework encompassing all nonlinear problems. The meeting addressed topic areas like identification of Markov processes from observations, modeling of nonlinear stochastic systems, stochastic integrals with respect to martingales and stochastic differential calculus, numerical synthesis of optimal systems and error performance on nonlinear systems via information theory methods, adaptive control, and optimal nonlinear filtering. Several applications on these different fields were reported, providing a bridge between the theory, algorithms, and real world problems. A general conclusion was perceived from the discussions held. It concerns the necessity of further work on simple, well defined examples that can be put in the perspective of different methods and approaches.

Chapter I briefly overviews the contents of the different contributions. Its Appendix gives a summary of each of the lectures and papers contributed, organizing them by topic areas. The remaining chapters of the book include all but three of the papers presented at the ASI. Except for the tutorial on manifolds, by Byrnes, the material for the other tutorials is not included in this volume. Under the general theme of **Nonlinear Stochastic Probems**, the book gives a hint to many different problems and areas. Through the work of the many authors included, it provides a guided entrance to the research being conducted today on these fields. Because of this, and aside the intrinsic value of each contribution, it is felt that the book will be of worth to graduate students, research workers in applied mathematics, and research engineers, working on the fields mentioned above.

ACKNOWLEDGEMENTS

The meeting was made possible by a grant from the Advanced Study Institute (ASI) program of NATO Scientific Affairs Division (Brussels, Belgium). Dr. Di Lullo and Dr. Graig Sinclair, administrators of the program, provided all required assistance.

A second grant was awarded by Instituto Nacional de Investigação Científica (INIC-Portugal). Dra. Luisa Pinto, the manager of the grants' division was ready to help whenever it proved necessary.

Further support was provided by Correios Telegrafos e Telefones de Lisboa e Porto (CTT/TLP), and by Instituto de Engenharia de Sistemas e Computadores (INESC).

For giveaways and logistic help, the following organizations are also acknowledged: TAP Air Portugal, Madeira Wine Association Limitada, Banco Totta & Açores, Cooperativa Agricola de Reguengos de Monsaraz, Comissão Regional de Turismo do Algarve, Offley Forrester Limitada, INESC.

One institution supported unconditionally and in various ways the whole organization of the ASI: Centro de Análise e Processamento de Sinais (CAPS). It would be meaningless to list the different instances where this support took place.

Numerous people directly helped with the preparation of the meeting, and the organization of the preprints and of the final proceedings. Orlando Serrenho designed the stationary used for the ASI correspondence. Maria Odete Subtil conceived and painted the motive used in the poster and leaflets announcing the meeting. José Leitão and João Lemos helped during the meeting with organizational details. Maria Manuela Veloso was of constant assistance throughout the preparation and during the meeting. Emília Martinho did all the secretarial work concerned with the ASI. Her expertise almost made it seem an easy task to organize the ASI. Carlos Belo and Maria Isabel Ribeiro, who also integrated the organizing committee, were tireless from the begining till the very last moment. They generously provided their time and effort. Without them it would have been unconceivable to have embarked on such a venture.

Finally, it is a pleasure to acknowledge the active participation of all the attendants of the ASI. Their knowledge and enthousiasm made it a rewarding experience to organize the Advanced Study Institute on Nonlinear Stochastic Problems.

Richard S. Bucy
Univ. Southern Calif.

José M. F. Moura
Inst. Superior Tecnico

CHAPTER I

INTRODUCTION

R. S. BUCY, JOSÉ M. F. MOURA
Overview of the ASI on Nonlinear Stochastic Problems
Appendix

OVERVIEW OF THE ASI ON NONLINEAR STOCHASTIC PROBLEMS

R. S. Bucy

Univ. South. California
Los Angeles Calif. USA

José M. F. Moura

Instituto Superior Técnico
P-1096 Lisboa PORTUGAL

The present book includes contributions presented at the Advanced Study Institute (ASI) on Nonlinear Stochastic Problems held in Algarve, Portugal, May 16-28, 1982.

The ASI span the fields of realization (spectral estimation, model identification, systems realization theory), control (adaptive, stochastic, optimal), new technologies and signal processing, nonlinear filtering (innovations approach, information and filtering, geometric approach, conections with quantum mechanics, implementation issues), stochastic processes (qualitative theory for stochastic systems on manifolds, stochastic differential equations, stochastic calculus, functionals of Brownian motion, etc.). During the ASI,besides theoretical aspects, also methods, implementation issues, and applications were discussed.

The first week of the ASI concentrated on realization, control applications and technological issues. The second week focused on stochastic processes theory and nonlinear filtering. The ASI was organized in a series of 23 lectures of 1-1/2 hour long, plus 5 of 45 min. each, complemented by the presentation of advanced research papers. Also, background material on certain topics was provided by six tutorials (45 min. each). Finally, three round table discussions of 1-1/2 hour duration took place on the general themes of applications, spectral

R. S. Bucy and J. M. F. Moura (eds.), Nonlinear Stochastic Problems, 3–5.

estimation and identification, nonlinear filtering and stochastic control.

Without pretending to describe the lectures and the advanced research papers, the Appendix, organized by topic areas, briefly overviews them. Its inclusion aims at providing the reader with a quick means for grasping the topics dealt with during the ASI.

A more detailed look at the ASI contents shows that the lectures of Burg and Haykin concentrate on spectral estimation, being complemented by the paper of Wagstaff and Berrou. Alengrin addresses fast algorithms in identification. The background for these topics were covered at the ASI by two tutorials given by Burg and Alengrin, respectively. The papers by Y-C Chen, Sarkar, and Teles discuss topics related to identification and modeling.

Lindquist and Picci, and Ruckebush addressed the topic of realization system theory. C. Byrnes gave a tutorial on calculus on manifolds, vector fields, etc. He also presented a contribution on geometric aspects of the convergence analysis of identification algorithms. The control area was addressed in the adaptive aspects by E. Mosca's lectures, and then by the papers of Byrnes, Mosca and Zappa, Yaz and Istefanopulos. Mosca gave a tutorial on the fundamentals of adaptive control. Stochastic control was the topic of one of L. Shepp's lectures. The second lecture was on positron emission reconstruction: the E-M algorithm. Algorithmic aspects in optimal control was the theme of Carro Incertis paper. Other aspects of optimal control were addressed by Helmes. The area of nonlinear filtering was the object of several tutorials and lectures, as well as of research papers. Bucy provided a tutorial covering Bucy's Representation Theorem, the Stratonovich-Kushner partial differential equation of nonlinear filtering, and the discrete nonlinear filtering theory. He presented at the ASI, two movies showing the evolution of nonlinear filters. Di Masi gave a tutorial on robust nonlinear filering. Mitter gave the third tutorial on this area, addressing the geometric approach to nonlinear filtering and stochastic control, exploring relations with Quantum Physics, the existence of sufficient statistics, finite dimensional nonlinear filters, etc. The lectures of Bucy were on relations between nonlinear filtering and information theory with particular emphasis on bounds evaluation. Those of Mitter addressed the topic of approximations in nonlinear filtering. Lo's lectures covered expansion methods to build nonlinear filters and the ones of Moura studied numerical implementation issues in nonlinear filtering when applied to phase modulation problems. Korezlioglu's paper discussed nonlinear filtering for Hilbert space valued processes. Di Masi's contribution presented generalized finite - dimensional filters in discrete time. Nithila's paper studied Volterra series and finite dimensional nonlinear filtering.

The qualitative theory of stochastic processes was the topic of Kliemann's talk. Protter discussed stochastic differential equations. Allinger spoke on the innovations approach, Hijab studied finite dimensional causal functionals of Brownian motion, Ustunel addressed several applications of infinite - dimensional stochastic calculus. Enchev discussed a stochastic differential rule, Maccone the topic of nonlinear time dependent Brownian motion. De La Rubia studied the drastic change of behavior on the structure of the solutions of certain nonlinear differential equations when driven by colored noise.

Dieulesaint lectured on new technologies and fast signal processing. He spoke on several acoustic surface wave devices. On the applications area, Brammer spoke on multiradar tracking networks, and on how engineering still uses, besides sounding theory, a lot of insight to design working algorithms in a real world of constrained computational power. Baram talked on reconstruction of two dimensional fields from sampled data in topographical applications, Barkana presented an application of design of perturbation signals to enhance resolution of nonlinear position detection schemes, Braumann addressed the field of probabilistic models in population dynamics, Grimble spoke on the design of real application algorithms to incorporate in ship positioning control, Graef talked on joint optimization of tramsmitter and receiver for cyclostationary signal processes, and Nardone spoke on passive underwater acoustics location systems.

APPENDIX

SHORT SYNOPSIS OF THE PAPERS

Papers signalled with a * refer to invited talks.

Spectral Estimation and High Resolution Techniques

J. P. Burg*: Estimation of Structured Covariance Matrices: A Generalization of the Burg Technique.

Covariance matrices for stationary time series. Block Toeplitz structure. Maximum likelihood covariance matrix for a zero mean multivariable Gaussian random process: basic equations. Analytic solutions. The Burg technique as the maximum likelihood solution for the problem of two random variables with equal variances. A fixed point algorithm that actualy solves the equations for moderated sized problems with reasonable computational ease. Convergence properties. Examples.

S. Haykin: Classification of Radar Clutter Using the Maximum-Entropy Method.

Development of an autoregressive model for radar clutter. The maximum-entropy spectral estimator (MESE). The multistage lattice (prediction error) filter, and the related harmonic-mean (Burg's) algorithm. Classification of radar clutter using the MESE for the detection of a moving target (e.g. aircraft) in the combined presence of a receiver noise and clutter.

R. A. Wagstaff & J. L. Berrou: A Simple Approach to High-Resolution Spectral Analysis.

A nonlinear high resolution technique for spectral analysis operating on the systems outputs rather than the imput sampled data. Application of the technique to obtain the undersea ambient noise spatial

R. S. Bucy and J. M. F. Moura (eds.), Nonlinear Stochastic Problems, 7–18.

spectrum from towed line array data and the frequency spectrum from time history data.

Identification of Parameters in Models

G. Alengrin*: Estimation of Stochastic Parameters for ARMA Models by Fast Filtering Algorithms.

Multivariable ARMA models. State space description. A review of the stochastic realization problem. Fast algorithms for estimation of stochastic parameters of ARMA models. A characterization of different parametrizations for the stochastic parameters.

Y-C. Chen: Convergence Study of Two Real-Time Parameter Estimation Schemes for nonlinear systems.

Comparison of algorithms based on the gradient method and on the least-squares method, when applied to the real-time parameter estimation of nonlinear systems. The systems dynamical equations are nonlinear. Through the study of this convergence behavior, especially for the problem of slowly varying systems, a more effective strategy results for choosing the gain factor.

T. K. Sarkar, D. K. Weiner, V. K. Jain, S. A. Dianat: Approximation by a Sum of Complex Exponentials Utilizing the Pencil of Function Method.

Approximation of a function by a sum of complex exponentials. The resulting nonlinear problem is linearized through the pencil - of - function method. The insensibility of the method to noise and its stability are illustrated through examples.

M. L. S. Teles: Minimax Estimation of Arma Systems.

A new method for the identification of parameters in models. The technique is based, not on least squares theory, but on a decision theoretical approach. Given a certain criterion, it uses Game theory to obtain the conditions for the existence of minimax decision rule. Simulation studies of parameter estimation for ARMA models excited by Gaussian and non Gaussian noise.

Systems Realization Theory

A. Lindquist, G. Picci: Stochastic Realization of Stationary Increment Processes.

State space representation of stationary increment processes. Nontrivial problems of probabilistic nature arise, notably related to semimartingale representation, requiring the use of the theory of conditional shift semigroups. Existence of forward and backward conditional derivatives. All this makes it possible to apply the stochastic realization theory for stationary processes developed in recent years both for linear and nonlinear Markovian representations. (This lecture was given at the ASI, but the paper is not included in the proceedings).

G. Ruckebusch: On the Structure of Minimal Markovian Representations.

Survey of the highlights of the (geometric) theory of Markovian representation as develloped recently by the work of several authors. Infinite dimensional representations. Markovian representations for strictly noncyclic (Gaussian) processes. Hardy space theory. A new concept of dimensional which is meaningful for both finite and infinite dimension Markovian representations is defined as some scalar inner function. Relations with the concept of dimension as introduced by Fuhrmann in his Hilbert space realization theory.

Nonlinear Systems Theory

C. I. Byrnes*: A Brief Tutorial on Calculus on Manifolds, with emphasis on Applications to Identification and Control.

Fundamental definitions of differentiable manifold functions, and vector fields, along with examples taken from identification, filtering and realization theory. Calculus of real valued functions on a manifold leading to the Morse theory. Examples in identification and in the stability analysis of a power system.

F. J. de la Rubia, J. G. Sanz, M. G. Velarde: Role of Multiplicative Non-White Noise in a Nonlinear Surface Reaction.

Nonlinear deterministic systems exhibiting multiple steady states on limit cycle oscillations can be drastically altered by an external noise. The paper studies the case of sustained oscillations induced by the noise, when no limit cycle is predicted from the deterministic equations. The model is relevant to surface catalytic reactions. It illustrates the influence of colored noise upon a simple nonlinear dissipative system.

Adaptive Control

C. I. Byrnes: Geometric Aspects of the Convergence Analysis of Identification Algorithms.

The relationship between the topology of a parameter set of candidate systems and the convergence analysis of recursive identification algorithms on this parameter set is studied, using the "ODE" approach of Ljung, together with classical differential-topological methods for the qualitative analysis of differential equations. The analysis suggests that the defining conditions for the candidate systems in a globally convergence identification scheme have the effect of forcing the parameter space$_*$X of candidates to be Euclidean.

E. Mosca : Multivariable Adaptive Regulators Based on Multistep Quadratic Cost Functionals.

Innovations representations (state space and ARMAX models). LQG regulation theory. Ljung's theory on convergence analysis. Convergence analysis of adaptive regulators. Single - step - ahead versus multi - step - ahead adaptive control. A basic multi - step - ahead regulation problem and its optimal solution. Implementational algorithm for unknown multivariable plants. Applications to pathological cases: Convergence analysis for multi - step - ahead minimum variance adaptive regulators.

E. Mosca, C. Zappa: Overparametrization, Positive Realness and Multistep Minimum Variance Adaptive Regulators.

Application of the "ODE" method of Ljung to show that even when the plant does not satisfy a positive realness condition, local convergence can be achieved by extending the time horizon of the controller. The algorithm behaves like an overparametrized LS+MV regulator. The MUSMAR approach has the advantage of requiring less parameters in the feedback gain vector. Results on simulation experiments.

E. Yaz, Y. Istefanopolus: Adaptive Receding Horizon Controllers for Discrete Stochastic Systems.

Devellopment of a control scheme which separates into an adaptive estimator and a certainty equivalent controller. The estimator (adaptive state estimator of Ljung) filters simultaneously the states and identifies the parameters of the system. The controller is designed via the receding horizon concept of Thomas and Barraud. It uses the state and parameter estimates as if they were the true values.

Stochastic Control

L. A. Shepp*: An Explicit Solution to a Problem in Nonlinear Stochastic Control Involving the Wiener Process Optimal Stopping.

Principle of smooth fit. Explicit solution to several new problems in stochastic control, among them the "finite fuel follower". Most problems of the above type are solved by first writing the Bellman equation and then applying numerical techniques. Thanks to the special properties of the problem studied (homogeneity of the Wiener process, the exponential discounting, the squared error criterion) we give the optimal tracking control explicitly.

Algorithms in Control

F. C. Incertis: Optimal Stochastic Control of Linear Systems with State and Control Dependent Noise: Efficient Computational Algorithms.

Efficient algorithm solution of the state - and - control dependent noise linear quadratic optimal control problem. The algorithms are particularly useful when the system matrix is diagonal - dominant (e.g. as is the case when dealing with many spatially quantized distributed parameter systems). Previous results on the computation of the square-root of a matrix and on the solution of a deterministic algebraic Riccati equation are generalized to the stochastic case. The algorithm implementation requires as basic operations only additions, products and inversions in the field of positive definite matrices.

Nonlinear Filtering

R. S. Bucy*: Joint Information and Demodulation.

Optimal Filtering viewed as mutual information optimization. Phase Lock Loop (PLL) Information loss. Rate distortion and filter error. Causal realizability and error bounding. Comparison of joint informations for the nonlinear filter and for the PLL. PLL discards out of phase measurements, suffering consequent loss of information and error performance.

G. B. Di Masi, W. J. Rungaldier, B. Barozzi: Generalized Finite-Dimensional Filters in Discrete Time.

Bayesian approach to nonlinear filtering problems in discrete-time. Generalized finite - dimensional filters. Existence. Recursive Implementation. Application to various situations.

H. Korezlioglu: Nonlinear Filtering Equations for Hilbert Space Valued Processes.

The integration of operator valued processes with respect to a Hilbert Space valued Brownian motion is shown to be equivalent to the integration of Hilbert-Schmidt operator valued processes with respect to a cylindrical Brownian motion. As a consequence, a more general filtering equation than those already known is obtained.

J. T-H. Lo, S-K. Ng: Optimal Orthogonal Expansion for Estimation I: Signal in White Gaussian Noise.

Expansion methods in nonlinear filtering. Systematic development of approximate recursive estimator which is optimal for the given structure and approaches the best estimate, when the order of the approximation increases. Projection of the minimal variance estimate into the Hilbert subspace of all Fourier - Hermite (FH) series, driven by the observations, with the same given index set. A system of linear algebraic equations for the FH coefficients results. The estimator consists of finitely many Wiener integrals of the observations and a memoryless nonlinear postprocessor. The postprocessor is an arithmetic combination of the Hermite polynomials.

J. T-H. Lo, S-K. Ng: Optimal Orthogonal Expansion for Estimation II: Signal in Counting Observations.

The approach of the previous paper is used to construct approximate estimators for the signal that modulates the rate of a counting process. A product - to - sum formula for the multiple Wiener integrals, with respect to the counting process martingale, is derived.The minimal variance estimate is projected onto the Hilbert subspace of all Fourier - Charlier (FC) series, driven by the counting observations, with the same given index set. A system of linear algebraic equations for the FC coefficients results. The estimator consists of finitely many Wiener integrals of the counting observations and a nonlinear postprocessor. The nonlinear postprocessor, however, is not memoryless.

S. K. Mitter: Approximations for Nonlinear Filtering.

Approximation schemes for optimal nonlinear filters. Analysis of the Stratonovich form of the Zakai equation for the unnormalized conditional density. Stochastic control interpretation of the Zakai equation: duality between estimation and control. Solution of the Zakai equation as a nonlinear least-squares problem. Impact of approximation schemes solving stochastic control problems on the solution of the Zakai equation. Conversely, perfectly observable, stochastic control problems can be converted to the solution of a linear parabolic equation. Insight on the extended Kalman - Bucy filter. Filtering with small process noise.

J. M. F. Moura: Phase Demodulation: A Nonlinear Filtering Approach.

Classes of phase modulation problems: cyclic and absolute. Long time or tracking behavior of phase demodulators. Short time or acquisition ability. Fokker-Planck theory of phase locked loops. Operator type solution. Continuous and discrete optimal nonlinear filtering. Phase demodulators as optimal nonlinear filters. Implementation and performance in the context of absolute phase tracking and acquisition.

M. T. Nithila: Volterra Series and Finite Dimensional Nonlinear Filtering.

Filtering of a class of nonlinear differential systems described by two subsystems connected with a polynomial link map. The first subsystem is linear, while the second is given by a finite Volterra series driven by polynomials of the state of the preceeding system. A finite dimensional representation of the optimal recursive filter for the given system structure. The solution relates to one of Markus and Willsky. It uses realization results on finite Volterra series of P. Crouch.

Stochastic Processes

D. F. Allinger: Causal Invertibility: An Approach to the Innovations Problem.

The innovations problem for stochastic differential equations (sde) is studied by means of a method called causal invertibility. Causal invertibility of transformations F and strong solutions of sde's. If and only if conditions for causal invertibility of a transformation F. Examples. Applications to sdes. Extensions.

R. Dominicis: Spectral Analysis of Nonlinear Semi-Markov Processes.

Spectral analysis of a time series associated to a homogeneous semi-Markov process having a finite number of states. A statistical test for verifying the semi-Markovian hypothesis. Determination of the asymptotic power.

O. Enchev: Differential Calculus for Gaussian Random Measures.

Link between stochastic integrals and notions from Quantum theory. Second quantization and canonical free fields. A rule for differential calculus.

O. Hijab: Finite Dimensional Causal Functionals of Brownian Motion.

Construction of functionals via ordinary differential equations. Finite dimensional functionals. Rank of a functional. The paper proves that a finite dimensional functional is smooth and of finite rank.

W. Kliemann: Transience, Recurrence and Invariant Measures for Diffusions.

Qualitative behavior of nonlinear differential systems dx(t)/dt=f(x(t),w(t)) where w(t) is a nondegenerate (in some sense) Markov process. If w is a diffusion process, the pair (x,w) is a diffusion, but everywhere degenerate. Connections between stochastic systems and associated deterministic control systems. Generalized concepts of weak and strong transience and recurrence, as well as existence and uniqueness of invariant measures, hence existence of stationary, ergodic solutions, are characterizable via this connection. Applications of the qualitative theory to stability theory of linear systems, stationary solutions and long term behavior of models in the natural sciences. Weak recurrence and controllability concepts for stochastic systems. Results on linear systems with Markovian noise.

C. Maccone: Eigenfunction Expansion for the Nonlinear Time Dependent Brownian Motion.

Nonlinear time dependent Brownian motions B(g(t)) are considered. It is proved that B(g(t)) is equivalent to a white-noise integral having zero mean and a time-dependent variance.An eigenfunction (Karhunen - Loeve) expansion is found. The eigenfunctions are Bessel functions of the first kind, with a suitable time function for the argument, multiplied by another time function. The eigenvalues are virtually determined by the zeroes of the Bessel function.

P. Protter: Stochastic Differential Equations with Feedback in the Differentials.

The theory of stochastic differential equations customarily assumes one is a priori given known coeffcients and known driving terms (e.g., white noise, Poisson white noise, etc.). Motivated by simple models in microeconomics, stochastic differential equations are investigated where one of the driving terms is not a priori given but evolves in a fashion depending on the paths of the solution. The paper shows that establishing the existence of solutions to equations of this type is equivalent to showing the existence of a solution of a "classical" equation. The difficulty arises in that the coefficients of this equation are not Lipschtzian, hence do not fall into the existing theory. The paper studies the existence and uniqueness of solutions.

K. Seitz: Transformation Properties of Stochastic Differential Equations.

Algebraic structure of a class of transformations generalyzing the Lyapounov transformation of linear time variant deterministic systems.

A. S. Ustunel: Some Applications of Stochastic Calculus on the Nuclear Spaces to the Nonlinear Stochastic Problems.

Applications of the infinite dimensional stochastic calculus to the following problems: Construction of the C infin.-flows of the finite dimensional diffusions with regular coefficients, a simple and short proof of the generalized Ito - Stratonovich formula, probabilistic solutions of the heat and Schrodinger's equations in $D'(R^d)$, diffusions depending on a parameter and the reduction of the unnormalized density equation of the filtering of nonlinear diffusions into a deterministic Cauchy problem depending on a parameter.

Technological Aspects and Signal Processing.

E. Dieulesaint*: Elastique waves Applied to Signal Processing.

Functions which can be performed with elastic waves propagating in solids: Delay, Pulse Compression, Filtering, Convolution, Spectral Analysis, Acoustic Wave Devices. (Lecture given at the ASI, but paper not included in the proceedings).

Applications

Y. Baram, M. Margalit: Reconstruction and Compression of Two Dimensional Fields from Sampled Data by Pseudo - Potential Functions.

Reconstruction of two dimensional fields from sampled data for purposes of smoothing, interpolation and compression. The underlying model is a finite sum of pseudo - potential elements. The local behavior of the data, along with the structure of the pseudo -

potential functions are used to convert the nonlinear parameter estimation problem into linear subproblems. Solution by standard techniques. Illustrative example: toppographical mapping from radar altimetry data.

A. Barkhana: Optimum Perturbation Signal That Makes Nonlinear Systems Approach to Linear Systems.

Optimum design of perturbation signals used in the study of nonlinear systems via linearizing techniques. The nonlinearity is described by a fifth order homogeneous polynomial. The physical nonlinear system has a built in low pass filter. The perturbation is periodic, continuous, well approximated by the first three terms of a Fourier sine series. Example with application of a gradient algorithm.

K. G. Brammer*: Stochastic Filtering Problems in Multiradar Tracking.

Linear and nonlinear aspects of the filter algorithms used for tracking of aircraft trajectories on the basis of measurements made by surveillance - type radars (track - while - scan). Review of the mathematical models for aircraft motion and radar measurements. Possible simplifications. Large number of tracked aircraft and limited computing facility: need for effective suboptimal filter algorithms. Multiradar networks with overlapping coverage, where the plot rate per track is proportional to the number of sensors seeing the same aircraft. Discussion of four methods. The mosaic method, one of its variants, a more sophisticated approach that uses all available plots from every radar seeing the aircraft, and a fourth method based on continuous monoradar tracking for each radar. Association, track discontinuities at switching points, smoothing. Although the original measurement model is nonlinear with a matrix which is state dependent, and the maneuver distribution is nonGaussian, the engineering approach employing reformulations, heuristic arguments and a series of simplifications results in a class of computationally austere linear tracking filters with good performances in many applications.

C. A. Braumann: Population extinction Probabilities and Methods of Estimation for Population Stochastic Differential Equation Models.

Population growth models are discussed (Malthusian, logistic, Gompertz). There are available methods of estimation, testing and prediction based on discrete time observations. Estimation of population extinction probabilities. By studying the effects on these probabilities, judgements of practical use can be made, eg. how "bad" is a fishing policy or what is its environmental impact.

P. T. K. Fung, Y. L. Chen, M. J. Grimble: Dynamic Ship Positioning Control Systems Design Including Nonlinear Thrusters and Dynamics.

Dynamic ship positioning control (eg. oil-rig drill ships, support vessels and survey vessels) include filters to remove the wave motion signal. A combined self-tuning filter and Kalman filter scheme is proposed. An optimal state estimate feedback control scheme is described. The ship dynamic equations are nonlinear. Significant nonlinearities are present in the thrust devices. Simulation results are presented. The self tuning filter maintains optimum performance in different weather conditions and compensates modelling errors introduced by the nonlinearities.

F. K. Graef: Joint Optimization of Transmitter and Receiver for Cyclostationary Random Signal Processes.

The paper deals with the joint optimization of transmitter and receiver in linear modulation systems. The channel introduces linear distortion and additive noise. Optimization is performed with respect to some mean square error criterion. The study addresses the problem of periodically varying filters and cyclostationary source signals (stochastic processes whose first and second order moments fluctuate periodically). A frequency domain characterization is used by introducing operator - valued spectral densities and transfer functions. The time averaged mean square transmission error is derived as a nonlinear functional of the spectral characteristics of the system. A specific example considers wide sense stationary source signals with arbitrary power spectra.

S. C. Nardonne, A. G. Lindgren, K. F. Gong: Fundamental Properties and Performance of Nonlinear Estimators for Bearings-Only Target Tracking.

The problem of estimating the position and velocity of an unaccelerated target from noise corrupted bearing measurements obtained by a single moving observation platform. The process is nonlinear and exhibits unusual observability characteristics that are geometry dependent. Attention is focused on the large - to - baseline geometries employing a symmetric observer maneuver strategy. Maximum likelihood estimation techniques are applied. Simulation results presented for the long range situation. Analytical results for the Cramer - Rao bound derived.

CHAPTER II

SPECTRAL ESTIMATION

ESTIMATION OF STRUCTURED COVARIANCE MATRICES
A GENERALIZATION OF THE BURG TECHNIQUE

John Parker Burg

Time & Space Processing, Inc., Santa Clara, CA 95051

ABSTRACT

Covariance matrices from stationary time series are Toeplitz. In this case and many other situations, one knows that the actual covariance matrix belongs to a particular subclass of covariance matrices. This paper discusses a rather straight-forward but computationally involved theory for estimating a covariance matrix of specified structure from vector samples of the random process. The theoretical foundation of the method is to assume that the random process is zero mean multivariate gaussian, and to find the maximum likelihood covariance matrix that has the specified structure. A major contribution of this paper is an ingenious and appealing fixed point algorithm that allows us to actually solve the equations for moderate sized problems with reasonable computational ease.

I. INTRODUCTION

In doing spectral analysis of a stationary time series, one modern approach is to use the "Burg technique" to estimate second order statistics from the raw time series data and then to use the maximum entropy method to generate the estimate of the power density spectrum. These two steps are independent in that one can use the Burg technique to estimate the autocorrelation function out to lag N and then use a conventional fourier transformation with a window function to get the spectral estimate, or, one can use the conventional lag product method of estimating the autocorrelation function followed by use of the maximum entropy method of spectral estimation. The Burg technique and

R. S. Bucy and J. M. F. Moura (eds.), Nonlinear Stochastic Problems, 21–39.

the maximum entropy method solve two separate but related problems.

The maximum entropy method of spectral estimation can be considered to be a generalization of the autoregressive method of spectral estimation. That is, if the second order statistics that are known about the spectrum consist of the first N+1 lags of the autocorrelation function and if the entropy of the time series is given by the integral of the logarithm of the spectrum, then the maximum entropy estimation procedure generates an Nth order all pole model as the functional form satisfying the variational extremum. However, the maximum entropy principle also tells us precisely how to do multi-channel and multi-dimensional spectral estimation using correlation information about the spectrum. Actually, in the much broader sense, the maximum entropy principle supplies us with a general approach to estimation theory in which one combines information with an extremum principle to select a possible solution to a problem.

In this paper, we shall be concerned with developing a similar generalization of the Burg technique. The approach again uses a variational principle combined with information to produce a feasible solution. The particular problem that we attack is simply stated. Given a set of vector samples from a random process, we wish to select a covariance matrix of specified structure that corresponds in a reasonable way to the given data. The solution formulation is to assume that the random process is zero mean multi-variate gaussian, that the vector samples are independent and to take as our solution the covariance matrix of specified structure that maximizes the probability of occurance of our vector samples. As we shall see, it is easy to write down the probability of the vector samples given that the covariance matrix is R. In fact, the information in the vector samples is neatly compressed into the sample covariance matrix S, so we just end up with a function, p(S,R), in the two matrices, S and R. R is constrained to be a covariance matrix of the proper structure while S is a random sample covariance matrix without any special structure.

If we momentarily disregard statistical considerations, the p(S,R) function gives us the desired variational formulation for the problem. That is, given the vector samples, we calculate the sample covariance matrix S and then solve for the constrained covariance matrix R that maximizes p(S,R). Aside from the many technical questions that one might ask, there is the subjective question, namely, why is the R that maximizes this function a "good" solution to the problem? The response to this question is that p(S,R) really comes from maximum likelihood considerations and thus should, in some sense, give us a reasonable answer, even if the process is not gaussian and the vector samples are not independent. After all, the process might be gaussian and the samples independent. In the final analysis, however, the p(S,R) variational principle will survive only if it works well in practice. And a practical principle must meet two criteria. It must work well on a large majority of meaningful situations and it must not be too difficult to compute. We will try to show that the p(S,R) principle has the first of these attributes to a high degree and that the fixed point algorithm presented herein helps greatly to reduce the problem of numerical computation.

II. DERIVATION OF THE VARIATIONAL PRINCIPLE

Suppose a column vector x is drawn from an N-dimensional Gaussian distribution with zero mean and covariance matrix R. Using a superscript T for the matrix transpose, the corresponding probability density function is

$$p(x) = (2\pi)^{-N/2} |R|^{-1/2} \exp(-x^T R^{-1} x/2). \tag{1}$$

Now, instead of a single vector sample, suppose that we have M independent vector samples, x_m, m = 1 to M. The probability density for this set of vectors follows from (1) as

$$p(x_m, m = 1 \text{ to } M) = (2\pi)^{-MN/2} |R|^{-M/2} \exp(-\sum_{m=1}^{M} x_m^T R^{-1} x_m/2). \tag{2}$$

We consider the situation where R is unknown except that it is a member of a certain family $\mathcal{R}$ of feasible covariances. This family is determined by the structure of the underlying source of the data vectors. For example, an important case is where $\mathcal{R}$ is the collection of all positive definite symmetric Toeplitz matrices, corresponding to a vector sample being N consecutive values from a sampled stationary time series.

Given the set of vector samples, x_m, m=1 to M, the R that belongs to $\mathcal{R}$ and which maximizes expression (2) is the "maximum likelihood" estimate of the covariance matrix. Since we are using (2) only as a function to be maximized, we do not change the problem if we maximize a strictly monotonic function of (2), for example, the natural logarithm of (2). Thus, taking the logarithm of (2), we get

$$-(MN/2) \log(2\pi) - (M/2) \log|R| - (1/2) \sum_{m=1}^{M} x_m^T R^{-1} x_m.$$

Dropping the leading constant term and dividing through by M/2, we define our objective function g(S,R) to be

$$g(S,R) = -\log|R| - (1/M) \sum_{m=1}^{M} x_m^T R^{-1} x_m. \tag{3}$$

Maximizing g(S,R) is clearly equivalent to maximizing expression (2). To simplify (3), we first state and then use the following matrix theorem.

The trace of a square matrix is defined to be the sum of the elements along the main diagonal of the matrix. Now if A is a r by s matrix and B is a s by r matrix, then both AB and BA are square matrices and thus their traces are defined. The matrix theorem is that their traces are equal, even if they are different sized matrices. We now note that because it is a scalar, i.e., a one by one matrix,

$$x^T R^{-1} x = \text{trace}(x^T R^{-1} x) = \text{trace}(A\ B),$$

where, $A = x^T R^{-1}$, a one by N matrix and B = x, a N by one matrix.

Our matrix theorem says that

$$x^T R^{-1} x = \text{trace}(A\ B) = \text{trace}(B\ A) \equiv \text{trace}(R^{-1} x\, x^T).$$

Note that $R^{-1} x\, x^T$ is a N by N matrix.

Using this result, (3) can be written as

$$g(S,R) = -\log|R| - \text{trace}(R^{-1} (1/M) \sum_{m=1}^{M} x_m x_m^T).$$

Defining the sample covariance matrix S to be

$$S = (1/M) \sum_{m=1}^{M} x_m x_m^T,$$

we arrive at the compact equation for g(S,R) of

$$g(S,R) = -\log|R| - \text{trace}(R^{-1} S). \tag{4}$$

This is our basic objective function. We wish to find the R that maximizes this function, given the sample covariance matrix S and given that R is constrained to have a particular structure.

III. THE NECESSARY CONDITIONS

The problem is to maximize g(S,R) over the matrices R belonging to a class $\mathcal{R}$. We shall assume that the class $\mathcal{R}$ is defined by a linear variety and is a subset of the class of symmetric positive definite matrices. A good example is the subset of Toeplitz matrices. With these assumptions, it is easy to characterize the solution of the problem in terms of the gradient of the objective function. Specifically, the gradient must be orthogonal to variations in $\mathcal{R}$. Geometrically, that is all there is to it. Of course, to make this more concrete, it is necessary to define an inner product on the space of matrices so that the notions of gradient and orthogonality have specific meanings. We shall define the inner product of two matrices C and D to be given by the trace of C transposed times D. Note that the inner product is symmetric, bilinear, and that the inner product of a non-zero matrix with itself is positive.

With this definition, we now need to find the gradient of g(S,R) with respect to R. We shall do this by deriving the variation of g in terms of the variation of R. We note that if R is a N by N symmetric matrix, we may have up to N(N-1)/2 independent variables! One is thus faced with the thought of having to deal with very large matrices for moderate sized problems. Fortunately, the large number of equations can still be treated in terms of just N by N matrices.

To derive the necessary conditions, we begin with some definitions and two matrix theorems. First, we define the variation of R to be

$$\delta R = \begin{bmatrix} \delta R(1,1) & \delta R(1,2) & - - - - - & \delta R(1,N) \\ \delta R(2,1) & \delta R(2,2) & - - - - - & \delta R(2,N) \\ - - - - & - - - - & - - - - - & - - - - \\ \delta R(N,1) & \delta R(N,2) & - - - - - & \delta R(N,N) \end{bmatrix}$$

where $\delta R(i,j)$ is the variation of the i,j th element of R.

Our first matrix theorem gives us an expression for the variation of the determinant of R in terms of the variation of

R. If $|R|$ is not zero, then

$$\delta|R| = |R| \operatorname{trace}(R^{-1}\delta R).$$

One derives this equation by noting that if the determinant of R is explicitly written out in terms of its elements, then the coefficient of R(i,j) in this expansion is the co-factor of R(i,j). In the inverse of R, the j,i th element is equal to the co-factor of the i,j th element of R divided by the determinant of R. In our above equation, we see that this is the coefficient of the variation of the i,j element of R. Now, noting that $\delta\log|R| = \delta|R|/|R|$, we have the important corollary that

$$\delta\log|R| = \operatorname{trace}(R^{-1}\delta R).$$

Our second useful matrix theorem gives us the variation of the inverse of R in terms of the variation of R. We first express the relation between the elements of R and R inverse by the matrix identity

$$R\,R^{-1} = I.$$

Taking the variation of this identity, we have

$$\delta R\,R^{-1} + R\,\delta(R^{-1}) = \delta I = 0, \text{ the null matrix.}$$

Our result is then

$$\delta(R^{-1}) = -R^{-1}\delta R\,R^{-1}.$$

Now we can derive the variation of g(S,R) easily as

$$\delta g(S,R) = -\delta\log|R| - \delta\operatorname{trace}(R^{-1}S) = -\operatorname{trace}(R^{-1}\delta R)$$

$$-\operatorname{trace}[\delta(R^{-1})S] = -\operatorname{trace}(R^{-1}\delta R - R^{-1}\delta R\,R^{-1}S) =$$

$$\operatorname{trace}(R^{-1}S\,R^{-1}\delta R - R^{-1}\delta R).$$

Our expression for the variation of g(S,R) is thus neatly

written as

$$\delta g(S,R) = \text{trace}\left[(R^{-1} S R^{-1} - R^{-1})\, \delta R\right].$$

The condition for maximation is that the gradient, $R^{-1} S R^{-1} - R^{-1}$, is orthogonal to changes in $\mathcal{R}$ space. That is, the variation of g is zero for any feasible variation of R. Thus the equation we shall solve is

$$\text{trace}\left[(R^{-1} S R^{-1} - R^{-1})\, \delta R\right] = 0. \quad (5)$$

It often happens that the structural constraint on the variation of R is satisfied by R itself. The toeplitz constraint is one case of this. We can then replace δR in equation (5) by R itself and the equation remains true. Our equation then says that

$$\text{trace}\,(R^{-1} S) = N. \quad (6)$$

If we substitute this into g(S,R), we have

$$g(S,R) = -\log|R| - N.$$

From this, we see that if R itself satisfies the structural constraints on the variation of R, then we can restate our variational principle as

Minimize the $\log|R|$ under the additional condition that (6) is true.

This is a very interesting and intuitive way of stating our variational principle. Condition (6) in effect places a scale factor constraint on R so that $\log|R|$ is bounded below. In fact, we shall show later that (6) gives us the minimum variance estimate of the optimum scale factor for R when we are dealing with a gaussian process. In the gaussian situation, we also note that the entropy of the random process is given by $\log|R|$, so this special case of our general variational principle is saying,

Choose the R that corresponds to the minimum entropy process under the auxiliary constraint that R is normalized by the minimum variance scale factor.

IV. SOME SIMPLE CASES OF COVARIANCE ESTIMATION AND THEIR SOLUTIONS

In this section, we shall formulate some relatively simple cases of covariance estimation and solve them using our variational principle. We assume that our data consisted of a set of N-dimensional vector samples, and that we have already formed the sample covariance matrix S. To simplify our discussions, we shall assume here that the sample covariance matrix is positive definite. If S is singular, then in some cases, R will need to be singular, which requires a more careful derivation of our basic equations. We point out here, however, that S being singular does not mean in general that R is singular. For example, for Toeplitz structures, we normally obtain a non-singular matrix for our estimate even if S is formed from a single vector sample.

One case of particular importance is when S satisfies the constraints that R must obey. As one would hope, R = S is a solution in this situation. One proves this immediately by noting that R = S makes the gradient identically zero. We shall show later that this is the only solution, which is what one would wish.

In some of the chosen examples below, we already "know" what the best answer should be. In other cases, one may not be so certain that the derived answer is the best. In this latter case, use of the variational principle may end up changing one's intuition about what answers do make the most sense.

The Unconstrained Case

The simplest problem to solve is when the R matrix is not constrained. Then the variation of R is arbitrary and the gradient must be identically zero. Then we have from equation (5) that

$$R^{-1} S R^{-1} - R^{-1} = 0,$$

which gives us immediately that R = S. It is of course nice to see that if R is arbitrary, then the sample covariance matrix S is the unique solution. It would be depressing if this were not the case.

Unknown Scale Factor Case

Suppose that we know the covariance matrix up to an unknown scale factor. An example of this is if one knows the shape of a spectrum as a function of frequency, but does not know the average power. This can occur if one passes white noise of unknown power through a known filter. For our general problem, we let R = aW, where W is a given positive definite symmetric matrix that is known to be proportional to the true covariance matrix and "a" is the unknown scale factor. In the white noise example, we note that the elements of W would be obtained from the autocorrelation of the impulse response of the filter out to lag N-1. Then, with R = aW, δR = δaW and have from equation (5) that

$$\text{trace of } [(aW)^{-1} S (aW)^{-1} - (aW)^{-1}] W = 0, \text{ or}$$

$$\text{trace}[(aW)^{-1} S - I] = 0, \text{ giving}$$

$$a = (1/N) \text{ trace}[W^{-1} S]. \tag{7}$$

One might not recognize that this unique solution is indeed the best answer since in our above example of white noise of unknown power passing through a known filter, the output power is normally estimated by a direct power average over our data samples. We shall now show in general that using the information provided by knowing W, formula (7) gives us the minimum variance estimate of "a".

Let us express the unique Cholesky decomposition of W in the form

$$W = G^{-1} G^{-T}, \text{ where}$$

G is lower triangular and its main diagonal consists of positive terms. Let us assume that x is one of our column vector samples and let us create the vector sample y by the linear transformation

$$y = G x.$$

We can now write

$$aI = aG W G^T = \text{Average value of } G x x^T G^T =$$

$$\text{Average value of } y y^T.$$

Thus, the vector sample y is made up of N independent random variables of uniform variance a. If x is multi-variant gaussian, so is y and in this case, the best estimate of the variance of the y variables is simply the average square value over all elements in all vector samples. Weighting each independent sample equally gives us the minimum variance estimate. We now derive our estimate for "a" in terms of S and W as

$a\,N$ = sample average of the trace of $y\,y^T$ =

sample average of the trace of $G\,x\,x^T G^T$ = trace$(G\,S\,G^T)$ =

trace$(S\,G^T G)$ = trace$(S\,W^{-1})$ = trace$(W^{-1} S)$, since $W^{-1} = G^T G$.

We now see that our variational principle has indeed led us to the best estimate of "a".

If W were already equal to our solution matrix R, then we see that "a" = 1 and (7) becomes (6). Thus, the optimally scaled R matrix does indeed satisfy (6) as discussed at the end of Section III.

The Burg Technique Case

One of the main features of the Burg technique is that the problem of estimating the reflection coefficients of a stationary time series is turned into one of estimating the covariance matrix of a pair of random variables whose individual variances are known to be equal. In this case, the structure of the two by two covariance matrix is of the form

$$R = \begin{bmatrix} a & b \\ b & a \end{bmatrix}, \quad \text{with} \quad \delta R = \begin{bmatrix} \delta a & \delta b \\ \delta b & \delta a \end{bmatrix}.$$

Equation (5) now tells us that the sum of the two diagonal elements of the gradient must be zero. Also, since the gradient matrix is always symmetric, we see that the gradient matrix must actually be diagonal in this present case. Thus the gradient is of the form

$$R^{-1} S R^{-1} - R^{-1} = \begin{bmatrix} c & 0 \\ 0 & -c \end{bmatrix}.$$

We now invoke a matrix theorem concerning transposes about the minor diagonal, i.e., flipping the matrix about the diagonal that runs upward to the right at 45 degrees. If we denote the transpose about the minor diagonal by a pre-superscript T, then it is easy to prove that if A B = C, then

$$({}^T B)\,({}^T A) = {}^T C.$$

Then we see that since $R^{-1} = {}^{T}R^{-1}$, we have

$$R^{-1} S R^{-1} - R^{-1} + R^{-1} ({}^{T}S) R^{-1} - R^{-1} = 0, \text{ the null matrix, or}$$

$$R^{-1} (S + {}^{T}S) R^{-1} = 2 R^{-1}, \text{ or that finally,}$$

$$R = (1/2)\ (S + {}^{T}S).$$

Thus, the estimated covariance matrix R is the minor diagonal symmetrized version of S.

The above use of the transpose about the minor diagonal gives us the following interesting and useful theorem. If the structure of R is such that R is equal to its minor diagonal transpose, then the inverse of R and the variation of R have this same property. Now, since transposing a matrix around either diagonal does not change its trace, we see that if R is also minor diagonal symmetric, then averaging the minor diagonal transposed form of equation (5) with itself, we have that R also satisfies

$$\text{trace of } [R^{-1}\ ((S + {}^{T}S)/2)\ R^{-1} - R^{-1}]\ \delta R = 0.$$

Thus, we can average S with its minor diagonal transpose and use this average in (5) to get the same answer for R. In fact, one can now note that if R is minor diagonal symmetric, then S can be replaced by the minor diagonal symmetrized matrix without any change to our objective function (4). Thus, we can start with this replacement and not change the functional form or numerical value of any of our equations. This is of more than passing significance.

One could hope that when there are three or more variables, whose variances are known to be equal, that the optimum estimate of their variance is also the average of the sample average. Unfortunately, this simple property does not extend beyond two variables. The reader can investigate for himself why this is so from a mathematical point of view. An intuitive feel for this fact can be developed if one supposes that the S matrix happens to indicate that the first variable is almost independent of the other variables, but that the other variables are strongly dependent. Then, weighting all the variables equally to estimate the variance would not seem to be the right thing to do. This line of reasoning only says that a straight average is not optimum. The variational principle gives us a solution to this problem, but it can not be written down explicitly.

Two Variables With Fixed Value Constraints

In addition to the above cases, there are some other situations in which an explicit solution to the variational principle exists. In particular, these situations involve pairs of variables in which the precise value of some of the second order statistics is known.

One Of The Variances Is Known. Let us assume that the variance of the first variable is known to be unity. Choosing unity is clearly as general as any other constant. So let

$$R = \begin{bmatrix} 1 & c \\ c & b \end{bmatrix} \quad \text{and} \quad S = \begin{bmatrix} A & C \\ C & B \end{bmatrix}.$$

Now since

$$\delta R = \begin{bmatrix} 0 & \delta c \\ \delta c & \delta b \end{bmatrix}, \text{ we see that } R^{-1} S R^{-1} - R^{-1} = \begin{bmatrix} \alpha & 0 \\ 0 & 0 \end{bmatrix},$$

which leads to

$$\begin{bmatrix} A-1 & C-c \\ C-c & B-b \end{bmatrix} = \alpha \begin{bmatrix} 1 & c \\ c & cc \end{bmatrix}.$$

Thus, $\alpha = A-1$ and we have $c = C/A$ and $b = B + (1-A)(C/A)^2$. Before trying to interpret this answer, one should ask if this R matrix is always non-negative definite, that is, is $b \geq 0$ and $b-cc \geq 0$? Clearly, the first inequality is true if the second is true, and the second is true if $b = B - CC/A = (AB-CC)/A \geq 0$. Thus if S is positive definite, so is R and if S is singular, so is R. b is zero if and only if B is zero. All of these properties are strongly desired.

Having $c = C/A$ is clearly reasonable and perhaps intuitive. To see that the solution for b is also reasonable, one can note that to do the linear least mean square prediction of the second variable from the first involves merely multiplying the first variable by c. One could estimate the resulting mean square prediction error from either S or R. Actually, b is such that both of these estimates are the same, that is,

$$\begin{bmatrix} -c & 1 \end{bmatrix} \begin{bmatrix} A & C \\ C & B \end{bmatrix} \begin{bmatrix} -c \\ 1 \end{bmatrix} = \begin{bmatrix} -c & 1 \end{bmatrix} \begin{bmatrix} 1 & c \\ c & b \end{bmatrix} \begin{bmatrix} -c \\ 1 \end{bmatrix} = b-cc. \quad (8)$$

One Variance and the Cross-Variance are Known. Suppose in addition to knowing that the first variable has unity variance, it is also known that the cross-variance is c. Using the above expressions for R and S, we note that

$$R^{-1} = 1/(b-cc)\begin{bmatrix} b & -c \\ -c & 1 \end{bmatrix} \text{ and } R^{-1}SR^{-1} - R^{-1} = \begin{bmatrix} \alpha & \gamma \\ \gamma & 0 \end{bmatrix}.$$

Only the bottom righthand equation needs to be solved and it is

$$b = c^2 + c^2 A - 2cC + B.$$

We note again that this b is such as to make the two prediction error estimates as shown in (8) equal. Since $b-cc \geq 0$, we also again have R to be non-negative definite.

The Cross-Variance is Known.

Here we start with

$$R = \begin{bmatrix} a & c \\ c & b \end{bmatrix} \text{ and } R^{-1}SR^{-1} - R^{-1} = \begin{bmatrix} 0 & \gamma \\ \gamma & 0 \end{bmatrix}.$$

Pre and post multiplying this last equation by R, we have

$$S - R = \gamma R \begin{bmatrix} 0 & 1 \\ 1 & 0 \end{bmatrix} R = \gamma \begin{bmatrix} a & c \\ c & b \end{bmatrix}\begin{bmatrix} c & b \\ a & c \end{bmatrix}, \text{ or}$$

$$\begin{bmatrix} A-a & C-c \\ C-c & B-b \end{bmatrix} = \gamma \begin{bmatrix} 2ac & ab+cc \\ ab+cc & 2bc \end{bmatrix}.$$

From the two equations on the main diagonal, we see that $A = a(1+2\gamma c)$ and $B = b(1+2\gamma c)$. Thus a has the same proportion to A as b has to B. To actually solve for a, we end up with a cubic equation in a that says that

$$\begin{bmatrix} -c/a & 1 \end{bmatrix}\begin{bmatrix} a & c \\ c & b \end{bmatrix}\begin{bmatrix} -c/a \\ 1 \end{bmatrix} = \begin{bmatrix} -c/a & 1 \end{bmatrix}\begin{bmatrix} A & C \\ C & B \end{bmatrix}\begin{bmatrix} -c/a \\ 1 \end{bmatrix}.$$

Thus, we see again that the solution says that the prediction error is the same whether we apply the optimum estimated filter to the sample covariance matrix or to the estimated covariance matrix.

<u>Both Variances are Known</u>. We can assume without loss of generality that both variances are unity. In working out the equation for c, we end up with a cubic that again says that the sample and estimated least mean square error in predicting one variable from the other are equal.

V. THE INVERSE ITERATION ALGORITHM

We shall now develop the general iterative process for determining a solution to the necessary conditions as given by (5). The approach used here is somewhat abstract, but it greatly simplifies the development since concepts are isolated from the complex computations that may be required to express the concepts in concrete form. To avoid notational complexity in our derivations, we shall consider here only the common case where R itself satisfies the conditions on the variation of R. When R is constrained to have particular numerical values, one can treat the problem by similar methods but using a reduced dimensionality.

<u>The Algorithm</u>. The iterative process that we consider here we shall term <u>inverse iteration</u>. At any stage, we begin with an approximation R_κ and a new approximation $R_{\kappa+1}$ is determined as follows:

1. Find $D_\kappa \in \mathfrak{R}$ so that $g(S-D_\kappa, R_\kappa)$ satisfied the necessary conditions.

2. Put $R_{\kappa+1} = R_\kappa + D_\kappa$.

Note that in step 1, we find the change (belonging to $\mathfrak{R}$) in our data S that makes the current approximation optimal. Then, in step 2 the approximation is updated by the negative of this virtual change in the data. That is why we have chosen to name the method inverse iteration. The main reason for using inverse iteration for this problem is that the gradient of the objective is linear with respect to S, so the problem implied by step 1 is a linear problem. Note that at each step of our iteration, our approximation satisfies our linear variety constraint. However, experience has shown that the new approximation may not be positive definite, a phenomenon that we shall deal with later.

<u>Improving Direction</u>. There are some basic properties of the inverse iteration process that are extremely important. The first is that the direction of change D_κ is an improving direction. That is, if R_κ is changed by adding a small amount of D_κ, the objective function will increase over what it was with R_κ. We shall now prove this.

What we wish to show is that the gradient of g has a positive inner product with the direction D determined by step 1. This means that small movement along D will yield a positive change in g. In mathematical terms, we wish to show that

$$\mathrm{Trace}\left[(R^{-1} S R^{-1} - R^{-1})D\right] > 0.$$

This can be rewritten as

$$\mathrm{trace}\left[(R^{-1}(S-D)R^{-1} - R^{-1})D\right] + \mathrm{trace}\left[R^{-1} D R^{-1} D\right].$$

The first term is zero by construction since this is the requirement that R is optimal for the data S-D. We now prove that the second term is positive.

Since R is positive definite, we again do the Cholesky decomposition as we did in section III, this time in the form

$$R^{-1} = G^T G.$$

Then, $\mathrm{trace}\left[R^{-1} D R^{-1} D\right] = \mathrm{trace}\left[G^T G D G^T G D\right] = \left[\mathrm{trace}\,(G D G^T)(G D G^T)\right].$

Since G and D are symmetric matrices, the matrix $G D G^T$ is symmetric and we are looking at its inner product with itself. Since D is not zero, this inner product is positive and we have proved that D is an improving direction.

We shall slip in here the promised proof that when S obeys the constraints on R, then R = S is the one and only solution. The proof is valid for the general type of constraint conditions.

We note that with S belonging to $\mathcal{R}$, S-R satisfies the constraints on the variation of R and thus

$$\mathrm{trace}\left[R^{-1}(S-R)\, R^{-1}(S-R)\right] = 0.$$

We have seen above that if R is positive definite, then this expression is positive unless S-R is equal to the null matrix and thus R = S is our one and only solution.

Study of Asymptotic Properties. Although D defines an improving direction, it is not yet clear how far to move in that direction. The basic algorithm moves a distance qD, with q = 1, but that may not always be appropriate. It would be best to move the distance that maximizes the objective function. Computational experience has shown that the best value for q is indeed often equal to unity, and we explore this here.

We find a quadratic approximation in q for our objective function and then find the value q = qmax that maximizes this approximation. Then we try to infer that qmax approaches 1 in the limit when D goes to zero. Although the complete estimate has been found in closed form, we have at this point an incomplete verification that the limiting optimal distance is 1.

Using the matrix theorems developed in section III, we derive the second order expansion of g(S,R+qD) as a function of q about q = 0 by first noting that

$$dg(S,R+qD)/dq = -\mathrm{trace}[(R+qD)^{-1}D] + \mathrm{trace}[(R+qD)^{-1}D(R+qD)^{-1}S].$$

The second derivative, evaluated at q = 0, can now be derived as

$$d^2g(S,R+qD)/d^2q\Big|_{q=0} = \mathrm{Trace}(R^{-1}DR^{-1}D) - 2\,\mathrm{trace}(R^{-1}DR^{-1}DR^{-1}S). \quad (9)$$

Having already shown that the first derivative evaluated at q = 0 is just the first term in (9), we can now write down the Taylor expansion out to second order as

$$g(S,R+qD) = g(S,R)+(q+q^2/2)\mathrm{trace}(R^{-1}DR^{-1}D)-q^2\,\mathrm{trace}(R^{-1}DR^{-1}SR^{-1}D).$$

Differentiating with respect to q and setting the result to zero gives us the maximum of this quadratic expression as

$$q_{max} = [\mathrm{trace}(R^{-1}DR^{-1}D)]/[2\,\mathrm{trace}(R^{-1}DR^{-1}SR^{-1}D) - \mathrm{trace}(R^{-1}DR^{-1}D)].$$

To estimate qmax, we shall use the fact that

$$\mathrm{trace}(R^{-1}SR^{-1}D) = \mathrm{trace}(R^{-1}DR^{-1}D) + \mathrm{trace}(R^{-1}D),$$

which follows from the definition of D. Now, although it is not quite legitimate, let us assume that the above equality holds when each trace has $R^{-1}D$ inserted. This yields

$$q_{max} = 1/[1+0(D)+E(D)],$$

where 0(D) means "of order D" and E(D) is two orders less than the error introduced by the above substitution. It is easy to see that E(D) is at most 0(1). It is conjectured that E(D) is 0(D). If so, then qmax goes to unity as q goes to zero.

The above result would be exact if the eigenvalues of $R^{-1}D$ were all equal. This is likely to be approximately true if $\mathcal{R}$ is robust since R would tend toward S and so would D. Thus $R^{-1}D$ would tend to the identity. Unfortunately, a precise statement in this regard has not been found.

VI. COMPUTATION OF THE INVERSE ITERATION

The iterative algorithm discussed in the previous section is based on repeatedly solving a linear problem in order to determine the new estimate of the covariance matrix R. The solution of this linear problem is, computationally, the most difficult part of the algorithm. By suitable formulation and use of available structure, however, a very efficient procedure can be developed.

Formulation and Duality. The necessary condition for a solution has been given by equation (5). It is repeated here (with slightly different notation) as that of finding an R that satisfies

$$\mathrm{trace}[(R^{-1}SR^{-1} - R^{-1})Q] = 0$$

for all Q in $\mathcal{R}$. The iterative process is deduced by assuming a trial solution R is given. A new trial R' is sought that satisfies

$$\mathrm{trace}[(R^{-1}SR^{-1} - R^{-1}R'R^{-1})Q] = 0 \tag{10}$$

for all Q in $\mathcal{R}$. This is linear in the unknown R'.

A problem of this type is a generalized projection problem. It can be solved by selecting a basis either in $\mathcal{R}$ or in the orthogonal complement of $\mathcal{R}$. We propose here to use a basis in $\mathcal{R}$ itself since in most applications, such as the case where $\mathcal{R}$ is the space of symmetric Toeplitz matrices, this will be of considerably less dimension than the dimension of the orthogonal complement.

Let the basis in $\mathcal{R}$ be Q_m, m = 1 to M, where we are assuming that the space is of dimension M. We note that the Q_m are symmetric matrices. Then we write an expansion for the unknown R' as

$$R' = \sum_{m=1}^{M} x_m Q_m. \tag{11}$$

This expansion is substituted into the basic linear condition (10) which leads to

$$\sum_{m=1}^{M} \mathrm{trace}[R^{-1} Q_m R^{-1} Q_j]\, x_m = \mathrm{trace}[R^{-1} S R^{-1} Q_j],$$

for j = 1 to M.

This is a system of M equations in the M unknowns x_m, m = 1 to M. Solution of this system yields the next approximation R' by using (11). If we define the matrix A by

$$A_{ij} = \mathrm{trace}[R^{-1} Q_i R^{-1} Q_j], \text{ and}$$

$$c_j = \mathrm{trace}[R^{-1} S R^{-1} Q_j],$$

the system takes the standard form

$$Ax = c. \tag{12}$$

Proof that A is Positive Definite. We can show that the matrix A defined above is symmetric and positive definite. Actually, the symmetry is seen immediately. Being positive definite is quite important for implementation, since it means that the efficient algorithms for solution of symmetric positive definite systems can be employed.

Consider

$$\sum_{i,j=1}^{M} x_i A_{ij} x_j = \sum_{i,j=1}^{M} x_i \mathrm{trace}[R^{-1} Q_i R^{-1} Q_j x_j] = \mathrm{trace}[R^{-1} B R^{-1} B],$$

where

$$B = \sum_{j=1}^{M} Q_j x_j.$$

We have shown before that when R is positive definite and B is not the null matrix, this expression is positive, showing that A is positive definite.

The Overall Procedure. The overall algorithm is outlined by the following steps:

1. Start with an initial positive definite approximation R, (We usually start with the identity matrix).

2. Using this R, calculate all the traces required to define the matrix A and solve the system (12). Then use (11) to obtain a tentative new approximation R'.

3. Evaluate the function g(S,R'). If this value is not larger than with the previous approximation or if the new approximation for R is not positive definite, then define D = R'-R and try an approximation of the form R +qD for q = 1/2. Keep cutting q by a factor of two until an approximation satisfying the above requirements is found. Set the new R equal to this value and go back to step 2.

There are of course numerous variations possible, especially regarding the step-size determination.

VII. SUMMARY AND CONCLUSIONS

The basic approach of maximum likelihood estimation of a structured covariance matrix is both basically sound and computationally feasible. In evaluating the procedure with simulated random vector data from a stationary time series, we have found that the maximum entropy spectra obtained from the autocorrelation functions estimated by the generalized Burg technique to be consistently better and often much better than those obtained by the usual Burg technique. In particular, there is no evidence of line splitting with the maximum likelihood procedure.

It is probable that the most important use of this new technique will be in estimating covariance matrices from multichannel and especially mult-dimensional data. Researchers have generalized the Burg technique to the multi-channel case in several ways using prediction error filter approaches. To me, none of these generalizations were totally satisfactory from a theoretical point of view and are certainly not on as sound a basis as the maximum likelihood approach presented here.

The most telling argument in favor of the maximum likelihood procedure, however, is that it shows us how to estimate multidimensional covariance matrices, a situation for which no one has offered a reasonable generalization of the Burg technique.

There is a great amount of research and testing to be done on this new estimation technique. If there is only one extremal to the variational principle, then this should be proven. It is already known that there is at least one solution in the space of positive definite matrices. There is also a need to see if this theory can be better related to general maximum entropy theory (Edwin T. Jaynes) and in particular to the work being done with minimum cross-entropy (John E. Shore).

CLASSIFICATION OF RADAR CLUTTER USING THE MAXIMUM-ENTROPY METHOD

Simon Haykin

Communications Research Laboratory
McMaster University
Hamilton, Ontario, Canada

ABSTRACT

In this paper, we demonstrate the usefulness of the maximum-entropy method in analysing the spectral (Doppler) content of clutter as encountered in a radar surveillance environment. Experimental results (based on recorded radar data) are included which show that it is indeed possible to extract Doppler-based features for deciding whether the clutter is due to reflections from the ground, weather disturbances, or migrating flocks of birds. In this way, it becomes possible to vector an aircraft around a hazardous (turbulent) area, and thereby avoid the possibility of danger to the safety of the aircraft.

1. INTRODUCTION

Radar clutter is defined [1] as "confused unwanted echoes on a radar display". The name is descriptive of the fact that such echoes tend to "clutter" the radar display and make the detection of wanted (target) echoes more difficult. An interesting property of radar clutter is that it can be modeled fairly closely as an autoregressive process of relatively low order (two to five) [2]. This suggests the use of the maximum-entropy method (MEM) [3] to analyse the spectral content of radar clutter. In this paper, we describe the application of the MEM to extract Doppler-based features that can be used to classify the different forms of clutter as encountered in a surveillance radar environment.

In Section 2, we briefly describe the power spectrum of

R. S. Bucy and J. M. F. Moura (eds.), Nonlinear Stochastic Problems, 41–54.

radar clutter. In Section 3 we describe an extension of the MEM that involves both temporal and spatial forms of averaging, thereby providing the basis of radar clutter classification. In Section 4 we present experimental results (based on recorded radar data) which demonstrate the ability of the MEM to distinguish between the different sources of radar clutter.

2. CLUTTER POWER SPECTRUM

In a radar surveillance environment, clutter arises due to reflections from (a) fixed or slowly moving objects such as trees, vegetation, hills, man-made structures, (b) sotrm clouds, precipitation, and other meteorological phenomena, and (c) migrating flocks of birds. Clutter targets are either isolated or composite [4]. An isolated clutter target is one which may be considered as a "point scatterer"; an example of an isolated clutter target is a water tower. The composite clutter target is more common and is characteristic of most ground clutter, meteorological echoes, and migrating flocks of birds. It consists of many individual scatterers within the coverage of the radar antenna beam. Extended composite radar clutter can limit radar detection capability since the total radar cross-section can be large. In this paper we consider only the effects of composite clutter.

In the case of composite radar clutter arising from a volume of randomly distributed scatterers, we find that as the scatterers "reshuffle", successive echoes from the same volume fluctuate. Obviously the rate of echo signal fluctuation is related to the "reshuffling rate" of the scatterers,, and as such it must contain useful information about the physical source of the scatterers or clutter. The instantaneous strength of the echo depends upon whether the scatterers are interfering constructively or destructively, and so the rate of fluctuation must depend on how fast the scatterers are moving relative to one another in the phase space, that is, across the constant-phase surfaces which are perpendicular to the radial direction of propagation from and to the radar.

Consider a point target or scatterer at distance r from the radar. The number of wavelengths contained in the two-way path, from the radar to the point target and back, equals $2r/\lambda$, where λ is the transmitted radar wavelength. The phase, θ, of the echo (relative to the transmitted signal) equals $4\pi r/\lambda$ radians. If the target is in motion with respect to the radar with radial velocity $v = dr/dt$, the phase θ changes proportionately. Since the rate of change of phase with respect to time represents angular frequency, we may write:

$$\omega = 2\pi f = \frac{d\theta}{dt} = \frac{4\pi}{\lambda}\frac{dr}{dt} = \frac{4\pi}{\lambda} v \tag{1}$$

or

$$f = \frac{2v}{\lambda} \tag{2}$$

This is the Doppler frequency shift resulting from the radial velocity v.

In coherent Doppler radar applications, the statistical function that most conveninetly characterizes the echo resulting from radar backscatter from a collection of point scatterers is its power spectrum [5]. This power spectrum is intimately connected with the velocities of the scatterers. In particular, for any number of scatterers the power spectrum, S(f), is a weighted image of the velocity distribution. If S(v)dv denotes that fraction of the received power returned by scatterers whose velocity components in the direction of the radar lie in the increment dv, then we may write

$$S(f)df = S(v)dv \tag{3}$$

where it is understood that the Doppler relation $f = 2v/\lambda$ applies. This shows that the power spectrum S(f) does indeed provide a picture of the radial velocity distribution within the volume of scatterers of interest.

In [2], it is shown that the power spectra due to various sources of radar clutter (ground targets such as wooded hills, sea echo, rain clouds) may be closely approximated by an all-pole rational function. This, in turn, suggests that radar clutter may be modelled as an autoregressive process of relatively low-order (two to five).

In a conventional moving-target indicator (MTI) radar, some form of filtering is used to remove clutter from the plan-position indicator (PPI) in order to improve the detection of moving targets (e.g., aircraft). However, in so doing, there is the distinct possibility of vectoring an aircraft into an area of stormy weather or dense bird activity, and thereby subjecting the aircraft to a hazardous situation. Rather than simply rejecting the clutter as an unwanted signal (which it is, from a target-detection viewpoint), it is much more desirable to analyze and classify the clutter according to some level of danger which it presents to air traffic. Such a classification should be possible by estimating the power spectrum of the clutter, which (as described in Section 2) represents the radial velocity distribution of the scatterers. Since the physical movement characteristics of ground, weather, and birds are different, so should be the Doppler spectra of their radar returns. For example, radar returns from stationary ground objects have a very small Doppler spectral spread, caused primarily by antenna scanning modulation and possibly vegetation movement due to wind. The power spectrum of ground clutter is thus centred at zero Doppler frequency and is concentrated in a very narrow frequency band. On the other hand, reflections from stormy clouds, rain, snow, and other meteorological conditions, have a wider Doppler spectral spread due to their more vigorous motion. In the case of strong winds, the overall motion can also result in an offset in the centre frequency of the power spectrum. This centre frequency offset can also be particularly apparent in the case of radar returns from migrating flocks of birds, which may have velocities as high as 50 knots in favourable winds. Wingbeat modulation and relative bird movements within the flock tend to cause the resulting clutter power spectrum to be even more widely spread than in the case of weather clutter.

We see that these three forms of clutter have more or less definite spectral characteristics: (1) ground clutter has a very narrow spectrum centred at zero Doppler, (2) weather clutter has a wider spread, with a possible shift in the centre frequency, and (3) clutter due to returns from birds is more widely spread in frequency with the centre frequency shifted noticeably away from zero Doppler. The position of the peak power in the spectrum, although a useful indicator, cannot be used alone as a basis for classification, since it depends on the velocity of the scatterers with respect to the radar, and can have quite a wide range of values (except for the case of ground clutter). On the other hand, the amount of spectral spread, caused by the intrinsic motion of the scatterers, should give a good indication of the source of clutter.

3. MULTISEGMENT MAXIMUM-ENTROPY METHOD

A data segment is defined as the set of time-contiguous radar samples returned from a single resolution cell in a single scan. The number of samples in this segment equals the number of hits per beamwidth. Spatial averaging involves using data segments from adjacent resolution cells, while temporal averaging uses data segments from the same resolution cell in successive scans. The difficulty in combining these data segments is the discontinuity in time between the pertinent data segments. Because of these time discontinuities, the segments cannot be simply combined to produce one long data record to be used as input for the usual MEM algorithm. Instead, the MEM is modified to operate individually on the data segments at each stage of the lattice filter used to compute the maximum-entropy spectral estimate, then average across the segments to produce one estimate of the reflection coefficient for application to the next stage. The MEM algorithm thus operates on multiple segments of data, disjoint in time; hence, the name "multisegment maximum-entropy method".

Suppose for example that we have available for analysis I disjoint segments of samples, denoted by $x_i(n)$ with $i = 1, 2, \ldots, I$ and $n = 1, 2, \ldots, N_i$. The total number of data samples is then

$$N = \sum_{i=1}^{I} N_i \tag{4}$$

We specify I forward prediction errors and I backward prediction errors for stage m of the lattice prediction-error filter in the MEM algorithm in terms of the corresponding sets of prediction errors at the previous stage, m−1, as follows, respectively [6]

$$f_{m,i}(n) = f_{m-1,i}(n) + \rho_m \, b_{m-1,i}(n-1)$$

$$b_{m,i}(n) = b_{m=1,i}(n-1) + \rho_m^* \, f_{m-1,i}(n) \tag{5}$$

where $n = m+1, m+2, \ldots, N_i$

$i = 1, 2, \ldots, I.$

and ρ_m is the reflection coefficient of stage m of the lattice filter.

The average value of the forward prediction error power and backward prediction error power, for stage m and data segment i, equals

$$P_{m,i} = \frac{1}{2}(P_{f,m,i} + P_{b,m,i})$$

$$= \frac{1}{2} \sum_{n=m+1}^{N_i} [|f_{m.i}(n)|^2 + |b_{m,i}(n)|^2] \tag{6}$$

By spatially averaging this result over all I data segments, we get

$$P_m = \frac{1}{I} \sum_{i=1}^{I} P_{m,i}$$

$$= \frac{1}{2I} \sum_{i=1}^{I} \sum_{n=m+1}^{N_i} [|f_{m,i}(n)|^2 + |b_{m,i}(n)|^2] \qquad (7)$$

Upon substituting Eqs. (5) into (7) and setting the derivative of P_m with respect to ρ_m equal to zero, we obtain an estimate for the reflection coefficient that is averaged both temporally and spatially, as shown by [2]

$$\hat{\rho}_m = \frac{-2 \sum_{i=1}^{I} \sum_{n=m+1}^{N_i} f_{m-1,i}(n)\, b^*_{m-1,i}(n-1)}{\sum_{i=1}^{I} \sum_{n=m+1}^{N_i} [|f_{m-1,i}(n)|^2 + |b_{m-1,i}(n-1)|^2]}, \qquad (8)$$

where m = 1, 2, ..., M

Equation (8) is the multisegment extension of the Burg formula or harmonic mean algorithm.

To compute the multisegment maximum-entropy spectral estimate, we proceed in the usual way [6], except that we use Eq. 8 as the estimate of the reflection coefficient, $\hat{\rho}_m$, for stage m, where m = 1, 2, ..., M, and M denotes the preassigned order of the prediction-error filter.

4. APPLICATION OF MULTISEGMENT MEM TO CLUTTER CLASSIFICATION

A prototype radar clutter classifier has been built for use with the new Canadian military Terminal Radar and Control System - Airport Surveillance Radars (TRACS-ASR) [7]. This prototype uses the multisegment MEM described above to estimate the power spectra of different sources of clutter. The purpose of the prototype is to demonstrate proof-in-principle of using maximum-entropy spectral estimates to separate out different sources of clutter. Thus, for the prototype to be as flexible as possible, its implementation is based on a 16-bit microcomputer kit, supplemented with a floating-point arithmetic chip and additional memory for program and data storage. The TRACS-ARS radar has 10-bit digital in-phase (I) and quadrature (Q) video channels, so that the prototype is designed to accept 4096 I, Q sample pairs at the range clock rate (about 1.25 MHz), then analyze this block of samples to produce a result. Within the limit of a total of

4096 samples, the physical area sampled is defined in terms of its location and its extent in range, bearing, and number of successive scans.

For the results described here, the sampling area consists of 10 consecutive ranges by 100 successive sweeps by 4 successive scans. Samples in range are taken at the sampling clock rate, while samples in azimuth for a given range are taken at the radar pulse repetition frequency, i.e., the radar sweep rate, where a sweep denotes the radar return corresponding to one pulse repetition period of the radar. The parameters of the radar yield about 20 hits/beamwidth, so that a data segment is defined as 20 successive sweep values at a given range; this constitutes the return from a single resolution cell as the antenna sweeps across it. As the bearing separation between the start of data segments at the same range increases beyond the antenna beamwidth, the segments represent returns from different resolution cells.

By analyzing the entire sampling area, we obtain a total of 81 spectra. Figure 1 shows a plot of some of these spectra for the case of ground clutter plus receiver noise. The original classification idea was to average the spectral widths of the pertinent set of 81 spectra to produce one estimate of the turbulence. However, Fig. 1 shows one immediate problem that can arise under certain conditions. Specifically, in those areas where there is no clutter, we have "white" (flat) spectra due to the noise alone. These "white" spectral contributions contaminate the spectral width estimate, and should therefore be eliminated. This is done by comparing the average input power of the samples with a preset threshold, and ignoring any data sets with lower power. Accordingly, in general, there are a total of 81 or less spectra per sampling area for computing the average spectral width. Figure 2 shows the resultant spectra after application of the power threshold test.

When analyzing weather clutter, we sometimes find that this clutter occurs "above" ground clutter, i.e. both weather and ground produce echoes. This phenomenon is illustrated in the spectra of Fig. 3. We see that two distinct groups of peaks are present: one at zero frequency due to the ground, and one at 0.15-0.20 due to the weather.

A similar phenomenon arises in the case of bird-plus-ground clutter. Figure 4 shows the spectra for such a case. Here again there appear to be two groups of peaks, but for birds the peak position is much more variable.

Note that the maximum-entropy spectral estimates shown plotted in Figs. 2 to 4 are all normalized with respect to the peak value of the spectral estimate. Also, the frequency is normalized with respect to the pulse-repetition frequency of the radar. The prediction-error filter order M equals 3.

In studying Figs. 2 to 4, there appear to be at least three features possible for classification:

(1) The first is the spectral width or turbulence. As anticipated, it seems to increase from ground to weather to bird clutter.

(2) The other feature is the spread of the peak position, which indicates the distribution of velocities within the sampling area, again an indication of turbulence. The standard deviation of the peak position increases from ground to weather to birds. One danger in calculating this standard deviation, as exemplified in Figure 3, is the lumping together of the two groups of peaks; one being due to ground clutter and the other to weather clutter. To avoid this difficulty, each spectrum is first classified as "ground" or "non-ground" based on the position of the main peak. Each spectrum contains 248 calculated points in frequency, numbered from 1 at -0.5 to 248 at +0.5. Zero frequency corresponds to a spectral array index of 124. A "ground" spectrum is one for which the main peak position index falls between 114 and 134. This index spread of 20, corresponding to ± 4 knots, allows for ground clutter peak offset due, for example, to vegetation movement. The standard deviation is calculated separately for each group, i.e. the "ground" and "non-ground" spectra. To take advantage of the information contained in the spectrum containing two strong peaks, the location of the secondary peak is recorded if its power is above -10 db relative to the main peak.

(3) Another useful feature that may be used to indicate the "goodness-of-fit" of the maximum-entropy spectral estimate is the ratio of the residual prediction-error power to the input power. This power ratio is denoted by α.

Table I shows a summary of these various features extracted from three representative data sets: one for each of ground, weather, and bird clutter. These figures confirm the visual observations made in Figs. 2 to 4. The standard deviation of the peak location increases from 1.2 for ground clutter to 4.3 for weather clutter to 15.9 for bird clutter ("non-ground" spectral group for the latter two). Similarly, the -10 dB width increases from 12.2 for ground to 31.7 for weather to 71.9 for bird clutter, as does the power ratio: 0.15 for ground, 0.261 for weather, and 0.353 for bird clutter. Based on the results of Table I, it indeed seems quite possible to distinguish the various forms of clutter based on one or more of these features.

TABLE I Summary of feature values for representative data sets for ground, weather, and bird clutter.

	MAIN PEAK LOC'N	SPECTRAL WIDTH AT -5 dB	SPECTRAL WIDTH AT -10 dB	SECONDARY PEAK LOC'N	POWER RATIO α
GROUND CLUTTER					
"GROUND"					
AVERAGE	124.4	6.5	12.2	0.0	.150
STD DEV	1.2	3.9	7.8	0.0	.063
# OF SPECT	81	81	81	0	81
"NON-GROUND"					
AVERAGE	0.0	0.0	0.0	0.0	.000
STD DEV	0.0	0.0	0.0	0.0	.000
# OF SPECT	0	0	0	0	0
WEATHER CLUTTER					
"GROUND"					
AVERAGE	123.4	9.9	29.3	170.6	.207
STD DEV	.5	2.6	21.8	4.4	.024
# OF SPECT	8	8	8	7	8
"NON-GROUND"					
AVERAGE	164.6	15.0	31.7	128.1	.261
STD DEV	4.3	4.4	12.9	3.6	.037
# OF SPECT	73	73	73	7	73
BIRD CLUTTER					
"GROUND"					
AVERAGE	123.1	26.7	60.1	175.2	.344
STD DEV	5.5	19.5	32.0	8.1	.093
# OF SPECT	34	34	34	23	34
"NON-GROUND"					
AVERAGE	157.5	33.7	71.9	150.1	.353
STD DEV	15.9	20.0	24.5	27.3	.069
# OF SPECT	40	40	40	25	40

Notes:

(1) The average and standard deviation figures for the peak locations are in terms of the spectral array index, where the index is 1 for frequency -0.5 and 248 for +0.5. The peak widths are given by the difference between the indices of the -5 and -10 dB points to the right and left of the main peak location.

(2) The number labelled "# of spect" indicates in how many of the 81 available spectra that particular feature was found.

(3) The spectra for each type of clutter are divided into "ground" and "non-ground" groups based on the main peak location. Main peaks with an index between 114 & 134 are "ground" spectra.

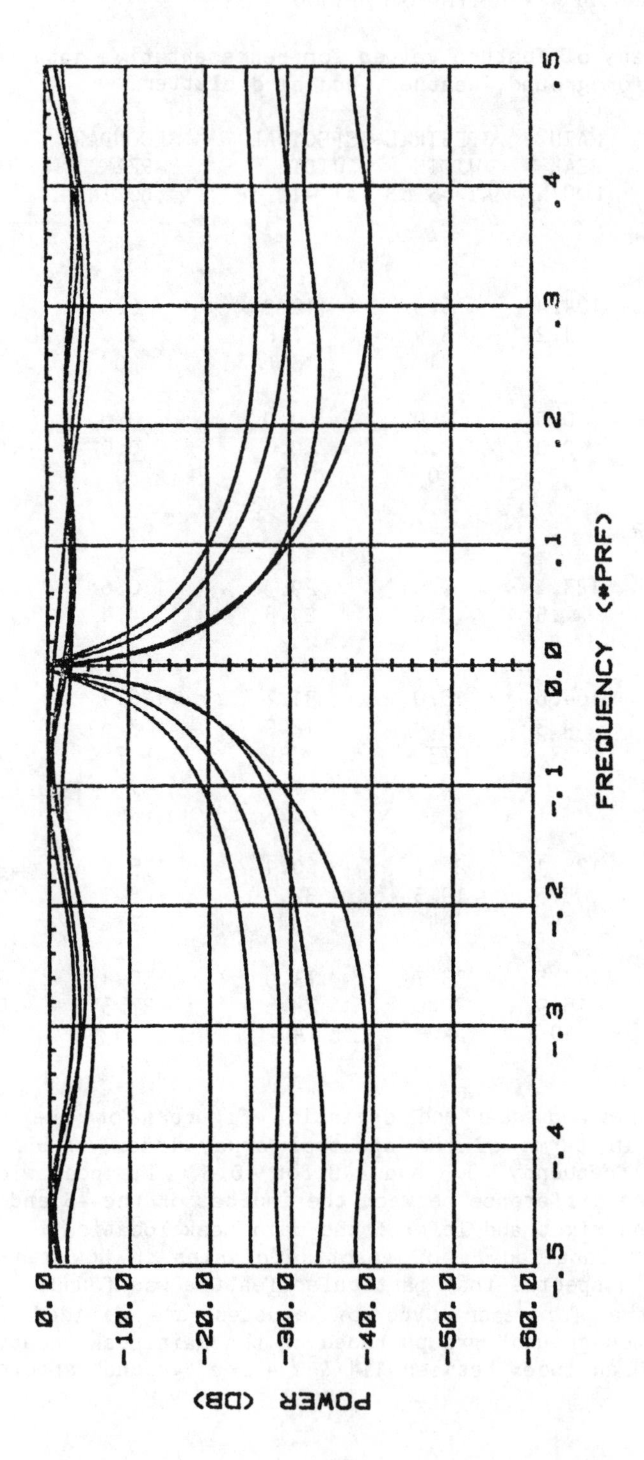

Figure 1: Some of the Doppler spectra produced in the analysis of data containing ground clutter and noise.

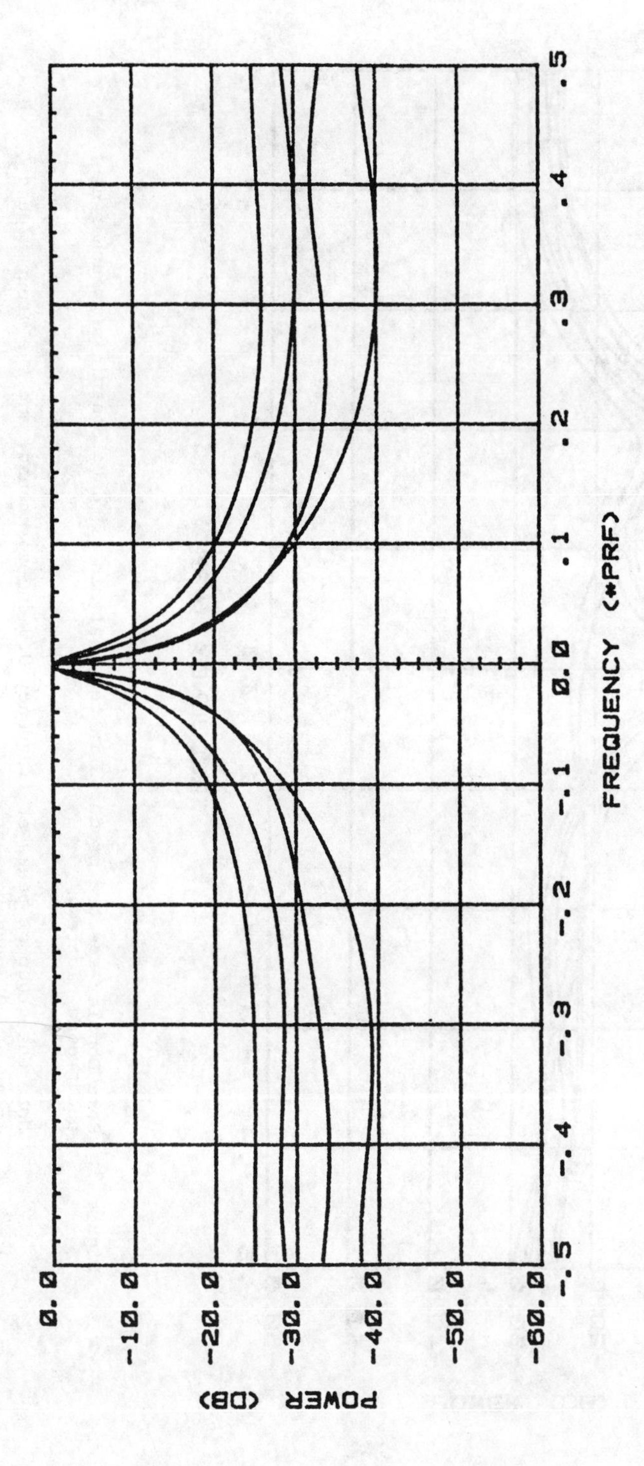

Figure 2: Doppler spectra produced in the analysis of data containing ground clutter and noise.

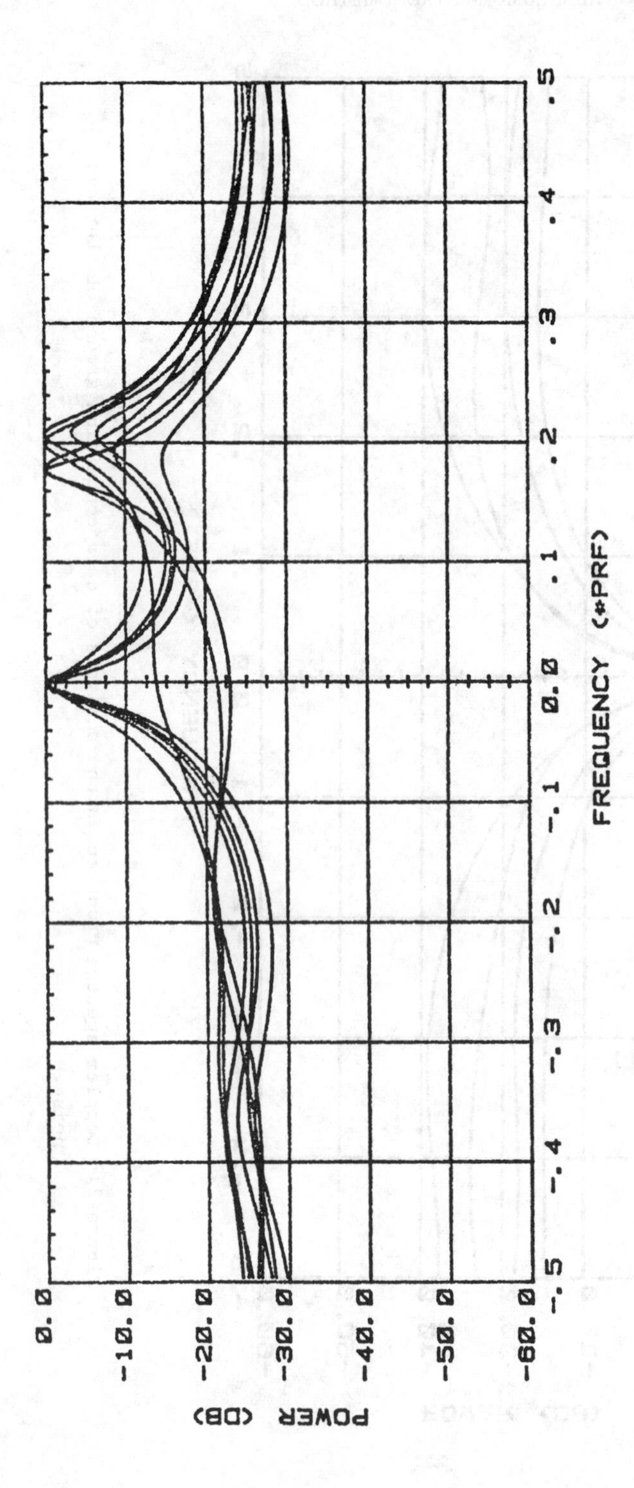

Figure 3: Some Doppler spectra produced in the analysis of data containing weather clutter and ground clutter.

Note the group of peaks at frequency 0.0 due to ground, and the group around frequency 0.2 due to weather.

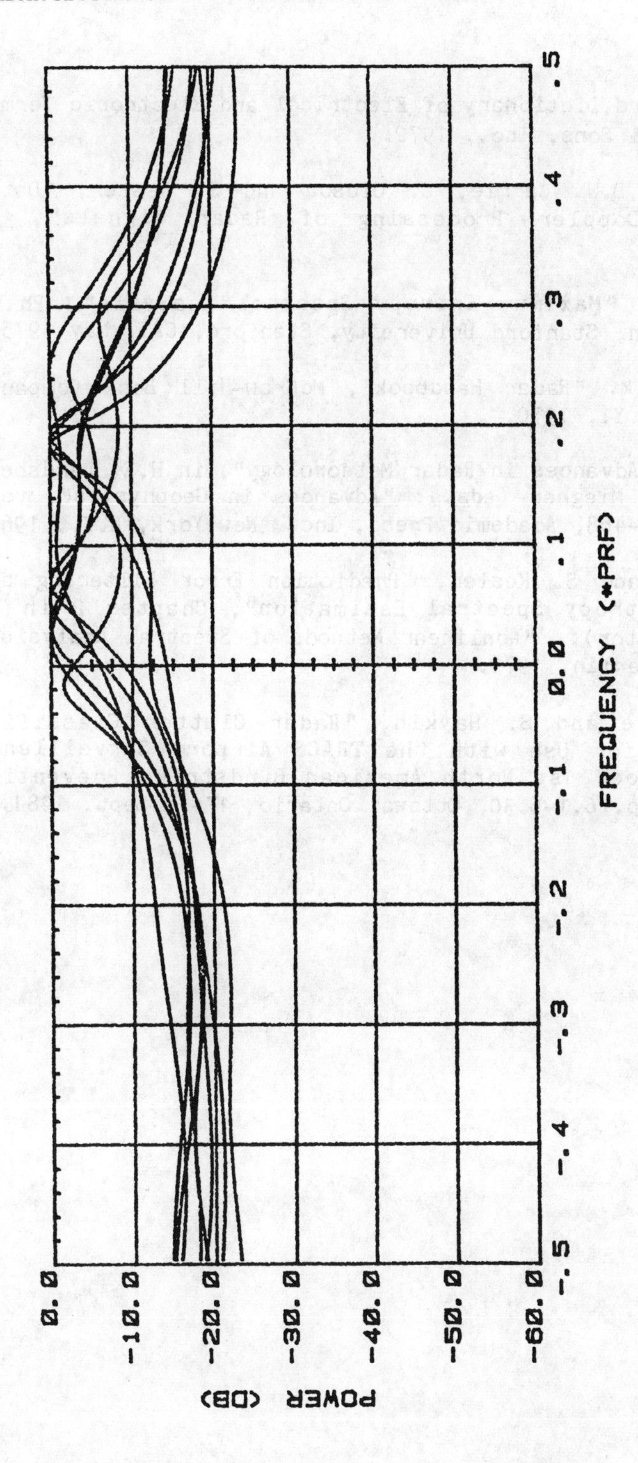

Figure 4: Some Doppler spectra produced in the analysis of data containing bird clutter and ground clutter.

Note the variance in the peak position and also the increased peak width as compared to the weather clutter in Figure 3.

REFERENCES

[1] IEEE Standard Dictionary of Electrical and Electronic Terms, John Wiley & Sons, Inc., 1972.

[2] S. Haykin, B.W. Currie, C. Gibson and S. Kesler, "Date-Adaptive Doppler Processing of Radar Signals", in preparation.

[3] J.P. Burg, "Maximum Entropy Spectral Analysis", Ph.D. Dissertation, Stanford University, Stanford, CA., May 1975.

[4] M.I. Skolnik, "Radar Handbook", McGraw-Hill Book Company, New York, N.Y., 1970.

[5] D. Atlas, "Advances in Radar Meteorology", in H.E. Landsberg and J. van Mieghen (eds.): "Advances in Geophysics", vol. 10, pp. 318-468, Academic Press, Inc., New York, N.Y., 1964.

[6] S. Haykin and S. Kesler, "Prediction Error Filtering and Maximum Entropy Spectral Estimation", Chapter 2 in S. Haykin (Editor): "Nonlinear Methods of Spectral Analysis", Springer, Berlin, 1979.

[7] B.W. Currie and S. Haykin, "Radar Clutter Classifier Prototype for Use with the TRACS Airport Surveillance Radar", Proc. 1st North American Birdstrike Prevention Workshop, pp. 6.1-6.30, Ottawa, Ontario, 17-18 Sept. 1981.

A SIMPLE APPROACH TO HIGH-RESOLUTION SPECTRAL ANALYSIS

R.A. Wagstaff and J.L. Berrou

NATO SACLANT ASW Research Centre
Viale San Bartolomeo, 400
19026 La Spezia, Italy

ABSTRACT

A non-linear high-resolution technique for spectral analysis is presented. This technique differs significantly from those presently in use in that it operates on the system outputs rather than the input sampled data. The technique has general application and can be used wherever a response function is known or can be assumed. It has been applied to obtain the undersea ambient noise spatial spectrum from towed line array data and the frequency spectrum from time history data. The technique is relatively insensitive to errors and can handle data with high dynamic ranges as a result of its non-linearity. Examples of improved resolution in both space and frequency are given.

INTRODUCTION

The Wagstaff-Berrou wideband (WB^2) technique for high-resolution spectral analysis presented in this paper was initially developed and extensively used to estimate the undersea ambient noise spatial spectrum from towed line-array data (1). The robustness of the technique and recent improvements in the algorithm make it an attractive candidate for other applications of high-resolution spectral analysis. The authors view it as a practical solution to a complex problem. Infact, we think of it as nothing more elaborate than common sense implemented in computer coding.

WB^2 has a basic approach similar to some of the constrained iterative restoration algorithms discussed by Shafer, Mersereau, and Richards (2). However, it differs significantly in that it is non-linear and is definable only in algorithm form. WB^2 is not at all similar to the high-resolution techniques discussed by Kay and Marple (3). Those techniques are designed to resolve signals. WB^2 on the other hand is designed to resolve colored noise. However, it fits well into this latter category of high-resolution techniques because it will give high-resolution if lines are present.

R. S. Bucy and J. M. F. Moura (eds.), Nonlinear Stochastic Problems, 55–62.

There are additional important ways in which WB^2 differs from other high-resolution methods. It uses very simple non-linear operations to solve the problems associated with the estimation process. The matrix multiplications are replaced by decibel additions and the matrix inversions are replaced by sign changes. Since the mathematics are performed in decibels and are relative rather than absolute, negative powers are not possible and oscillation does not occur. As a result, the technique can handle data with a high dynamic range, estimate the low levels with as much confidence as the high levels, and do so while keeping only four significant figures in the calculations. The simple mathematics on decibel quantities is also responsible for the robustness of the technique. For example, two or three dead hydrophones from a 40 element array have very little effect on the final result and no effect on the rate of convergence.

THE TECHNIQUE

The assumptions inherent in the WB^2 technique are:

- separate sources within the same bin are statistically independent
- the system response functions are either known or can be estimated
- the parameter being estimated is colored noise unless the data indicate otherwise
- the parameter can be estimated from the system output data and a knowledge of the system response functions.

The basic concept of WB^2 is very simple. The implementation is also simple. However, since it is an algorithm rather than a set of mathematical formulas, it is difficult to define in explicit terms.

For simplicity, the WB^2 technique will be explained in terms of a sonar system which has a line array. The system outputs utilized by WB^2 are the set of beam levels from the beamformer. The system response functions required by WB^2 are the set of beam patterns of the arrays. These two sets are all that are required by WB^2 to estimate the spatial spectrum of the acoustic field.

The procedure by which WB^2 obtains a high-resolution spectral estimate is the following. First a guess of the spectrum is made. This "guess" can be nearly anything, including a flat spectrum. Next, this "guess" is convolved with the beam patterns of the array to get estimated beam levels. These are then compared with the measured levels. A fraction of the difference for each beam is stored in accumulators corresponding with each point in the estimated spatial spectrum covered by the beam. Since the beams overlap in their spatial coverage, each point will fall within the mainlobe coverage area of several beams. These accumulated differences are then added to the original "guess" to produce an improved estimate of the spectrum.

This addition and subtraction process is done in terms of decibel quantites. For example, if the calculated beam level 3 dB is too low; some of the spectral points within the beam coverage area will have a fraction of 3 dB entered into their accumulators and added to their levels when generating the next spectral estimate. Since this is done in dB and not power, it is non-linear. This permits the calculations to be done with only four significant figures. It also permits WB^2 to handle data with a high dynamic range and estimate the low level components with the same confidence as the high level components.

This convolution - comparison - modification process is continued until the estimated beam levels are in acceptable agreement with the measured ones. Less than 20 iterations are usually adequate. The final spectral estimate is the high-resolution spatial spectrum. A logic flow chart is included in Fig. 1.

WB^2 was explained in terms of estimating the spatial spectrum from line array beam data. The example could well have been the frequency spectrum from the output of a spectrum analyzer. After all, the beam patterns of the array are simply spatial filters. Furthermore, it could have been any other type of processor for which outputs and response functions are available.

RESULTS

Some characteristics of WB^2 can be illustrated by the modelled spatial spectrum in Fig. 2 which consists of a two step platform with a line component at the edge of a step. The top curve is the result obtained by convolving the array beam pattern with the spatial spectrum. This is equivalent to taking a 64 point spatial FFT which has 24 zero padding. The step function in this figure is the WB^2 estimate. The final curve at 0 dB gives the errors between the measured and calculated beam levels. The discontinuities in the WB^2 estimate are not as sharp as the true spectrum but the levels are generally correct within a decibel. The apparent differences between the estimated and true line component is an artifact of the plot. The power is contained in many resolution cells of the beam containing the line. The total level is obtained by adding the power in each resolution cell. The plotted level could be increased simply by decreasing the number of resolution cells, thereby forcing each cell to contain more power. These results illustrate the ability of the WB^2 technique to handle a line in a variable background of white noise.

The results in Fig. 3 are from a towed line array of 40 hydrophones in excellent condition. The top curve is the beamformer output levels. The peak between beams 20 and 25 is the noise received from the towship. The beams outside the range 22 to 42 are virtual beams which do not receive acoustic power on their mainlobes. They receive it only through sidelobes extending into acoustic space. These are the beams produced by the FFT beamformer which have phase shifts greater than that corresponding to an endfire beam. The dynamic range of these data exceeds 40 dB. The curve immediately below the beam data is the WB^2 high-resolution estimate of the spatial spectrum. The towship noise component has been well resolved. Several other sources which are not obvious in the beam data have also been resolved. The error curve at the bottom of the plot indicates the WB^2 result is excellent (less than a dB) along most areas in "acoustic space" (between beams 22 and 42). Agreement outside this range is not expected since only the mainlobe response is used in the calculations and not the sidelobes. The acoustic power on the virtual beams is from sidelobes.

The data in Fig. 4 are when the same array was degraded by three faulty hydrophones out of 40. The dynamic range of the beam data has been reduced considerably due to the bad hydrophones. The towship noise line is not as well resolved as before. Four other low-level sources have been resolved as well. This curve has been normalized by the noise beamwidth to illustrate the gain due to WB^2. The confidence curve indicates that the calculated beam levels are generally within 1 dB of the measured beam levels in real acoustic space. These results demonstrate that the WB^2 technique is relatively insensitive to errors in the data and to errors due to using the incorrect (ideal rather than actual) beam response patterns.

Kay and Marple (3) did an excellent review of eleven of the more commonly used methods for spectrum estimation. In discussing the characteristics of each method they found it useful to utilize each to calculate the frequency spectrum based on the same 64-point real time process containing three sinusoids and colored noise. Figure 5 contains the true spectrum (upper left-hand corner) and the results from the eleven methods (Fig. 16 of (3) reproduced with permission and without modification). A common characteristic of the methods in Fig. 6 seems to be the failure to adequately handle discrete lines and colored noise simultaneously. Some cannot handle colored noise at all.

The WB^2 estimate of the spectrum in Fig. 5a is given in Fig. 6. The scale at the bottom of this plot is beam number because the data normally used by the authors comes from a towed line-array. However, beam patterns are analogous to filter response patterns. They are interchangeable. The locations and levels of the three line components have been estimated very well. The level of the lowest line component appears about 3 dB below the level of the original line. This, again, is an artifact of the plot. The total level of the bin is obtained by summing the power in each resolution cell. WB^2 got 20 dB for this line, the exact value. The same is true for the two lines closely spaced.

The colored noise in the process of Fig. 5a is reasonably well estimated by WB^2, considering that only 64 points of the time-series were used. Sixty-four points are not enough to adequately characterize the colored noise; an infinite series is required. If a feature is not contained in the input data, WB^2 will not get it out. It would, however, faithfully reproduce the colored noise if the data stream were sufficiently long to characterize it. Even so, WB^2 produces a very good estimate of the original spectrum. For this process, the WB^2 result seems to be a better estimate than the other results in Fig. 5.

SUMMARY

WB^2, a relatively robust and simple technique for high resolution spectral analysis, has been presented. The technique is sufficiently general that it can be applied whenever a system response function is reasonably well known or can be estimated. It utilizes additions and subtractions of decibel quantities to replace matrix multiplication and inversion operations. As a result of its non-linearity, it requires as few as four significant figures in the calculations, can handle extremely high dynamic ranges while still representing low-level sources well, and is relatively insensitive to errors. The examples of frequency and spatial spectra demonstrate that the technique successfully estimates arbitrary combinations of line components and white or colored noise.

REFERENCES

1. Wagstaff, R.A.: 1978, "Iterative technique for ambient noise horizontal directionality estimation from towed line-array data", J. Acoust. Soc. Am. 63, pp 863-869.

2. Shafer, R.W., Mersereau, R.M. and Richards, M.A.: 1981, "Constrained Iterative Restoration Algorithms", Proc. IEEE, Vol. 69, No. 4, pp 432-460.

3. Kay, S.M. and S.L. Marple, Jr.: 1981, "Spectrum Analysis - A Modern Perspective", Proc. IEEE, Vol. 69, No. 11, pp. 1386-1419.

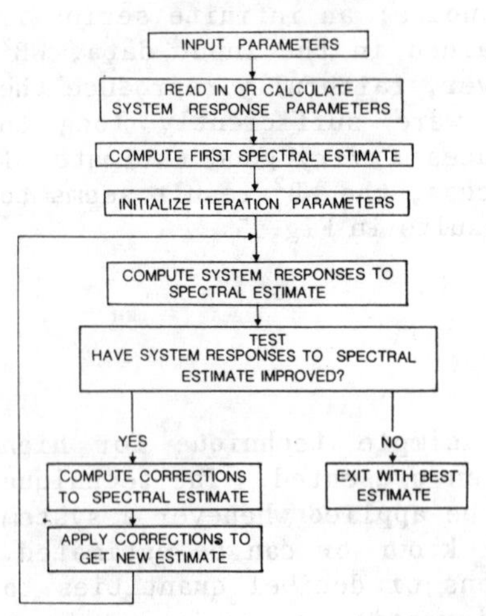

Fig. 1
Flow chart of the WB^2 high-resolution spectral analysis method.

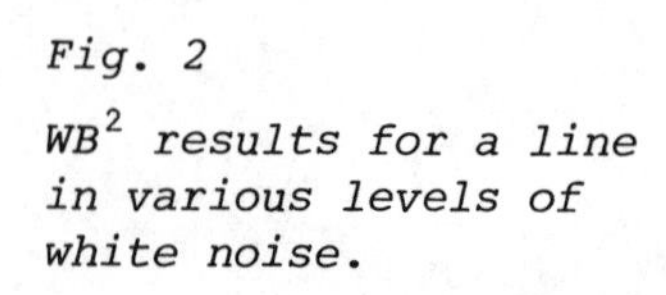

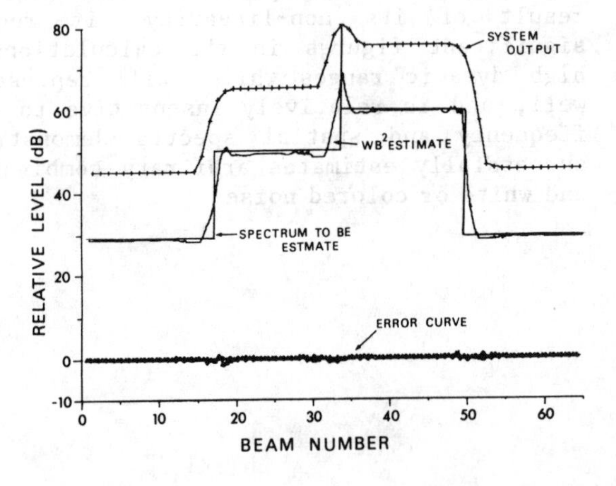

Fig. 2
WB^2 results for a line in various levels of white noise.

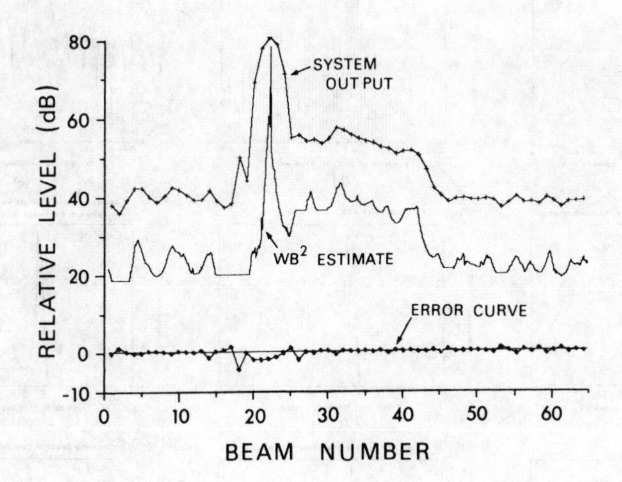

Fig. 3 *WB^2 results for a towed array system in perfect condition with 40 hydrophones. Dominant source is towship noise.*

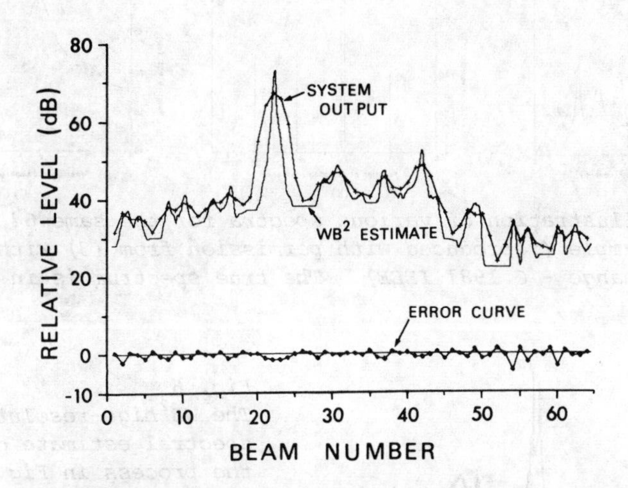

Fig. 4 *WB^2 results for a towed array system degraded by three faulty hydrophones out of 40. Dominant source is towship noise.*

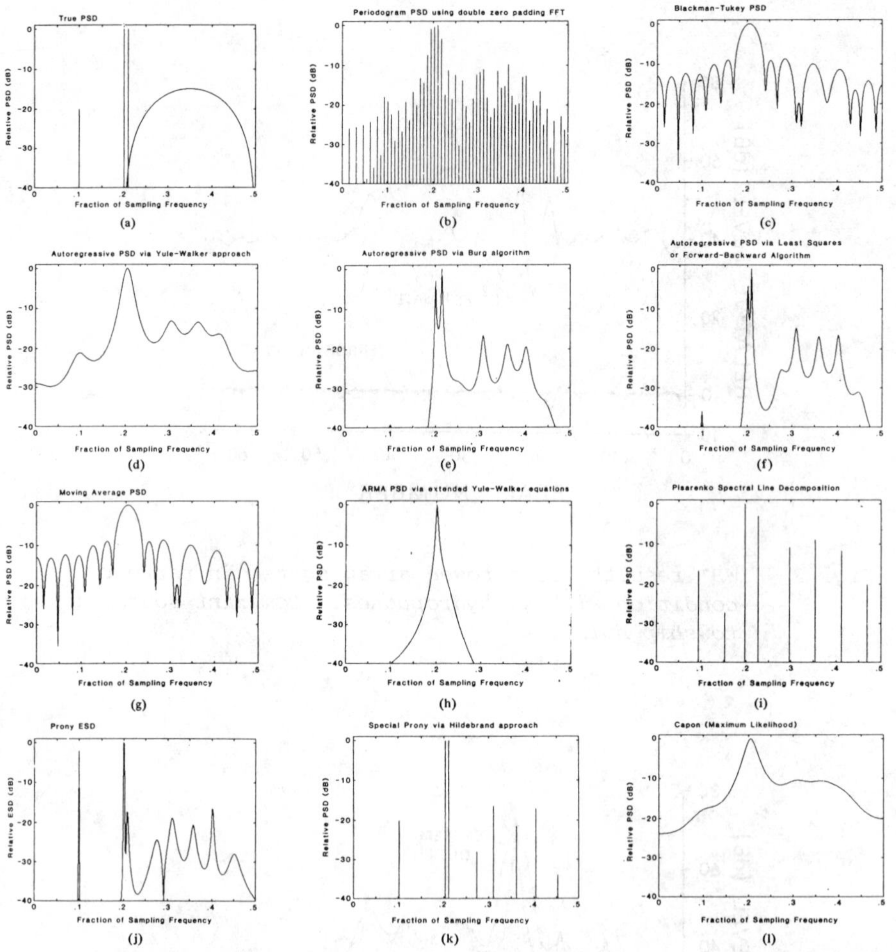

Fig. 5 *Illustration of various spectra for the same 64 point sample (reproduced with permission from (3) without change - C 1981 IEEE). The true spectrum is in (a).*

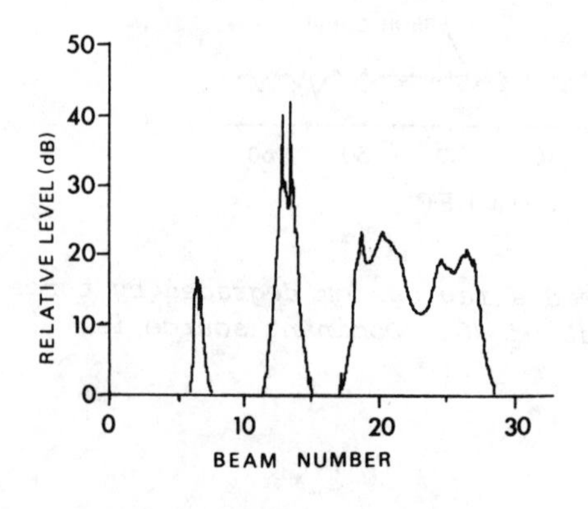

Fig. 6
The WB^2 high-resolution spectral estimate of the process in Fig.5a. The same 64 datapoints were used by WB^2 as were used to get the other results in Fig.5.

CHAPTER III

IDENTIFICATION

ESTIMATION OF STOCHASTIC PARAMETERS FOR ARMA MODELS BY FAST FILTERING ALGORITHMS

G. Alengrin

Université de Nice - Laboratoire de Signaux et Systèmes
Equipe de Recherche associée au CNRS, n°835
41, Bd Napoléon III - 06041 NICE CEDEX, France

Abstract : The aim of this paper is to present some remarks about the parametrization of multivariable ARMA models. First a canonical state-space model is introduced, displaying two different kinds of parameters. Then parameters of the stochastic part of the model are estimated by fast filtering algorithms. Finally a method is given to derive explicitely all sets of stochastic parameters leading to equivalent wide sense time series.

PREFACE

The purpose of this paper is to study representations of vectorial rational stationary time series.
Two kinds of representation are generally used, either a state-space representation or the so-called auto-regressive moving average (ARMA) model. In the following we show the connections between these two representations and we point out the usefulness of particular state-space representations for the estimation of ARMA parameters.
The multivariable ARMA model is written as :

$$\mathscr{A}(z^{-1})\, y(k) = \mathscr{B}(z^{-1})\, e(k)$$

defining a stationary gaussian time series $\{y(k)\}$, with zero mean. $e(k)$ is a white gaussian noise vector.
Dimensions of the observed process $y(k)$ and of the noise process $e(k)$ are :

$$\dim(y) = p \qquad \dim(e) = p.$$

$\mathscr{A}(z^{-1})$ and $\mathscr{B}(z^{-1})$ are polynomial matrices of the unit lag operator z^{-1} :

$$\mathscr{A}(z^{-1}) = \Sigma\, A_i\, z^{-i} \qquad \mathscr{B}(z^{-1}) = \Sigma\, B_i\, z^{-i}$$

R. S. Bucy and J. M. F. Moura (eds.), Nonlinear Stochastic Problems, 65–78.

1. PARAMETRIZATION OF AN ARMA MODEL

1.1 Relationship between state-space and ARMA model

Assume that the time series $\{y(k)\}$ is generated by the following linear stationary dynamical system :

$$\begin{aligned} x(k+1) &= F\, x(k) + K\, \tilde{e}(k) \\ y(k) &= H\, x(k) + \tilde{e}(k) \end{aligned} \tag{1.1}$$

with : $\dim(y) = p$, $\dim(\tilde{e}) = p$, $\dim(x) = n$, (F,H) completely observable, $\tilde{e}(k)$ a white gaussian noise vector.

Moreover $\operatorname{rank}(H) = p$ with $H = \begin{pmatrix} h_1 \\ h_2 \\ \vdots \\ h_p \end{pmatrix}$.

In this section we shall develop an ARMA model displaying the structural parameters and a minimal number of characteristic parameters.

Consider, as in [1], the ordered set of vectors
$h_1', h_2', \ldots, h_p', F'h_1', \ldots, F'h_p', (F')^2 h_1', \ldots$

The vector selection is achieved by scanning this set from the left and by keeping a vector if it is linearly independent of its antecedents.

Let n_i be the smallest integer such that $(F')^{n_i} h_i'$ is a linear combination of its antecedents. n_i is the ith Kronecker invariant and then

$$h_i F^{n_i} = \sum_{k=1}^{i-1} \sum_{j=o}^{\alpha_k} a_{ijk}\, h_k F^j + \sum_{k=i}^{p} \sum_{j=o}^{\beta_k} a_{ijk}\, h_k F^j \tag{1.2}$$

with $\begin{aligned} \alpha_k &= \min(n_i, n_k - 1) \\ \beta_k &= \min(n_i - 1, n_k - 1) \end{aligned}$ and $\sum_{i=1}^{p} n_i = n$

$\{n_i, a_{ijk}\}$ are fundamental invariants of the system.

Defining, as in [2], generalized parameters from the relationship:

$$h_i F^{n_i} = \sum_{k=1}^{i-1} \sum_{j=o}^{n_i} \overline{a_{ijk}}\, h_k F^j + \sum_{k=i}^{p} \sum_{j=o}^{n_i} \overline{a_{ijk}}\, h_k F^j \tag{1.3}$$

with :

$$\left.\begin{aligned} \overline{a_{ijk}} &= o \quad \text{if } j > n_k - 1 \\ \overline{a_{i,n_i,k}} &= o \quad \text{if } k \geq i \\ \overline{a_{ijk}} &= a_{ijk} \quad \text{otherwise} \end{aligned}\right\} \tag{1.4}$$

it is easy to show that the multivariable ARMA model associated with the system (1.1) is given by :

$$\sum_{j=o}^{n_m} A_j \, y(k-j) = \sum_{j=o}^{n_m} D_j \, \tilde{e}(k-j) \qquad (1.5)$$

where $n_m = \max(n_i)$ is the greatest Kronecker index of the system, A_j and D_j being defined as follows :

$$A_o = \begin{pmatrix} 1 & 0 \ldots\ldots & 0 \ldots & 0 \\ \overline{-a_{2,n_2,1}} & 1 \ldots\ldots & 0 \ldots & 0 \\ \overline{-a_{3,n_3,1}} & \overline{-a_{3,n_3,2}} & 1 \ldots & 0 \\ \vdots & & \vdots & 1 \end{pmatrix} \quad A_j = -\begin{pmatrix} \overline{a_{1,n_1-j,1}} \ldots & \overline{a_{1,n_1-j,p}} \\ \overline{a_{2,n_2-j,1}} \ldots & \overline{a_{2,n_2-j,p}} \\ \vdots & \vdots \\ \overline{a_{p,n_p-j,1}} \ldots & \overline{a_{p,n_p-j,p}} \end{pmatrix} \qquad (1.6)$$

with $\overline{a_{ijk}}$ as in (1.4) and the additional condition $\overline{a_{ijk}} = o$ if $j < o$.

Moreover $D_o = A_o$ and D_j takes the general form :

$$D_j = \begin{pmatrix} \beta_{1,n_1-j,1} \ldots\ldots & \beta_{1,n_1-j,p} \\ \beta_{2,n_2-j,1} \ldots\ldots & \beta_{2,n_2-j,p} \\ \vdots & \vdots \\ \beta_{p,n_p-j,1} \ldots\ldots & \beta_{p,n_p-j,p} \end{pmatrix} \quad \text{with } \beta_{ijk} = o \text{ if } j < o$$

The left side of the ARMA model (1.5) is thus set by a minimal number of characteristic parameters a_{ijk} deduced from the deterministic part of the system (F,H).

By analogy, we shall call stochastic parameters the elements of matrices D_j and of the covariance matrix of $\tilde{e}(k)$.

From (1.5) it is also easy to infer the standard ARMA model :

$$\sum_{j=o}^{n_m} A_j \, y(k-j) = \sum_{j=o}^{n_m} B_j \, e(k-j) \qquad (1.7)$$

by defining $e(k) \triangleq A_o \, \tilde{e}(k)$.

Then we have B_o = identity matrix and

$$B_j = \begin{pmatrix} b_{1,n_1-j,1} \ldots\ldots & b_{1,n_1-j,p} \\ b_{2\ n_2-j,1} \ldots\ldots & b_{2,n_2-j,p} \\ \vdots & \vdots \\ b_{p,n_p-j,1} \ldots\ldots & b_{p,n_p-j,p} \end{pmatrix} \quad \text{with } b_{ijk} = o \text{ if } j < o \qquad (1.8)$$

1.2 Canonical state-space representation of ARMA models

From the above considerations, it is now possible to derive a canonical state-space representation displaying directly the dynamical parameters a_{ijk} and the stochastic parameters b_{ijk}. The following result can then be proved [3] :

A canonical state-space representation of the ARMA model (1.7) is given by

$$\begin{cases} \zeta(k+1) = J\,\zeta(k) + A\,y(k) + B\,e(k) \\ A_o\,y(k) = C\,\zeta(k) + e(k) \end{cases} \tag{1.9}$$

where :

$$J = \begin{pmatrix} J_1 & & & 0 \\ & J_2 & & \\ & & \ddots & \\ 0 & & & J_p \end{pmatrix} \text{ with } J_i = \begin{pmatrix} 0 & \cdots & \cdots & 0 \\ 1 & \ddots & & \vdots \\ 0 & \ddots & \ddots & \vdots \\ 0 & \cdots & 0\;1 & 0 \end{pmatrix} \quad C = \begin{pmatrix} C_1 & & & 0 \\ & C_2 & & \\ & & \ddots & \\ 0 & & & C_p \end{pmatrix}$$

and dim $J_i = (n_i \times n_i)$ with $C_i = (0 \;\cdots\; 0\; 1)$

and dim $C_i = (1 \times n_i)$

Matrix A is made from the deterministic parameters a_{ijk}, and matrix B from the parameters b_{ijk} of the stochastic part.

One can easily verify the above result by deriving the input-output relationship from (1.9), using the two following properties:

$$J_i^m = 0 \qquad \forall\, m \geq n_i \tag{1.10}$$

$$C_i\, J_i^{\ell} = (0 \;\cdots\; 0\; 1\; 0 \;\cdots\; 0) \qquad \forall\, \ell < n_i \tag{1.11}$$

$$\uparrow \text{ column } n_i - \ell$$

We shall now consider the estimation of parameters a_{ijk} and b_{ijk}, given the covariance of the time series $\{y(k)\}$.
There exist several estimation algorithms for the determination of the structural parameters n_i and of the parameters a_{ijk} . See for example [4],[5]. These algorithms are equivalent to a deterministic realization algorithm for a canonical structure and may be considered as the first step of the estimation procedure.

1.3 Canonical representation of the M.A. part

Once the structural and dynamical parameters have been determined, one can define a "residual" process

$$\epsilon(k) = \mathscr{A}(z^{-1})\, y(k) \tag{1.12}$$

which is a stationary process and, clearly a multivariable M.A. process as :

$$\epsilon(k) = \mathscr{B}(z^{-1})\, e(k) \tag{1.13}$$

A simplified canonical stationary state-space representation for the process is given by :

$$\begin{cases} \xi_2(k+1) = J\,\xi_2(k) + B\,e(k) \\ \epsilon(k) = C\,\xi_2(k) + e(k) \end{cases} \tag{1.14}$$

where J,B and C have been defined in section 1.2.
J and C are completely identified by the structural parameters n_i.

From the residual process (1.12), we can obtain estimates of the covariance matrices :

$$\Lambda(i)= E[\epsilon(k+i)\ \epsilon'(k)]$$

And the problem of estimating the elements of B is set in terms of a stochastic realization problem.
From the model (1.14) we obtain :

$$\Lambda(i)= C\ J^{i}\Pi_o\ C'+ C\ J^{i-1}B\ \tilde{Y}_o \quad i > o \tag{1.15}$$

$$\Lambda(o)= C\ \Pi_o\ C'+ \tilde{Y}_o \tag{1.16}$$

where $\tilde{Y}_o= E[e(k)\ e'(k)]$ and $\Pi_o= E[\xi_2(k)\ \xi_2'(k)]$

Then
$$\begin{pmatrix} \Lambda(1) \\ \vdots \\ \Lambda(n_m) \end{pmatrix} = \mathcal{O}(J\ \Pi_o\ C'+ B\ \tilde{Y}_o)$$

with $\mathcal{O} \triangleq \begin{pmatrix} C \\ C\ J \\ \vdots \\ C\ J^{n_m-1} \end{pmatrix}$ the observability matrix.

Properties (1.10) and (1.11) give $\mathcal{O}'\mathcal{O}= I_{p\times p}$

so that we define :

$$G \triangleq J\ \Pi_o C'+ B\ \tilde{Y}_o= \mathcal{O}' \begin{pmatrix} \Lambda(1) \\ \vdots \\ \Lambda(n_m) \end{pmatrix} \tag{1.17}$$

The known quantities J,C,G "realize" the covariance $\Lambda(i)=C\ J^{i-1}G \quad i>o.$

2. A REVIEW OF THE STOCHASTIC REALIZATION PROBLEM [7]

2.1 Problem statement

For the residual process $\epsilon(k)$, the stochastic realization problem can be stated as follows :
Given the covariance $\Lambda(i)$ of the stationary process $\epsilon(k)$, determine a markovian representation of the form :

$$\left\{ \begin{aligned} \xi(k+1) &= J\ \xi(k)+ B_1 u(k)+ B_2 v(k) \\ \eta(k) &= C\ \xi(k)+ u(k) \end{aligned} \right. \tag{2.1}$$

such that $\eta(k)$ has the same covariance as $\epsilon(k)$. u(k) and v(k) are white gaussian noises with covariances :

$$E\left[\begin{pmatrix} u(k) \\ v(k) \end{pmatrix}(u'(k)\ v'(k))\right] = \begin{pmatrix} R_1 & 0 \\ 0 & R_2 \end{pmatrix}.$$

To each markovian representation (2.1), corresponds a set of matrices such that :

$$\left\{\begin{array}{l} P-J\ P\ J' = B_1 R_1 B_1' + B_2 R_2 B_2' \\ G-J\ P\ C' = B_1 R_1 \\ \Lambda(o)-C\ P\ C' = R_1 \\ \begin{pmatrix} R_1 & 0 \\ 0 & R_2 \end{pmatrix} \geqslant o \quad P > o \end{array}\right. \qquad (2.2)$$

Definition : We define by $\mathscr{P}$ the set of all symetric positive definite matrices P satisfying (2.2). It is well known [7] that $\mathscr{P}$ is a closed bounded convex set.
As a consequence of the positive real lemma, the set $\mathscr{P}$ associated with $\Lambda(i)$ is non empty, $\mathscr{P}$ has a maximal element P^* and a minimal element P_* i.e. $P - P_* \geqslant o \ \forall\, P \in \mathscr{P}$.

2.2 Algorithms for the determination of P_* and P^*

Algorithm 2.1 [7] : The minimal element P_* of $\mathscr{P}$ can be computed by generating the sequence :

$$\left\{\begin{array}{l} \Pi_{n+1} = J\ \Pi_n\ J' + (G-J\ \Pi_n\ C')(\Lambda(o)-C\ \Pi_n\ C')^{-1}\ (G'-C\ \Pi_n\ J') \\ \Pi_o = o \end{array}\right. \qquad (2.3)$$

Then $P_* = \lim_{n\to\infty} \Pi_n$

Algorithm 2.2 [7] : The maximal element P^* of $\mathscr{P}$ can be computed by generating the sequence :

$$\left\{\begin{array}{l} \Omega_{n+1} = J'\ \Omega_n\ J + (C'-J'\Omega_n\ G)(\Lambda(o)-G'\Omega_n\ G)^{-1}\ (C-G'\Omega_n\ J) \\ \Omega_o = o \end{array}\right. \qquad (2.4)$$

Then $P^* = \lim_{n\to\infty} \Omega_n^{-1}$

Consider the mapping [8]

$$\Gamma(P) = -P + J\ P\ J' + (G-J\ P\ C')(\Lambda(o)-C\ P\ C')^{-1}\ (G'-C\ P\ J') \qquad (2.5)$$

Then we have the following result :

Theorem 2.1 [8] :
Let $\Gamma(\cdot)$ be given by (2.5), then $\mathscr{P}$ is characterized by :

$$\mathscr{P} = \{P \mid P=P'\ ,\ \Gamma(P) \leqslant o\}$$

We can easily check that :

$$\Gamma(P) = -B_2 R_2 B_2'\ .$$

In our case of estimation of an ARMA model we are looking for markovian representations with a single noise sequence, then $B_2 = o$. Solutions to take into account are thus the P matrices belonging to the set : $\mathscr{P}_o = \{P \in \mathscr{P} \mid \Gamma(P) = o\}$.

We shall call these solutions equilibrium solutions of (2.3).
The solution $P_* \in \mathscr{P}_o$ leads to the well known steady-state Kalman filter :

$$\begin{cases} \xi_*(k+1) = J\,\xi_*(k) + B_*\,e_*(k) \\ \eta(k) = C\,\xi_*(k) + e_*(k) \end{cases} \qquad (2.6)$$

with $R_* = \Lambda(o) - C\,P_*\,C'$

and $B_* = (G - J\,P_* C')R_*^{-1}$
uniquely determined from (2.2).

Furthermore the dynamical matrix $\Phi_* = J - B_* C$ is asymptotically stable.
This representation gives the minimal phase representation of the M.A. process with spectral factor :

$$\mathscr{B}_*(z^{-1}) = C(z\,I - J)^{-1} B_* + I.$$

Consider now the dual problem defined by the transformation $(C, J, G, \Lambda(o)) \longrightarrow (G', J', C', \Lambda(o))$ and corresponding to the dual operator $\bar{\Lambda}(i) = \Lambda(-i)$. To the stochastic realization of $\bar{\Lambda}(i)$ is associated a backward stationary representation :

$$\begin{cases} \bar{\xi}(k-1) = J'\,\bar{\xi}(k) + \bar{B}_1 \bar{u}(k) + \bar{B}_2 \bar{v}(k) \\ \eta(k) = G'\,\bar{\xi}(k) + \bar{u}(k) \end{cases} \qquad (2.7)$$

with state covariance $\bar{P}$.
Then we have the following result :

Theorem 2.2 [7] :
If $P > o$ is a solution of (2.2), then $\bar{P} = P^{-1}$ is solution of the dual problem and conversely.
Then $(\bar{P}_*)^{-1} = P^*$.

$\bar{P}_*$ is thus given as an equilibrium point of equation (2.4), and by duality P^* is an (unstable) equilibrium point of equation (2.3). $P^* \in \mathscr{P}_o$ and consequently we can give another markovian representation for the M.A. process as :

$$\begin{cases} \xi^*(k+1) = J\,\xi^*(k) + B^*\,e^*(k) \\ \eta(k) = C\,\xi^*(k) + e^*(k) \end{cases} \qquad (2.8)$$

with $R^* = \Lambda(o) - C\,P^*\,C'$
$B^* = (G - J\,P^*\,C')(R^*)^{-1}$

this representation gives the spectral factor :

$$\mathscr{B}^*(z^{-1}) = C(z\,I - J)^{-1}\,B^* + I$$

with all the roots of $\det(\mathscr{B}^*(z^{-1})) = o$ unstables.

In the following section we examine new fast algorithms for the determination of the two extremal representations (2.6) and (2.8).

3. FAST ALGORITHMS FOR THE DETERMINATION OF STATISTICAL PARAMETERS

3.1 Algorithm for minimal phase representation

In the last section we have shown that the determination of the M.A. parameters of an ARMA model could be carried out by using stochastic realization algorithms leading to extremal realizations. But the Riccati equations (2.3) or (2.4) require at each step of the recurrence the updating of $\frac{n(n+1)}{2}$ non-linear equation.

Furthermore, for our problem, it is not necessary to know the value of Π_n or Ω_n at each step, since we need only to know B_* and R_* (or B^* and R^*).

In this section we derive non-Riccati algorithms for the direct determination of the statistical parameters B and R with a reduced number of non-linear equations.
These algorithms are strongly connected with results of [9],[10] and are called "fast" algorithms.
Consider the canonical state-space representation defined in section 1 :

$$\begin{cases} \xi_2(k+1) = J\,\xi_2(k) + B\,e(k) \\ \epsilon(k) = C\,\xi_2(k) + e(k) \end{cases} \qquad (3.1)$$

For this model, the filtering equations are :

$$\hat{\xi}_2(k+1/k) = J\,\hat{\xi}_2(k/k-1) + B(k)\,\nu(k) \qquad (3.2)$$

where $\bar{\nu}(k) \triangleq \epsilon(k) - C\,\hat{\xi}_2(k/k-1)$.

Define the state covariance of (3.2) at time k as

$$P(k) \triangleq E[\hat{\xi}_2(k/k-1)\,\hat{\xi}_2'(k/k-1)]$$

then we have : $$B(k) = G(k)[R^{\nu}(k)]^{-1} \qquad (3.3)$$

$$\text{with}\ \begin{cases} G(k) = G - J\,P(k)\,C' \\ R^{\nu}(k) = \Lambda(o) - C\,P(k)\,C' \triangleq E[\nu(k)\,\nu'(k)] \end{cases} \qquad (3.4)$$

with initial conditions :

$$G(o) = G \quad \text{and} \quad R^{\nu}(o) = \Lambda(o) \qquad (3.5)$$

Furthermore, from equation (3.2) we obtain :

$$P(k+1) = J\,P(k)J' + B(k)\,R^{\nu}(k)\,B'(k). \qquad (3.6)$$

Define now the increment :

$$\delta P(k+1) = P(k+1) - P(k) \qquad (3.7)$$

$$\text{with}\quad \delta P(1) = G\,\Lambda^{-1}(o)G' = L(o)\,M(o)\,L'(o) \qquad (3.8)$$

$$\begin{cases} L(o) = G & \dim L(o) = (n\times p) \\ M(o) = \Lambda^{-1}(o) & \dim M(o) = (p\times p) \end{cases} \qquad (3.9)$$

Then we derive the algorithm by relating the factorized form of the increment at time k

$$\delta P(k) = L(k-1)\,M(k-1)\,L'(k-1) \qquad (3.10)$$

with the factorized form at time k+1. We obtain the following equations, derived in [3],[6].

Algorithm 3.1 [6] : The matrices G(k) and $R^{\nu}(k)$ defined in (3.4) are obtained recursively by solving the set of equations :

$$G(k)= G(k-1)-J\ L(k-1)\ M(k-1)\ L'(k-1)\ C' \tag{3.11}$$
$$L(k)= J\ L(k-1)- G(k-1)[R^{\nu}(k-1)]^{-1} C\ L(k-1) \tag{3.12}$$
$$R^{\nu}(k)= R^{\nu}(k-1)- C\ L(k-1)\ M(k-1)\ L'(k-1)\ C' \tag{3.13}$$
$$M(k)= M(k-1)+ M(k-1)\ L'(k-1)C'[R^{\nu}(k)]^{-1} C\ L(k-1)\ M(k-1) \tag{3.14}$$

with initial conditions (3.5) and (3.9).

Remark : The matrix B(k-1) defined by (3.3) is explicitely calculated in (3.12).

Algorithm 3.1 has been obtained by factorizing the same matrix as in Algorithm 2.1, then one can use the same convergence results together with (3.4) to see that :

$$\begin{cases} B_* = G(\infty)[R^{\nu}(\infty)]^{-1} \\ R_* = R^{\nu}(\infty) \end{cases} \tag{3.15}$$

Thus this algorithm gives the minimal phase representation (2.6) of the M.A. part of the process and from this representation we can easily determine the parameters of matrices B_j (1.8).

Equations (3.11)-(3.14) consist of 2np elements for L(k) and G(k) and p(p+1) elements for M(k) and $R^{\nu}(k)$. Thus 2np + p(p+1) equations have to be updated at each step. This result can be compared with the $\frac{n(n+1)}{2}$ equations for Algorithm 2.1.

Furthermore, by the structural properties of J this matrix has p null rows, then there exist p^2 invariant elements in G(k) equation (3.11) so that finally only (2n+1)p equations are to be solved.

Remark : In the scalar case (p=1), algorithms similar to Algorithm (3.1) have been derived in [11],[12]. The structure of these algorithms allows a parallel calculation of each component of the gain vector B(k).

3.2 Dual algorithm

By duality we can now study the backward filter

$$\begin{cases} \bar{\xi}(k-1)= J'\ \bar{\xi}(k)+ \bar{B}(k)\ \bar{\nu}(k) \\ \epsilon(k) = G'\ \bar{\xi}(k)+ \bar{\nu}(k) \end{cases} \tag{3.16}$$

the backward gain is given by :

$$\bar{B}(k) = \bar{G}(k)[\bar{R}(k)]^{-1} \tag{3.17}$$

with $\bar{P}(k) = E[\bar{\xi}(k)\ \bar{\xi}'(k)]$, $\bar{G}(k)= C'- J'\ \bar{P}(k)\ G$ (3.18)

and $\bar{R}(k) \triangleq E[\bar{e}(k)\ \bar{e}'(k)] = \Lambda(o)- G'\ \bar{P}(k)\ G$ (3.19)

If we define the increment

$$\delta\bar{P}(k)= \bar{P}(k-1)- \bar{P}(k)$$

we obtain, using the same development as in [3] the following algorithm :

Algorithm 3.2 The matrices $\bar{G}(k-1)$ and $\bar{R}(k-1)$ defined in (3.18) are obtained recursively by solving the set of equations :

$$\bar{G}(k-1) = \bar{G}(k) - J' L(k) M(k) L'(k) G \tag{3.20}$$
$$L(k-1) = (J' - \bar{B}(k) G') L(k) \tag{3.21}$$
$$\bar{R}(k-1) = \bar{R}(k) - G' L(k) M(k) L'(k) G \tag{3.22}$$
$$M(k-1) = M(k) + M(k) L'(k) G [\bar{R}(k-1)]^{-1} G' L(k) M(k) \tag{3.23}$$

with initial conditions :

$$\bar{G}(o) = L(o) = C' , \quad M(o) = [\bar{R}(o)]^{-1} = [\Lambda(o)]^{-1}.$$

At the convergence we have :

$$\begin{cases} \bar{B}_* = \bar{G}(-\infty)[\bar{R}(-\infty)]^{-1} \\ \bar{R}_* = \bar{R}(-\infty) \end{cases}$$

so that we can give the following backward representation of the process :

$$\begin{cases} \bar{\xi}_*(k-1) = J' \bar{\xi}_*(k) + \bar{B}_* \bar{e}_*(k) \\ \eta(k) = G' \bar{\xi}_*(k) + \bar{e}_*(k) \end{cases} \tag{3.24}$$

which corresponds to the extremal representation associated with P^* as $(\bar{P}_*)^{-1} = P^*$.

3.3 Doubling algorithm

For the discrete problem, it can be derived from [7] that the extremal elements P_* and P^* are related by

$$(P^* - P_*)^{-1} = \Phi_*' (P^* - P_*)^{-1} \Phi_* + C' R_*^{-1} C \tag{3.25}$$

Assuming that the minimal representation Φ_*, R_* has been determined, we can solve equation (3.25) by the following doubling algorithm :

$$S(2^{k+1}) = (\Phi_*'^2)^k S(2^k) (\Phi_*^2)^k + S(2^k) \tag{3.26}$$

with $S(1) = C' R_*^{-1} C$.

At the convergence of the algorithm we have thus :

$$P^* = P_* + S^{-1}(\infty).$$

4. DETERMINATION OF ALL STOCHASTIC REALIZATIONS FOR THE M.A. PROCESS

In the last section we have determined two sets of parameters for the M.A. process, one corresponding to the minimal state covariance P_*, the other to the maximal state covariance P^*. The aim of this section is to find other sets of parameters. From section 2 all these sets are characterized by matrices P belonging to $\mathcal{P}_o$ (equilibrium solutions). We give a method to find all the elements of $\mathcal{P}_o$ and consequently all stochastic realizations for the M.A. process.

Consider a stationary stochastic realization

$$\begin{cases} \xi_2(k+1) = J \xi_2(k) + B e(k) \\ \epsilon(k) = C \xi_2(k) + e(k) \end{cases} \tag{4.1}$$

with state covariance $\Pi_o \triangleq E[\xi_2(k) \xi_2'(k)]$ and $R \triangleq E[e(k) e'(k)]$

then we have the following lemma :

Lemma 4.1 :
Equilibrium solutions P of (2.3) ($P \in \mathscr{P}_o$) are connected to equilibrium solutions of the algebraic Riccati equation :

$$\Sigma = \Phi \Sigma \Phi' - \Phi \Sigma C'(C \Sigma C' + R)^{-1} C \Sigma \Phi' \tag{4.2}$$

by the relationship $P = \Pi_o - \Sigma$ (4.3)

with $\Phi \triangleq J - B C$. (4.4)

Proof : From equations (3.2),(3.3),(3.4) it is easy to derive the steady-state Kalman-filter for model (4.1) and to obtain for the stationary covariance $P \triangleq E[\hat{\xi}_2(k) \hat{\xi}_2'(k)]$ the relationship $\Gamma(P) = o$ with $\Gamma(P)$ defined by (2.5).
Model (4.1) may also be written as

$$\xi_2(k+1) = (J - B C)\, \xi_2(k) + B\, \epsilon(k) \tag{4.5}$$

Define the stationary error covariance as :

$$\Sigma \triangleq E[(\xi_2(k) - \hat{\xi}_2(k))(\xi_2'(k) - \hat{\xi}_2'(k))] = \Pi_o - P$$

then the steady-state Kalman-filter for (4.5) is :

$$\hat{\xi}_2(k+1) = \Phi\, \hat{\xi}_2(k) + B\, \epsilon(k) + \Phi \Sigma C'(C \Sigma C' + R)^{-1} \nu(k)$$

and Σ satisfies algebraic Riccati equation (4.2).

By this lemma it is then possible, using results on equilibrium solutions of algebraic Riccati equation [13],[14],[15], to solve the problem of finding all markovian representations of the M.A. process.

Consider the Hamiltonian matrix $\mathcal{H}$ associated with the filtering problem :

$$\mathcal{H} = \begin{pmatrix} (\Phi')^{-1} & (\Phi')^{-1} M \\ 0 & \Phi \end{pmatrix} \tag{4.6}$$

with $M = C'R^{-1}C$.

Lemma 4.2 :
The characteristic polynomial of $\mathcal{H}$ is given by :

$$\det(\lambda I - \mathcal{H}) = \frac{(-1)^n \lambda^n}{\det(\Phi)} \det(\lambda I - \Phi) \det(\tfrac{1}{\lambda} I - \Phi) \tag{4.7}$$

Proof : The block triangular structure of $\mathcal{H}$ yields to :

$$\begin{aligned}\det(\lambda I - \mathcal{H}) &= \det(\lambda I - (\Phi')^{-1}) \cdot \det(\lambda I - \Phi) \\ &= \det[-\lambda (\Phi')^{-1} (\tfrac{1}{\lambda} I - \Phi')] \det(\lambda I - \Phi)\end{aligned}$$

then we obtain equation (4.7).

This result gives a factorization of the characteristic polynomial in the following form

$$\det(\lambda I - \mathcal{H}) = \alpha\, D(\lambda) \cdot D(\tfrac{1}{\lambda})$$

where $D(\lambda) = \det(\lambda I - \Phi)$.

The eigenvalues of $\mathcal{H}$ consist then of the eigenvalues of Φ and their inverses. This result holds for all Φ; particularly when $\Phi = \Phi_*$ $D(\lambda)$ is a Hurwitz polynomial.
Thus the factorization of the characteristic polynomial of $\mathcal{H}$ can be carried out in several ways as

$$\det(\lambda I - \mathcal{H}) = \alpha\ \Delta(\lambda) \cdot \Delta\left(\frac{1}{\lambda}\right) \tag{4.8}$$

$\Delta(\lambda)$ being a polynomial of degree n, with m roots chosen among the eigenvalues of Φ_*.
The remaining (n-m) roots are then the inverses of the remaining (n-m) eigenvalues of Φ_*.

Theorem 4.1 :
Each equilibrium point $\Sigma.$ of the algebraic Riccati equation (4.2) is given by the relationship :

$$(-\Sigma.,I)\ \Delta(\mathcal{H}) = 0 \tag{4.9}$$

where $\Delta(\mathcal{H})$ is a matrix polynomial built from the factorization (4.8), and $\Delta(\lambda)$ is the characteristic polynomial of $\Phi - K.C$
where

$$K. \overset{\Delta}{=} \Phi\ \Sigma. C'(C\ \Sigma.\ C' + R)^{-1} \tag{4.10}$$

Outline of the proof :
For the continuous case the proof is given in [15]. Chosing the transformation matrix :

$$T = \begin{pmatrix} 0 & I \\ I & \Sigma. \end{pmatrix}$$

we get

$$T^{-1}\mathcal{H}T = \begin{pmatrix} -\Sigma.(\Phi')^{-1}M + \Phi & -\Sigma.(\Phi')^{-1} - \Sigma.(\Phi')^{-1}M\ \Sigma. + \Phi\ \Sigma. \\ (\Phi')^{-1}M & (\Phi')^{-1} + (\Phi')^{-1}M\ \Sigma. \end{pmatrix} \tag{4.11}$$

By assumption $\Sigma.$ is an equilibrium point of (4.2) then it can be checked that the upper right matrix is equal to zero. Furthermore the upper left matrix is equal to $\Phi - K.C$, with K. defined in (4.10), and the lower right matrix is $(\Phi' - C'\ K!)^{-1}$.

Then

$$T^{-1}\mathcal{H}\ T = \begin{pmatrix} \Phi - K.C & 0 \\ (\Phi')^{-1}M & (\Phi' - C'K!)^{-1} \end{pmatrix} \tag{4.12}$$

It follows that :

$$\Delta(\mathcal{H}) = T \begin{pmatrix} \Delta(\Phi - K.C) & 0 \\ N_1 & \Delta[(\Phi' - C'K!)^{-1}] \end{pmatrix} T^{-1}$$

and $\Delta(\Phi - K.C) = 0$ according to the Cayley-Hamilton theorem :
Then

$$\begin{pmatrix} -\Sigma. & I \\ I & 0 \end{pmatrix} \Delta(\mathcal{H}) = \begin{pmatrix} 0 & 0 \\ N_1 & N_2 \end{pmatrix} T^{-1}$$

which completes the proof.

From the above result, it is now possible to construct all the markovian representations of the M.A. process.
The number of different representations is equal to the number of combinations of eigenvalues in (4.8). For example if all the eigenvalues of Φ_* are distinct and real, there exists 2^n different representations [15].
Furthermore the number of representations is finite if and only if Φ_* is a cyclic matrix [16].
Finally, different cases are examined in [15].

Remark : In the scalar case (p=1) it is not necessary to determine the Hamiltonian matrix. Owing to the particular structure of matrix $\Phi = J - B\,C$ the elements of vector B are easily calculated from the coefficients of polynomial $\Delta(\lambda)$. Then the covariance R is determined from the spectral equality :

$$\mathscr{B}_*(z^{-1})\ R_*\ \mathscr{B}_*(z) = \mathscr{B}(z^{-1})\ R\ \mathscr{B}(z)$$

setting z=1 for example.

To conclude this section, we note that the above method allows us to find all the sets of statistical parameters leading to equivalent time series.

REFERENCES

[1] Popov, V.M., Invariant description of linear time-invariant controllable systems, SIAM J. of Control, vol.10, n°2, pp 252-264, May 1972

[2] Salut, G., Identification optimale des systèmes linéaires stochastiques, Thèse de Doctorat d'Etat, Univ. Paul-Sabatier,Toulouse, June 1976

[3] Favier, G., G. Alengrin & M. Orsini, Le filtre de Kalman et les algorithmes de factorisation, Rapport DRET-CETHEDEC, Jan. 1981

[4] Tse, E., H.L. Weinert, Structure determination and parameter identification for multivariable stochastic linear systems, IEEE Trans. Aut. Control, vol. AC-20, n°5, pp 603-612, Oct. 1975

[5] Guidorzi, R.P., Canonical structures in the identification of multivariable systems, Automatica, vol. 11, pp 361-374, 1975

[6] Favier, G., Stochastic realization algorithms for identification of multivariable ARMA models, Rapport interne, LASSY, Feb. 1980

[7] Faurre, P., Réalisations markoviennes de processus stationnaires, Rapport LABORIA, n°13, March 1973

[8] Pavon, M., Stochastic realization and invariant directions of the matrix Riccati equation, SIAM J. of Control, vol.18, n°2, pp 155-180, March 1980

[9] Lindquist, A., A new algorithm for optimal filtering of discrete-time stationary processes. SIAM J. of Control, vol.12, n°4, pp 736-746, nov.1974

[10] Morf, M., G.S. Sidhu, T. Kailath, Some new algorithms for recursive estimation in constant, linear, discrete-time systems, IEEE Trans. on Aut. Control, vol. AC-19, n°4, pp 315-323, Aug. 1974

[11] Favier, G., A comparison of stochastic realization algorithms for identification of ARMA models, 2nd Int. Conf. on Information Sciences & Systems, Patras, Greece, July 1979

[12] Alengrin, G., Parameter identification for autoregressive moving average models, 2nd Int. Conf. on Information Sciences & Systems, Patras, Greece, July 1979

[13] Bucy, R.S., P.D. Joseph, Filtering for stochastic processes with applications to guidance, *Interscience*, New York, 1968

[14] Willems, J.C., Least squares stationary optimal control and the algebraic Riccati equation, IEEE Trans. Aut. Control, vol. AC-16, n°6, pp 621-634, Dec. 1971

[15] Rodriguez-Canabal, J.M., The geometry of the Riccati equation, Ph.D., thesis, Univ. of Southern California, June 1972

[16] Ruckebush, G. Representations markoviennes de processus gaussiens stationnaires, Thèse de 3° cycle, Univ. Paris VI, May 1975.

CONVERGENCE STUDY OF TWO REAL-TIME PARAMETER ESTIMATION SCHEMES FOR NONLINEAR SYSTEM

Yung-chang Chen

Department of Electrical Engineering
National Central University, Chung-li, Taiwan, R.O.C.

ABSTRACT

In this paper, gradient method and least-square method are used to identify the parameters of a nonlinear system. The nonlinear system is described by nonlinear state differential equation, and model adjustment technique is used for identification. The convergence behavior of the whole scheme is then investigated, under the circumstances where the output data is corrupted with measurement noise. Comparison of the two methods are also presented.

1. PROBLEM FORMULATION

The process or plant under consideration is represented by the following nonlinear state equation

$$\dot{\underline{x}}_D = \underline{f}\left[\underline{x}_D, \underline{p}_D, \underline{u}(t)\right] \tag{1}$$

$$\underline{y}_D = C\underline{x}_D + \underline{v}(t) \tag{2}$$

where $\underline{x}_D$ is an n-dimensional state vector, $\underline{p}_D$ is an m-dimensional parameter vector to be estimated, $\underline{u}$ is a p-dimensional input vector, $\underline{y}_D$ is a q-dimensional output vector, and $\underline{v}(t)$ is a q-dimensional noise vector. It is also assumed that the n-dimensional vector function $\underline{f}$ is differentiable with respect to $\underline{x}_D$ and $\underline{p}_D$. An adjustable system model

$$\dot{\underline{x}} = \underline{f}\left[\underline{x}, \underline{p}, \underline{u}(t)\right] \tag{3}$$

$$\underline{y} = C\underline{x} \tag{4}$$

R. S. Bucy and J. M. F. Moura (eds.), Nonlinear Stochastic Problems, 79–86.

will be used to study the convergence behavior in estimating the unknown parameter $\underline{p}_D$. $\underline{v}(t)$ is assumed to be stationary, zero mean, Gaussian noise vector. $\underline{v}(t_i)$ and $\underline{v}(t_j)$ are statistically independent if $t_i \neq t_j$. $v_i(t)$ and $v_j(t)$ are also statistically independent if $i \neq j$. We have

$$E\{\underline{v}\} = \underline{0} \tag{5}$$

$$E\{\underline{v}(t_i)\underline{v}^T(t_j)\} = 0 \text{ for } i \neq j \tag{6}$$

$$E\{\underline{v}(t_i)\underline{v}^T(t_i)\} = NI \tag{7}$$

with

$$N = \begin{bmatrix} \sigma_1^2 & 0 & \cdots\cdots & 0 \\ 0 & \sigma_2^2 & & \cdot \\ \cdot & 0 & \cdot & \cdot \\ \cdot & & \cdot & 0 \\ 0 & \cdots\cdots & 0 & \sigma_q^2 \end{bmatrix} \tag{8}$$

2. SENSITIVITY EQUATIONS [4] [5]

From Eq.(3), the sensitivity functions $\frac{\partial x_i}{\partial p_j}$, $i = 1,2,\cdots,n$ and $j = 1,2,\ldots,m$ can be obtained from the following sensitivity equations:

$$\frac{\partial^2 x_i}{\partial p_j \partial t} = \sum_{\beta=1}^{n} \frac{\partial f_i}{\partial x_\beta} \cdot \frac{\partial x_\beta}{\partial p_j} + \frac{\partial f_i}{\partial p_j} \tag{9}$$

If the changes of the parameters in time are sufficiently small, Eq.(9) can be approximated by

$$\frac{d}{dt}\left(\frac{\partial x_i}{\partial p_j}\right) = \sum_{\beta=1}^{n} \frac{\partial f_i}{\partial x_\beta} \cdot \frac{\partial x_\beta}{\partial p_j} + \frac{\partial f_i}{\partial p_j} \tag{10}$$

In a compact matrix form

$$\dot{S} = F_x S + F_p \tag{11}$$

where

$$S \overset{\Delta}{=} \begin{bmatrix} S_{11} & S_{12} & \cdots & S_{1m} \\ S_{21} & S_{22} & \cdots & S_{2m} \\ \cdot & & & \\ \cdot & & & \\ \cdot & & & \\ S_{n1} & S_{n2} & \cdots & S_{nm} \end{bmatrix}_{nxm} = \begin{bmatrix} \frac{\partial x_1}{\partial p_1} & \cdots & \frac{\partial x_1}{\partial p_m} \\ \vdots & & \\ \frac{\partial x_n}{\partial p_1} & \cdots & \frac{\partial x_n}{\partial p_m} \end{bmatrix}_{nxm} \tag{12}$$

$$F_x \triangleq \begin{bmatrix} \frac{\partial f_1}{\partial x_n} \cdots \frac{\partial f_1}{\partial x_n} \\ \vdots \\ \frac{\partial f_n}{\partial x_1} \cdots \frac{\partial f_n}{\partial x_n} \end{bmatrix}_{nxm} \qquad F_p \triangleq \begin{bmatrix} \frac{\partial f_1}{\partial p_1} \cdots \frac{\partial f_1}{\partial p_m} \\ \vdots \\ \frac{\partial f_n}{\partial p_1} \cdots \frac{\partial f_n}{\partial p_m} \end{bmatrix}_{nxm} \tag{13}$$

If the parameter vector takes on its nomial value $\underline{p}_o$, the nomial state trajectory solution $\underline{x}_o = \underline{x}(t,\underline{p}_o)$ is obtained directly from state differential equation. On the other hand, the change of state trajectory solution is given by

$$\begin{aligned} \underline{\Delta x}(t,\underline{p}) &= \underline{x}(t,\underline{p}) - \underline{x}_o(t,\underline{p}_o) = S(t,\underline{p}_o)\,(\underline{p} - \underline{p}_o) \\ &= S(t,\underline{p}_o)\cdot \underline{\Delta p} \end{aligned} \tag{14}$$

3. GRADIENT METHOD [2]

The error criterion for identifying the system parameters is chosen to be

$$J_k = E\{\, F\left[\underline{p}(k)\right] \} \tag{15}$$

and $F\left[\underline{p}(k)\right] = \tfrac{1}{2}\,\underline{e}^T(k) R\, \underline{e}(k)$ (16)

Eq.(16) is known as stochastic error criterion function with

$$\underline{e}(k) = C\left[\underline{x}_D(k) - \underline{x}(k)\right] + \underline{v}(k) \tag{17}$$

and R being a positive-definite weighting matrix. By taking gradient on $F\left[\underline{p}(k)\right]$, the following result is obtained

$$\nabla_{\underline{p}}\, F\left[\underline{p}(k)\right] = -\, S^T(k) C^T R\, \underline{e}(k) \tag{18}$$

Thus, we obtain a parameter-adjustment scheme of gradient type with $\rho(k)$ as gain factor.

$$\underline{p}(k+1) = \underline{p}(k) + \rho(k) S^T(k) C^T R\, \underline{e}(k) \tag{19}$$

3.1 Convergence Analysis

Define $\underline{p}_D - \underline{p}(k) \triangleq \underline{\delta p}(k)$. From Eq.(19), we have

$$\underline{\delta p}(k+1) = \underline{\delta p}(k) - \rho(k)S^T(k)C^T R\,\underline{e}(k) \tag{20}$$

Under the assumptions that $\underline{p}(k)$ is sufficiently in the vicinity of $\underline{p}_D$, and each iteration is small and update sampling period is large enough, we have

$$\underline{\Delta x}(k) = \underline{x}_D(k) - \underline{x}(k) \approx S(k)\cdot\underline{\delta p}(k) \tag{21}$$

From Eqs.(17), (20), (21), we have

$$\underline{\delta p}(k+1) = \left[I-\rho(k)S^T(k)C^T RCS(k)\right]\underline{\delta p}(k) - \rho(k)S^T(k)C^T R\underline{v}(k) \tag{22}$$

This is a forced, linear, and time-varying finite difference equation, and the forcing function is of stochastic nature; hence Eq.(22) is referred to as stochastic identification-error system. In order to find out conditions under which the parameter estimation scheme converges in mean-square sense, let's calculate the upper bound on $E\{\|\underline{\delta p}(k+1)\|^2\}$.
From Eq.(22)

$$\begin{aligned} E\{\|\underline{\delta p}(k+1)\|^2\} = {} & E\{\|\underline{\delta p}(k)\|^2\} - 2\rho(k)E\{\underline{\delta p}^T(k)Q\,\underline{\delta p}(k)\} \\ & +\rho^2(k)E\{\underline{\delta p}^T(k)Q^2\underline{\delta p}(k)\} \\ & -\rho(k)E\{\underline{v}^T(k)\left[S^T(k)C^T R\right]^T\left[I-\rho(k)Q\right]\underline{\delta p}(k)\} \\ & -\rho(k)E\{\underline{\delta p}^T(k)\left[I-\rho(k)Q\right]^T S^T(k)C^T R\underline{v}(k)\} \\ & +\rho^2(k)E\{\underline{v}^T(k)\left[S^T(k)C^T R\right]^T S^T(k)C^T R\underline{v}(k)\} \end{aligned} \tag{23}$$

where $Q \triangleq S^T(k)C^T RCS(k)$.

Since $\underline{v}(k)$ and $\underline{\delta p}^T(k)\left[I-\rho(k)Q\right]^T S^T(k)C^T R$ are statistically independent,

$$E\{\underline{\delta p}^T(k)\left[I-\rho(k)Q\right]^T S^T(k)C^T R\underline{v}(k)\} = 0 \tag{24}$$

and

$$E\{\underline{V}^T(k)\left[S^T(k)C^T R\right]^T\left[I-\rho(k)Q\right]\underline{\delta p}(k)\} = 0 \tag{25}$$

Furthermore, it can be shown that

$$E\{\underline{v}^T(k)\left[S^T(k)C^T R\right]^T S^T(k)C^T R\underline{v}(k)\} = \sigma_1^2 E\{d_1\} + \sigma_2^2 E\{d_2\} + \cdots + \sigma_q^2 E\{d_q\} \tag{26}$$

where $d_1, d_2, \cdots, d_q$ are the diagonal elements of

$\left[S^T(k)C^TR\right]^T S^TC^TR$ and $d_1, d_2, \cdots, d_q \geq 0$

and from conditional probability theory,

$$E\{\underline{\delta p}^T(k)Q^2\underline{\delta p}(k)\} = E\{\underline{\delta p}^T(k)E\{Q^2/\underline{\delta p}(k)\}\ \underline{\delta p}(k)\} \tag{27}$$

So, from Eqs.(23), (24), (25), (26) and (27)

$$\begin{aligned} E\{\|\underline{\delta p}(k+1)\|^2\} &\leq E\{\|\underline{\delta p}(k)\|^2\} - 2\rho(k)\lambda_{min} E\{\|\underline{\delta p}(k)\|^2\} \\ &+ \rho^2(k)\lambda^2_{max} E\{\|\underline{\delta p}(k)\|^2\} \\ &+ \rho^2(k)\left[\sigma_1^2 E\{d_1\} + \sigma_2^2 E\{d_2\} + \cdots + \sigma_q^2 E\{d_q\}\right] \\ &= (1 - 2\rho(k)\lambda_{min} + \rho^2(k)\lambda^2_{max}) E\{\|\underline{\delta p}(k)\|^2\} \\ &+ \rho^2(k)\left[\sigma_1^2 E\{d_1\} + \sigma_2^2 E\{d_2\} + \cdots + \sigma_q^2 E\{d_q\}\right] \end{aligned} \tag{28}$$

where λ_{min}, $\lambda_{max} \geq 0$ is the minimum and the maximum eigenvalue of $E\{Q^2/\underline{\delta p}(k)\}$, respectively.
If the gain factor $\rho(k)$ is chosen so that

$$\begin{aligned} &(a)\ \sum_{k=0}^{\infty} \rho(k) \to \infty \\ &(b)\ \rho(k) \to 0 \text{ as } k \to \infty \\ &(c)\ \sum_{k=0}^{\infty} \rho^2(k) < \infty \end{aligned} \tag{29}$$

then by use of Venter's theorem $E\{\|\underline{\delta p}(k)\|^2\} \to 0$ as $k \to \infty$.
The above three conditions can be satisfied by, e.g.

$$\rho(k) = \frac{1}{k^p} \quad \text{with } \tfrac{1}{2} < p < 1. \tag{30}$$

It is recommended that the whole identification process be started by choosing a small gain factor, and after while, choosing gain factor according to Venter's theorem.

4. LEAST-SQUARE METHOD

From the definition of error criterion and sensitivity matrix in the foregoing sections, we have

$$F\left[\underline{p}(k)+\underline{\Delta p}(k)\right] = \tfrac{1}{2}\{\underline{e}(k)-CS(k)\underline{\Delta p}(k)\}^T R\{\underline{e}(k)-CS(k)\underline{\Delta p}(k)\} \tag{31}$$

It follows

$$\frac{\partial F[\underline{p}(k) + \underline{\Delta p}(k)]}{\partial \underline{\Delta p}(k)} = - S^T(k)C^T R\{\underline{e}(k)-CS(k)\underline{\Delta p}(k)\} \tag{32}$$

Setting Eq.(32) equal to zero, we obtain the following necessary condition for minimizing $F[\underline{p}(k)]$

$$\underline{\delta p}^*(k) = \left[S^T(k)C^T RCS(k)\right]^+ S^T(k)C^T R\underline{e}(k) \tag{33}$$

where $[\ ]^+$ denote the generalized inverse of matrix. Thus, the identification algorithm is given by

$$\underline{p}(k+1) = \underline{p}(k) + \left[S^T(k)C^T RCS(k)\right]^+ S^T(k)C^T R\underline{e}(k) \tag{34}$$

From matrix theory, we have three cases
(a) If dim $\underline{e}$ < dim $\underline{p}$ and CS(k) is of full rank, then

$$\underline{p}(k+1) = \underline{p}(k) + \left[CS(k)\right]^T \left[CS(k)S^T(k)C^T\right]^{-1} e(k) \tag{35}$$

(b) If dim $\underline{e}$ = dim $\underline{p}$, and CS(k) is of full rank,

$$\underline{p}(k+1) = \underline{p}(k) + \left[CS(k)\right]^{-1} \underline{e}(k) \tag{26}$$

(c) If dim $\underline{e}$ > dim $\underline{p}$, and CS(k) is of full rank,

$$\underline{p}(k+1) = \underline{p}(k) + \left[S^T(k)C^T RCS(k)\right]^{-1} S^T(k)C^T R\underline{e}(k) \tag{37}$$

4.1 Convergence Analysis

Case (a)
(1) stochastic identification-error system

$$\underline{\delta p}(k+1) \approx \{I- \rho(k)S^T(k)C^T[CS(k)S^T(k)C^T]^{-1} CS(k)\}\underline{\delta p}(k)$$
$$-S^T(k)C^T[CS(k)S^T(k)C^T]^{-1} \underline{v}(k) \tag{38}$$

(2) upper bound on $E\{\| \underline{\delta p}(k+1)\|^2\}$

Let $G = CS(k)$, $F = CS(k)S^T(k)C^T$, $Z = G^T F^{-1} G$, we have

$$E\{\| \underline{\delta p}(k+1)\|^2\} \le \{1-2\rho(k)\lambda_{min}+\rho^2(k)\lambda_{max}^2\}E\{\| \underline{\delta p}(k)\|^2\}$$
$$+\rho^2(k)\left[\sigma_1^2 E\{d_1\} +\sigma_2^2 E\{d_2\} +\cdots+ \sigma_q^2 E\{d_q\}\right] \tag{39}$$

where $\lambda_{min}, \lambda_{max} \ge 0$ are the minimum and the maximum eigenvalue of matrix $T = E\{Z \mid \underline{p}(k)\}$, and $d_1, d_2, \cdots, d_q$ are the diagonal elements of the matrix $\left[G^T F^{-1}\right]^T G^T F^{-1}$.

Case (b)

(1) stochastic identification-error system

$$\underline{\delta p}(k+1) \approx (1-\rho(k))\, \underline{\delta p}(k) - \rho(k)\left[CS(k)\right]^{-1} \underline{v}(k) \tag{40}$$

(2) upper bound on $E\{\|\underline{\delta p}(k+1)\|^2\}$

$$E\{\|\underline{\delta p}(k+1)\|^2\} \le (1-\rho(k))^2 E\{\|\underline{\delta p}(k)\|^2\} + \rho^2(k)\left[\sigma_1^2 E\{d_1\} + \sigma_2^2 E\{d_2\} + \cdots + \sigma_q^2 E\{d_q\}\right] \tag{41}$$

where $d_1, d_2, \cdots, d_q$ are the diagonal elements of the matrix $\{\left[CS(k)\right]^{-1}\}^T \left[CS(k)\right]^{-1}$.

Case (c)

(1) stochastic identification-error system

Let $A \triangleq S^T(k)C^T RCS(k)$, $B \triangleq S^T(k)C^T R$

Thus,

$$\underline{\delta p}(k+1) = (1-\rho(k))\, \underline{\delta p}(k) - \rho(k)A^{-1}B\underline{v}(k) \tag{42}$$

(2) upper bound on $E\{\|\underline{\delta p}(k+1)\|^2\}$

$$E\{\|\underline{\delta p}(k+1)\|^2\} \le (1-\rho(k))^2 E\{\|\underline{\delta p}(k)\|^2\} + \rho^2(k)\left[\sigma_1^2 E\{d_1\} + \sigma_2^2 E\{d_2\} + \cdots + \sigma_q^2 E\{d_q\}\right] \tag{43}$$

where $d_1, d_2, \cdots, d_q$ are the diagonal elements of the matrix $(A^{-1}B)^T A^{-1}B$.

4.2 Practical Consideration on Least-Square Method

The behavior of case (a) is similar to that of gradient method, but case (b) and case (c) is better under stochastic condition, and much superior under deterministic condition if we do it properly.

When we update parameter in k-th iteration by $\underline{\delta p}^*$ and if it is quasi-constant in time, then we have

$$S(k+1) = S(k) + \Delta T\left[F_x S(k) + F_p\right] \tag{44}$$

$$\begin{aligned}\underline{\delta x}(k+1) &= \underline{\delta x}(k) - S(k)\underline{\delta p}^* + \Delta T\left[F_x(\underline{\delta x}(k) - S(k)\underline{\delta p}^*) + F_p(\underline{\delta p} - \underline{\delta p}^*)\right]\\ &= \underline{\delta x}(k) - S(k)\underline{\delta p}^* + \Delta T F_x \underline{\delta x}(k) - \Delta T F_x S(k)\underline{\delta P}^* + \Delta T F_p \underline{\delta p} - \Delta T F_p \underline{\delta p}^*\end{aligned} \tag{45}$$

Multiplying Eq.(44) by ($\underline{\delta p} - \underline{\delta p}^*$), we get

$$S(k+1)\left[\underline{\delta p}-\underline{\delta p}^*\right]=\left[S(k)+\Delta T(F_x S(k)+F_p)\right]\left[\underline{\delta p}-\underline{\delta P}^*\right]$$

$$=S(k)\underline{\delta p}-S(k)\underline{\delta p}^*+\Delta TF_x S(k)\underline{\delta p}-\Delta TF_x S(k)\underline{\delta p}^*$$

$$+\Delta TF_p\underline{\delta p}-\Delta TF_p\underline{\delta p}^* \qquad (46)$$

i) deterministic condition

In case (b) and case (c), if we set $\rho(k) = 1$, and use the sensitivity equation

$$\dot{S} = F_x S + F_p \quad , \; S(kT) = 0,$$

the convergence behavior will be supremely good, because it processes as if the identification started from the beginning.

ii) stochastic condition

The tracking ability of case (b) and case (c) is better than that of gradient method, because each has the term $(1-\rho)^2$ instead of the burden of the term $(1-\rho\,\lambda_{min} + \rho^2\lambda_{max}{}^2)$ in gradient method.

CONCLUSION

In this paper, algorithms of gradient method and least-square method for real-time parameter estimation of nonlinear system are studied and their convergence behaviors are compared. Through the deep insight of their convergence behaviors, especially for slowly-varying systems, we have a more effective strategy to choose the gain factor.

REFERENCES

(1) Eykhoff, P., "System Identification", Wiley-Interscience, 1974
(2) Paul M. Frank, "Introduction to System Sensitivity Theory", 1978
(3) Graybill, "Introduction to Matrices with Application in Statistics"
(4) Menahem Sidar, "Recursive Identification and Tracking of Parameter for Linear and Nonlinear System", Int. J. Control, Vol. 24, No. 3, pp. 361-378, 1976
(5) M. R. Matausek and S. S. Stanknvic, "Robust Real-Time Algorithm for Identification of Nonlinear Time-Varying System", Int. J. Control, Vol. 31, No. 1, 79-94, 1980
(6) Raman K. Mehra, "Optimal Input Signal for Parameter Estimation in Dynamic System-Survey and New Results", IEEE Transaction on Automatic Control, Vol. 19, No. 6, December 1974

APPROXIMATION BY A SUM OF COMPLEX EXPONENTIALS UTILIZING THE PENCIL OF FUNCTION METHOD.

Tapan K. Sarkar[1]
Donald K. Weiner[2]
Vijay K. Jain[3]
Soheil A. Dianat[1]

Abstract: The approximation of a function by a sum of complex exponentials, in general, is a nonlinear optimization problem. The nonlinear problem, however, is linearized through the pencil-of-function method. This method differs from the classical Wiener least squares approach where an integrated squared error is minimized. Among the advantages of this method are its natural insensitivity to noise in the data and explicit determination of the signal order. Examples are presented to illustrate the stability of this technique when noise is present in the data.

1. INTRODUCTION

The approximation of an arbitrary real function x(t) which is nonzero for $t \geq 0$, by a sum of complex exponentials often leads to a nonlinear least squares problem [1]. Recently Golub [2] has presented a class of least squares problems in which the variables separate. Also a number of methods are available based on the classical Wiener approach of minimizing an integrated squared error [3]. In this paper, we apply a different method called the pencil-of-function method, which decouples the approximation problem [4-5]. A continuous version of the pencil-of-function method is available in [6]. In this presentation we apply the pencil-of-function method to y(p), which is the discretized version of x(t). Finally examples are presented to illustrate the stability of this method in the calculation of exponentials, particularly for noisy data.

[1]Dept. of Electrical Engineering, Rochester Inst. of Tech. Rochester, NY 14623.

[2]Dept. of Electrical Eng. Syracuse University, Syracuse, NY 13210

[3]Dept. of Electrical Eng. Univ. of South Florida, Tampa FL 33620.

R. S. Bucy and J. M. F. Moura (eds.), Nonlinear Stochastic Problems, 87–100.

2. DEFINITION OF THE PROBLEM

The problem of interest is to approximate an arbitrary square integrable function $x(t)$ by a sum of complex exponentials, i.e.

$$x(t) \approx \sum_{i=1}^{\ell} \sum_{k=1}^{m_i} \hat{A}_{ik} \cdot \{t\}^{k-1} \cdot \exp\{s_i t\} \tag{1}$$

where

$$\sum_{i=1}^{\ell} m_i = n \tag{2}$$

Here s_i are referred to as the poles of the system and m_i is the order of the pole s_i. n is the total number of poles counting the multiplicities of repeated poles.

In most problems, however, we do not deal with continuous functions but with sampled sequences. So we will assume that the sequence $\{y(p)\}$ has been sampled at a sufficiently fast rate from the continuous function $x(t)$ so that no information has been lost. Thus for a sampled data system we have

$$y(p) = \sum_{i=1}^{\ell} \sum_{k=1}^{m_i} A_{ik} \{p\}^{k-1} \{z_i\}^p \tag{3}$$

where

$$z_i = \exp\{s_i \Delta t\} \tag{4}$$

where Δt is the sampling interval. Here z_i are referred to as the z-domain poles.

The problem of approximating $\hat{A}_{ik}$ and s_i in (1) or A_{ik} and z_i in (3) is a nonlinear approximation problem. However we can linearize this nonlinear problem through the introduction of the T operator.

3. PROPERTIES OF THE T OPERATOR

The T operator is defined to be an operator similar to a reverse integral operator. It is defined on the space of squared integrable sequences in the following way:

$$T[Y] \triangleq T\{y(p)\} \triangleq \{ \sum_{k=p}^{\infty} y(k)\} \tag{5}$$

So for example

$$T\{0,1,0,0,\ldots\} = \{1,1,0,0,0,\ldots\} \tag{6}$$

and

$$T\{p(0.5)^p\} = \{2p(0.5)^p + 2(0.5)^p\} \tag{7}$$

From the definition of T in (5) we can prove the following:

(a) T is a linear operator, since

$$T[\alpha Y + \beta D] = \alpha T[Y] + \beta T[D] \tag{8}$$

where α and β are scalar quantities, and [Y] and [D] are two sampled data sequences as defined in (5).

(b) A constant multiplication of a unit impulse remains unchanged by the application of the T operator, i.e.

$$T[a\delta] \triangleq T\{a,0,0,\ldots\} = aT\{1,0,0,\ldots\} = aT[\delta] \tag{9}$$

(c) For simple poles (i.e. $m_i = 1$ for all i) we have

$$T[Y] = T\{y(p)\} = T\{\sum_{i=1}^{n} A_i(z_i)^p\} = \{\sum_{i=1}^{n} A_i \frac{(z_i)^p}{1-z_i}\} \tag{10}$$

In particular we have

$$T\{A(z)^p\} = \{\frac{A(z)^p}{1-z}\} \tag{11}$$

(d) Successive operations of T are obtained in the following way:

$$T^0 = 1$$

$$T^k[Y] = T^{k-1}[TY] \tag{12}$$

and thus we can generate a polynomial in T. One such polynomial, which will be of fundamental importance, is $[1-(1-z)T]^k$.

(e) Also, it can be shown that

$$(1-z)T\{(p)^i(z)^p\} = (p)^i(z)^p + z\sum_{k=0}^{i-1}\binom{i}{k}T\{(p)^k(z)^p\} \tag{13}$$

where

$$\binom{i}{k} = \frac{i!}{k!(i-k)!} \tag{14}$$

Properties (c) and (e) demonstrate that the application of the operator T on the sequence $\{y(p)\}$ preserves the poles of the sequence.

4. PROPERTIES OF THE $[1-(1-z)T]^k$ POLYNOMIALS

From property (c) it is clear that if $\{y(p)\}$ be a sequence having a pole z repeated with multiplicity m, then the multiplicity of z in the sequences

$$[1-(1-z)T]^k\{y(p)\} \text{ is } m-k,\ k=0,1,2,\ldots,m-k \tag{15}$$

Furthermore,

$$[1-(1-z)T]^k\{y(p)\} = \{0\} \qquad \text{for } k \geq m \tag{16}$$

Geometrically, the above property asserts that the operator $[1-(1-z)T]^k$ projects [Y] into a lower dimensional hyperplane in z of order m-k. Hence, it follows that, if the operator

$$\prod_{i=1}^{n} [1-(1-z)T]^{m_i} \tag{17}$$

is applied to [Y] which is given by

$$[Y] = \{y(p)\} = \{\sum_{i=1}^{n} \sum_{k=1}^{m_i} A_{ik}(p)^{k-1}(z_i)^p\} \tag{18}$$

then the result would be the null sequence $\{0\}$. We have thus linearized the problem by first attempting to estimate the poles z_i. After the poles z_i have been computed, the residues A_{ij} at the poles can be obtained by a least squares procedure.

5. DETERMINATION OF THE POLES z_i

In order to determine the poles z_i, we define the following operations. Let

$$[Y_2] \triangleq T[Y_1] \triangleq T[TY_0] = T^2[Y_0] \tag{19}$$

and

$$[D_i] \triangleq [1-(1-z)T][Y_i] = [1+dT][Y_i] \tag{20}$$

where

$$d \triangleq -(1-z) \tag{21}$$

From (20) it is seen that

$$[D_i] = [Y_i]+d[T][Y_i]=[T]\{Y_{i-1}+d[T][Y_{i-1}]\} = [T][D_{i-1}] \tag{22}$$

The sequence $[D_i]$ is called a pencil-of-functions parameterized by the scalar parameter d [3-7]. It is a linear combination of functions $[Y_i]$ and $[Y_{i+1}]$ through the parameter d. The pencil of functions contain very important characteristics from a system identification point of view. When the set $[D_i]$ of n+1 functions becomes linearly dependent then (d+1) becomes a system pole. Now we define the gram matrix $[\underline{H}]$ whose ith row and jth column are defined as

$$[\underline{H}] = (h_{ij} \triangleq \langle D_i, D_j \rangle = \langle [1+dT][Y_i], [1+dT][Y_j] \rangle)$$

$$= [d^*\{d\langle Y_{i+1}, Y_{j+1}\rangle + \langle Y_i, Y_{j+1}\rangle + d\langle Y_{i+1}, Y_j\rangle + \langle Y_i, Y_j\rangle\}] \quad (23)$$

where

$$\langle Y_i, Y_j \rangle = \sum_{p=0}^{\infty} \{y_i(p)\}\{y_j(p)\} \quad (24)$$

and d* denotes the complex conjugate of d. The underbar distinguishes a matrix from a sequence.

The significance of the gram matrix $[\underline{H}]$ lies in the fact that the set $[D_1]$, $[D_2]$, $[D_3]\ldots[D_k]$ are linearly independent if and only if $\det[\underline{H}]>0$, whereas it is linearly dependent if and only if $\det[\underline{H}]=0$. Thus in order that there are n poles in the sequence $[D_i]$ it is necessary and sufficient that $\det[\underline{H}]$ be positive or zero according as $k<n$, or $k\geq n$. [This is clear from (16).]

After some very tedious algebraic manipulations it can be shown that

$$\det[\underline{H}] = \sum_{i=1}^{k} \sum_{j=1}^{k} (d^*)^{i-1}(d)^{j-1} Q_{ij} \quad (25)$$

where Q_{ij} are the diagonal cofactors of the matrix $[\underline{G}]$ whose elements are defined as

$$[G] = [g_{ij} \triangleq \langle Y_i, Y_j \rangle = \sum_{p=0}^{\infty} \{y_i(p)\}\{y_j(p)\}] \quad (26)$$

An on line procedure for determining (26) can be given by the following equations:

$$f_1(k) = y_1(k)$$

$$f_i(k) = \sum_{\nu=k}^{m-1} f_{i-1}(\nu)$$

$$g_{ij} = (-1)^{i+j} \sum_{k=0}^{M} [f_i(k) f_j(k) - \sum_{\nu=1}^{i} (k-m)^{i-\nu} f_\nu(M) f_j(k)$$
$$- \sum_{\nu=1}^{j} (k-M)^{j-\nu} f_\mu(M) f_i(k) + \sum_{\nu=1}^{i} \sum_{\nu=1}^{j} f_\nu(M) f_\mu(M) (k-M)^{i+j-\nu-\mu}]$$

and
$$f_i(M) = \sum_{\nu=0}^{M-1} f_{i-1}(\nu)$$

Here it is assumed that $f_i(p)=0$ for $p>M$. This assumption is reasonable for practical problems.

When the order of the approximation k in (25) becomes equal to the order of the system the sets $[D_1]$, $[D_2]$,...$[D_n]$ becomes linearly dependent and hence

$$\det[\underline{H}] = 0 = \sum_{i=1}^{k} \sum_{j=1}^{k} (d^*)^{i-1}(d)^{j-1} Q_{ij} \ . \tag{27}$$

since

$$Q_{ij}^2 + \det[\underline{G}]_n Q_{ij} Q_{jj} = Q_{ii} Q_{jj} \tag{28}$$

and

$$\det[\underline{G}] = 0 \text{ when } \det[\underline{H}] = 0 \ , \tag{29}$$

(27) can be rewritten as

$$\left\{ \sum_{i=1}^{n+1} \sqrt{Q_{ii}} (d^*)^{n-i+1} \right\} \cdot \left\{ \sum_{j=1}^{n+1} \sqrt{Q_{jj}} (d)^{n-j+1} \right\} = 0 \tag{30}$$

From (30) the poles z_i can be obtained from the solution of the following polynomial equation:

$$\sum_{i=1}^{n+1} \sqrt{Q_{ii}} (-1+z)^{n-i+1} = 0 \tag{31}$$

and the poles s_i are obtained from z_i through the transformation

$$s_i = \frac{1}{T} \ln [z_i]$$

Since the function to be approximated is considered to be real there will also be a complex conjugate set of poles s_i^* as indicated by (30).

Once the poles s_i or z_i are determined the unknown constants A_{ij} can be determined from a least squares procedure.

6. ERROR ANALYSIS IN THE PRESENCE OF NOISE

When the signal x(t) or the sequence $\{y(p)\}$ does not contain any noise then the poles z_i given by (31) are the exact poles of the system. However if noise is present then they no longer represent the system poles. The objective of this present section is to find the error in the location of the poles due to additive noise.

Let $\{r(p)\}$ be the noisy sequence and let there exist an approximant of $\{r(p)\}$ and call it $\{y(p)\}$, which provides the best exponential approximation of the signal in the given subspace. The assumption made here is that the optimum approximant is unique.

For convenience let us call

$$r(p) = y(p) + \varepsilon e(p) \tag{32}$$

We now introduce the error sequence in the normalized form

$$[\{r(p) - y(p)\}] = \varepsilon\{e(p)\} \tag{33}$$

where ε is a non-negative real number chosen to establish the equality

$$\sum_p \{e(p)\}^2 = \sum_p \{y(p)\}^2 \tag{34}$$

clearly ε represents the fractional error in the representation and is therefore a measure of the degree of approximation of $\{r(p)\}$. If $\{y(p)\}$ is the best approximant of $\{r(p)\}$, then $\{e(p)\}$ must be orthogonal to $\{y(p)\}$. Thus, in particular, $\varepsilon=0$ characterizes the perfect representation.

Now consider the following sequences:

$$[R_1] = \{r_1(p)\} = \{r(p)\}$$

$$[R_2] = \{r_2(p)\} = T\{r_1(p)\}$$

$$\vdots$$

$$[R_{k+1}] = \{r_{k+1}(p)\} = T\{r_k(p)\} = T\{y_k(p) + \varepsilon e_k(p)\} \tag{35}$$

In (35) $\{y(p)\}$ and $\{e(p)\}$ are not known explicitly, nor is ε for that matter. We begin with the computation of the $(n+1)\times(n+1)$ gram matrix $[\underline{B}]$ for the sequence $[R_1], [R_2], \ldots, [R_n]$, i.e.

$$[\underline{B}] = [b_{ij} \triangleq \langle\{r_i(p)\}, \{r_j(p)\}\rangle]$$

$$= \begin{bmatrix} y_1(p) + \varepsilon e_1(p) \\ \cdot \\ \cdot \\ y_{n+1}(p) + \varepsilon e_{n+1}(p) \end{bmatrix} \cdot \begin{bmatrix} \{y_1(p) + \varepsilon e_1(p) \ldots y_{n+1}(p) + \varepsilon e_{n+1}(p)\} \end{bmatrix} \tag{36}$$

Since $\{y(p)\}$ is a sequence of order n, the corresponding $[Y_1], \ldots, [Y_{n+1}]$ sequences are linearly dependent. Thus there exists then a nonsingular matrix $[\underline{P}]$ such that

$$[\underline{Y}] \triangleq \begin{bmatrix} \{y_1(p)\} \\ \vdots \\ \{y_{n+1}(p)\} \end{bmatrix} = [\underline{P}] \begin{bmatrix} \{\bar{y}_1\} \\ \vdots \\ \{\bar{y}_n\} \\ 0 \end{bmatrix}, \triangleq [\underline{P}][\bar{\underline{Y}}] \tag{37}$$

From the theory of orthogonal transformation the sequences $\{\bar{y}_1(p)\}, \{\bar{y}_2(p)\}, \ldots, \{\bar{y}_n(p)\}$ must form an orthonormal set.

Similarly we obtain

$$[\underline{E}] = \begin{bmatrix} \{e_1(p)\} \\ \vdots \\ \{e_{n+1}(p)\} \end{bmatrix} = [\underline{P}] \begin{bmatrix} \{\bar{e}_1(p)\} \\ \vdots \\ \{\bar{e}_{n+1}(p)\} \end{bmatrix} \triangleq [\underline{P}][\bar{\underline{E}}] \tag{38}$$

Observe in (38) the sequences $\{\bar{e}_1(p)\}, \ldots, \{\bar{e}_{n+1}(p)\}$ are not orthonormal to each other. Substitution of (37) into (36) yields

$$\begin{aligned} [\underline{B}] &= [\underline{P}][\bar{\underline{Y}}+\varepsilon\bar{\underline{E}}][\bar{\underline{Y}}+\varepsilon\bar{\underline{E}}]^T[\underline{P}]^T \\ &= [\underline{P}]\{[\bar{\underline{Y}}][\bar{\underline{Y}}]^T+\varepsilon[\bar{\underline{E}}][\bar{\underline{Y}}]^T+\varepsilon[\bar{\underline{Y}}][\bar{\underline{E}}]^T+\varepsilon^2[\bar{\underline{E}}][\bar{\underline{E}}]^T\}[\underline{P}]^T \end{aligned} \tag{39}$$

Observe that

$$[\bar{\underline{Y}}][\bar{\underline{Y}}]^T = \left[\begin{array}{c|c} [I]_{n\times n} & 0 \\ \hline 0 & o \end{array}\right]_{(n+1)\times(n+1)} \tag{40}$$

where [I] is the n×n identity matrix and

$$[\bar{\underline{E}}][\bar{\underline{Y}}]^T = \begin{bmatrix} \{\bar{e}_1(p)\}\cdot\{\bar{y}_1(p)\} & \ldots & \{\bar{e}_1(p)\}\cdot\{\bar{y}_n(p)\} \\ \\ \{\bar{e}_{n+1}(p)\}\cdot\{\bar{y}_1(p)\} & \ldots & \{\bar{e}_{n+1}(p)\}\cdot\{\bar{y}_n(p)\} \end{bmatrix} \tag{41}$$

By utilizing the theorem on determinant expansion as presented in Appendix A, we get

$$\begin{aligned} \det[\underline{B}] &= (\det[\underline{P}])^2 \cdot \det \left| \left[\begin{array}{c|c} I_{n\times n} & \varepsilon\{\bar{e}_{n+1}(p)\}\cdot\{\bar{y}_1(p)\} \\ \hline 0 & \varepsilon^2\{\bar{e}_{n+1}(p)\}\cdot\{\bar{y}_{n+1}(p)\} \end{array}\right] \right| \\ &\quad + \text{terms of the order of } \varepsilon^3 \text{ and higher} \\ &= \varepsilon^2 \cdot \|\{e_{n+1}(p)\}\|^2 \cdot (\det[\underline{P}])^2 + \Theta(\varepsilon^3) + \ldots \end{aligned} \tag{42}$$

Thus it is clear that $\sqrt{\det[\underline{B}]}$ is of the same order of magnitude as the error measure ε, i.e.

$$\det[\underline{B}] = \Theta(\varepsilon) \tag{43}$$

In particular, a necessary and sufficient condition for perfect approximation of a signal or order n is that $\det[\underline{B}]$ must vanish.

In order to find the error in the location of the poles first the error in the diagonal cofactors of the gram matrix $[\underline{B}]$ has to be estimated. The diagonal cofactors are given by:

$$D_{ii} = \det\left[\begin{bmatrix} \{y_1(p)+\varepsilon e_1(p)\} \\ \cdot \\ \{y_{i-1}(p)+\varepsilon e_{i-1}(p)\} \\ \{y_{i+1}(p)+\varepsilon e_{i+1}(p)\} \\ \vdots \\ \{y_{n+1}(p)+\varepsilon e_{n+1}(p)\} \end{bmatrix} \cdot \begin{matrix} \{y_1(p)+\varepsilon e_1(p)\}..\{y_{i-1}(p)+\varepsilon e_{i-1}(p)\} \\ \{y_{i+1}(p)+\varepsilon e_{i+1}(p)\}\ldots \\ \ldots\{y_{n+1}(p)+\varepsilon e_{n+1}(p)\} \end{matrix}\right] \tag{44}$$

Let $[\underline{S}_i]$ be the matrix which orthonormalizes the vectors, from $\{y_1(p)\}\ldots\{y_{i-1}(p)\}\{y_i(p)\}\ldots\{y_{n+1}(p)\}$ to $\{\bar{y}_i(p)\}\ldots$ $\{\bar{y}_{i-1}(p)\}\{\bar{y}_{i+1}(p)\}\ldots\{\bar{y}_{n+1}(p)\}$. So we have

$$\begin{bmatrix} \{y_1(p)\} \\ \vdots \\ \{y_{i-1}(p)\} \\ \{y_{i+1}(p)\} \\ \vdots \\ \{y_{n+1}(p)\} \end{bmatrix} = [\underline{S}_i] \cdot \begin{bmatrix} \{\bar{y}_1(p)\} \\ \vdots \\ \{\bar{y}_{i-1}(p)\} \\ \{\bar{y}_{i+1}(p)\} \\ \vdots \\ \{\bar{y}_{n+1}(p)\} \end{bmatrix} \triangleq [\underline{S}_i][\bar{\underline{Y}}_i] \tag{45}$$

$$\begin{bmatrix} \{e_1(p)\} \\ \vdots \\ \{e_{i-1}(p)\} \\ \{e_{i+1}(p)\} \\ \vdots \\ \{e_{n+1}(p)\} \end{bmatrix} = [\underline{S}_i] \cdot \begin{bmatrix} \{e_1(p)\} \\ \vdots \\ \{e_{i-1}(p)\} \\ \{e_{i+1}(p)\} \\ \vdots \\ \{e_{n+1}(p)\} \end{bmatrix} \triangleq [\underline{S}_i][\bar{\underline{R}}_i] \tag{46}$$

Observe that the vectors $\{e_1(p)\}\ldots\{\bar{e}_{n+1}(p)\}$ are not a set of orthonormal vectors. Substitution of (45) and (46) into (44) yields

$$D_{ii} = \det\left|[\underline{S}_i][\underline{\bar{Y}}_i+\varepsilon\underline{\bar{E}}_i][\underline{\bar{Y}}_i+\varepsilon\underline{\bar{E}}_i]^T[\underline{S}_i]^T\right|$$

$$= (\det[\underline{S}_i])^2\cdot\det\left|[\underline{\bar{Y}}_i][\underline{\bar{Y}}_i]^T+\varepsilon[\underline{\bar{E}}_i][\underline{\bar{Y}}_i]^T+\varepsilon[\underline{\bar{Y}}_i][\underline{\bar{E}}_i]^T+\varepsilon^2[\underline{\bar{E}}_i][\underline{\bar{E}}_i]^T\right| \tag{47}$$

By utilizing

$$[\underline{\bar{Y}}_i][\underline{\bar{Y}}_i]^T = [I]_{n\times n}$$

and applying the theorem on determinant expansion [as presented in Appendix A] we obtain

$$D_{ii} = Q_{ii} + 2\,(\det[\underline{S}_i])^2\cdot\left[\sum_{\substack{j=1\\ i\neq j}}^{n+1}\sum_{p=0}^{\infty}\{\bar{y}_j(p)\}\cdot\{e_j(p)\}\right] \tag{48}$$

where Q_{ii} are the diagonal cofactors of the gram matrix formed by the sets $[Y_1]$, $[Y_2]\ldots[Y_{n+1}]$ and is given by (26). The underlying assumption in (48) is of course we have the right order of the system n, i.e.

$$\det[\underline{G}] = 0\,. \tag{49}$$

Observe that (48) implies continuous differentiability of the diagonal cofactors D_{ii} with respext to the error ε, so that

$$\lim_{\varepsilon\to 0} D_{ii} = Q_{ii} \quad \text{for all } i \tag{50}$$

Let z_i' be the approximate poles that are obtained by utilizing the diagonal cofactors D_{ii} instead of Q_{ii} in (31). So the poles z_i' are the roots of the following equation

$$\sum_{i=1}^{n+1}\sqrt{D_{ii}}\,(-1+z_i')^{n+1-i} = 0 \tag{51}$$

Substitution of (48) into (51) illustrates that the approximate poles z_i' are related to the exact poles z_i by the relationship

$$z_i' = z_i + \varepsilon F_i + \Theta(\varepsilon^2) \tag{52}$$

where F_i are certain constants. The approximate poles z_i' are again a continuous function of ε. So the error in the approximation $(z_i'-z_i)$ is a continuous function of ε and approaches zero uniformly as $\varepsilon\to 0$. Hence it is expected that as the order of the approximation is increased, the lower order poles would remain relatively stable.

7. EXAMPLES

As an example consider the transient response of a conducting pipe tested at the ATHAMAS-I Electromagnetic Pulse (EMP) simulator. The conducting pipe is 10m long and 1m in diameter. Hence the true resonance of the pipe is expected to be in the neighborhood of 14MHz. Also the pipe has been excited in such a way that it is reasonable to expect only odd harmonics at the scattered fields. The data which have been measured are the integral of the electric field and hence is available in terms of a voltage. Thus, in addition to the frequencies of the conducting pipe one should also observe a very dominant low frequency pole. The same transient data as depicted in Fig. 1 is used for analysis. The results for a fifth and a seventh order system are as follows:

For $n = 5$, the poles in radians/sec are

$$(-0.0029 \pm j0.083) \times 10^9 \quad (=13.33\text{MHz})$$
$$(-0.0428 \pm j0.217) \times 10^9 \quad (=35.20\text{MHz})$$
$$(-0.0098 \qquad) \times 10^9 \quad (= 1.56\text{MHz})$$

For $n = 7$, the poles in radians/sec are

$$(-0.0058 \pm j0.084) \times 10^9 \quad (=13.40\text{MHz})$$
$$(-0.0270 \pm j0.219) \times 10^9 \quad (=35.10\text{MHz})$$
$$(-0.0270 \pm j0.550) \times 10^9 \quad (=87.60\text{MHz})$$
$$(-0.0012 \qquad) \times 10^9 \quad (= 0.19\text{MHz})$$

It is interesting to observe that the real pole due to the integrator has been obtained. This pole is a very dominant pole as the data were recorded after having passed through an integrator. The above results display a dynamic range of approximately 1000:1 for the values of poles of the conducting pipe.

Next the data were differentiated to get rid of the undesirable dominant pole of the integrator. The differentiation was done numerically. For a fourth and sixth order system the above results have been recalculated as follows:

For $n = 4$, the poles in radians/sec are

$$(-0.0026 \pm j0.086) \times 10^9 \quad (=13.70\text{MHz})$$
$$(-0.0480 \pm j0.235) \times 10^9 \quad (=37.47\text{MHz})$$

For $n = 6$, the poles in radians/sec are

$$(-0.005 \pm j0.983) \times 10^9 \quad (=13.23\text{MHz})$$
$$(-0.034 \pm j0.221) \times 10^9 \quad (=35.59\text{MHz})$$
$$(-0.071 \pm j0.406) \times 10^9 \quad (=65.9\text{ MHz})$$

Here a good approximation to the poles has been obtained with only four poles. Also, there seems to be a good agreement in the pole locations obtained from the original integrated data and the numerically differentiated data. Observe that indeed the poles are occurring approximately at odd harmonics of the fundamental.

In Figure 2, the true numerically differentiated data are plotted against the reconstructed response of a sixth order system. The plot has been normalized to unity amplitude. There is a close agreement even in the very early times of the two waveforms.

8. CONCLUSION

The pencil of function method has been applied to the approximation of an arbitrary function by a sum of complex exponentials. The important features of this technique are the natural insensitivity to noise and the continuous dependence of the pole locations to the error ε. Finally examples have been presented to illustrate the stability of the pole locations yielded by the pencil of function method.

9. REFERENCES

[1] R. N. McDonough and W.H. Huggins, "Best Lease Squares Representation of Signals by Exponentials," IEEE Trans, AC-13, No. 4, pp. 405-412, August 1968.

[2] G.H. Golub and V. Pereyera, "The Differentiation of Pseudo Inverse and Nonlinear Least Squares Problems Whose Variables Separate," SIAM J. of Numerical Analysis, vol. 10, No. 2, pp. 413-432, April 1973.

[3] T.K. Sarkar, J. Nebat, V.K. Jain and D.D. Weiner, "A Comparison of the Pencil of Function Method with Prony's Method, Weiner Filters and Other Identification Techniques," IEEE Trans. on Antennas and Propagation, vol. AP-30, January 1982, pp. 89-99.

[4] V.K. Jain and R.D. Gupta, "Identification of Linear Systems Through a Grammian Technique," Intl. J. of Control, vol. 12, pp. 421-431, 1970.

[5] V.K. Jain, "Filter Analysis by Use of Pencil of Functions: Part I and II," IEEE Trans. on Circuits and Systems, vol. CAS-21, pp. 574-579, September 1974.

[6] T.K. Sarkar, J. Nebat, D.D. Weiner and V.K. Jain, "Suboptimal Approximation/Identification of Transient Waveforms from Electromagnetic Systems by Pencil-of-Function Method," IEEE Trans. on Antennas and Propagation, vol. AP-28, Nov. 1980, pp. 928-933.

10. APPENDIX A

Let $\underline{A}$ and $\underline{B}$ be two square n×n matrices. Let the columns of $\underline{A}$ and $\underline{B}$ be represented by A_i and B_i for i=1,2,...,n. Denote

$$C^1_{\nu_i} = A_i \quad \text{if} \quad \nu_i = 1$$

$$= B_i \quad \text{if} \quad \nu_i = 0$$

and the matrix constituted by the columns of

$$C^1_{\nu_1}, C^2_{\nu_2}, C^3_{\nu_3}, \ldots, C^n_{\nu_n} \quad \text{as} \quad \underline{C}_{\nu_1,\nu_2,\ldots,\nu_n}.$$

Then

$$\det\left|[\underline{A}+\underline{B}]\right| = \sum_{m=0}^{n} \sum \det\left|\underline{C}_{\nu_1,\nu_2,\ldots,\nu_n}\right|$$

where in the second summation exactly m of the ν_i's equal 1 and the rest equal 0.

This is a standard result of determinant expansion.

ACKNOWLEDGEMENT

Grateful acknowledgement is made to Dr. Henry Mullaney for his great interest in this work. This work has been supported by the Office of Naval Research under Contract N00014-79-C-0598.

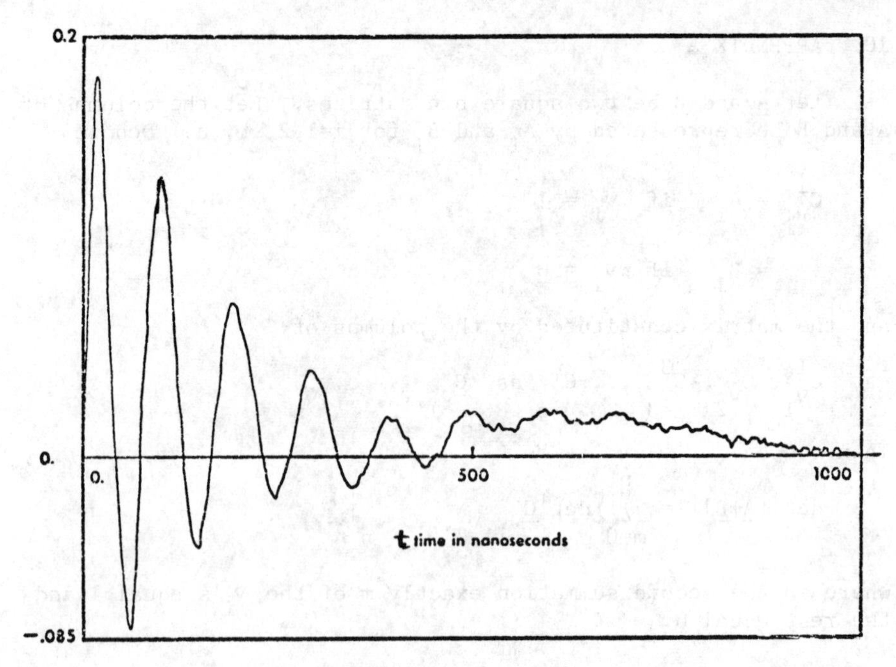

Figure 1. Transient Response of a Conducting Pipe Measured at the ATHAMAS-1 EMP Simulator.

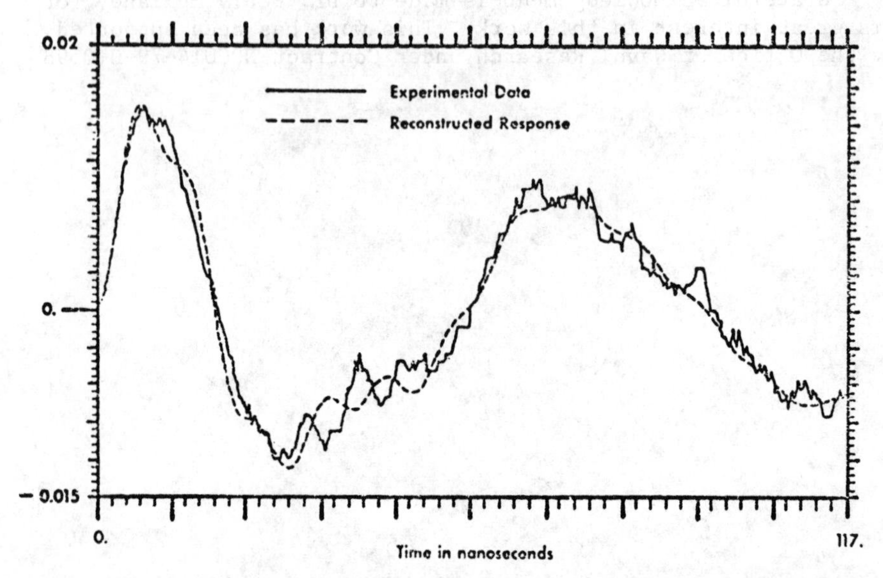

Figure 2. True Response vs. Reconstructed Response of a Sixth Order System for a 10m long 1m Diameter Conducting Pipe.

MINIMAX ESTIMATION OF ARMA SYSTEMS

M.L.S. Teles

Instituto de Física, Universidade de São Paulo,
Caixa Postal 20516
01000 - SÃO PAULO, SP - BRASIL

Abstract. The identification of a dynamic system with mixed-autoregressive — moving average models raises a nonlinear estimation problem which is solved by algorithms, usually based on least squares theory. This paper presents a new method based on a decision theoretical approach for the identification problem. When a certain criterion is chosen, game theory gives the conditions for the existence of a minimax decision rule, which is fundamental for robust estimation. Simulation studies are presented, starting from time series previously generated by autoregressive-moving average systems excited by gaussian and non-gaussian noise.

1. INTRODUCTION

Among the possible models to describe a fixed dynamic system subject to stationary stochastic disturbances, the mixed autoregressive-moving average (ARMA) is important because its parcimony in the number of parameters.

Since the estimation problem is then nonlinear, there are practical difficulties. Most of the very many techniques found in the literature, concerning identification and system parameter estimation, are based on least squares theory, and work properly when the noise distribution is gaussian [1, 2, 3]. If this is not the case, such methods loose their basic safety.

In order to obtain robust estimators, relative to noise distribution, one may rely on the minimax estimation, which consists of applying the Minimax Principle [4] to the chosen loss function, that is, the identification criterion. A more general formulation

R. S. Bucy and J. M. F. Moura (eds.), Nonlinear Stochastic Problems, 101–107.

results from setting the problem as a statistical game. Then Game Theory gives conditions for existence of the estimators.

In section 2, the identification of a linear system, fixed and of finite dimension is formally stated as a statistical game. In Section 3, one shows how the minimax estimates of an ARMA model can be calculated. Simulation results are shown in Section 4.

2. PROBLEM STATEMENT

Let S be a linear time invariant and finite dimensional system, which admits an ARMA (autoregressive-moving average) representation as follows:

$$A^0(z^{-1})y_{t+1} = C^0(z^{-1})\varepsilon_{t+1} \tag{2.1}$$

where y_{t+1} is the scalar output at time t+1, and ε_{t+1} is a real stochastic process defined in a probability space $(E, \mathcal{F}, P)$, adapted to the sequence of sub-σ-algebras $(\mathcal{Y}_i, i \in \mathbb{N})$ where $\mathcal{Y}_i$ is generated by observations up to time i. $\mathcal{Y}_0$ is the "initial condition", the available information at time i=0. Let

$$E\{\varepsilon_{t+1} / \mathcal{Y}_t\} = 0 \qquad \text{a.s.} \tag{2.2}$$

and

$$E\{\varepsilon^2_{t+1} / \mathcal{Y}_t\} < \infty \qquad \text{a.s.}$$

A^0 and C^0 are stable polynomials in z^{-1} (unit delay operator) and have no common factors.

As the objective here is to estimate system parameters, it is supposed that the orders of A^0 and C^0 are known: n and m respectively:

$$A^0(z^{-1}) = 1 + a^0_1 z^{-1} + a^0_2 z^{-2} + \ldots + a^0_n z^{-n} \tag{2.3}$$

$$C^0(z^{-1}) = 1 + c^0_1 z^{-1} + c^0_2 z^{-2} + \ldots + c^0_m z^{-m} \tag{2.4}$$

In innovations form [9], (2.1) becomes:

$$y_{t+1} = E[y_{t+1} / \mathcal{Y}_t] + \varepsilon_{t+1} \tag{2.5}$$

where

$$E[y_{t+1} / \mathcal{Y}_t] = \left[1 - \frac{A^0(z^{-1})}{C^0(z^{-1})}\right] y_{t+1} \tag{2.6}$$

Let

$\Theta \triangleq \{\theta \,|\, \theta=[a_1, a_2, \ldots, a_n, c_1, c_2, \ldots, c_m]^T\}$ - parameter space

Z - space of all possible observation sets $z_t=\{y_t, y_{t-1}, \ldots, y_1\}$

D - space of functions $d: Z \rightarrow R$ such that $E[L(\theta,d(z_t))]<k<\infty$ for all $\theta \in \Theta$.

$L(\theta,a)$ - loss function (identification criterion) associated to the action $d(z_t) = a$ when θ is the true parameter ($L:\Theta \times \mathcal{A} \rightarrow R$)

$\mathcal{A}$ - action space, the image set of functions $d \in D$.

$R(\theta,d) \triangleq E[L(\theta,d(z_t)]$ - risk function

The triple (Θ, D, R), associated to the experiment which produces $z_t \in Z$, defines the statistical game: nature chooses a parameter $\theta \in \Theta$ and the statistician, based on z_t, chooses an action $d(z_t) \in \mathcal{A}$. As this choice may be in probability (mixed strategies), it is convenient to define D^* as the sets of all probability distributions in D:

$D^* \triangleq \{\delta(d) \;/\; d \in D$ and δ is a probability distribution$\}$.

The minimax decision rule for the statistician is obtained as the solution of the problem:

$$\inf_{\delta \in D^*} \sup_{\theta \in \Theta} R(\theta,\delta) \tag{2.7}$$

where

$$R(\theta,\delta) = E[R(\theta,d(z_t)]= \int R(\theta, d(z_t)\, d\delta(z_t) \tag{2.8}$$

3. MINIMAX ESTIMATION

Let the loss function be based on the so-called incremental mean square error [8]:

$$L(\theta,d(z_t))=E_\theta[(y_{t+1} -d(z_t))^2-(y_{t+1} -E_\theta[y_{t+1} \;/\; \mathcal{Y}_t])^2 \;/\; \mathcal{Y}_t] \tag{3.1}$$

Here the decision rule for the statistician is the output prediction at time t+1, based on information $\mathcal{Y}_t$.

Using known properties of $E_\theta[y_{t+1} \;/\; \mathcal{Y}_t]$ (3.1) becomes:

$$L(\theta,d(z_t))=(d(z_t)-E_\theta[y_{t+1} \;/\; \mathcal{Y}_t])^2 \tag{3.2}$$

Because $L(\theta, d(z_t))$ is a convex function of $d(z_t)$ for all $\theta \in \Theta$, class D of pure strategies is essentially complete for the game (Θ, D, R) [4]. So, there is no need to consider mixed strategies with this loss function.

Now suppose a particular case when noise is amplitude-limited, that is, there exists $v>0$ such that

$$-v \leq \varepsilon_k \leq v \quad , \qquad k=1, 2, \ldots, t \tag{3.3}$$

From (3.3) and (2.5) it follows that:

$$y_k - v \leq E[y_k \,/\, \mathcal{Y}_{k-1}] \leq y_k + v \qquad k=1, 2, \ldots, t \tag{3.4}$$

and consequentely,

$$\alpha_1(z_t) \leq E[y_{t+1} \,/\, \mathcal{Y}_t] \leq \alpha_2(z_t) \tag{3.5}$$

with $\alpha_1(z_t)$ and $\alpha_2(z_t)$ being unknown quantities.

Using criterion (3.1) and under hypothesis (3.3), one can show [5] that the minimax decision rule is simply:

$$d^*(z_t) = \frac{\alpha_1(z_t) + \alpha_2(z_t)}{2} \tag{3.6}$$

where $\alpha_1(z_t)$ and $\alpha_2(z_t)$ are respectively the solutions of problems a) and b) below:

$$\text{a)} \ \min_{\theta \in \Theta} \ E_\theta[y_{t+1} \,/\, \mathcal{Y}_t] \longrightarrow \alpha_1(z_t)$$

$$y_k - v \leq E_\theta[y_{t+1} \,/\, \mathcal{Y}_t] \leq y_k + v, \qquad k = 1, 2, \ldots, t \tag{3.7}$$

$$\text{b)} \ \max_{\theta \in \Theta} \ E_\theta[y_{t+1} \,/\, \mathcal{Y}_t] \longrightarrow \alpha_1(z_t)$$

$$y_k - v \leq E_\theta[y_{t+1} \,/\, \mathcal{Y}_t] \leq y_k + v \ , \quad k = 1, 2, \ldots, t \tag{3.8}$$

Because $E_\theta[y_{t+1} \,/\, \mathcal{Y}_t]$ is nonlinear in θ, two difficulties arise: first, a) and b) are nonlinear programming problems, and consequentely not easy to solve. On the other side, solutions of a) and b) give only the output prediction as seen in (3.6), and not the required parameter estimates.

Now consider the particular case when $C^0(z^{-1}) = 1$, that is, the proposed model is pure autoregressive (AR). The linearity of $E_{\theta}[y_{t+1} / Y_t]$ relative to θ transforms a) and b) into linear programming problems. Let θ_1 and θ_2 be the arguments of the solutions of a) and b) respectively. From (3.3) and the mentioned linearity one gets the optimal $\hat{\theta}$ and y_{t+1} values:

$$y_{t+1}(z_t) = E_{\hat{\theta}}[y_{t+1} / Y_t] \qquad (3.9)$$

$$\hat{\theta} = \frac{\theta_1 + \theta_2}{2} \qquad (3.10)$$

After these considerations, it seems convenient to estimate ARMA models by means of an alternative procedure, envolving only minimax estimation of AR models. Since such problem decomposition is quite usual in identifying ARMA models [1, 2, 3], several methods may be proposed.

The algorithm described below was inspired in the two-stage least squares method [1]. It consists of two estimation problems: the first yields a "long AR" model, which is used to estimate noise sequence, and the second yields the required estimates of the ARMA model:

Step 1. Obtain a minimax estimate $\hat{\alpha} \triangleq [\hat{\alpha}_1\ \hat{\alpha}_2, \ldots, \hat{\alpha}_p]^T$ for $\alpha \triangleq [\alpha_1\ \alpha_2, \ldots, \alpha_p]^T$, the parameter vector of the following "long AR" model of order p (p>> max (m,n)):

$$(1+ \alpha_1 z^{-1} + \alpha_2 z^{-2} + \ldots + \alpha_p z^{-p})\, y_{t+1} = \varepsilon_{t+1}(\alpha) \qquad (3.11)$$

Step 2. Using $\hat{\alpha}$, obtain an estimate $\{\varepsilon_k(\hat{\alpha})\}$, k=1, 2, ..., t of the sequence $\{\varepsilon_k(\alpha)\}$ from:

$$\varepsilon_k(\hat{\alpha}) = (1+\hat{\alpha}_1 z^{-1} + \ldots + \hat{\alpha}_p z^{-p})\, y_k \qquad (3.12)$$

Step 3. Obtain minimax estimates $\hat{A}$ and $\hat{C}$ for the polynomials $A^0(z^{-1})$ and $C^0(z^{-1})$, using the model:

$$A(z^{-1})\, y_{t+1} = C(z^{-1})\, \varepsilon_{t+1}(\hat{\alpha}) \qquad (3.13)$$

Steps 1 and 3 envolve minimax estimation of AR models, thus requiring linear programming.

4. SIMULATION

In order to experimentally check the above procedure, a few tests were performed. For each given ARMA system, $A^0(z^{-1})\, y_t = C^0(z^{-1})\varepsilon_t$, a pseudo-random sequence $\{\varepsilon_t\}$ was generated and $\{y_t\}$ computed through the system equation. Minimax estimates are shown below, (sample size = 250)

1. System: $y_t - 0.5 y_{t-1} = \varepsilon_t + 0.5\, \varepsilon_{t-1}$
   Noise distribution: uniform (-0.5, 0.5)
   Minimax estimates: $\hat{a}_1 = -0.502$; $\hat{c}_1 = 0.465$

2. System: $y_t - 0.5 y_{t-1} = \varepsilon_t + 0.5 \varepsilon_{t-1}$
   Noise distribution: gaussian N(0,1) truncated on (-1,1)
   Minimax estimates: $\hat{a}_1 = -0.526$; $\hat{c}_1 = 0.463$

3. System: $y_t + 0.64 y_{t-2} = \varepsilon_t - 0.25\, \varepsilon_{t-2}$
   Noise distribution: uniform (-1,1)
   Sample size: 250
   Minimax estimates: $\hat{a}_1 = -0.004$; $\hat{a}_2 = 0.631$; $\hat{c}_1 = 0.018$;
   $\hat{c}_2 = -0.230$

4. System: $y_t + 0.64 y_{t-2} = \varepsilon_t - 0.25\, \varepsilon_{t-2}$
   Noise distribution: gaussian N(0,1) truncated on (-1,1)
   Minimax estimates: $\hat{a}_1 = -0.003$; $\hat{a}_2 = 0.664$; $\hat{c}_1 = -0.007$;
   $\hat{c}_2 = -0.228$

5. System: $y_t = \varepsilon_t + 0.5\, \varepsilon_{t-1}$
   Noise distribution: uniform (-1,1)
   Minimax estimate: $\hat{c}_2 = 0.504$

6. System: $y_t = \varepsilon_t + 0.5\, \varepsilon_{t-1}$
   Noise distribution: gaussian N(0,1) truncated on (-1,1)
   Minimax estimate: $\hat{c}_1 = 0.490$

REFERENCES

[1] Durbin, J. (1961). "The Fitting of Time Series Models". Revue Inst. de Stat., 28, 3, p. 233.

[2] Graupe, D. and Krause, D.J. (1975), "Identification of Auto-

regressive-Moving Average Parameters of Time Series. IEEE Trans. Aut. Control, AC-20, p. 104.

[3] Mayne, D.Q. and Firoozan, F. (1978). "Linear Estimation of ARMA Systems". Proc. 7th IFAC Congress, Helsinki, Finland.

[4] Ferguson, T.S. (1967) "Mathematical Statistics - A Decision Theoretic Approach". Ac. Press Inc.

[5] Teles, M.L.S. and Castrucci, P.B.L. (1982). "On the Minimax Approach for Dynamic System Identification", to appear in Preprints 6th IFAC Symposium on Ident. & System Par. Estimation, Washington, D.C. USA.

[6] Blackwell, D. and Girshick, M.A. (1954). "Theory of Games and Statistical Decisions". John Wiley & Sons, Inc.

[7] Box, G.E.P. and Jenkins, G.M. (1970). "Time Series Analysis, Forecasting and Control". Holden Day, Inc.

[8] Sebald, A.V. and Haddad, A.H. (1977) "Robust State Estimation in Uncertain Systems: Combined Detection - Estimation with Incremental MSE Criterion". IEEE Trans. Aut. Control, AC-22, p. 821.

[9] Ljung, L. (1976). "On the Consistency of Prediction Error Identification Methods". System Identification: Advances and Case Studies. R.K. Mehra & D.G. Lainiotis (eds.). Ac. Press.

[10] Rajbman, N.S. (1978). "Minimax Identification. Automatica, Vol. 14, p. 129. Pergamon Press.

CHAPTER IV

SYSTEM THEORY

ON THE STRUCTURE OF MINIMAL MARKOVIAN REPRESENTATIONS

Guy Ruckebusch

Schlumberger-Doll Research
Ridgefield, Connecticut 06877, USA

ABSTRACT : The theory of Markovian Representation is set forth in an abstract Hilbert space setting. This enables us to completely characterize the set of all (finite or infinite-dimensional) Markovian Representations of a given (stationary) Gaussian process. The paper discusses when Markovian Representations can be studied by means of functional models in the Hardy space. It is shown that it is the case for all the Minimal Markovian Representations of a strictly noncyclic process. Hence, new results on their structure can be derived.

1. INTRODUCTION

Let $y = \{y(t), t \in Z\}$ be a (real or complex) centered stationary Gaussian vector-process with a rational spectral density. A vector-process $x = \{x(t), t \in Z\}$ is said to be a *Markovian Representation* (MR) of y when x and y satisfy the following model for all t:

$$x(t+1) = Fx(t) + Lw(t), \qquad y(t) = Hx(t) + Jw(t), \tag{1}$$

where $w = \{w(t), t \in Z\}$ is a centered Gaussian unitary white noise process taking its values in an arbitrary (real or complex) euclidian space; (F,H,J,L) are matrices of appropriate dimensions with F asymptotically stable (all its eigenvalues are inside the unit disc). The above assumptions imply that x is a centered, stationary, purely nondeterministic Gaussian Markovian process. It is said to be a Minimal Markovian Representation (MMR) when there exists no other MR of y with a smaller dimension.

The classical *Stochastic Realization Problem* is to find all the MMRs of a given process y (if there is any). This problem has been extensively studied in the literature (cf. the references listed in [7] and [13]) and can be considered now as completely solved. In particular, a well-known result is that y has a finite-dimensional MR [i.e a model of type (1)] if and only if it has a rational spectral density. The theory has been extended in several directions and research has recently focused on two nontrivial

R. S. Bucy and J. M. F. Moura (eds.), Nonlinear Stochastic Problems, 111–122.

generalizations of model (1) : the study of the (necessarily infinite-dimensional) MRs of a Gaussian process with a nonrational spectral density and the study of the MRs of a non-Gaussian process (F and H are then replaced in (1) by two arbitrary nonlinear functions). Only the first topic will be covered in this paper. The second one is far from having reached the same maturity. Interestingly enough, it is shown in [6] that much of the material of the linear theory can still be of use in a nonlinear framework.

It has been demonstrated in the previous works of Lindquist-Picci and the author that the Hilbert space is the natural setting to define the concept of Markovian Representation without any dimensionality restriction. To naturally introduce our abstract approach, let us first translate the model (1) into geometric terms.

Let Y_∞ be the Gaussian space [10] generated by the process y. Since y is stationary, Y_∞ is endowed with a unitary operator V defined as the time-shift of the coordinates of y (cf. [12]). Let Y_p and Y_f be the "past" and "future" (closed) subspaces of Y_∞ generated respectively by the components of $\{y(t), t < 0\}$ and $\{y(t), t \geqslant 0\}$. Clearly, Y_p is invariant under V^{-1} while Y_f is invariant under V. Let H be a Gaussian space endowed with a unitary operator U. For any subspace X of H, we set

$$X_- = \bigvee_{n \leqslant 0} U^n(X) \quad \text{and} \quad X_+ = \bigvee_{n \geqslant 0} U^n(X). \tag{2}$$

(H,U) is said to be a dilation of (Y_∞, V) when Y_∞ is a subspace of H and the restriction of U to Y_∞ coincides with V. The geometric equivalent of model (1) is then given by the following.

PROPOSITION 1 [13]. *A process $x = \{x(t),\ t \in Z\}$ satisfies the dynamical equations (1) if and only if there exists a dilation (H,U) of (Y_∞, V) such that U shifts the coordinates of x and the Gaussian space X generated by $x(0)$ satisfies the three conditions :*

$$X \text{ is finite-dimensional}; \tag{3}$$

$$\bigcap_n U^n(X_- \vee Y_p) = \{0\}; \tag{4}$$

$$(X_+ \vee Y_f) \perp (X_- \vee Y_p) \mid X \quad \text{i.e.} \quad [(X_+ \vee Y_f) \ominus X] \perp [(X_- \vee Y_p) \ominus X]. \tag{5}$$

It follows from Proposition 1 that an MR could be defined as a subspace X included in a dilation of (Y_∞, V) and satisfying (3-5). In this paper, we shall assume that the pair (H,U) is fixed [it can be interpreted as a stationary flow of information which is available to construct the MRs of y]. The abstract formulation of the concept of Markovian Representation will now proceed from two generalizations. First, we shall drop the conditions (3) and (4) on X [note that this will enable us from now on to speak of infinite-dimensional representations, since (5) still makes sense in such a case]; secondly, we shall impose the minimal assumptions on the 4-tuple (H, U, Y_p, Y_f). In particular, the probabilistic context can be given up in favor of an abstract Hilbert space setting.

To achieve the greatest generality as well as clarity, the results of the paper will be presented in four parts, each corresponding to increasingly stringent assumptions on the 4-tuple (H, U, Y_p, Y_f). These assumptions have been chosen to ensure that the same theoretical framework applies for both stationary discrete-time processes and continuous-time processes with stationary increments (cf. [13]). Due to space limitations, some parts of the theory will not be covered (for example, the study of the links between the MRs of a process and the factorizations of its spectral density, for which we refer to [15]). To facilitate the task of the reader, we have left out only the proofs that can be found in [13].

2. BASIC GEOMETRIC THEORY

The framework to be used throughout this paper is the following. We consider a *fixed* (real or complex) *Hilbert space H endowed with a unitary operator U. We distinguish in H two subspaces* Y_p *and* Y_f *which satisfy the following assumption :*

$$\text{(A1)} \qquad U^{-1}(Y_p) \subset Y_p \quad \textit{and} \quad U(Y_f) \subset Y_f.$$

In the sequel we shall designate by Y_∞ the smallest subspace including both Y_p and Y_f. Note that (A1) does not necessarily imply that Y_∞ is a reducing subspace for U (in fact, this assumption can be dispensed with until section 3).

DEFINITION 1 - *A subspace X is called an MR* [of the 4-tuple (H,U,Y_p,Y_f)] *if*

$$(X_+ \vee Y_f) \perp (X_- \vee Y_p) \mid X \qquad [X_+ \text{ and } X_- \text{ are defined as in (2)}]. \tag{6}$$

With X is associated its transition operator F defined by $F = E^X U|X$.

In Definition 1, E^X (also denoted $E[\cdot|X]$) designates the orthogonal projector onto X and U|X stands for the restriction of U to X. Clearly, F is a contraction on X and, hence, U can be interpreted as a (non-necessarily minimal) unitary dilation of F (cf. [16]). Note also that F generalizes the matrix F of model (1).

Let us fix the terminology. An MR X is said to be internal when X is included in Y_∞. The fundamental concepts of Minimality, Observability and Constructibility are defined below by analogy with System Theory (cf. [4]). In this respect, the next proposition should come as no surprise.

DEFINITION 2 - *An MR X is called a Minimal Markovian Representation (MMR) if it contains no other MR as a proper subspace. For any MR X, we define its observable part* X_o *and its constructible part* X_c *by*

$$X_o = \bar{E}\,[Y_f|X] \quad \text{and} \quad X_c = \bar{E}\,[Y_p|X]. \tag{7}$$

X is said to be an Observable Markovian Representation (OMR) *when* $X = X_o$ *and a Constructible Markovian Representation* (CMR) *when* $X = X_c$.

PROPOSITION 2 [13] - *For any MR X,* X_o *is an OMR and* X_c *is a CMR. Moreover, X is an MMR if and only if* $X = X_o$ *and* $X = X_c$.

We now turn to the problem of constructing the MRs. The following Lemma is analogous to the well-known *Duality Principle* of System Theory [4].

LEMMA 1 [13] - *X is an MR of* (H,U,Y_p,Y_f) *if and only if X is an MR of* (H,U^{-1},Y_f,Y_p).

Thus, the Observability and the Constructibility can be seen as dual concepts. The set of all MRs is characterized by the following theorem, which is stated in an equivalent form in [7].

THEOREM 1 [13] - *A subspace X is an MR if and only if*

$$X = \bar{E}\,[H_+|H_-] \tag{8}$$

[the bar over E stands for closure], *where* H_+ *and* H_- *are two subspaces including* Y_f *and* Y_p *and invariant under U and* U^{-1} *respectively.*

From Lemma 1, note that we could exchange the roles of H_+ and H_- in (8). The structure of the OMRs is given by the following theorem (note that the structure of the CMRs could be inferred using Lemma 1).

THEOREM 2 [13] - *A subspace X is an OMR if and only if*

$$X = \bar{E}\,[Y_f|H_-], \tag{9}$$

where H_- *is a subspace including* Y_p *and invariant under* U^{-1}. *Moreover, the set of all MRs included in an OMR X with the transition operator F is exactly the set of invariant subspaces of* F^* (defined as the adjoint operator of F) *that include* X_c.

REMARK - From Proposition 2 the OMR X defined by (9) will be an MMR if and only if X is constructible. Clearly, a necessary condition is : $E[Y_f|H_-] \cap Y_p^{\perp} = \{0\}$, which is equivalent to

$$H_- \perp (Y_f \cap Y_p^{\perp}). \tag{10}$$

However, note that (10) is not sufficient to guarantee the minimality of X in general. We shall see in Section 5 that this is in fact the case under stronger assumptions on the 4-tuple (H,U,Y_p,Y_f).

Theorem 2 implies that an OMR [a CMR] X contains only one MMR, namely X_c [X_o]. Since Y_p and Y_f are a CMR and an OMR respectively, it follows that the subspaces

$$Y_* = \bar{E}\,[Y_f|Y_p] \quad \text{and} \quad Y^* = \bar{E}\,[Y_p|Y_f] \tag{11}$$

are the only two MMRs included in Y_p and Y_f respectively. These two MMRs have important properties; in particular, any internal MMR can be shown [13] to be a subspace of $(Y_* \vee Y^*)$, called the *frame space* in [7]. Let F_i and F_j be the transition operators of Y_* and Y^* respectively. It can be shown [13] that F_i and F_j satisfy

$$F_i = E^{Y_p} U|Y_* \quad \text{and} \quad F_j^* = E^{Y_f} U^{-1}|Y^*. \tag{12}$$

We terminate this section by presenting some "isomorphism" theorems between MMRs. We first introduce some new terminology. Let H_1 and H_2 be two Hilbert spaces and let M be an operator from H_1 to H_2. We say that M is a *weak-affinity* if M is one-to-one, closed, with a dense domain and a dense range. We say that M is a *quasi-affinity* [16] if M is furthermore bounded. Let F_1 and F_2 be two operators in H_1 and H_2 respectively. We say that F_1 is a *weak-affine* [*quasi-affine transform*] of F_2 when there exists a weak-affinity [quasi-affinity] M from H_1 to H_2 such that $M F_1 = F_2 M$. Clearly, if F_1 is a weak-affine transform of F_2, the converse is also true. To emphasize this symmetry, we say that F_1 and F_2 are *weakly-similar.* Finally, we say that F_1 and F_2 are *quasi-similar* when each operator $F_i(i=1,2)$ is a quasi-affine transform of the other.

PROPOSITION 3 [13] - *Let X be any MMR with the transition operator F. Then F and F^* are quasi-affine transforms of F_i and F_j^* respectively, the corresponding quasi-affinities being $E^{Y_p}|X$ and $E^{Y_f}|X$ respectively.*

In particular, Proposition 3 implies that, for any MMR X, we have $Y_* = \bar{E}[X|Y_p]$ and $Y^* = \bar{E}[X|Y_f]$. These results relate Y_* and Y^* to the "predictor spaces" of X (cf. [7],[13]). Between any two MMRs, only a weaker form of Proposition 3 holds. Note that this is to be related to an analogous result in the Realization Theory in the Hilbert space [3].

THEOREM 3 [13] - *Let X and $\dot{X}$ be any two MMRs with the transition operators F and $\dot{F}$ respectively. Then F and $\dot{F}$ are weakly-similar.*

3. MARKOVIAN REPRESENTATIONS AND INNOVATION SUBSPACES

In this section, the 4-tuple (H,U,Y_p,Y_f) will be assumed to satisfy not only (A1) but also :

(A2) $$\bigvee_n U^n(Y_p) = \bigvee_n U^n(Y_f) = Y_\infty;$$

(A3) $$\bigcap_n U^n(Y_p) = \bigcap_n U^n(Y_f) = \{0\}.$$

These assumptions are satisfied if Y_p and Y_f are the past and future spaces associated with a stationary Gaussian process which is both regular and regular after "time-reversal" (cf. [12]). Note also that (A1), (A2) and (A3) imply together that Y_p and Y_f are respectively an *incoming* and an *outgoing* subspace of Y_∞ in the terminology of Lax-Phillips [5].

In this section we shall study the geometric equivalent of the white noise process w of model (1). In our setting, the concept of white noise process is naturally replaced by the concept of *wandering subspace* [16, p. 4]. We say that a subspace G of H is wandering (for U) when $U(G) \perp G$. We then define the subspace generated by G, denoted G_∞, as the subspace containing all the subspaces $U^n(G)$ for all n.

DEFINITION 3 - *For any MR X, its innovation subspace W and its co-innovation subspace $\tilde{W}$ are defined respectively by*

$$U(X_- \vee Y_p) = W \oplus (X_- \vee Y_p) \quad \text{and} \quad (X_+ \vee Y_f) = \tilde{W} \oplus U(X_+ \vee Y_f). \tag{13}$$

Note that (13) implies that W and $\tilde{W}$ are two wandering subspaces of H. We define the proper innovation subspace I and the proper co-innovation subspace $\tilde{J}$ as the innovation subspace of Y_* and the co-innovation subspace of Y^* respectively.

The relationships between the Markovian Representations and their innovation subspaces are explained in the following.

PROPOSITION 4 [13] - *For any MR X, its innovation and co-innovation subspaces W and $\tilde{W}$ satisfy, respectively,*

$$(X_+ \vee Y_f) = X \oplus W_+ \quad \text{and} \quad (X_- \vee Y_p) = X \oplus U^{-1}(\tilde{W}_-). \tag{14}$$

If X is an OMR [a CMR], then the correspondence between X and W [$\tilde{W}$] is one-to-one; more precisely, we have

$$X = \bar{E}\,[Y_f | W_+^{\perp}] \quad \text{and} \quad X = \bar{E}\,[Y_p | U^{-1}(\tilde{W}_-^{\perp})] \tag{15}$$

[for any subspace G of H, $G^{\perp}$ denotes the orthogonal complement of G in H]. *If X is internal (not necessarily observable or constructible), then the correspondence between X and the pair $(W, \tilde{W})$ is one-to-one; more precisely the following decomposition holds:*

$$Y_\infty = X \oplus W_+ \oplus U^{-1}(\tilde{W}_-). \tag{16}$$

Note that equation (16) strenghtens the analogies with Scattering Theory (cf. [3],[7]).

The purely nondeterministic condition (4) of the Introduction can be generalized as follows.

DEFINITION 4 - *An MR X is called regular, or coregular, according to whether*

$$\bigcap_n U^n(X_- \vee Y_p) = \{0\} \quad \text{or} \quad \bigcap_n U^n(X_+ \vee Y_f) = \{0\}. \tag{17}$$

Using the Wold decomposition (cf. [2],[16]), it can be shown that the regularity and the coregularity of X are respectively equivalent to $(X_- \vee Y_p) = U^{-1}(W_-)$ and $(X_+ \vee Y_f) = \tilde{W}_+$. It could also be shown [13] that an MR X with the transition operator F is regular [coregular] if and only if F^n [F^{*n}] converges to zero strongly. From (A3), note that Y_* is regular and Y^* is coregular. More precisely, the innovation subspace I of Y_* and the co-innovation $\tilde{J}$ of Y^* satisfy, respectively,

$$Y_p = U^{-1}(I_-) \quad \text{and} \quad Y_f = \tilde{J}_+. \tag{18}$$

PROPOSITION 5 - *All regular [coregular] MRs X satisfy*

$$U^{-1}(W_-) = X \oplus U^{-1}(\tilde{W}_-) \qquad [\tilde{W}_+ = X \oplus W_+]. \tag{19}$$

When X is regular and coregular the two above equations coincide. If X is regular [coregular], then its constructibility [observability] is equivalent to

$$W_- = \tilde{W}_- \vee I_- \qquad [\tilde{W}_+ = W_+ \vee \tilde{J}_+]. \tag{20}$$

PROOF : The two equations of (19) follow from (14) and the remarks after Definition 4. Thus, it remains only to show (20). Let us assume that X is coregular. From (15) and (18), X will be observable if and only if $\overline{E}[\tilde{W}_+|W_+^\perp] = \overline{E}[\tilde{J}_+|W_+^\perp]$, which is equivalent to $\tilde{W}_+ = W_+ \vee \tilde{J}_+$. To complete the proof, note that the constructibility condition follows by duality.

Note that Proposition 5 implies that the correspondence between X and the pair $(W,\tilde{W})$ is one-to-one if X is regular or coregular. In the sequel we shall need the following Lemma which is but a trivial consequence of Proposition I 2.1 of [16].

LEMMA 2 - *The innovation and co-innovation subspaces of a regular and coregular MR have the same (finite or infinite) dimension. Furthermore, this dimension is invariant in the set of all internal (regular or coregular) MRs.*

The following theorem gives a necessary and sufficient condition on the 4-tuple (H,U,Y_p,Y_f) to ensure that all MMRs are regular and coregular.

THEOREM 4 [13] - *All MMRs are regular [coregular] if and only if Y^* is regular [Y_* is coregular].*

4. MARKOVIAN REPRESENTATIONS AND FUNCTIONAL MODELS

In this section, the 4-tuple (H,U,Y_p,Y_f) will be assumed to satisfy not only (Al), (A2) and (A3) but also :

(A4) *U has a finite multiplicity in H assumed complex and separable.*

Note that (A4) is satisfied if H is the Gaussian space generated by a finite-dimensional stationary Gaussian process. Assuming U of finite multiplicity will guarantee that the innovation and co-innovation subspaces of any MR will be both finite-dimensional. This fact will enable us to obtain simple functional models of the MRs by using the H^2-theory of finite-dimensional Hilbert spaces.

Let us recall some basic elements of Hardy space theory which will be needed hereafter (for more details we refer to [2] or [16]). Let N be a finite-dimensional Hilbert space . We designate by $L^2(N)$ the Hilbert space of all (equivalence classes of) measurable N-valued square integrable functions on the unit circle C. We denote by $H^2(N)$ the subspace of $L^2(N)$ of all functions whose negatively indexed Fourier coefficients vanish. All functions in $H^2(N)$ have analytic continuations into the disc, and whenever convenient we shall assume that the functions have been continued. Let N_1 and N_2 be two finite-dimensional Hilbert spaces. We designate by $B(N_1,N_2)$ the finite-dimensional space of all operators from N_1 into N_2. $B(N_1,N_2)$ can be endowed with an Hilbertian structure and we denote by $H^\infty(B(N_1,N_2))$ the subspace of $H^2(B(N_1,N_2))$ of all essentially bounded functions on C. A function Q of $H^\infty(B(N_1,N_2))$ is said to be inner when Q maps C into the set of unitary operators from N_1 onto N_2. Thus, N_1 and N_2 have the same dimension and we shall refer to the dimension of Q (denoted dimQ) as to the common dimension of N_1 and N_2. Given any inner function Q in

$H^\infty(B(N_1,N_2))$ we define a subspace $H(Q)$ of $H^2(N_2)$ and an operator $S(Q)$ (called the restricted shift associated to Q in [2]) on $H(Q)$ by

$$H(Q) = H^2(N_2) \ominus Q H^2(N_1) , \qquad S(Q)f = E[\chi f \mid H(Q)] \quad (f \in H(Q)) , \tag{21}$$

where χ denotes the identity operator on C, i.e. $\chi(z) = z$.

Let us elucidate the relationships between Q and S(Q). First, S(Q) is completely determined (up to a unitary equivalence) by the purely contractive part Q^o of Q.(In fact Q^o can be identified to the *characteristic function* of S(Q) as defined by Sz.-Nagy and Foias in [16].) An inner function Q in $H^\infty(B(N_1,N_2))$ is said to be purely contractive if there is no subspace of N_1 where the restriction of the operator Q(0) is an isometry. If Q is an inner function in $H^\infty(B(N_1,N_2))$, it follows from Proposition V 2.1 of [16] that there exist uniquely determined decompositions $N_1 = N_1^o \oplus N'_1$ and $N_2 = N_2^o \oplus N'_2$, so that Q can be written as the direct sum operator $Q = Q^o \oplus Q'$, where Q^o is a purely contractive inner function in $H^\infty(B(N_1^o,N_2^o))$ [by definition it is the purely contractive part of Q] and Q' is a constant unitary operator from N'_1 onto N'_2. Note that Q and Q^o have the same determinant up to a unit factor. Although the knowledge of all Q^o is needed to completely characterize S(Q), the knowledge of detQ [det stands for determinant] is sufficient to determine the spectrum of S(Q). Since detQ is a scalar inner function, it can be uniquely factored as the product of a Blaschke product b and a singular inner function s [16, p. 101]. We define the support of detQ as the union of the closure of the zero set of b and the closed support of the singular measure defining s. From Proposition 4.2 of [1], it could be shown that the point spectrum of S(Q) consists of the zero set of b in the open unit disc while the continuous spectrum of S(Q) consists of the intersection of the support of detQ with the unit circle (note that the residual spectrum of S(Q) is empty). Finally, S(Q) satisfies detQ(S(Q)) = 0, which shows that S(Q) is an operator of the class C_0 introduced by Sz.-Nagy and Foias [16, p. 123]. This implies that both $S(Q)^n$ and $S(Q)^{*n}$ converge to zero strongly.

In Section 5 we shall consider the following problem : "Let Q_1 and Q_2 be two inner functions associated through (21) with the operators $S(Q_1)$ and $S(Q_2)$. What is a necessary and sufficient condition on Q_1 and Q_2 to ensure that $S(Q_1)$ is a quasi-affine transform of $S(Q_2)$?". When Q_1 and Q_2 have the same dimension, the answer is provided by the following result from [9] : "Q_1 and Q_2 must be *quasi-equivalent* [11] or, equivalently, they must have the same *invariant factors* [11]. Moreover, this implies that $S(Q_1)$ and $S(Q_2)$ are in fact quasi-similar". We recall that an invariant factor is defined up to a unit multiplicative constant. It will be called trivial if it is equal to 1. Now the complete solution to the above problem is given by the following.

LEMMA 3 - *Let Q_1 and Q_2 be two inner functions. They have the same nontrivial invariant factors (and, therefore, the same determinant up to a unit factor) if and only if $S(Q_1)$ is a quasi-affine transform of $S(Q_2)$. In such a case $S(Q_1)$ and $S(Q_2)$ are in fact quasi-similar.*

PROOF : As the lemma is true when $\dim Q_1 = \dim Q_2$, let us suppose without loss of generality that $\dim Q_1 < \dim Q_2$. Up to a unitary equivalence, we may assume that Q_1 and Q_2 are inner in $H^\infty(B(N_1,N_1))$ and $H^\infty(B(N_2,N_2))$ respectively, with N_1 being a subspace of N_2. We then define an inner function Q in $H^\infty(B(N_2,N_2))$ as the direct sum operator $Q = Q_1 \oplus \mathrm{id}$, where id denotes the constant identity operator in $(N_2 \ominus N_1)$. Q and Q_1 have the same purely contractive part and, therefore, S(Q) and $S(Q_1)$ are unitarily equivalent. Thus, $S(Q_1)$ is a quasi-affine transform of $S(Q_2)$ if and only if $S(Q)$ is a quasi-affine transform of $S(Q_2)$. This in turn amounts to $S(Q)$ [and, therefore, $S(Q_1)$] being quasi-similar to $S(Q_2)$ and Q having the same invariant factors as Q_2. Since Q and Q_1 have the same nontrivial invariant factors, it follows that Q_1 and Q_2 also have the same nontrivial invariant factors. To complete the proof of the lemma we have just to remark that $\det Q_1$ and $\det Q_2$ are equal to the product of their invariant factors (up to a unit multiplicative constant).

We are now in a position to use all this body of knowledge to study the functional models associated with the regular and coregular MRs. The functional models will be obtained using the concept of *Fourier Representation* of a wandering subspace (cf. [16]). Let W be such a subspace of H, its Fourier Representation is a unitary operator Φ from W_∞ onto $L^2(W)$ defined on the generators by $\Phi U^n a = a\chi^n$ for all n in Z and a in W. Clearly, $\Phi U = \chi\Phi$ on W_∞. Moreover, Φ maps W_+ onto $H^2(W)$. Let X be a regular and coregular MR associated with its innovation and co-innovation subspaces W and $\tilde{W}$. We denote by Φ and $\tilde{\Phi}$ the Fourier Representations of W and $\tilde{W}$ respectively. From (19), W_+ is included in $\tilde{W}_+$ and Lemma V 3.2 of [16] implies that there exists an inner function Q in $H^\infty(B(W,\tilde{W}))$ such that $\tilde{\Phi} = Q\Phi$ on $W_\infty = \tilde{W}_\infty$. Since $\tilde{\Phi}(W_+) = QH^2(W)$, we have $\tilde{\Phi}(X) = H(Q)$. Now the very definitions of F and $S(Q)$ imply that $\tilde{\Phi}F = S(Q)\tilde{\Phi}$ on X, from which it follows that F and $S(Q)$ are unitarily equivalent. Conversely, let X be an MR whose transition operator F is unitarily equivalent to $S(Q)$, where Q is some inner function. Since $S(Q)$ is in the class C_o, it follows by unitary equivalence that F is also in C_o. Thus, F^n and F^{*n} converge to zero strongly and, therefore, X is regular and coregular. Hence, we have proven the following.

PROPOSITION 6 - *An MR X is regular and coregular if, and only if, its transition operator F is unitarily equivalent to the restricted shift associated with some inner function.*

Following the terminology of [7], the inner function Q associated with X will be called the *structural function* of X. Remarking that $(\det Q)H^2(\tilde{W}) \subset QH^2(W)$, it follows from Theorem III 3.8 of [2] that X is finite-dimensional if and only if detQ (as well as Q) is a rational function. In such a case the dimension of X is equal to the number of zeros (multiplicities counted) of detQ. The following proposition shows that, for infinite-dimensional regular and coregular MRs, detQ can be used as a "generalized" dimension function. It is to be noted that this concept is closely related [15] to the degree theory of strictly noncyclic functions introduced by Fuhrmann in [2]. In the proposition we shall say that a scalar inner function v *divides* a scalar inner function u if there exists a scalar inner function w such that u = vw.

PROPOSITION 7 - *Let X be a regular and coregular MR with the structural function Q. If X_1 is an MR included in X, then X_1 is regular and coregular and its structural function Q_1 is such that $\det Q_1$ divides $\det Q$. If $\det Q_1 = \det Q$ up to a unit factor, then $X = X_1$.*

PROOF : If X_1 is included in X, its regularity and coregularity follow immediately from (17). Let F and F_1 be the transition operators of X and X_1 respectively. The definitions of F and F_1 imply that $F_1^n = E^{X_1} F^n | X_1$ for all $n \geq 0$. It follows from Proposition 1.3 of [1] that there exist two invariant subspaces A and B of F such that $A = X_1 \oplus B$. Let W and $\tilde{W}$ be the two innovation subspaces of X with their respective Fourier Representations Φ and $\tilde{\Phi}$. The decomposition $\tilde{W}_+ = X \oplus W_+$ implies that A and B are both orthogonal to W_+. It is easy to check that $(A \oplus W_+)$ and $(B \oplus W_+)$ are two subspaces of $\tilde{W}_+$ invariant under U. From the Wold decomposition it follows that there exist two wandering subspaces G and K in W_∞ $(= \tilde{W}_\infty)$ such that $G_+ = A \oplus W_+$ and $K_+ = B \oplus W_+$. This implies that the subspaces generated by G and K are equal to W_∞ and, hence, W, $\tilde{W}$, G and K have all the same finite dimension from Lemma 2. Clearly we have the decomposition $G_+ = X_1 \oplus K_+$. Let Φ_g and Φ_k be the Fourier Representations of G and K. Since $K_+ \subset G_+$, there exists an inner function Q_2 in $H^\infty(B(K,G))$ such that $\Phi_g = Q_2 \Phi_k$ on W_∞. But, as in the proof of Proposition 6, this implies that F_1 is unitarily equivalent (through Φ_g) to $S(Q_2)$. From the definition of the structural function Q_1 of X_1 it follows that Q_1 and Q_2 have the same purely contractive part and hence $\det Q_1 = \det Q_2$. Remembering that $G_+ \subset \tilde{W}_+$ and $W_+ \subset K_+$, we infer the existence of two inner functions P in $H^\infty(B(G,\tilde{W}))$ and R in $H^\infty(B(W,K))$ such that $\tilde{\Phi} = P\Phi_g$ and $\Phi_k = R\Phi$ (both equations being valid on W_∞). This and the definition of Q_2 imply that $\tilde{\Phi} = PQ_2R\Phi$ on W_∞. From the definition of Q it follows that $Q = PQ_2R$. Taking the determinants, we obtain that $\det Q_1$ (equal to $\det Q_2$) divides $\det Q$. Thus, the first part of the proposition is proved. Suppose now that $\det Q$ divides $\det Q_1$. This implies that $\det P$ and $\det R$ are unit constants and, therefore, P and R are constant unitary operators. From Lemma V 3.2 of [16], it follows that $G_+ = \tilde{W}_+$ and $K_+ = W_+$, from which it results that $X = X_1$.

In the next section we shall study the set of all regular and coregular MMRs. Thus, we shall suppose that the assumptions of Theorem 4 are satisfied. Under (A4) they can be made more precise.

LEMMA 4 - *Y^* is regular if and only if Y_* is coregular.*

PROOF : From (A3) the transition operators F_i of Y_* and F_j of Y^* are such that F_i^n and F_j^{*n} converge to zero strongly. This implies that F_i and F_j are both completely nonunitary (cf. [2],[16]). Proposition 3 asserts that F_j is a quasi-affine transform of F_i and it follows from Proposition III 4.6 of [16] that, if one of the operators F_i or F_j is in the class C_o, so is the other. The Lemma results after remarking that Y^* is regular if and only if F_j is in C_o while Y_* is coregular if and only if F_i is in C_o.

In the following a Gaussian process will be called *strictly noncyclic* (cf. [7],[14]) when Y^* is regular. To justify the terminology, note that a Gaussian process can be shown [15] to be strictly noncyclic if and only if all its spectral factors are strictly noncyclic functions in the sense of [2].

5. STRUCTURE OF THE MMRs OF STRICTLY NONCYCLIC PROCESSES

In this section, the 4-tuple (H,U,Y_p,Y_f) will be assumed to satisfy not only (A1) through (A4) but also :

(A5) *Y^* is regular or, equivalently, Y_* is coregular.*

Thanks to this assumption all MMRs are regular and coregular and, therefore, they all have structural functions as defined in Section 4. The set of these minimal structural functions is described by the following.

THEOREM 5 - *The structural functions of any two MMRs have the same nontrivial invariant factors.*

PROOF : Let X be any MMR with the transition operator F and the structural function Q. From Proposition 3, F is a quasi-affine transform of the transition operator F_i of Y_*. The structural function Q_i of Y_* is such that S(Q) is a quasi-affine transform of $S(Q_i)$. But Lemma 3 implies that Q and Q_i have the same nontrivial factors. As X is arbitrary, the theorem follows by transitivity.

Theorem 5 has several interesting consequences. First, the transition operators of any two MMRs are quasi-similar (cf. [8],[14]). This follows from the quasi-similarity of the two associated restricted shifts. Secondly, the structural fuctions of *internal* MMRs have the same dimension and have therefore the same invariant factors [8]. Finally, if X and $\dot{X}$ are any two MMRs, their structural functions have the same determinant up to a unit factor [14]. This implies that the transition operators of X and $\dot{X}$ have the same spectrum. Moreover, X and $\dot{X}$ have the same "dimension" as defined in Section 4. This can be seen as a generalization to strictly noncyclic processes of a well-known fact for processes with a rational spectral density.

For strictly noncyclic processes Theorem 2 can be particularized to characterize the set of all MMRs. Note that Theorem 5 and the following proposition have been established in [8], but under the unnecessary assumption that the MMRs are real and internal.

PROPOSITION 8 - *A subspace X is an MMR if and only if*

$$X = \bar{E}[Y_f | H_-] , \tag{22}$$

where H_- is a subspace including Y_p, invariant under U^{-1} and orthogonal to $(Y_f \cap Y_p^\perp)$.

PROOF : The necessity follows from the remark after Theorem 3. For the sufficiency, let X be the OMR defined by (22). We denote by F its transition operator and by X_c its constructible part. We first show that the transition operator F_j of Y^* is a quasi-affine transform of F. For that, let M denote the restriction to X of the orthogonal projector onto Y_f. From Definition 1 we have $MF^* = F_j^*M$. By taking the adjoints, we see that F_j will be a quasi-affine transform of F if M is a quasi-affinity. That M is one-to-one follows from the observability of X. It remains to show that M(X) is dense in Y^*. But this follows from the two following remarks : first, M(X) is included in Y^* since H_- (and therefore X) is orthogonal to $(Y_f \cap Y_p^\perp)$; secondly, $M(X_c)$ is a subspace of M(X) [since $X_c \subset X$] which is dense in Y^* [since X_c is an MMR, cf. Proposition 3]. Thus, we have shown that F_j is a quasi-affine transform of F. But, F_j belongs to the class C_o and Proposition 5.2 of [1] implies that F must also belong to C_o. This establishes that X is both regular and coregular. Therefore, X has a structural function, say Q. If Q_j denotes the structural function of Y^*, the above reasoning implies that $S(Q_j)$ is a quasi-affine transform of S(Q). From Lemma 3 it follows that $\det Q_j = \det Q$ (up to a unit factor). Let Q_c denote the structural function of X_c [Q_c exists since X_c is an MMR]. Theorem 5 implies that $\det Q_j = \det Q_c$ (up to a unit factor). Thus, we have obtained that X_c is a subspace of X such that $\det Q_c = \det Q$ (up to a unit factor). It follows from Proposition 7 that X is identical to the MMR X_c, which concludes the proof.

It should be pointed out that the map $H_- \to X$ is not one-to-one (except for internal MMRs) and, therefore, is not a proper parametrization of the set of all MMRs. A proper parametrization is derived in [13] and applied to the construction of all finite-dimensional MMRs.

ACKNOWLEDGMENT - The author would like to thank Professor E.A. Nordgren for his help in establishing the proof of Proposition 7.

REFERENCES

[1] Douglas,R.G., *Canonical Models.in Topics in Operator Theory,* Math. Surveys No. 13, American Math. Society, Providence, 1974, pp. 161-218.

[2] Fuhrmann,P.A., *Linear Systems and Operators in Hilbert Space,* Mc Graw-Hill, 1981.

[3] Helton,J.W., *Systems with Infinite-Dimensional State Space : The Hilbert Space Approach,* Proc. IEEE, Trans. Autom. Cont., vol. 64, No. 1, January 1976, pp. 145-160.

[4] Kalman,R.E., *Lectures on Controllability and Observability,* C.I.M.E. Summer Course, 1968, Cremonese, Roma, 1969, pp. 1-149.

[5] Lax,P.D., and Phillips,R.S., *Scattering Theory,* Academic Press, 1967.

[6] Lindquist,A., Mitter,S., and Picci,G., *Toward a Theory of Nonlinear Stochastic Realization, in Feedback and Synthesis of Linear and Nonlinear Systems,* Eds. D. Hinrichsen and A. Isidori, Springer Verlag.

[7] Lindquist,A., Pavon,M., and Picci,G., *Recent Trends in Stochastic Realization Theory, in Harmonic Analysis and Prediction Theory. The Masani Volume,* Eds. V. Mandrekar and H. Salehi, North Holland.

[8] Lindquist,A., and Picci,G., *On a Condition for Minimality of Markovian Splitting Subspaces,* Systems and Controls Letters, to be published.

[9] Moore,B.,III, and Nordgren,E.A., *On Quasi-Equivalence and Quasi-Similarity,* Acta Sci. Math., vol. 34, 1973, pp. 311-316.

[10] Neveu,J., *Processus Aléatoires Gaussiens,* Presses de l'Université de Montréal, 1968.

[11] Nordgren,E.A., *On quasi-equivalence of matrices over* H^∞, Acta Sci. Math., vol. 34, 1973, pp. 301-310.

[12] Rozanov,Yu.A., *Stationary Random Processes,* Holden Day, 1967.

[13] Ruckebusch,G., *Théorie Géométrique de la Représentation Markovienne,* Ann. Inst. Henri Poincaré, vol. 16,No. 3, 1980, pp. 225-297.

[14] Ruckebusch,G., *Markovian Representation Theory and Hardy spaces,* Proc. IEEE, International Symposium on Circuits and Systems, Rome, May 1982.

[15] Ruckebusch,G., *Markovian Representations and Spectral Factorizations of Stationary Gaussian Processes, in Harmonic Analysis and Prediction Theory - The Masani Volume,* Eds. V. Mandrekar and H. Salehi, North Holland.

[16] Sz.-Nagy,B., and Foias,C., *Harmonic Analysis of Operators on Hilbert Space,* North Holland, 1970.

A BRIEF TUTORIAL ON CALCULUS ON MANIFOLDS, WITH EMPHASIS ON APPLICATIONS TO IDENTIFICATION AND CONTROL

Christopher I. Byrnes

Division of Applied Sciences
and Department of Mathematics
Harvard University
Cambridge, MA 02138

Abstract: In this tutorial, fundamental definitions of differentiable manifold, functions, and vector fields are given along with examples taken from identification, filtering, and from realization theory. Calculus of real-valued functions on a manifold, leading to the Morse Theory, is also discussed accompanied by examples in identification and in the stability analysis of a power system.

0. WHAT CAN MANIFOLDS DO FOR PROBLEMS OF IDENTIFICATION, FILTERING, AND ADAPTIVE CONTROL?

0.1 Manifolds

Let us start out with an intuitive definition of manifold: A manifold M is a topological space which is (locally) Euclidean in a neighborhood of each of its points. Three examples of manifolds are

Figure 0.1. The real plane $\mathbb{R}^2$.

R. S. Bucy and J. M. F. Moura (eds.), Nonlinear Stochastic Problems, 123–150.

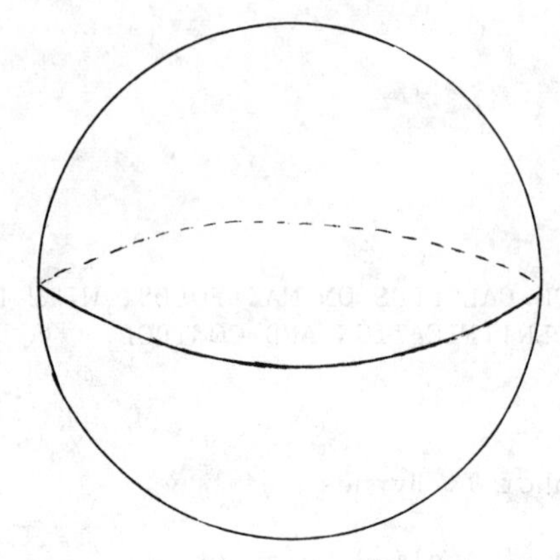

Figure 0.2. The sphere S^2.

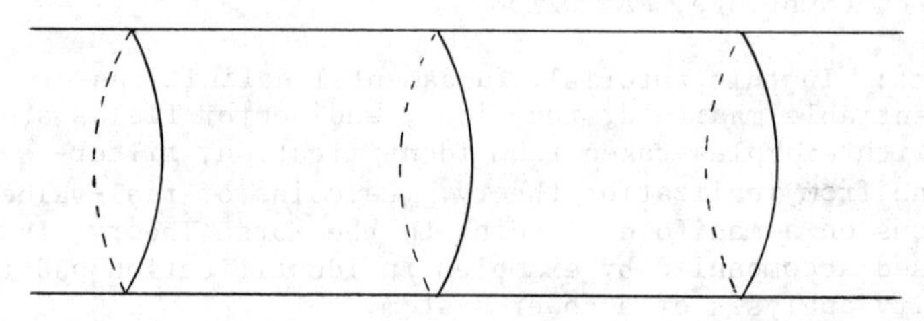

Figure 0.3. The (infinite cylinder) $S^1 \times \mathbb{R}$.

Thus, for any point $p \in M$, for $M = \mathbb{R}^2$, S^2, or $S^1 \times \mathbb{R}$, there is a sufficiently small open set U containing p which is homeomorphic to a disk. This much of the precise definition of smooth manifold will suffice at present.

The geometry of and analysis on manifolds have already made some deep contributions to deterministic control theory, as can be measured by the solution of outstanding classical problems - which may not use "manifolds" in the problem statement, but whose best (or only) solution or partial solution to date have come as an application of this theory. The basis for these successes, as well as for the potential applications in stochastic control can be traced to the following roles played by manifolds:

0.2. Calculus

Manifolds provide a setting for differential/integral calculus on nonlinear spaces. For example, consider the Hamiltonian system

$$\frac{dx}{dt} = \frac{\partial H}{\partial y} , \quad \frac{dy}{dt} = -\frac{\partial H}{\partial x} \tag{0.1}$$

where $(x,y) \in \mathbb{R}^n \times \mathbb{R}^n$ and $H(x,y) \in \mathbb{R}$. Even though (0.1) is defined on $\mathbb{R}^{2n}$, conservation of energy implies that the solution to (0.1) with initial condition (x_o,y_o) will evolve, in fact, on the hypersurface

$$H(x,y) = H(x_o,y_o) = \text{constant} \tag{0.2}$$

In the interest of integrating (0.1) explicitly by quadrature (i.e. in finding the "action-angle coordinates" for (0.1), see [2]), we are led to study a differential equation on a non-linear (in general) hypersurface in $\mathbb{R}^{2n}$ even though we started with an equation (0.1) on Euclidean space. Moreover, it was an historically important observation, due to Gibbs, in the development of ergodic theory that one can integrate smooth functions (with growth constraints) on any surface (0.2) of constant energy not containing an equilibrium point of (0.1). The hypersurfaces not containing equilibria are always manifolds, as we shall see.

0.3. Parameterized Systems

Manifolds often parameterize the set of linear systems having certain discrete (or continuous) invariants constrained. For example, the set of miniphase systems with r zeroes and n poles,

$$X_{n,r} = \{p(s)/q(s): p,q \text{ are monic, } \deg(p)=r\leq n=\deg(q),$$
$$\text{and } p(s)=0 \text{ implies } |s|<1\}$$

is an open subset of $\mathbb{R}^{n+r}$. Explicitly, if $p_o+\ldots+s^r/q_o+\ldots+s^n$ is an element of $X_{n,r}$ it corresponds to the point

$$(p_o,\ldots,p_{r-1},q_o,\ldots,q_{r-1}) \in \mathbb{R}^{r+n}$$

where, however, $p_o,\ldots,p_{r-1}$ satisfy the (discrete) Routh-Hurwitz inequalities. Thus, $X_{n,r} \subset \mathbb{R}^{r+n}$ is an open subset.

Åström-Söderström prove uniqueness of the maximum likelihood estimates for a recursive identification algorithm on $X_{n,r}$, introduced in [3], by doing calculus on this "manifold of systems"; i.e. by analyzing the extrema of the function

$$V_{g_o} : X_{n,r} \to \mathbb{R}^+$$

defined by

$$V_{g_o}(g) = \frac{1}{2\pi i} \int_{|s|=1} \frac{g_o(s)g_o(s^{-1})}{g(s)\, g(s^{-1})} \frac{ds}{s} \qquad (0.3)$$

In [3], it was shown that, for g_o stable and minimum phase, the extrema of V_{g_o} coincide with the maximum likelihood estimates. Uniqueness is proved in [4] by proving the uniqueness of the critical points of the differentiable function V_{g_o} on the parameter manifold $X_{n,r}$. $X_{n,r}$ is in fact a Euclidean space [14], [15], but many of the parameter spaces arising in identification [19], [24], or [25], although they are manifolds, are not Euclidean.

0.4. Realization Theory

Manifolds can and do arise as state-spaces for nonlinear input-output systems ([8], [28], [31] and [39]). A very pertinent example from filtering would be the input-output pair consisting of an observation process driving a black box with output some conditional statistic which is to be filtered. In nonlinear filtering, this input-output pair is nonlinear thus a finite-dimensional filter, if it were to exist, would necessarily be a nonlinear system. To date, this use of nonlinear realization theory has only produced (conjectured) necessary conditions for the existence of finite dimensional filters. For the cubic sensor problem [13]

$$dx_t = dw_t$$

$$dy_t = x_t^3\, dt + dv_t$$

through the work of many people [26], [40], (S.K. Mitter, unpublished), the ansatz suggested in [12] has been verified, yielding the theorem that no statistic $s(x_t)$ can be propagated by a finite dimensional filter. That is, there does not exist a system

$$dz_t = f(z_t)dt + g(z_t)dy_t$$

$$s(x_t) = h(z_t)$$

defined on a finite-dimensional manifold M.

This work was partially supported by Air Force Office of Scientific Research under Grant No. AFOSR-81-0054, the National Science Foundation under Grant No. ECS-81-21428, and NASA under Grant No. NSG-2265.

1. LOCAL COORDINATES AND DIFFERENTIABLE MANIFOLDS

1.1. Topological Manifolds

In many problems it becomes useful, or even necessary, to study the extrema of a function defined on a set X, say a subset of some $\mathbb{R}^N$. A question that then arises is: Can one use calculus to study, say, max-min problems on X? If X is a differentiable manifold, then the answer is of course yes.

Definition 1.1. A topological space M is a topological manifold provided there exists an open cover (U_α) of M and homeomorphisms

$$h_\alpha : U_\alpha \to \mathbb{R}^n \tag{1.1}$$

of each U_α with $\mathbb{R}^n$.

For example, S^2 is a topological manifold. For, S^2 is covered by 2 open sets

$$U_1 = S^2-\{\text{north pole}\}, \quad U_2 = S^2-\{\text{south pole}\}$$

and stereographic projection from the north pole defines a homeomorphism

$$h_1 : U_1 \simeq \mathbb{R}^2$$

while stereographic projection from the south pole gives rise to a homeomorphism

$$h_2 : U_2 \simeq \mathbb{R}^2$$

(see Figure 1.1).

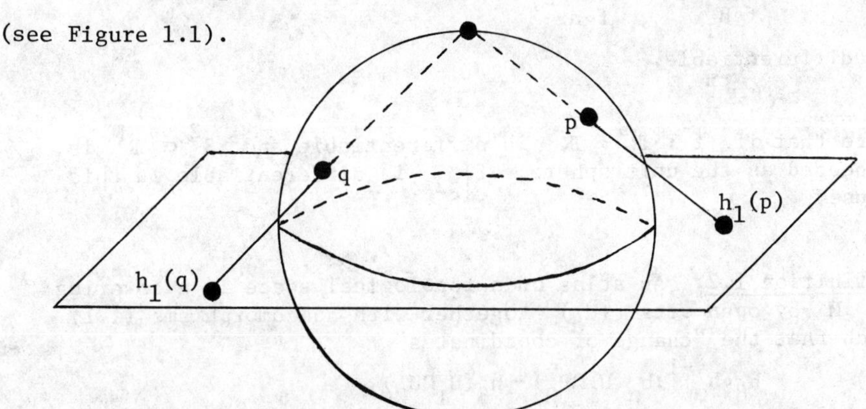

Figure 1.1. The stereographic projection h_1.

Finally, we note that the infinite cylinder (Figure 0.3) $S^1 \times \mathbb{R}$ is also a topological manifold, covered by two "Euclidean" open sets

$$U_1 = S^1 \times \mathbb{R} - \{\text{north pole}\} \times \mathbb{R}\ ,\quad U_2 = S^1 \times \mathbb{R} - \{\text{south pole}\} \times \mathbb{R}$$

1.2. Differentiable Manifolds

The covering (U_1, U_2) of S^2 introduced in Section 1.1 has a further property which is crucial to the existence of a well-defined calculus on S^2. Namely, the change of coordinates from (U_1, h_1) to (U_2, h_2):

$$h_2 \circ h_1^{-1} : U_1 \cap U_2 \to U_1 \cap U_2$$

is differentiable. Indeed

$$h_2 \circ h_1^{-1}(x_1, \ldots, x_n) = (x_1, \ldots, x_n)/||x||^2$$

In particular, if

$$f : h_1(U_1 \cap U_2) \to \mathbb{R}$$

is differentiable then

$$f(h_2(h_1^{-1}(x)))\quad ,\quad x \in h_2(U_1 \cap U_2)$$

is also differentiable, so that there is a well-defined notion of differentiable function on S^2, relative to the covering $(U_1; h_1)$, $(U_2; h_2)$. That is,

$$f : S^2 \to \mathbb{R}$$

is differentiable if, and only if, each of

$$f \circ h_1^{-1}\quad ,\quad f \circ h_2^{-1}$$

is differentiable.

Note that if $f : \mathbb{R}^3 \to \mathbb{R}$ is differentiable and $S^2 \subset \mathbb{R}^3$ is embedded as the unit sphere, $f|_{S^2}$ is differentiable in this sense.

Definition 1.2. An atlas on a topological space is a covering of M by open sets (U_α) together with homeomorphisms (1.1) such that the "change of coordinates"

$$h_\beta \circ h_\alpha^{-1} : h_\alpha(U_1 \cap U_2) \to h_\beta(U_1 \cap U_2)$$

is infinitely differentiable; i.e. is C^∞.

In particular, each $h_\beta \circ h_\alpha^{-1}$ has a differentiable inverse $h_\alpha \circ h_\beta^{-1}$. If M is connected this implies the n in (1.1) is independent of α. This integer is called <u>the dimension</u> of M.

Given an atlas on M, we can define differentiability with respect to that atlas so that calculus on manifolds, with an atlas, is underway. However, it would be extremely inconvenient to have to use local coordinates (U_α, h_α) coming from that atlas alone. Let us say that two atlases $\{(U_\alpha, h_\alpha)\}$, $\{(V_i, k_i)\}$ are compatible if, say, each k_i is a differentiable function - with respect to $\{(U_\alpha, h_\alpha)\}$ -

$$k_i : V_i \cap U_\alpha \to \mathbb{R}^n$$

whenever $V_i \cap U_\alpha \neq \emptyset$. Thus, we can speak of a maximal atlas, viz. the union of all atlases compatible with a given one.

<u>Definition 1.3</u>. A differentiable n-manifold is a topological space together with a maximal atlas of coordinate charts (U_α, h_α) satisfying (1.1)

To check that M is a differentiable n-manifold, however, one need only produce a single atlas. Thus, $\mathbb{R}^2$ or S^2, and $S^1 \times \mathbb{R}$ are all differentiable 2-manifolds. Finally, we have

<u>Definition 1.4</u>. A function $f : M \to \mathbb{R}$ is differentiable if and only if

$$f \circ h_\alpha^{-1} : \mathbb{R}^n \to \mathbb{R}$$

is differentiable for all coordinate charts (U_α, h_α) in an atlas for M.

<u>Exercise 1.5</u>. The "height function"

$$f : S^1 \to \mathbb{R}$$

in Figure 1.4, is differentiable.

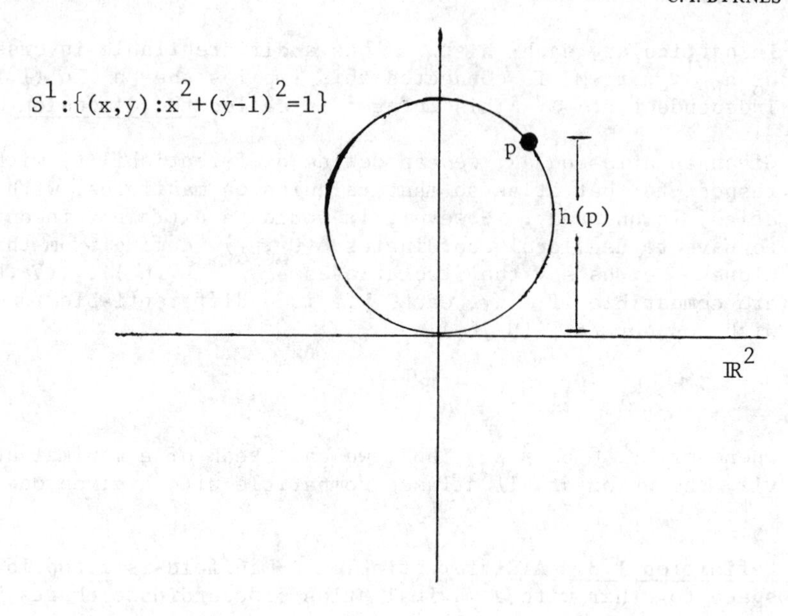

Figure 1.4.

1.3. Constructions with Manifolds

Calculus on open subsets of $\mathbb{R}^n$ is of course classical, but it is still important to note that an open subset U of $\mathbb{R}^n$ is a differentiable n-manifold since U can be covered by ε-balls.

Proposition 1.6. An open subset of a differentiable n-manifold is a differentiable n-manifold.

Proof. See e.g. [7].

In general, we strongly recommend the introductory books [33], [38] as well as [7], [23], [37].

Now certain closed subsets of $\mathbb{R}^n$ turn out to be manifolds, as a consequence of the Implicit Function Theorem. For example, consider the locus of $f^{-1}(0)$ of a single differentiable function

$$f : \mathbb{R}^2 \to \mathbb{R}$$

According to the Implicit Function Theorem (e.g. [38]), $f^{-1}(0)$ is a smooth 1-manifold provided the Jacobian of f,

$$Jf = \left[\frac{\partial f}{\partial x}, \frac{\partial f}{\partial y}\right]$$

is maximal rank - in this case, never vanishes - along $f^{-1}(0)$. Thus, if

$$f(x,y) = x^2 + y^2 - 1$$

then $Jf(x,y) \neq 0$ on $f^{-1}(0)$, for

$$Jf(x,y) = [2x,2y] = 0 \Rightarrow (x,y) = (0,0)$$

This shows, again, that the unit circle

$$S^1 = \{(x,y) : x^2 + y^2 = 1\}$$

As a second example, consider a Hamiltonian system

$$\frac{dx}{dt} = \frac{\partial H}{\partial y}, \quad \frac{dy}{dt} = -\frac{\partial H}{\partial x} \tag{1.2}$$

with Hamiltonian (or energy) function $H : \mathbb{R}^{2n} \to \mathbb{R}$. Suppose that the hypersurface (of constant energy)

$$M : H(x,y) = c$$

contains no equilibrium. Then M is a manifold of dimension n-1, for

$$JH(x,y) = \left[\frac{\partial H}{\partial x}, \frac{\partial H}{\partial y}\right] \neq 0$$

precisely when (x,y) is not an equilibrium point of (1.2). By the Implicit Function Theorem, such an M is a manifold (compare, Section 0.2). It can also be shown that M is "orientable" and therefore possesses a volume form, with respect to which one can develop the integral calculus (see [38]).

One can also use the Implicit Function Theorem to prove that functions are differentiable, since

Proposition 1.7. A function $f : M \to \mathbb{R}$ is differentiable if, and only if, its graph,

$$gr(f) = \{(x,f(x))\} \subset M \times \mathbb{R}$$

is a differentiable manifold.

Exercise 1.8. Show that the graph of the height function

$h : S^1 \to \mathbb{R}$ is a submanifold of the "infinite cylinder" $S^1 \times \mathbb{R}$.

Remark 1.9. Using precisely this technique, D. Delchamps [18] has proved that the unique positive definite solution of the Riccati equation

$$AP + PA' - PBB'P + C'C = 0$$

is a differentiable (in fact, real analytic) function of the parameters $(a_{ij}, b_{k\ell}, c_{mn})$ in a minimal triple (A,B,C).

Of course, this hints at another construction for manifolds, viz.

Proposition 1.10. If M, N are differentiable manifolds with atlases $\{(U_\alpha,h_\alpha)\}$, $\{(V_\beta,k_\beta)\}$, then $M \times N$ is a differentiable manifold, with atlas $\{(U_\alpha \times V_\beta, h_\alpha \times k_\beta)\}$, of dimension

$$\dim M \times N = \dim M + \dim N$$

In the next section, we shall look at some examples of manifolds arising in system theory, using the facts that open subsets, and certain closed subsets, of manifolds are again manifolds.

2. EXAMPLES: MANIFOLDS AS PARAMETER SPACES FOR SYSTEMS

2.1 Observable Systems

Consider the system

$$x(t+1|t) = Fx(t|t-1) + K\varepsilon(t)$$
$$y(t) = Hx(t|t-1) + \varepsilon(t)$$

where F is $n \times n$, K $n \times p$, H $p \times n$ and (F,H) is observable, i.e.

$$\text{rank } [H^t, (HF)^t, \ldots, (HF^{n-1})^t] = n \tag{2.1}$$

Then, (F,K,H) may be regarded as point

$$(F,K,H) \in \mathbb{R}^N, \quad N = n^2+2np$$

by using the entries of (F,K,H) as coordinates for an N-vector. Thus,

$$\tilde{\mathscr{A}}^n_p = \{(F,K,H) : (F,H) \text{ is observable}\}$$

is a subset of $\mathbb{R}^N$ and is, in fact, an N-manifold since

Proposition 2.1. $\tilde{\mathscr{A}}^n_p$ is an open, dense subset of $\mathbb{R}^N$.

Proof. To say (F,K,H) is not in $\tilde{\mathscr{A}}^n_p$ is to say that for every choice of n columns $i_1,\ldots,i_n$ of the observability matrix (2.1), the corresponding $n \times n$ determinant

$$\det(H^t,\ldots,(HF^{n-1})^t)_{i_1 \ldots i_n} = 0 \qquad (2.2)$$

That is, $\tilde{\mathscr{A}}^n_p$ is the complement of the union of a finite number $\binom{N}{n}$ of zero sets of functions (2.2). Thus, $\tilde{\mathscr{A}}^n_p$ is open. In fact, $\tilde{\mathscr{A}}^n_p$ is dense since the functions (2.2) are polynomials and

Lemma 2.2. The zero set of a nonzero polynomial

$$p : \mathbb{R}^n \to \mathbb{R}$$

has no interior points.

2.2. Observable Systems with Fixed Observability Indices

This set of systems, which appears in the work of Alengrin [1], is also a manifold. In fact, if $n = n_1 + \ldots + n_p$, $n_i \geq 0$, $n_i \in \mathbb{Z}$ and

$$\tilde{\mathscr{A}}^n_p(n_i) = \{(F,K,H) : (F,H) \text{ has observability indices } (n_i)\}$$

then, $\tilde{\mathscr{A}}^n_p(n_i)$ is canonically a product manifold $\mathbb{R}^N \times GL(n,\mathbb{R})$, where

$$N = n^2 + np + p - \sum_{n_i \geq n_n} (n_i - n_j + 1)$$

and $GL(n,\mathbb{R}) \subset \mathbb{R}^{n^2}$ is the open subset of invertible $n \times n$ matrices (see [10], [14]).

2.3. Miniphase Systems

The parameter space $X_{n,r}$ introduced in Section 0.3 is a differentiable manifold, as an open subset of $\mathbb{R}^r \times \mathbb{R}^n$. And, the "objective function" (0.3) is a differentiable function on $X_{n,r}$.

2.4. Partial Realizations.

Consider the set

$$C_n = \{(Q_n, Q_{n-1}\zeta) : Q_n, Q_{n-1} \text{ monic, coprime polynomials of degrees } n, n-1 \text{ and } \zeta \in \mathbb{R}\}$$

C_n arises in the Chebychev recursion relations for partial realizations, see e.g. [22]. Using the coefficients $Q_{n,i}$, $Q_{n-1,j}$ of $Q_n(s)$ and $Q_{n-1}(s)$ and using γ as the coördinates of a vector $\underline{Q} \subset \mathbb{R}^{2n}$, we see

$$C_n \subset \mathbb{R}^{2n}$$

Proposition 2.3. C_n is an open dense subset of $\mathbb{R}^{2n}$, in particular C_n is a manifold.

Proof. The proof reposes on the existence of the resultant [32], $\mathrm{Res}(Q_n, Q_{n-1})$ which is a polynomial in the coefficients $Q_{n,i}$, $Q_{n-1,j}$ and which satisfies

$$\mathrm{Res}(Q_n, Q_{n-1}) = 0 \iff Q_n, Q_{n-1} \text{ have a common root.}$$

C_n is the complement of an algebraic set, $\mathrm{Res}^{-1}(0)$, in $\mathbb{R}^{2n}$ and the proposition follows from Lemma 2.2.

A similar argument implies that $\mathrm{Rat}(n)$ is a manifold, see [9], [14].

3. DIFFERENTIABLE FUNCTIONS

3.1. Real-Valued Functions

Suppose M is a differentiable manifold, and $f : M \to \mathbb{R}$ is a differentiable function. As in calculus, the extrema of f always give rise to critical points, i.e.

Definition 3.1. $p \in M$ is a critical point for f provided $h_\alpha(p)$ is a critical point for $f \circ h_\alpha^{-1}$, for some coordinate chart (U_α, h_α) containing p.

Note that if $h_\alpha(p)$ is a critical point for $f \circ h_\alpha^{-1}$ then $h_\beta(p)$ is a critical point for $f \circ h_\beta^{-1}$, by the chain rule applied to the identity

$$f \circ h_\beta^{-1} = (f \circ h_\alpha) \circ (h_\alpha^{-1} \circ h_\beta)$$

noting that $h_\alpha^{-1} \circ h_\beta$ and $h_\beta^{-1} \circ h_\alpha$ are differentiable inverses. Thus, although the values $\partial f/\partial h_\alpha^i$ change in the h_β coordinates, their simultaneous vanishing does not.

In particular, if p is a critical point with respect to one coordinate chart, then p is a critical point with respect to every coordinate chart. It follows, for example, that if $p \in M$ is a minimum for f, p is a critical point. More is true:

Definition 3.2. If p is a critical point, the Hessian of f with respect to a chart

$$h_\alpha = (h_\alpha^1, \ldots, h_\alpha^n) : U_\alpha \to \mathbb{R}^n$$

is the symmetric matrix of second partials

$$H_p(f \circ h_\alpha^{-1}) = \left[\frac{\partial^2 (f \circ h_\alpha^{-1})}{\partial h_\alpha^i \partial h_\beta^j} \right]$$

If $H_p(f \circ h_\alpha)^{-1}$ is nonsingular, then p is a nondegenerate critical point. In this case, the number, $i_p(f)$, of negative eigenvalues of $H_p(f \circ h_\alpha^{-1})$ is called the index of the critical point p.

As before, Definition 3.2 is well-defined since, although the components of $H_p(f \circ h_\alpha^{-1})$ may change in a different coordinate system, the nonsingularity of $H_p(f \circ h_\alpha^{-1})$ does not change, as

$$H_p(f \circ h_\alpha^{-1}) = J(h_\beta^{-1} \circ h_\alpha)^t H_p(f \circ h_\beta^{-1}) J(h_\beta^{-1} \circ h_\alpha) \tag{3.1}$$

where $J(h_\beta^{-1} \circ h_\alpha)$ is the Jacobian matrix of change of coordinates $h_\beta^{-1} \circ h_\alpha$. (3.1) also shows that $i_p(f)$ is invariant. Since the calculations of i_p are done in a coordinate chart, the standard results from advanced calculus on $\mathbb{R}^n$ apply, e.g.:

Second Derivative Test 3.3. $p \in M$ is a minimum for f if $i_p(f) = 0$, p is a maximum if $i_p(f) = \dim M$.

It is well-known that a differentiable function

$$f : [0,1] \to \mathbb{R} \tag{3.2}$$

has a maximum and a minimum. The min-max principle asserts that, if f has 2 minima, then f must have a maximum lying between them. If f satisfies $f^{(i)}(0) = f^{(i)}(1)$ for each i, then the min-max principle implies that if f has two minima, f must have at least two maxima. Such principles are, in fact, consequences of the Morse inequalities which relate the number and nature of the critical points of f to the topology of M (see [6], [33]).

Definition 3.4. $f : M \to \mathbb{R}$ is a Morse function if all of its critical points are nondegenerate, and if

$$f^{-1}(-\infty,c] \text{ is compact for all } c \in \mathbb{R}$$

The fundamental theorem of elementary Morse theory is The Morse Inequalities, which are valid for any Morse function f:

Theorem 3.5. (Morse) If $c_j(f)$ is the number of critical points of f having index j, then

$$b_o(M) \leq c_o(f)$$

$$b_1(M)-b_o(M) \leq c_1(f)-c_o(f)$$

$$\vdots$$

$$b_{n-1}(M)-\ldots+(-1)^n b_o(M) \leq c_{n-1}(f)-\ldots+(-1)^n c_o(f)$$

$$b_n(M)-\ldots+(-1)^{n+1} b_o(M) = c_n(f)-\ldots+(-1)^{n+1} c_o(f)$$

where the $b_i(M)$ are topological invariants of M - its Betti numbers.

For example, if f in (3.2) satisfies $f^{(i)}(0) = f^{(i)}(1)$ then f is a differentiable function on S^1, via the definition

$$f(e^{2\pi i\theta}) = f(\theta)$$

The Betti numbers of S^1 are $b_o(S^1) = 1 = b_1(S^1)$.

Suppose $c_o(f) = 2$; i.e. that f has 2 minima. The Morse inequalities then give

$$1 = b_o(S^1) \leq c_o(f) = 2$$

$$0 = b_1(S^1)-b_o(S^1) \leq c_1(f)-c_o(f) = c_1(f)-2$$

Thus, $c_o(f) = 2 \Rightarrow c_1(f) \geqslant 2$.

3.2 Example

The objective function (0.3) is defined on $X_{n,r}$. Thus a point $p(s)/q(s) \in X_{n,r}$ is a critical point of V_{g_o} if, and only if,

$$\frac{\partial V_{g_o}}{\partial p_i} = , i=0, \dots, r-1 ; \frac{\partial V_{g_o}}{\partial q_i} = 0 , i=0 , \dots , n-1.$$

According to [4], V_{g_o} has a unique critical point, viz. g_o, which satisfies $H_{g_o}(V_{g_o}) > 0$. Moreover [15], $V^{-1}(-\infty,c]$ is compact.

Comparing with the Morse inequalities, we must have $b_o(X_{n,r}) = 1$ reflecting the fact that $X_{n,r}$ is connected and

$$b_1(X_{n,r}) = \dots = b_{n+r}(X_{n,r}) = 0 ,$$

implying that $X_{n,r}$ is contractible. In fact, by Milnor's Theorem (4.8) we know that $X_{n,r}$ is in fact $\mathbb{R}^{n+r}$; see [14], [15]. This exemplifies the "Euclidean Ansatz": If X is a parameter manifold of systems which admits a globally convergent recursive identification algorithm, then X is Euclidean space.

3.3. Example

Let $M = T^2$ and consider the function on M,

$$V(\theta_1,\theta_2) = -\alpha_{12}\cos(\theta_1-\theta_2)-\alpha_{13}\cos(\theta_1)-\alpha_{23}\cos(\theta_2)$$

V is in fact the potential function for the load flow equations for a 3-bus network (see [5]) and has critical points precisely at the load flow for zero power injection into the system. The coefficients α_{ij} are coupling parameters. An outstanding problem for several decades has been to compute the number of critical points of V of indices 0, 1, or 2.

For example, (0,0) is a critical point. To see this, choose the open set

$$U = \{(\theta_1,\theta_2) : -\pi/2 < \theta_i < \pi/2\}$$

with chart

$$h(\theta_1,\theta_2) = (\theta_1,\theta_2)$$

Then

$$\left.\frac{\partial V}{\partial h^1}\right|_{(0,0)} = \alpha_{12}\sin(\theta_1-\theta_2) + \alpha_{13}\sin(\theta_1)\Big|_{(0,0)} = 0$$

$$\left.\frac{\partial V}{\partial h^2}\right|_{(0,0)} = -\alpha_{12}\sin(\theta_1-\theta_2) + \alpha_{13}\sin(\theta_2)\Big|_{(0,0)} = 0$$

Moreover,

$$H_{(0,0)}(V\circ h^{-1}) = \begin{bmatrix} \alpha_{12}+\alpha_{13} & -\alpha_{12} \\ -\alpha_{12} & \alpha_{12}+\alpha_{13} \end{bmatrix} > 0 .$$

Similarly $(0,\pi)$, $(\pi,0)$, and (π,π) are critical points of V, lying in other charts. As it turns out, their indices can change as a function of (α_{ij}) so that understanding their nature is not so simple, but is possible since V is Morse for almost all α_{ij}.

Note, however, that the torus T^2 has Betti numbers

$$b_o(T^2) = 1,\ b_1(T^2) = 2,\ b_2(T^2) = 1$$

Thus, if V has only the points

$$(0,0),\ (0,\pi),\ (\pi,0),\ (\pi,\pi)$$

as critical points, we must have

$$c_o(V) = 1,\ c_1(V) = 2,\ c_2(V) = 1 \tag{3.2}$$

V can, however, possess more critical points; e.g., in [41] it is shown that when $\alpha_{ij} = 1$, V has 6 critical points. In [5] it is shown that V can only have either 4 or 6 critical points and we deduce from the Morse inequalities and the fact that V is even, $V(\theta) = V(-\theta)$, the only possible distributions of critical points are (3.2) or

$$c_o(V) = 1,\ c_1(V) = 3,\ c_2(V) = 2 \tag{3.3}$$

In particular [5], in either case there is always a unique stable load flow (for sufficiently small power injections).

3.4 Differentiable Functions (general case)

Suppose M,N are differentiable manifolds with atlases (U_α,h_α), (V_β,k_β). A function

$$f : M \to N$$

is differentiable provided

$$k_\beta \circ f \circ h_\alpha^{-1} : \mathbb{R}^m \to \mathbb{R}^n$$

is differentiable whenever it is defined. The case $N = \mathbb{R}$ agrees with our previous definition and the case $M = N = \mathbb{R}^n$ arises in the consideration of differentiable changes of coordinates. For manifolds of systems, changes of coordinates often arise in practice as a change in the system parameterization; e.g. from a time-domain description to a frequency domain description, or from ARMA parameters to Markov-parameters, etc.

Definition 2.4. A diffeomorphism $f : M \to M'$ is a differentiable function with a differentiable inverse.

For a nontrivial, but explicit, example of a diffeomorphism, the reader is referred to [16] where it is shown that the Chebychev parameter space C_n (Section 2.4) and an open dense submanifold of the sequence space $S(2n,n)$ (see [11]) arising in the partial realization problem are diffeomorphic. Using this diffeomorphism and earlier work by Chebychev and Stieltzes, it is possible to then answer (in the negative) several of the questions raised by Kalman in [30]. For example, it is shown that there does not exist an open dense subset of the space of sequences $(\gamma_1,\ldots,\gamma_{2n})$, possessing a stable partial realization, for which the sequence $(\gamma_1,\ldots,\gamma_{2n-2})$ possesses a stable partial realization. Thus, stability is not preserved or detectable through partial realization - and this failure manifests itself on an open set of systems.

4. VECTOR FIELDS AND FLOWS ON MANIFOLDS

4.1. Vector Fields on $\mathbb{R}^n$

An ordinary differential equation on $\mathbb{R}^n$

$$\frac{dx}{dt} = f(x) \quad , \quad x \in \mathbb{R}^n \tag{4.1}$$

is determined by the "field of vectors"

$$x \longmapsto f(x) \quad \mathbb{R}^n \tag{4.2}$$

(4.1) also determines a collection of integral curves $\gamma_t(x_o)$, viz. $\gamma_t(x_o)$ is the solution of (4.1), with initial condition x_o, at time t. Supposing that $\gamma_t(x_o)$ exists for all $t \in \mathbb{R}$ and all x_o $\mathbb{R}^n$ (i.e., that (4.1) has no finite escape time), this collection of integral curves determines a "flow" on $\mathbb{R}^n$, viz.

$$\Phi : \mathbb{R} \times \mathbb{R}^n \to \mathbb{R}^n \tag{4.3}$$

$$\Phi(t,x) = \gamma_t(x)$$

Note that Φ satisfies

(i) $\Phi(0,\cdot)$ is the identity

(ii) $\Phi(t,\cdot)$ is 1-1 (uniqueness of solutions)

(iii) $\Phi(t_1+t_2,x)=\Phi(t_1,\Phi(t_2,x))$ (time invariance of (4.1))

If f is infinitely differentiable (C^∞), then Φ is infinitely differentiable (C^∞) and is therefore a diffeomorphism of $\mathbb{R}^n$. Moreover, given a C^∞ function Φ satisfying (i)-(iii), we can recover a C^∞ vector field f generating Φ by differentiating the flow lines of Φ:

$$f(x) = \frac{d}{dt}\Phi(t,x)\Big|_{t=0}$$

Thus, any of (4.1), (4.2), (4.3) constitute an ordinary differential equation on $\mathbb{R}^n$. (In the case of "finite escape time," we must work with "local flows", see [ ]).

It is clear from definition (4.3), that flows can be defined on manifolds in an unambiguous way.

Example 4.1. The van der Pol oscillator

$$\frac{dx}{dt} = y - (x^3-x)$$
$$\frac{dy}{dt} = -x \tag{4.4}$$

defines a vector field on $\mathbb{R}^2$ with a limit cycle and an unstable equilibrium at (0,0). It is intuitively clear that (4.4) should have an unstable equilibrium at ∞, too, and in this way (4.4) gives rise to a flow on S^2:

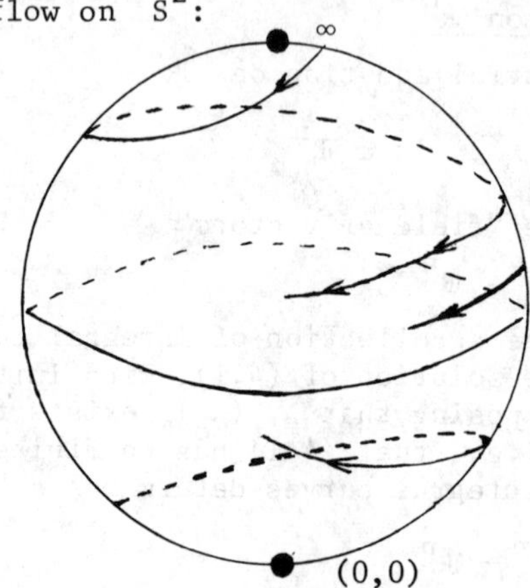

Figure 4.1. The van der Pol flow on S^2.

Example 4.2. The control system

$$\frac{d\theta}{dt} = u(t)\theta \qquad 0 \leqslant \theta \leqslant 2\pi$$

can be regarded as a time invariant system

$$\begin{aligned} \frac{d\theta}{dt} &= u(\tau)\theta \\ \frac{d\tau}{dt} &= 1 \end{aligned} \tag{4.5}$$

on $S^1 \times \mathbb{R}$. For $u(t) = t$, an integral curve of the corresponding flow on the infinite cylinder is sketched in Figure 4.2.

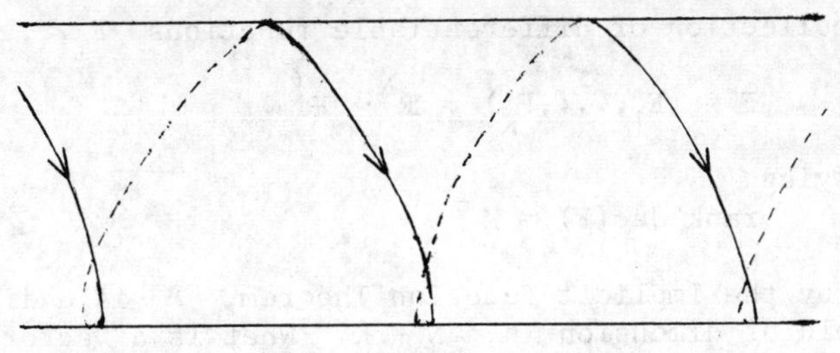

Figure 4.2. The flow (4.5) on $S^1 \times \mathbb{R}$.

Example 4.3. Consider the "height function" h, as in Figure 4.1, on S^2 instead of S^1. Then

$$h : S^2 - \{\text{north pole}\} \to \mathbb{R}$$

has a gradient, grad h, which gives a vector field on $\mathbb{R}^2$. In fact, -grad h has a stable equilibrium point at 0 and an unstable equilibrium "at ∞". Thus, grad h induces a flow on all of S^2, which is illustrated in Figure 4.3

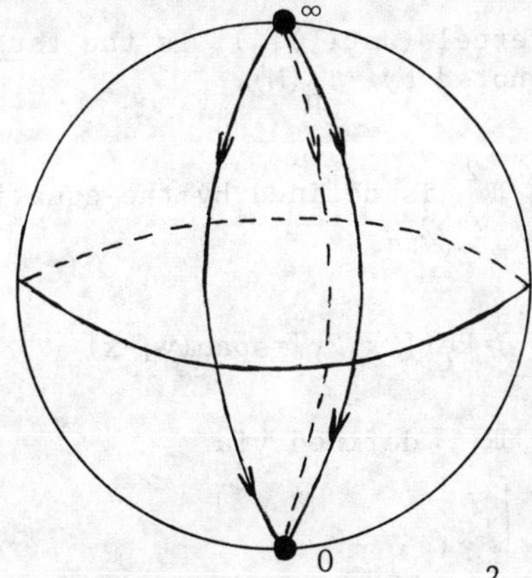

Figure 4.3. A gradient flow on S^2.

4.2. The Tangent Space to a Submanifold of $\mathbb{R}^N$

Although the definition ((4.3) and (i)-(iii) following) of a C^∞ flow Φ makes sense if $\mathbb{R}^n$ is replaced by a differentiable manifold M, it is desirable to have an analogue of (4.2) for noneuclidean manifolds in order to define ordinary differential equations. After all, to know Φ is to have explicit formulas for the solutions of (4.1) for all x_o and this is rarely possible. This leads to the construction of the tangent space to a manifold M at a point $p \in M$.

Suppose $M \subset \mathbb{R}^N$ is defined as the zero set

$$M = F^{-1}(0)$$

of a collection of differentiable functions $F_1, \ldots, F_K$

$$F = (F_1, \ldots, F_K) : \mathbb{R}^N \to \mathbb{R}^K$$

satisfying

$$\text{rank } \mathrm{Jac}(F) = K \tag{4.6}$$

Then, by the Implicit Function Theorem, M is a differentiable manifold of dimension $n = N - K$. What is a vector field on M? A very natural definition is to consider vector fields

$$f : U \to \mathbb{R}^N, \tag{4.7}$$

where U is an open, $M \subset U \subset \mathbb{R}^N$ of M, which are tangent to M itself.

<u>Definition 4.4.</u> A vector field f on U is <u>tangent to</u> M if for each $p \in M$, the N-vector $f(p)$ satisfies

$$f(p) \in \mathrm{Kernel}(\mathrm{Jac}(F)|_p). \tag{4.8}$$

The vector space Kernel $(\mathrm{Jac}(F)|_p)$ is the <u>tangent space to</u> M <u>at</u> p, which is denoted by $T_p(M)$.

<u>Example 4.5.</u> $S^1 \subset \mathbb{R}^2$ is defined by the equation

$$F(x,y) = x^2+y^2-1$$

Thus,

$$T_{(x,y)}(S^1)=\ker[2x, 2y]=\mathrm{span}(y,-x)$$

For example,

$$f : S^1 \to \mathbb{R}^2 \quad \text{defined via}$$

$$f(x,y) = \begin{bmatrix} y \\ -x \end{bmatrix}$$

is a vector field on S^1.

The condition (4.8) generalizes the observation (for $K = 1$) that, since the gradient of F points "in the direction of most change," grad F is always perpendicular to $T_p(M)$.

It is a theorem of Whitney that every (paracompact) manifold can be imbedded as a closed subset of $\mathbb{R}^{2n}$ defined locally by n equations. From this theorem it is possible to show that M can be defined globally as the zeroes of n functions, thus the definitions given above are in fact general.

4.3. Differential Equations and Control Systems on Manifolds

If $M \subset \mathbb{R}^N$ is the zero set of $F = (F_1,\ldots,F_K)$, a differential equation on M is an equation

$$\frac{dx}{dt} = f(x) \quad , \quad x \in \mathbb{R}^N \tag{4.9}$$

where $f : M \to \mathbb{R}^N$ is tangent to M and the initial condition x_o lies in M.

Proposition 4.6. Suppose the solution x_t to (4.9) exists in $\mathbb{R}^N$ at time t. If $x_o \in M$, then $x_t \in M$.

This can be proved for sufficiently small t by choosing a coordinate chart (U_α, h_α) about x_o - using the Implicit Function Theorem - and rewriting (4.9) in these new coordinates. For t large, we partition $[0,t]$ into sufficiently small intervals, applying the previous argument to each interval.

For example, the equations

$$\frac{dx}{dt} = f(x) + \sum_{i=1}^{m} u_i(t)g_i(x) \tag{4.10a}$$

$$y = h(x) \tag{4.10b}$$

define a control system on M, provided f and the g_i are tangent to M, and the output h is a differentiable function on M with values, say, in $\mathbb{R}^p$. If f, g_i, h are real analytic and there is no finite escape time, then analogues of the state-space isomorphism theorem hold in the presence of nonlinear controllability and observability [8], [27], [31], [39]. There is conversely, also a well-developed realization theory (see [28], [39]) which requires the consideration of noneuclidean M in the construction of minimal realization for nonlinear response maps. The reader is referred to [12] for an overview of how this theory makes contact with problems of nonlinear filtering.

An important aspect of the qualitative study of differential equations on manifolds, first noted by Poincaré in a series of papers appearing about one hundred years ago, is the strong influence which the topology of M has on the equilibrium structure of a vector field. Let us say that an equilibrium point p,

$$f(p) = 0$$

is <u>simple</u> if $\text{Jac } f$ is invertible at p and that f is <u>nondegenerate</u> if all its equilibria are simple. A first theorem in this direction is already hinted at in Examples 4.1 and 4.3:

<u>Proposition 4.7</u>. (Poincaré, 1885) Every nondegenerate vector field on S^2 has at least 2 equilibria.

Note, for example, in contract that the vector field

$$f(x,y) = \binom{y}{-x}$$

on S^1 has no zeroes. These observations are, in fact, very crude corollaries of the Poincaré-Hopf Index Theorem, see [35], which is more explicit and holds in greater generality.

In particular, there is no nondegenerate vector field on S^2 having just one equilibrium which is a globally attracting sink. This has implications in identification, especially in the existence of globally convergent recursive identification algorithms on parameter manifolds (see [14]). In this direction, the key topological result seems to be ([34], also [42]):

<u>Theorem 4.8</u>. (Milnor) If a manifold M admits a vector field with just one equilibrium, which is locally and globally attracting, then M is diffeomorphic to $\mathbb{R}^n$.

4.4. <u>The Tangent Bundle of an Abstract Manifold</u>

Since it is not always convenient to think of certain manifolds as embedded in $\mathbb{R}^N$, it is desirable to have constructions of vector fields, tangent spaces, etc., which proceed directly from the definition 1.3.

Let $x_o \in M$ and consider a coordinate chart (U_α, h_α) with $x_o \in U_\alpha$. A vector field on M should correspond, via h_α, with a differential equation

$$\dot{z}_\alpha = f(z_\alpha) \quad , \quad z_\alpha = h_\alpha(x) \tag{4.11}$$

This corresponds, on $U_\alpha \subset M$, to a pair $(x, f_\alpha(x)) \in U_\alpha \times \mathbb{R}^n$ where $f_\alpha(x) = f \circ h_\alpha(x)$. If $x_o \in U_\beta$, then a vector field would correspond, via the h_β coordinates, to a pair $(x, f_\beta(x))$. Now, $f_\alpha(x)$ will not coincide, in general, with $f_\beta(x)$ but, since a vector field on M (say, as defined in 4.2) is a globally defined object, there must be a definite relation between the two.

Indeed, the vector field $(x, f_\beta(x))$ corresponds on $U_\alpha \cap U_\beta$ to (4.11) under the change of coordinates $h_\beta \circ h_\alpha^{-1}$. Explicitly,

$$f_\beta(x) = \mathrm{Jac}(h_\beta \circ h_\alpha^{-1})(f_\alpha(x)),\ x \in U_\alpha \cap U_\beta$$

A vector field would then be prescribed by the collection of pairs

$$(x, f_\alpha(x)) \in U_\alpha \times \mathbb{R}^n$$

subject to the constraint

$$(x, f_\alpha(x)) = (x, \mathrm{Jac}(h_\beta \circ h_\alpha^{-1}) f_\beta(x)),\ x \in U_\alpha \cap U_\beta \tag{4.12}$$

Put another way, the collection of tangent vectors should itself be a manifold - the "tangent bundle" $T(M)$ - with an atlas given by the neighborhoods

$$V_\alpha = U_\alpha \times \mathbb{R}^n \tag{4.13}$$

and coordinates

$$H_\alpha = (h_\alpha ; Jh_\alpha) \tag{4.14}$$

with coordinate changes

$$H_\beta \circ H_\alpha^{-1} = (h_\beta \circ h_\alpha^{-1} ; J(h_\beta \circ h_\alpha^{-1})) \tag{4.15}$$

on the intersection $V_\alpha \cap V_\beta$.

TM can be constructed as a set as

$$TM = \cup\, U_\alpha \times \mathbb{R}^n /\!\sim$$

where $(x, f_\alpha(x)) \sim (x, f_\beta(x)$ if, and only if, (4.12) is satisfied. We topologize M by using the images of the subsets $U_\alpha \times \mathbb{R}^n$ as basic open sets.

In this light (4.13)-(4.14) make TM as topological manifold, while (4.15) guarantees that TM is a differentiable manifold, with atlas (4.13)-(4.14).

One can check that the projection

$$\pi[m,v] = m\ ,$$

which maps TM onto M, is smooth and that $\pi^{-1}(m)$ is naturally a linear space of dimension n. $\pi^{-1}(m)$ is defined to be the tangent space $T_p(M)$.

Definition 4.9. A differentiable function $f : M \to TM$ is a vector field if, and only if, $\pi \circ f = id_M$.

Example 4.10. $T\mathbb{R}^2 \simeq \mathbb{R}^2 \times \mathbb{R}^2$, since we can take a single coordinate chart $(U_\alpha, h_\alpha) = (\mathbb{R}^2, id)$. Furthermore, the projection $\pi : \mathbb{R}^2 \times \mathbb{R}^2 \to \mathbb{R}^2$ is just the projection on the first factor. Therefore, if

$$f : \mathbb{R}^2 \to \mathbb{R}^2 \times \mathbb{R}^2$$

satisfies $\pi \circ f = id$, then f is given by

$$f(p) = (p, \tilde{f}(p));$$

i.e., f corresponds to "an honest" vector field $\tilde{f}$ on $\mathbb{R}^2$.

Example 4.11. $TS^1 \simeq S^1 \times \mathbb{R}$, by a diffeomorphism which is a linear map on the tangent space. To see this, take the standard atlas

$$U_\alpha = S^1 - \{\text{north pole}\}, \; U_\beta = S^1 - \{\text{south pole}\}$$

with h_α, h_β the stereographic projections as in Figure (1.1). Thus,

$$TS^1 = (U_\alpha \times \mathbb{R}) \cup (U_\beta \times \mathbb{R}) /\sim$$

In particular, if $z_\alpha = h_\alpha(x)$ and $z_\beta = h_\beta(x)$, then

$$(z_\alpha, 1+z_\alpha^2) \sim (z_\beta, 1+z_\beta^2) \quad \text{for} \quad x \in U_\alpha \cup U_\beta$$

since

$$z_\beta = h_\beta \circ h_\alpha^{-1}(z_\alpha) = z_\alpha/||z_\alpha||^2 = 1/z_\alpha$$

Thus

$$f(x) = \begin{cases} (x, 1+z_\alpha^2) & x \in U_\alpha \\ (x, 1+z_\beta^2) & x \in U_\beta \end{cases}$$

is a well defined function

$$f : S^1 \to TS^1$$

satisfying

$$\pi \circ f(x) = x$$

Moreover, the vector $f(x) \in T_x(S^1)$ is never zero. One can therefore define a diffeomorphism, linear on $T_x(S^1)$,

$$F : TS^1 \to S^1 \times \mathbb{R}$$

via

$$F[x,v] = (x,\alpha(v))$$

where $\alpha(v)$ is uniquely defined by the equation

$$v = \alpha(v)f(x)$$

since $f(x) \neq 0$ and $T_x(S^1)$ is 1-dimensional.

In fact, $f(x)$ correspond to the vector field $\partial/\partial\theta$, which is the restriction of $\begin{bmatrix} y \\ -x \end{bmatrix}$ to S^1, under the assignment

$$x = \tan^{-1}(\theta).$$

Example 4.12. $TS^2 \neq S^2 \times \mathbb{R}^2$. If this were true, one could find two vector fields f_1, f_2 or S^2 which would be everywhere linearly independent. f_1, f_2 would of course be nondegenerate, but since f_1 would never be zero this would contradict the Poincaré-Hopf Theorem (Proposition 4.7).

5. BIBLIOGRAPHY

[1] G. Alengrin, "Estimation of Stochastic Parameters for ARMA Models by Fast Filtering Algorithms," in Nonlinear Stochastic Problems (R.S. Bucy, J.F. Moura, eds.), D. Reidel, Dordrecht, 1983 (this volume, pp. 65-78)

[2] V.I. Arnold, Mathematical Methods in Classical Mechanics, Springer-Verlag, New York, 1978.

[3] K.J. Åström and T. Bohlin, "Numerical Identification of Linear Dynamical Systems from Normal Operating Records," in Theory of Self-Adaptive Control Systems (P.H. Hammond, ed.), Plenum, N.Y., 1966.

[4] K.J. Åström and T. Söderström, "Uniqueness of the Maximum Likelihood Estimates of the Parameters of an ARMA Model," IEEE Trans. Aut. Control 19 (1974) 769-773.

[5] J. Baillieul and C.I. Byrnes, "Geometric Critical Point Analysis of Electrical Power Systems," IEEE Trans. Circuits and Systems, to appear in 1982.

[6] R. Bott, Morse Theory and Its Application to Homotopy Theory, Als Manuskript vervielfältigt im Mathematischen Institut der Universität Bonn, 1960.

[7] F. Brickell and R.S. Clark, Differentiable Manifolds, Van Nostrand Reinhold, London, 1970.

[8] R.W. Brockett, "System Theory on Group Manifolds and Coset Spaces," SIAM J. Control 10 (1972) 265-284.

[9] R.W. Brockett, "Some Geometric Questions in the Theory of Linear Systems," IEEE Trans. Aut. Control 21 (1976) 449-455.

[10] R.W. Brockett, "The Geometry of the Set of Controllable Linear Systems," Research Reports of the Aut. Cont. Labs. Fac. Eng., Nagoya Univ. 24 (1977).

[11] R.W. Brockett, "The Geometry of Partial Realization Problems," Proc. of 1978 IEEE Conf. on Decision and Control, 1048-1052.

[12] R.W. Brockett, "Nonlinear Systems and Nonlinear Estimation Theory," in Stochastic Systems: The Mathematics of Filtering and Identification and Applications (M. Hazewinkel and J.C. Willems), D. Reidel, Dordrecht, 1981.

[13] R.S. Bucy and J. Pages, "A Priori Bounds for the Cubic Sensor Problem," IEEE Trans. Aut. Control 23 (1978) 88-91.

[14] C.I. Byrnes, "Geometric Aspects of the Convergence Analysis of Identification Algorithms," in Nonlinear Stochastic Problems (R.S. Bucy and J.F. Moura, eds.), D. Reidel, Dordrecht, 1983. (this volume, pp. 163-186)

[15] C.I. Byrnes and P.S. Krishnaprasad, "On the Euclidean Ansatz for Stable and Miniphase Systems," to appear.

[16] C.I. Byrnes and A. Lindquist, "The Stability and Instability of Partial Realizations," Systems and Control Letters 2 (1982), to appear.

[17] J.M.C. Clark, "The Consistent Selection of Local Coordinates in Linear System Identification," in Proc. of the J.A.C.C., Purdue, 1976.

[18] D.F. Delchamps, "A Note on the Analyticity of the Riccati Metric," in Algebraic and Geometric Methods in Linear System Theory (C.I. Byrnes and C.F. Martin, eds.), Amer. Math. Soc., Providence, 1980.

[19] M. Diestler, "The Properties of the Parameterization of ARMAX Systems and Their Relevance for Structural Estimation and Dynamic Specification," paper presented at the 4th World Congress of the Econometric Society, Aix-en-Provence, 1980.

[20] M. Diestler and E.J. Hannan, "Some Properties of the Parameterization of ARMA Systems with Unknown Order," J. of Multivariate Analysis 11 (1981).

[21] B.F. Doolin and C.F. Martin, Elements of Global Differential Geometry, preliminary version.

[22] W.B. Gragg and A. Lindquist, "On the Partial Realization," to appear.

[23] V. Guillemin and A. Pollack, Differential Topology, Prentice-Hall, Englewood Cliffs, 1974.

[24] E.J. Hannan, "System Identification," in Stochastic Systems: The Mathematics of Filtering and Identification and Applications (M. Hazewinkel and J.C. Willems, eds.), D. Reidel, Dordrecht, 1981.

[25] M. Hazewinkel, "Fine Moduli Spaces: What Are They and What Are They Good For," in Geometrical Methods for the Theory of Linear Systems (C.I. Byrnes and C.F. Martin, eds.), D. Reidel, Dordrecht, 1980.

[26] M. Hazewinkel and S.I. Marcus, "Some Results and Speculations on the Role of Lie Algebras in Filtering," in Stochastic Systems: The Mathematics of Filtering and Identification and Applications, D. Reidel, Dordrecht, 1981.

[27] R. Hermann, "On the Accessibility Problem in Control Theory," in Nonlinear Differential Equations and Nonlinear Mechanics (J.P. LaSalle and S. Lefschetz, eds.), Academic Press, N.Y., 1963.

[28] B. Jakubczyk, "Existence and Uniqueness of Nonlinear Realizations," Astérisque 75-76 (1980) 141-147.

[29] R.E. Kalman, "Global Structure of Classes of Linear Dynamical Systems," preprint of a lecture given at the NATO-ASI on Geometric and Algebraic Methods for Nonlinear Systems, London, 1971.

[30] R.E. Kalman, "On Partial Realizations, Transfer Functions, and Canonical Forms," Acta Polytechnica Scandinavica Helsinki MA31 (1979) 9-32.

[31] A.J. Krener and R. Hermann, "Nonlinear Controllability and Observability," IEEE Trans. Aut. Control 22 (1977) 728-740.

[32] S. Lang, Algebra, Addison-Wesley, 1965.

[33] J.W. Milnor, Morse Theory, Princeton Univ. Press, Princeton, 1963.

[34] J.W. Milnor, "Differential Topology," in Lectures in Modern Mathematics Vol. II (T.L. Saaty, ed.), Wiley, N.Y., 1964.

[35] J.W. Milnor, Topology from the Differentiable Viewpoint, Univ. Press of Virginia, Charlottesville, 1965.

[36] S.K. Mitter, "On the Analogy Between Mathematical Problems of Nonlinear Filtering and Quantum Physics," Ricerche di Automatica X (1979) 163-216.

[37] I.M. Singer and J.S. Thorpe, Lecture Notes on Elementary Geometry and Topology, Springer-Verlag, N.Y., 1981.

[38] M. Spivak, Calculus on Manifolds, Benjamin, N.Y., 1965.

[39] H.J. Sussmann, "Existence and Uniqueness of Minimal Realizations of Nonlinear Systems," Math. System Theory 10 (1975) 1476-1487.

[40] H.J. Sussmann, "Rigorous Results on the Cubic Sensor Problem," in Stochastic Systems: The Mathematics of Filtering and Applications (M. Hazewinkel and J.C. Willems, eds.), D. Reidel, Dordrecht, 1981.

[41] C.J. Tavora and O.J.M. Smith, "Equilibrium Analysis of Power Systems," IEEE Trans. Power Apparatus and Systems 91 (1972) 1131-1137.

[42] F.W. Wilson, Jr., "The Structure of the Level Surfaces of a Lyapunov Function," J. of Diff. Eqns. 3 (1967) 323-329.

ROLE OF MULTIPLICATIVE NON-WHITE NOISE IN A NONLINEAR SURFACE CATALYTIC REACTION

F.J. de la Rubia, J. Garcia Sanz and M.G. Velarde

U.N.E.D.-Fisica Fundamental, Apdo.Correos 50 487,
Madrid, Spain

Time-dependent oscillations induced by a non-white noise in a model system of a surface catalyzed reaction are described. The role of finite correlation time and variance of the noise is discussed in various illustrative cases.

1. INTRODUCTION: NON LINEAR DETERMINISTIC PROBLEM

Non linear deterministic systems exhibiting multiple steady states or limit cycle oscillations can be drastically altered by an external noise|1-4|. An obvious consequence of noise might be a state of chaos, not to be confused with the appearance of strange attractors in purely deterministic systems |5,6|. Other possibilities include the case of sustained oscillations induced by the noise when no limit cycle is predicted from the deterministic equations. In this note we study the latter case in a model problem|7| supposedly relevant to a surface catalyzed reaction like the oxidization of CO on a Pt surface. Evidence of the role of noise in such problems exists in the literature|8|.

The model refers to the following pair of non linearly coupled ordinary differential equations

$$d\Theta_1/dt=\alpha_1(1-\Theta_1-\Theta_2)-\gamma_1\Theta_1-\Theta_1\Theta_2(1-\Theta_1-\Theta_2)^2=f(\Theta_1,\Theta_2) \quad (1.a)$$

$$d\Theta_2/dt=\alpha_2(1-\Theta_1-\Theta_2)-\gamma_2\Theta_2-\Theta_1\Theta_2(1-\Theta_1-\Theta_2)^2=g(\Theta_1,\Theta_2) \quad (1.b)$$

where Θ_i (i=1,2) accounts for the ratio of chemisorbed molecules on a surface to the coverage or number of available adsorption sites, $\Theta_i \geq 0, 0 \leq \Theta_1+\Theta_2 \leq 1$. The other parameters α_i and γ_i are non negative quantities which incorporate both dynamic and static features of

R. S. Bucy and J. M. F. Moura (eds.), Nonlinear Stochastic Problems, 151–159.

the surface reaction problem, like reaction rates, partial pressure effects, impurity concentrations, temperature dependence of reaction rates, etc.

According to the values of the parameters α_i and γ_i the system (1) has one or three steady states or limit cycle behavior. Here for illustration we shall restrict consideration to the values $\alpha_1=0.016$, $\gamma_1=0.001$, $\gamma_2=0.002$, and reasonable values of α_2 for which the system possesses a single steady state, stable or not. Specifically we take two values $\alpha_2=0.0295$ and $\alpha_2=0.0270$.

Experimentally, the most relevant quantity in a surface reaction is the *production rate*, $r(t)=\Theta_1\Theta_2(1-\Theta_1-\Theta_2)^2$. Oscillations in either of the Θ_i yield oscillations in $r(t)$.

2. STOCHASTIC PROBLEM

We consider a noise $\eta(t)$ such that the actual (fluctuating) value of α_2 is

$$\alpha_2 = \alpha_2^d + \eta(t) \tag{2}$$

where α_2^d is the given value of the parameter. The noise, $\eta(t)$, is assumed to be bounded ($\alpha_2 \geqq 0$), colored with finite correlation time and such that $E\{\alpha_2\} = \alpha_2^d$. As a matter of fact we pose

$$\eta(t) = (2a/\pi) \text{ arc tan } \beta_t \tag{3}$$

where β_t is the Ornstein-Uhlenbeck process that obeys the equation |9|

$$d\beta = -\rho\beta dt + \sigma dW_t \tag{4}$$

and

$$\beta_t \sim N(0, \sigma^2/2\rho) \tag{5.a}$$

$$E\{\beta_t\beta_{t'}\} = (\sigma^2/2\rho)\exp(-\rho|t-t'|) \tag{5.b}$$

The constant a delineates the domain of the noise, $\eta(t)$. Thus we have $\alpha_2^d - a \leqq \alpha_2 \leqq \alpha_2^d + a$.

The above given properties yield $\eta(t)$ as a stationary Markov process with distribution, normalized in the interval $(-a,a)$, |9|

$$P(\eta) = (\pi^{\frac{1}{2}}/2a\varepsilon)\{\cos^2(\pi\eta/2a)\}^{-1}\exp\{-\varepsilon^{-1}\tan^2(\pi\eta/2a)\} \tag{6}$$

where $\varepsilon = \sigma/\rho^{\frac{1}{2}}$. From Eq. (6) it follows that P has a single maximum at $\eta=0$ provided $\varepsilon \leqq 1$, whereas if $\varepsilon > 1$ there is a minimum at the origin and $\eta_\pm = \pm(2a/\pi)\text{arc cos}(1/\varepsilon)$ define maxima of $P(\eta)$. Thus we assume $\varepsilon \leqq 1$.

From Eq.(6) we get

$$E\{\eta\} = \int_{-a}^{+a} \eta \, P\{\eta\} \, d\eta = 0 \tag{7.a}$$

$$Var\{\eta\} = E\{\eta^2\} - E^2\{\eta\} = (4a^2/\pi^{5/2}) \, I(\varepsilon) \tag{7.b}$$

with

$$I(\varepsilon) = \int_{-\infty}^{+\infty} \text{arc tan}^2 (\varepsilon x) \exp(-x^2) \, dx \tag{7.c}$$

Then the variance of the noise is governed by ε. We also have the correlation

$$C(\tau) = E\{\eta(0) \, \eta(\tau)\} = E\{\eta^2\} \exp(-\rho\tau) \tag{7.d}$$

with a correlation time that scales with $1/\rho$.

As a consequence of the above given properties it follows that the triplet $\{\Theta_1, \Theta_2, \eta\}$ defines a Markov process with a unique invariant measure in the relevant portion of phase space $\{\Theta_1 \geqq 0, \Theta_2 \geqq 0, \Theta_1 + \Theta_2 \leqq 1\}$. Moreover the system is ergodic and a numerical simulation of the process is perfectly all right |10,11|.

3. THE ROLE OF EXTERNAL NOISE : DISCUSSION OF RESULTS

It is clear that the role of fluctuations is measured by $\delta = (Var\alpha_2)^{\frac{1}{2}}/E\{\alpha_2\}$. For illustration we take $\varepsilon = 0.2, 0.5$ and 0.9 which correspond to $\delta = 0.1, 0.2$ and 0.3, respectively. On the other hand the correlation time of the noise is taken much smaller than a characteristic time or period of the deterministic system, $\tau \ll T$.

3.1. CASE OF A STABLE FOCUS

For $\alpha_2^d = 0.0295$ the system (1) possesses a single steady state which is a stable focus. Thus asymptotically all initial conditions must decay to the steady state and a constant value of the production rate, $r(t)$, is predicted.

Fig. 1 depicts the effect of noise upon the steady state of (1), for $\varepsilon = 0.2$ and two correlation times $\tau=1$ (dotted line) and $\tau = 50$ (solid line). The steady and constant value of Θ_1 is also given for reference. As a matter of fact this value was the initial condition for our numerical simulations. As it can be seen the noise induces a marked time-dependence in Θ_1, and thus in the production rate (Fig.2). A study of the power spectrum of the signal shows a major frequency component around $T = 10^3$ which corresponds to a characteristic constant of the steady state (inverse of the imaginary part of the linear eigenvalue around the steady state).

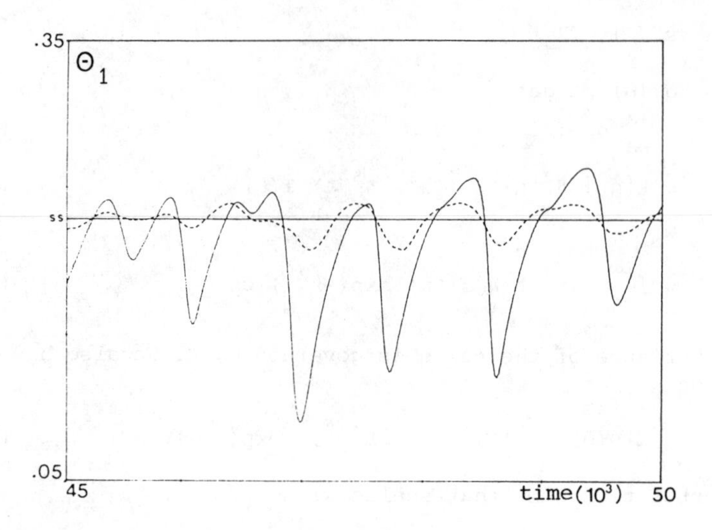

Fig. 1. Θ_1 *vs* time(in units of 10^3) for ε=0.2 and τ=1(broken line)and τ=50(solid line).The deterministic system has a single steady state which is a *stable* focus(ss level line).

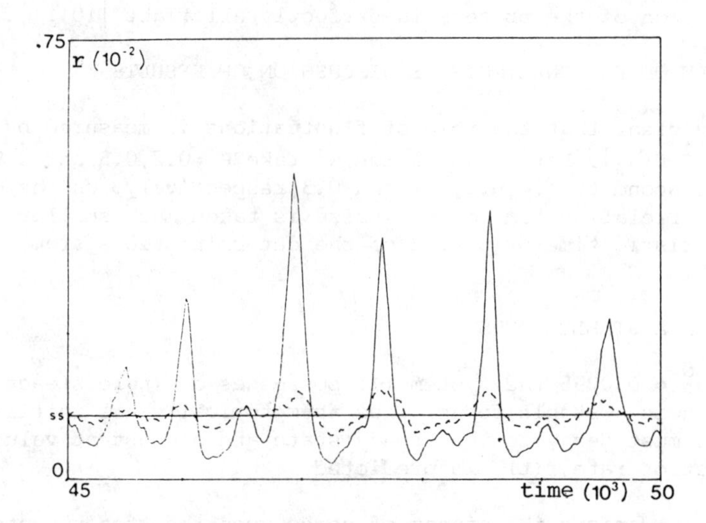

Fig. 2. Production rate,r(in units of 10^{-2})*vs* time(in units of 10^3) for ε=0.2 and τ=1(broken line)and τ=50(solid line). The level line ss corresponds to the constant rate at steady state in the deterministic result.

It also appears that the finite correlation time does sustain a time-dependence(limit cycle-like)production rate in a parameter range where in the noise-free case the deterministic system has a constant r(t). Lastly, a change in the variance of the noise also induces a tendency to an increase in the amplitude of the time-dependent signal thus providing peaks in the production rate that can reach values an order of magnitude higher than the expected value at the steady state. There is no doubt that such effects are clearly outside any reasonable error bar to be expected in an experiment.

3.2. CASE OF AN UNSTABLE FOCUS AND LIMIT CYCLE OSCILLATION

For $\alpha_2^d = 0.0270$ the deterministic system (1) has an unstable focus with limit cycle behavior in its neighborhood.The period of either of the Θ_i and,consequently,of the production rate,r(t), is $T \sim 10^3$.

Figures 3 and 4 illustrate the role played by the noise in Θ_1 and r(t), respectively.Note the striking similarity that exists between Figures 2 and 3. Again an increase in the variance of the noise induces an increase in the amplitude of the oscillations in the production rate.

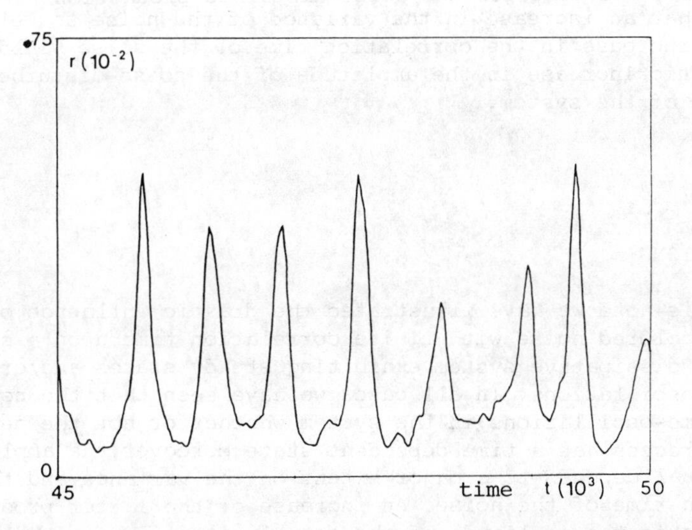

Fig.3. Production rate,r(in units of 10^{-2}) *vs* time(in units of 10^3) for ε=0.2 and τ=50 << T=10^3. T is the period of the predicted limit cycle behavior in the deterministic description.Notice the similarity of this time-dependent noise-disturbed rate with the result given in the preceding Fig.2.

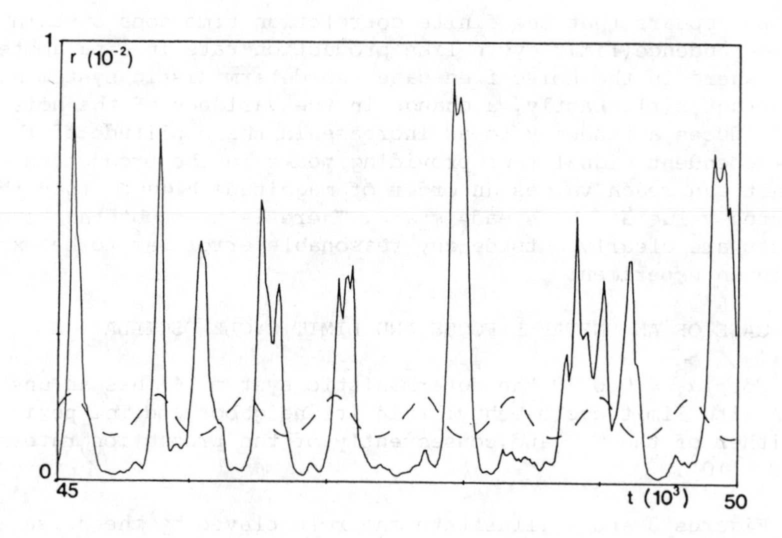

Fig.4. Production rate,r, for the case of a noise-disturbed limit cycle(like in the preceding fig.3)when ε=0.9 and τ=50. The broken line accounts for the deterministic limit cycle and the solid line for the noise-disturbed production rate. Note that an increase in the variance of the noise as well as an increase in the correlation time of the noise yield a drastic increase in the amplitude of the noise-disturbed output of the system.

4. CONCLUSIONS

In this note we have illustrated the drastic influence of a non-white,colored noise with finite correlation time upon a simple non linear dissipative system exhibiting steady states and/or limit cycle oscillations. In all cases we have seen that the noise induces time-oscillations in the system whether or not the deterministic process has a time-dependent state.Moreover,the amplitude of the signal depends to a major extent on the variance and the correlation time of the noise. An increase of the latter produces an increase in the amplitude of the actual signal. The signal, modified by the noise shows peaks which in the power spectrum have a major component around a characteristic time constant of the deterministic system(for instance, the period of the deterministic system).The height of the peaks may eventually reach values an order of magnitude higher than the expected value for the noise-free state. All these features cannot appear in the case of white,delta correlated noise |12| , however large the variance is taken.One is tempted to conclude that a non-white noise in a non linear system rather plays the expected role of a time-delay added to the deterministic dynamics of the process .

Further characteristic properties of the specific system discussed in this note are the following. Fig. 5 shows the steady values of the relevant quantities when there is a steady state in (1).In the abscissa we have the bifurcation parameter,α_2(scaled with 10^{-2}, which corresponds to the physically relevant range). The introduction of a noise in the bifurcation parameter adds a new time constant as the noise has finite correlation time,τ . Increasing this correlation time we increase the time interval that α_2(now a fluctuating quantity)spends in the neighborhood of its deterministic value. Both the residence time at a given value and the deviation from the deterministic steady state are related to the actual noise introduced in the system. The system,however, has its own dynamics (1) and the inspection of Fig. 5 clearly justifies that an increase in the correlation time corresponds to an increase in the amplitude of the stochastic signal which in fact means a dramatic decrease of the minima in $\Theta_1(t)$ rather than an increase in its maxima(past α_2^d =0.034 the curve is rather flat whereas at $\alpha_2^d \leqq$ 0.034 there is a steep slope).The decrease in Θ_1 and Θ_2 yields the appearance of enhanced peaks in the production rate or just the appearance of peaks when no variation in time is expected according to the deterministic analysis of the problem (see Fig.6).

As a final remark we could say that the above given comments are not so restrictive that they only apply to the model problem (1). Rather we think that a colored noise ,with finite correlation time induces a time-dependence of its own.However, the interaction of the noise with the specific dynamics of the system triggers indeed results that drastically depend on this dynamics and the parameter range displayed by the system's operation.

ACKNOWLEDGMENTS

The authors acknowledge M. Ehrhardt for his help in proving the existence and unicity of invariant measures. Fruitful discussions with S. Grossmann and W. Kliemann are also acknowledged. This research has been sponsored by the Stiftung Volkswagenwerk.

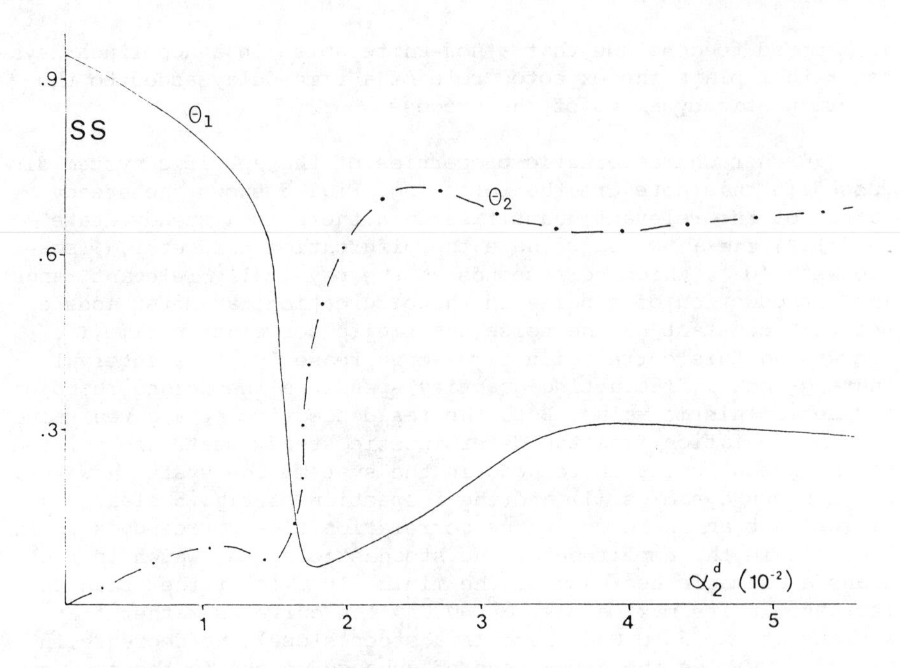

Fig.5. Steady state values of Θ_1 and Θ_2 *vs* the bifurcation parameter in the deterministic case. Upon variation of α_2 we have either stable node, stable focus or unstable focus. The region between 0.01 and 0.02 roughly corresponds to stable node. Around 0.02 there is a change to stable focus and somewhere in the middle of 0.02 and 0.03 there is the cross-over from stable to unstable focus. Later on the system takes again on stable focus or node as the bifurcation parameter increases.

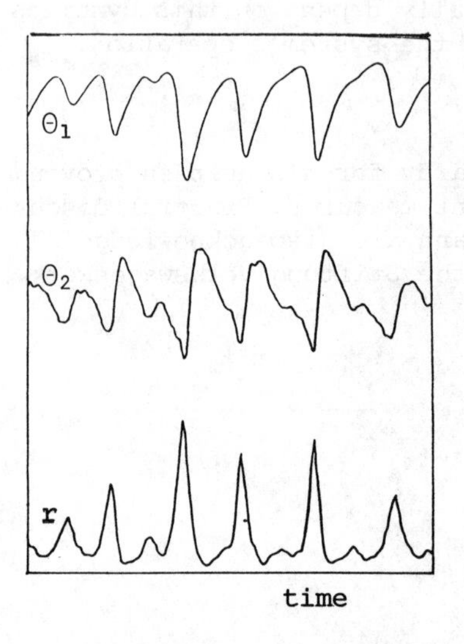

Fig.6. Noise-disturbed values of Θ_1, Θ_2 and the production rate, r, in the case of a single steady state of the deterministic system. Note that the maxima of r(t) almost coincide with the minima of Θ_i (i=1,2). Case ε =0.2 and τ =50.

REFERENCES

1. G. Nicolis and I. Prigogine,SELF-ORGANIZATION IN NON-EQUILIBRIUM SYSTEMS, John Wiley, New York,1977.For the pioneering work see R. Landauer,J. Appl. Phys. *33*,2209(1962)
2. L. Arnold and R. Lefever(editors),STOCHASTIC NONLINEAR SYSTEMS, Springer-Verlag,Berlin,1981, and references therein
3. See, for instance, F.J. de la Rubia and M.G. Velarde,Phys. Lett. A *69*,304(1978)
4. H. Haken, SYNERGETICS (2nd. edition),Springer-Verlag,Berlin,1977
5. M.G. Velarde, in Proceedings Summer School on Thermodynamic Stability(Bellaterra,1981), Springer-Verlag,Berlin,to appear
6. M.G. Velarde, in T. Riste(editor),NONLINEAR PHENOMENA AT PHASE TRANSITIONS AND INSTABILITIES, Plenum Press,New York,1982
7. C.G. Takoudis,L.D. Schmidt and R. Aris,Surface Sci.*105*,325(1981)
8. J.E. Turner, B.C. Sales and M.B. Maple,Surface Sci.*103*,54(1981)
9. L. Arnold,STOCHASTIC DIFFERENTIAL EQUATIONS,John Wiley,New York, 1974
10. M. Ehrhardt,Invariant probabilities for systems in random environment,Bremen University, Forschungsschwerpunkt Dynamische Systeme, Report nº 60, 1981
11. L. Arnold and W. Kliemann,in A.T. Bharucha-Reid(editor),PROBABILISTIC ANALYSIS AND RELATED TOPICS,vol.3,Academic Press, New York,1981
12. J.M. Sancho, M. San Miguel,H. Yamazaki and T. Kawakubo,Physica, to be published.

CHAPTER V

ADAPTIVE / STOCHASTIC CONTROL

GEOMETRIC ASPECTS OF THE CONVERGENCE ANALYSIS OF IDENTIFICATION ALGORITHMS

Christopher I. Byrnes

Division of Applied Sciences and Department of Mathematics
Harvard University
Cambridge, MA 02138

Abstract: The relationship between the topology of a parameter set of candidate systems and the convergence analysis of recursive identification algorithms on this parameter set is studied using, for example, the "ODE approach" of Ljung together with classical differential-topological methods for the qualitative analysis of differential equations. Combined with recent studies ([4], [8], [10], [12], [19] and [23]) on the topology of certain spaces of systems this analysis suggests that, topologically, the defining conditions for the candidate systems in a globally convergent identification scheme have the effect of forcing the parameter space X of candidates to be Euclidean. This interpretation of such defining properties as stable, miniphase, constant controllability indices, etc., is supported by examples from the literature (esp. [1], [3], [16]).

0. INTRODUCTION

Let X = a parameter set of candidate systems,
M = a set of measurements, say input-output pairs (u^N, y^N), and consider the possibly time-dependent functions

$$A_t : X \times M \to X \times M \quad , \quad t \in \mathbb{N} \text{ or } \mathbb{Z}$$

or

$$\tilde{A}_t : X \times M \to X \times M \quad , \quad t \in \mathbb{R}^+ \text{ or } \mathbb{R}$$

which define discrete-time, or continuous-time, algorithms (or flows) on $X \times M$, i.e.

$$A(x,m) = (x_t, m^1).$$

R. S. Bucy and J. M. F. Moura (eds.), Nonlinear Stochastic Problems, 163–186.

The convergence of identification algorithms to a system "x" $\in X$ is intimately related to the global convergence of such flows. If the algorithm A is time-dependent, then one can replace M by $M \times R$ obtaining a time-independent flow in the usual way, so I shall always consider time-invariant flows. By global convergence is meant that there exists $x_o \in X$ such that A_t (or $\widetilde{A}_t$) converges asymptotically (or uniformly asymptotically) to $\{x_o\} \times M$. I assume $M \simeq \mathbb{R}^n$, so M will be suppressed.

Even for discrete-time linear systems, the analysis of continuous-time flows appears in a crucial way in several successful convergence proofs. An important method, for example, for convergence analysis is the "ODE approach" of Ljung ([24], [25]) which associates to a recursive identification algorithm a flow on the parameter set X in such a way that global convergence of the flow to an equilibrium point $x \in X$ implies, under reasonable hypotheses, convergence of the recursive algorithm. There is also a converse to this assertion, so the convergence question can be stated as:

Question 0.1. Are the properties (e.g. stability, miniphase, etc.) defining the parameter set X such that X will admit a globally convergent flow?

This question arises also in the proof, by Åström-Söderström [3], of the uniqueness of maximum likelihood estimates under certain constraints on the set of candidate systems. In [2] it is shown that, under suitable statistical properties (e.g. ergodicity) the maximum likelihood method leads to a deterministic function

$$V : X \to \mathbb{R}$$

whose global minima coincide with the maximum likelihood estimates. In general, the existence of such an "objective function" can also be phrased as:

Question 0.2. Are the properties defining the paramater set such that X admits a real-valued function V with a unique minimum and no other critical points?

If X is a manifold, then choosing a gradient flow of V leads from Question 0.2 to Question 0.1. As these questions suggest, on X one has a topology - i.e. a notion of convergence - already defined. And, the major theme pursued here will be the interpretation of various hypothesis on candidate systems in terms of the topology of X. Indeed, as the examples arising in the literature imply, X can be taken to be a manifold or, at worse, a "manifold with edges" [15] and the relation between the topology of X and the analysis of flows and functions is then classical (see, e.g. the tutorial [7] and references cited therein). In this context, the work by Brockett, Byrnes, Delchamps, Duncan and Krishnaprasad which is surveyed here combined with the work of Ljung suggests that, whenever a globally convergent algorithm exists, the parameter space X must be a Euclidean space. Several supporting examples from the literature are discussed in this paper.

I would like to thank Roger Brockett, David Delchamps and P.S. Krishnaprasad for illuminating discussions on topology, identification, and adaptive control.

This research was partially supported by Air Force Office of Scientific Research under Grant No. AFOSR-81-0054, the National Science Foundation under Grant No. ECS-81-21428, and NASA under Grant No. NSG-2265.

1. AN OBJECTIVE FUNCTION ON THE SPACE OF MINIPHASE SYSTEMS

Following Åström-Bohlin [2], one can construct a real-valued function V on the space of miniphase systems having local extrema at precisely the maximum likelihood estimates for the ARMA system identification algorithm introduced in [2].

Using the calculus of residues, Åström-Söderström [3] show that V has a unique minimum, and no other critical points, thereby proving the uniqueness of the maximum likelihood estimates. In this section I present this construction together with a proof of the result [10] needed in the next section to show that uniqueness of the critical point of V implies that the space of miniphase systems is, in fact, Euclidean.

Set

$$X_n = \{\text{miniphase systems of degree} \leq n\} \tag{1.1}$$

Thus, $p(s)/q(s) \in X_n$ if and only if

$$q(s) = q_0 + q_1 s + \ldots + s^n \, , \; p(s) = p_0 + p_1 s + \ldots + s^r \tag{1.2}$$

where $p(s)$ has all of its $r(\leq n)$ zeroes satisfying $|s| < 1$. In particular,

$$X_n = \cup X_{n,r} \qquad r = 0,\ldots,n$$

where $p(s)/q(s) \in X_{n,r}$ if, and only if, $\deg(p) = r$.

For example,

$$X_{1,0} = \{1/s + a : |a| < 1\}$$

so that, as a topological space, $X_{1,0}$ may be parameterized by the interval $(-1,1)$. More generally, it follows from Hermite's Theorem [20] that $X_{n,o}$ is an open subset of $\mathbb{R}^n$ defined by polynomial inequalities in the (q_i). Noting that $X_{n,o}$ is path-connected and that

$$X_{n,r} = \mathbb{R}^n \times X_{r,o}$$

one knows the

Lemma 1.1. $X_{n,r}$ is a connected $(n+r)$-dimensional manifold.

Suppose one is interested in either identifying a "true system" g_o or in finding a "best n-th order fit" for a time series and suppose the order of g_o and the number of zeroes, say (n,r), is known. The proof of uniqueness of a maximum likelihood estimate of g_o given in [3] rests on the introduction of a measure, using only an input-output knowledge of g_o, of how far an estimate $g \in X_{n,r}$ is from g_o, see Figure 1.1.

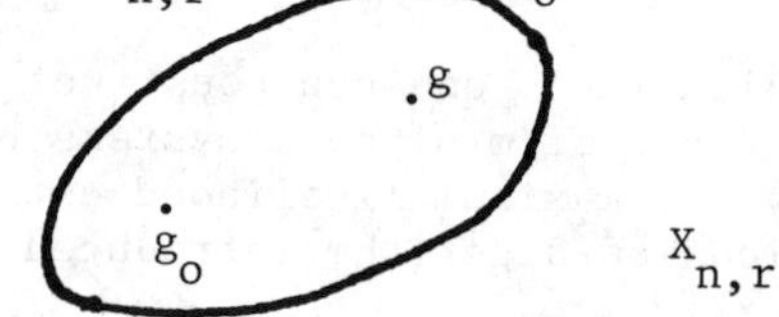

Figure 1.1. Depicting an estimate g of g_o.

Explicitly, Åström-Bohlin define an objective function

$$V_{g_o} : X_{n,r} \to \mathbb{R}^+$$

in the following way: Drive g_o by zero mean, unit variance white noise ε_t which gives rise to an output process η_t. Since g_o is stable, η_t is a stationary process. Driving g^{-1} with η_t one obtains, since g is minimum phase, a stationary output process γ_t. Thus, γ_t possesses an invariant probability measure with respect to which it has finite variance which satisfies

$$\mathrm{Var}(\gamma_t) = \mathrm{Var}(\gamma_{t+\tau}), \quad \tau \in \mathbb{Z}$$

Define, then, $V_{g_o}(g)$ via

$$V_{g_o}(g) = \frac{1}{2\pi}\mathrm{Var}(\gamma_o) \tag{1.3}$$

Alternatively, one may compute

$$V_{g_o}(g) = \frac{1}{2\pi i} \int_{|z|=1} \frac{g_o(z)g_o(z^{-1})}{g(z)\, g(z^{-1})} \frac{dz}{z} \tag{1.4}$$

Using the calculus of residues, it can be proven that

Theorem 1.2. [3]. V_{g_o} satisfies the properties:

(i) $V_{g_o}(g) \geq 1$, $V_{g_o}(g) = 1 \Leftrightarrow g_o = g$;

(ii) grad $V_{g_o}(g) = 0 \Leftrightarrow g_o = g$;

(iii) $D^2 V_{g_o}(g_o) > 0$.

At this point, it seems plausible that, in fact, V_{g_o} is a Lyapunov function for the flow on $X_{n,r}$ which is associated to the maximum likelihood estimation algorithms by Ljung's "O.D.E. method" ([24], [25]). This would account not only for uniqueness of the maximum likelihood estimate in $X_{n,r}$, but for convergence as well. This leads to the following question

Question 1.3. If, along trajectories of a flow $\tilde{A}_t$, the inequality

$$\frac{dV_{g_o}}{dt} \leq 0 \quad , \quad \frac{dV_{g_o}}{dt}(g) = 0 \Leftrightarrow g = g_o \tag{1.5}$$

holds, does $\tilde{A}_t(g) \to g_o$ uniformly?

That not every smooth function on a smooth manifold, satisfying (i)-(iii) of Theorem 1.2, is a Lyapunov function can be easily demonstrated. Consider $M = \mathbb{R}^2 - \{1,1\}$ and the smooth function

$$V : M \to \mathbb{R}^+$$

defined via

$$V(x,y) = x^2 + y^2 + 1 \tag{1.6}$$

Evidently, V satisfies

(i) $V(x,y) \geq 1$, $V(x,y) = 1 \Leftrightarrow (x,y) = (0,0)$;

(ii) grad $V(x,y) = [2x,2y] = 0 \Leftrightarrow (x,y) = (0,0)$;

$$\text{(iii)} \quad D^2V(0,0) = \begin{bmatrix} 2 & 0 \\ 0 & 2 \end{bmatrix} > 0$$

However, (i)-(iii) does not imply that $-\text{grad}V$ has $(0,0)$ as a global sink! For example, choosing $(2,2)$ as an initial condition one has

$$\gamma_t = -(\text{grad}V)_t(2,2) \not\rightarrow (0,0) \tag{1.7}$$

Indeed, γ_t does not exist for all $t > 0$ but has a finite escape time, viz $t_o = \ln 2/2$.

The reason for this phenomenon is illustrated in Figure 2. For example, for any initial condition (x,y) satisfying $V(x,y) \leq 3/2$ $\gamma_t = -(\text{grad}V)_t(x,y) \to (0,0)$. The absence of finite escape time can be verified by explicit calculation, but is also implied by the compactness of $V^{-1}(-\infty,3/2]$. In contrast, for any initial condition (x,x) satisfying $V(x,x) > 1 + \sqrt{2}$ the trajectory escapes from M at time $t_o = (\ln x)/2$.

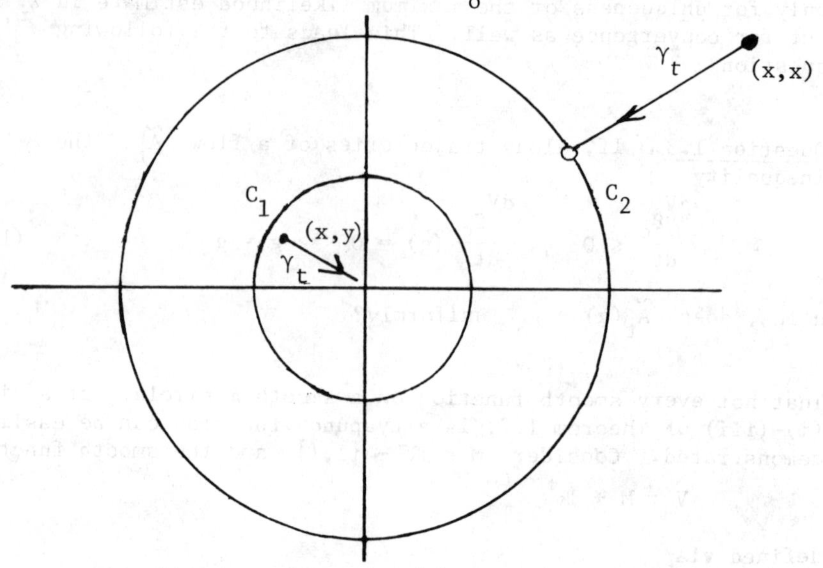

$C_1 = \{(x,y) : V(x,y) \leq 3/2\}$

$C_2 = \{(x,y) : V(x,y) \leq 1 + \sqrt{2}\}$

Figure 1.2. Depicting the lack of global convergence of -gradV.

Thus, conditions (i)-(iii) do not suffice to imply global convergence of flows satisfying (1.4) and, in addition, an objective function must satisfy

(iv) $V^{-1}(-\infty,c]$ is compact, $\forall c$.

Definition 1.4. $V : M \to \mathbb{R}$ is an objective function provided V satisfies

(i) V has a unique minimum at $m \in M$;

(ii) V has no other critical points except m;

(iii) $D^2V(m) > 0$;

(iv) $V^{-1}(-\infty,c]$ is compact for all $c \in \mathbb{R}$.

In the language of the tutorial [7], V is a Morse function on M with no other critical points aside from a minimum. In the next section, I will prove that - for topological reasons alone - the failure of $V(x,y)$ to be an objective function is, in fact, independent of V. That is,

Nota Bene 1.5. There does not exist any objective function V defined on $M = \mathbb{R}^2 - \{(1,1)\}$.

Now consider the objective function defined in (1.3)-(1.4). If V_{g_o} has compact sublevel sets, then V_{g_o} must have singularities on the boundary points

$$(q(s),p(s)) \in \mathbb{R}^n \times \partial X_{r,o} \subset \mathbb{R}^n \times \mathbb{R}^r \tag{1.8}$$

That this occurs is clear from (1.4), compare [3]. The presence of such singularities is not, however, enough a priori to conclude compactners. This has been proved by Byrnes and Krishnaprasad in [10]:

Theorem 1.6. $V_{g_o} : X_{n,r} \to \mathbb{R}^+$ has compact sublevel sets.

Combining theorems 1.2 and 1.6, one obtains an affirmative answer to Question 1.3, viz:

Corollary 1.7. $V_{g_o} : X_{n,r} \to \mathbb{R}^+$ is an objective function.

Sketch of a Proof of 1.6. Consider the "autoregressive" case. If $g(s) \in X_{n,o}$ then

$$g(s) = \frac{1}{q_o+q_1 s+\ldots+s^n}$$

and to $g(s)$ one assigns the vector

$$\tilde{g} = (q_{n-1},\ldots,q_o,1) \in \mathbb{R}^{n+1}$$

Denoting by c_k the Fourier coefficient

$$c_k = \frac{1}{2\pi i} \int_{|z|=1} z^k g_o(z) g_o(z^{-1}) \frac{dz}{z}$$

one forms the Toeplitz matrix, for $m \geqslant 1$,

$$C_m = \begin{bmatrix} c_o & c_1 & \cdots & c_{m-1} \\ c_1 & c_o & c_1 & \cdot \\ c_2 & c_1 & \cdot & \cdot \\ & & \cdot & \cdot & \cdot \\ c_{m-1} & & - & c_o \end{bmatrix}$$

By Bochner's Theorem, $C_{n+1} > 0$. Indeed,

$$V_{g_o}(g) = \langle\tilde{g}, C_{n+1}\tilde{g}\rangle$$

Thus, if $\lambda_{min} = \min\{\text{eigenvalues of } C_{n+1}\}$,

$$V_{g_o}^{-1}(-\infty,c] = \{\tilde{g} : ||g||^2 \leqslant \frac{c}{\lambda_{min}} + 1 ,\ g \text{ Hurwitz}\} \qquad (1.9)$$

Modulo the condition that $g(s)$ be Hurwitz, the sublevel set (1.9) is clearly compact in $\mathbb{R}^{n+1}$ and hence in the affine hyperplane $c_{n+1} = 1$. This technicality can be analyzed through Rouché's Theorem and the singularities of V_{g_o} in (1.8), from which the theorem follows in the autoregressive case.

The cases $r \geqslant 1$ can be analyzed through a similar argument involving compact quadratic forms on Hilbert space, see [10].

Q.E.D.

2. MILNOR'S THEOREM AND THE TOPOLOGY OF OBJECTIVE FUNCTIONS

In this section, the topological implications of the existence of an objective function

$$V : X \to \mathbb{R}$$

on a parameter space X are presented. In particular, it is shown using Morse-theoretic reasoning (see [27], [7]) that the space $X_{n,r}$ of miniphase systems is Euclidean [10].

First of all, I shall prove

Proposition 2.1. There exists no objective function

$$V : \mathbb{R}^2 - \{(1,1)\} \to \mathbb{R}$$

Proof. Suppose such a

$$V : \mathbb{R}^2 - \{(1,1)\} \to \mathbb{R}$$

is given, and consider the vector field

$$\frac{dx}{dt} = -\text{grad}V(x) \quad , \quad x \in \mathbb{R}^2 - \{(1,1)\} \tag{2.1}$$

Let $\gamma : [0,1] \to \mathbb{R}^2 - \{(1,1)\}$ be a continuous closed curve

$$\gamma(0) = \gamma(1) = (0,0)$$

If $x \in \gamma$ is fixed, there exists a time T such that the solution x_T to (2.1) at time T with initial condition x lies in the open ball

$$B_{1/2}(0,0) = \{x : ||x|| < 1/2\}$$

By continuity of the function

$$x \longmapsto x_T ,$$

there exists an open neighborhood U of x in γ such that for all $x' \in U$

$$(x')_T \in B_{1/2}(0,0)$$

Since γ is compact, γ is covered by finitely many such neighborhoods $U_1,\ldots,U_N$ and hence there exists a finite time T' such that the function

$$x \longmapsto x_{T'} \quad , \quad x \in \gamma$$

maps γ to a curve $\gamma' \subset B_{1/2}(0,0)$. Indeed, there exists a jointly continuous map

$$\Phi : [0,T'] \times [0,1] \to \mathbb{R}^2 - \{(1,1)\}$$

satisfying

$$\Phi(0,t) = \gamma(t)$$

$$\Phi(1,t) = \gamma'(t)$$

Since $B_{1/2}(0,0) \subset \mathbb{R}^2 - \{(1,1)\}$ is contractible, γ' and hence γ is contractible in $\mathbb{R}^2 - \{(1,1)\}$. Since γ is arbitrary, this implies that $\mathbb{R}^2 - \{(1,1)\}$ is simply-connected, contrary to fact. For example, the circle of radius $\sqrt{2}$ centered about $(1,1)$ is not contractible to a point in $\mathbb{R}^2 - \{(1,1)\}$. Therefore, there exists no such V.

Q.E.D.

The proof of this theorem provides an elementary criterion for the nonexistence of globally convergent flows:

Lemma 2.2. ([8], [12]). If M is a manifold which admits a vector field with a unique locally and globally attracting sink, then M is path-connected and simply-connected. In particular, if $V : M \to \mathbb{R}$ is an objective function, then M is a path-connected, simply-connected manifold.

In particular, one qualitative difference between the space $X_{n,r}$ of miniphase systems and the manifold $\mathbb{R}^2 - \{(1,1)\}$ is that $X_{n,r}$ is simply-connected. Much more, however, is true.

Note that the proof of Proposition 2.1 uses only the compactness of γ, so that one can prove more generally that each i-sphere in X is contractible. Thus, the higher homotopy groups $\pi_i(X)$, for $i \geq 1$, are all zero. By Whitehead's Theorem ([29], pg. 512), this implies further that X must be contractible. Using a deeper argument, Milnor has shown (see also [31]).

Milnor's Theorem 2.3. [26]. If X^n is an n-manifold which admits a vector field with a unique locally and globally attracting sink, then X^n is diffeomorphic to $\mathbb{R}^n$.

Remark 2.4. If $V : X^n \to \mathbb{R}$ is an objective function then the conclusion that X^n is diffeomorphic to $\mathbb{R}^n$ follows from Morse Theory. Milnor's Theorem allows one, for example, to handle cases where V has a degenerate critical point.

Corollary 1.7 and Milnor's Theorem imply

Corollary 2.4. The space $X_{n,r}$ of miniphase systems with r finite zeroes is diffeomorphic to $\mathbb{R}^{n+r}$.

3. OBJECTIVE FUNCTIONS ON SPACES OF RATIONAL FUNCTIONS

As a second example, consider

$$\mathrm{Rat}(n)=\{f(s)/g(s):\deg g=n,\ \deg f\leqslant n-1,\ (f,g)=1 \text{ and } g \text{ monic}\}$$

For example,

$$\mathrm{Rat}(1) = \{a/s + b : a \neq 0\}$$

Rat(1) may be parameterized, via (a,b), as an open dense subset of $\mathbb{R}^2$; viz. $\mathbb{R}^2 - \{b\text{-axis}\}$, as in Figure 3.1

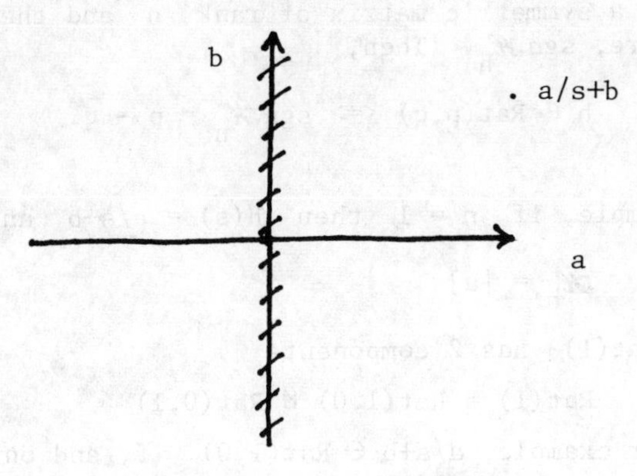

Figure 3.1. Depicting Rat(1).

In general (see the tutorial [7]), Rat(n) is an open dense subset of $\mathbb{R}^{2n}$.

Question 3.1. Does there exist an objective function on Rat(n)?

That the answer to Question 3.1 is always in the negative was proved by Brockett in [4]:

Theorem 3.2. [4]. Rat(n) is always disconnected. In fact,

$$\mathrm{Rat}(n) = \bigcup_{\substack{p,q\geqslant 0\\ p+q=n}} \mathrm{Rat}(p,q)$$

is a decomposition of Rat(n) into n + 1 disjoint path-connected open sets.

To explain the notation, expand $h \in \mathrm{Rat}(n)$ into a Laurent series,

$$h(s) = \sum_{i=1}^{\infty} \ell_i s^{-i}$$

and form the Hankel matrix

$$\mathcal{H}_h = \begin{bmatrix} \ell_1 & \ell_2 & \cdots & \ell_n \\ \ell_2 & & \cdot & \\ \cdot & & & \\ \cdot & & & \\ \cdot & & & \\ \ell_n & & & \ell_{2n-1} \end{bmatrix}$$

$\mathcal{H}_h$ is a symmetric matrix of rank n and therefore has a signature, $\operatorname{sgn}\mathcal{H}_h$. Then,

$$h \in \operatorname{Rat}(p,q) \iff \operatorname{sgn}\mathcal{H}_h = p - q$$

For example, if $n = 1$ then $h(s) = a/s+b$ and

$$\mathcal{H}_h = [a]$$

Thus, Rat(1) has 2 components

$$\operatorname{Rat}(1) = \operatorname{Rat}(1,0) \uplus \operatorname{Rat}(0,1)$$

and, for example, $a/s+b \in \operatorname{Rat}(1,0)$ if, and only if, $a > 0$.

Note that in particular, in order to define objective functions on parameter sets $X \subset \operatorname{Rat}(n)$, one must restrict the properties of the candidate systems g under consideration at least so far as to place g in the same path-component as g_o. For example, a more refined version of Question 3.1 is

<u>Question 3.1'</u>. Does there exist an objective function on Rat(p,q)?

As it turns out, the situation for $n = 1$ generalizes in the following sense. Recall that the signature of the Hankel matrix $\mathcal{H}_h$ can also be computed by a classical index formula. If $h(s)$ is rational, its Cauchy index is calculated as follows: Suppose the graph of h is as depicted in Figure 3.2.

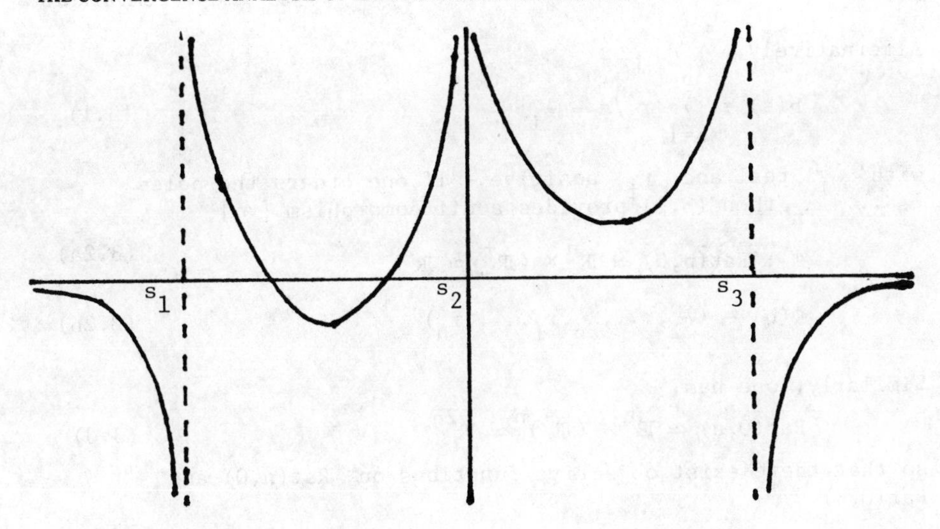

Figure 3.2. The graph of $h(s) \in \mathrm{Rat}(n)$.

Then at a pole (say s_1) where the value of $h(s)$ jumps from $-\infty$ to $+\infty$ one assigns a "local contribution" of 1; that is,

$$i_h(s_1) = 1$$

At a pole (say s_3), where $h(s)$ jumps from $+\infty$ to $-\infty$ one assigns a local contribution of -1. Finally, at a pole of even order (say s_2) where $h(s)$ does not change sign, one assigns a local contribution of 0. The Cauchy Index of h, $CI(h)$, is then defined as the sum

$$CI(h) = \sum_{s \text{ a pole}} i_h(s)$$

of the local contributions over all real poles. $CI(h)$ satisfies the following remarkable identity ([20], [21])

Theorem 3.3. (Hermite-Hurwitz). For $h \in \mathrm{Rat}(n)$,

$$\mathrm{sgn}(\mathscr{H}_h) = CI(h)$$

Using the Hermite-Hurwitz Theorem, one sees that $h \in \mathrm{Rat}(n,0)$ if, and only if,

$$n = \sum_{s \text{ a pole}} i_h(s) = n \cdot 1$$

Therefore, h has all n poles real with local contribution 1.

Alternatively,

$$h(s) = \sum_{i=1}^{n} r_i/s - \lambda_i \tag{3.1}$$

with λ_i real and r_i positive. If one orders the poles $\lambda_1 < \ldots < \lambda_n$, then (3.1) provides a diffeomorphism [4]

$$\Phi : \mathrm{Rat}(n,0) \simeq \mathbb{R}^n \times (\mathbb{R}^+) \simeq \mathbb{R}^{2n} \tag{3.2a}$$

$$\Phi(h) = (\lambda_1,\ldots,\lambda_n,r_1,\ldots,r_n) \tag{3.2b}$$

Similarly, one has

$$\mathrm{Rat}(0,n) \simeq \mathbb{R}^n \times (\mathbb{R}^-)^n \simeq \mathbb{R}^{2n} \tag{3.3}$$

so that there exist objective functions on $\mathrm{Rat}(n,0)$ and $\mathrm{Rat}(0,n)$.

Example 3.4. ($n = 2$). According to Theorem 3.2,

$$\mathrm{Rat}(2) = \mathrm{Rat}(2,0) \uplus \mathrm{Rat}(1,1) \uplus \mathrm{Rat}(0,2)$$

Consider, for $\theta \in [0,2\pi]$ the transfer function

$$g_\theta(s) = \frac{(\cos\theta) + \sin\theta}{s^2+1}$$

Since $g_\theta(s)$ has no real poles,

$$\mathrm{CI}(g_\theta) = 0 \quad , \quad \text{or} \quad g_\theta \in \mathrm{Rat}(1,1) \quad , \quad \forall\theta \in [0,2\pi]$$

Note that $\theta \longmapsto g_\theta$ is periodic, so that g_θ is a closed curve in $\mathrm{Rat}(1,1)$. Indeed, as $0 \leqslant \theta \leqslant 2\pi$ the zero of g (s) moves in a closed curve, viz. $\mathbb{R} \cup \{\infty\}$, about the pole $+ i$ of $g_\theta(s)$ in the extended plane $\mathbb{C} \cup \{\infty\}$. Heuristically, if one were to retract $g_\theta(s)$ to a point, say $g(s) \in \mathrm{Rat}(1,1)$, this zero would have to cancel a pole during the retraction. This would imply that g_θ is not contractible in $\mathrm{Rat}(2)$, i.e. in $\mathrm{Rat}(1,1)$.

This argument can be made rigorous yielding

Lemma 3.6. $\mathrm{Rat}(1,1)$ is not simply-connected.

This also follows from the explicit identity

$$\mathrm{Rat}(n-1,1) \simeq \mathrm{Rat}(1,n-1) \simeq S^1 \times \mathbb{R}^{2n-1}, \; n > 1$$

derived by Brockett in [4], but the argument generalizes in the following way. Consider, for $\min(p,q) \geq 1$,

$$g_\theta(s) = \frac{(\cos\theta)s+\sin\theta}{s^2+1} + \sum_{i=1}^{p-1} + \frac{1}{s-i} + \sum_{j=p+1}^{n-2} - \frac{1}{s-j}$$

$g_\theta(s) \in Rat(p,q)$, for θ $[0,2\pi]$, and in fact $g_\theta(s)$ is noncontractible (see [8]). Combining this fact with Lemma 2.3, one obtains

Theorem 3.7. [8]. There exists a globally convergent vector field (or an objective function) on $Rat(p,q)$ if, and only if, $\min(p,q) = 0$.

In particular, some hypothesis - e.g. stable, miniphase, positive definiteness of the Hankel $\mathscr{H}_h$ - is needed to prescribe a "good candidate set" X. Alternatively, if X is not Euclidean a choice of overparameterization

$$\Phi : \tilde{X} \to X$$

can also lead to Euclidean structure $\tilde{X}$. This point of view seems to play a role in recent work by Mosca and Zappa ([28], this proceedings) and is in fact quite classical in differential analysis and topology.

4. MULTIVARIABLE PARAMETER SPACES

In this section, I will briefly discuss the topology of certain parameter spaces of multi-input, multi-output systems.

Consider the system

$$x(t+1|t) = Fx(t|t-1) + K\varepsilon(t) \quad (4.1a)$$

$$y(t) = Hx(t|t-1) + \varepsilon(t) \quad (4.1b)$$

with no observed input, whose entire state space $\mathbb{R}^n$ is observable from the output strings $y_t \in \mathbb{R}^p$, and with identical driving noise and observation noise $\varepsilon(t)$ an $\mathbb{R}^p$-valued white Gaussian noise. Thus (4.1a-b) defines a point in the open, dense subset

$$\mathscr{A}_p^n = \{(F,K,H) : (F,H) \text{ is observable}\} \subset \mathbb{R}^{n^2+2np}$$

If

$$n = n_1+\ldots+n_p \quad , \quad n_i \in \mathbb{N}$$

then let

$$\tilde{\mathscr{A}}^n_p(n_i) = \{(F,K,H) : (F,H) \text{ has observability indices } (n_i)\}$$

Associated to the relation (4.1a-b) between (y,ε) is the ARMA representation

$$\sum_{j=0}^{n_m} A_j y(k-j) = \sum_{j=0}^{n_m} D_j \varepsilon(k-j) \qquad (4.2)$$

where $n_m = \max\{n_i\}$, and $D_o = A_o$. The autoregressive part, i.e. the left-handed side of (4.2), depends only on (F,H) while the moving average part, except for D_o, reflects the stochastic parameters of the system. Moreover, by realization theory the parameters (A_j, D_j) determine (F,K,H) up to a change of basis, i.e. modulo transformations of the form

$$(F,K,H) \longmapsto (TFT^{-1}, TK, HT^{-1}) \qquad (4.4)$$

Alternatively, if

$$\tilde{\mathscr{A}}^n_T(n_i) = \{H(sI-F)^{-1}K : (F,K,H) \in \tilde{\mathscr{A}}^n_p(n_i)\}$$

then the ARMA coefficients (A_j, D_j) parameterize the set $\mathscr{A}^n_p(n_i)$.

Now, assuming the (n_i)'s are known, it may be possible to estimate the deterministic parameters (A_j) as in [14], [30] and the identification algorithm outlined by Alengrin ([1], these proceedings) consists in first estimating the (A_j) and then determining the stochastic parameters (D_j) in terms of a stochastic realization of the stochastic, or moving average, part of (4.2). According to the "Euclidean Ansatz" pursued here, one would expect that $\mathscr{A}^n_p(n_i)$ is Euclidean.

For example, consider $\mathscr{A}^4_2(2,2)$. $(F,K,H) \in \tilde{\mathscr{A}}(2,2)$ if, and only if, the first four columns of the matrix

$$[h_1^t \; h_2^t \; F^t h_1^t \; F^t h_2^t \; \dots \; (F^t)^3 h_2^t]$$

are linearly independent. Denoting this 4×4 submatrix by T one has that $(F,K,H) \in \tilde{\mathscr{A}}(2,2)$ if, and only if,

$$TFT^{-1} = \begin{bmatrix} 0 & 0 & 1 & 0 \\ 0 & 0 & 0 & 1 \\ * & * & * & * \\ * & * & * & * \end{bmatrix}, \quad HT^{-1} = \begin{bmatrix} 1 & 0 & 0 & 0 \\ 0 & 1 & 0 & 0 \end{bmatrix}$$

and $\tilde{K} = TK$ is arbitrary. In the ARMA parameters one has

$$A_0 = \begin{bmatrix} 1 & 0 \\ 0 & 1 \end{bmatrix}, \quad A_1 = \begin{bmatrix} * & * \\ * & * \end{bmatrix}, \quad A_2 = \begin{bmatrix} * & * \\ * & * \end{bmatrix}$$

with the "stochastic parameters" D_0, D_1, D_2 containing certain fixed 0's and 1's (in D_0 in this case) and 8 free parameters in D_1 and D_2. In particular, $\mathscr{A}(2,2) \simeq \mathbb{R}^{16}$ with Euclidean coordinates given by the free parameters in the (A_j,D_j). In general,

Theorem 4.1. (A_j,D_j) provide Euclidean coordinates for $\mathscr{A}_p^n(n_i)$. In particular,

$$\mathscr{A}_p^n(n_i) \simeq \mathbb{R}^N, \quad N = np+p^2 - \sum_{n_i \geq n_j} (n_i-n_j+1).$$

Proof. That $\mathscr{A}_p^n(n_i)$ is Euclidean follows as above. As for the dimension N, one first uses Brockett's formula [5]

$$\dim \tilde{\mathscr{A}}_p^n(n_i) = n^2+np+p^2- \sum_{n_i \geq n_j} (n_i-n_j+1) \tag{4.5}$$

for the dimension of $\tilde{\mathscr{A}}_p^n(n_i)$ and then the State-Space Isomorphism Theorem of realization theory. Explicitly, (A_j,D_j) determines (F,K,H) uniquely up to a change of basis (4.3). Moreover, if (F',K',H') is another realization of (A_j,D_j) then the transformation T which relates (F,K,H) and (F',K',H') is unique since

$$T[H^t,F^tH^t,\ldots,(F^{n-1})^tH^t] = [(H')^t,\ldots,((F')^{n-1})^tH^t]$$

and (F,H) is observable. Therefore,

$$\dim \mathscr{A}_p^n(n_i) = \dim \tilde{\mathscr{A}}_p^n(n_i) - n^2$$

since $n^2 = \dim GL(\mathbb{R}^n)$.

Q.E.D.

Defining the set

$$\mathscr{A}_p^n = \{(A_j,D_j):(A_j,D_j) \text{ admits a realization } (F,K,H) \in \tilde{\mathscr{A}}_p^n\}$$

one sees then that $\mathscr{A}_p^n$ is the disjoint union of subsets

$$\mathscr{A}_p^n = \cup \mathscr{A}_p^n(n_i) \tag{4.6}$$

each of which is a Euclidean space, but in general of different dimensions.

Example 4.2. (n = 4, p = 2) In Figure 4.1, the decomposition of $\mathscr{A}^4_2$ into the three Euclidean pieces $\mathscr{A}^4_2(n_i)$ is depicted, where the superscripts indicate the dimensions of the particular Euclidean spaces. Of course, the appearance of a boundary is due in part to the planar approximation to 16-dimensional space.

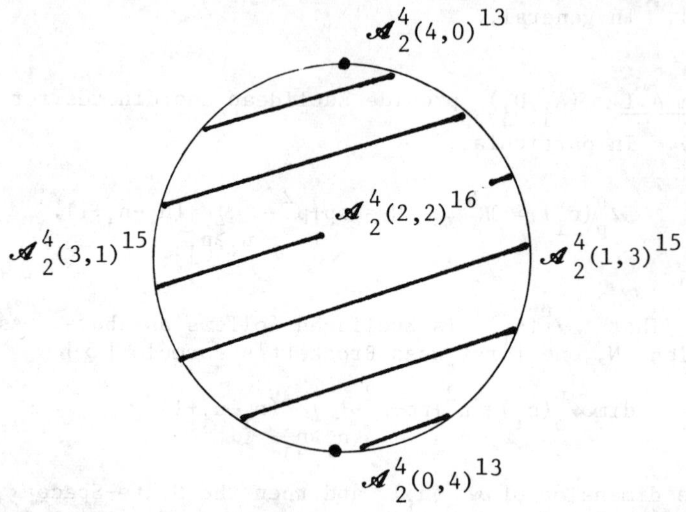

Figure 4.1. Depicting the decomposition of $\mathscr{A}^4_2$ into the $\mathscr{A}^4_2(n_i)$.

In fact, the $\mathscr{A}^n_p$ are manifolds and the $\mathscr{A}^n_p(n_i)$ submanifolds, see Remark 4, but this is not needed for the analysis of the $\mathscr{A}^n_p(n_i)$. Of course, as Example 4.2 indicates, all the $\mathscr{A}^4_2(n_i)$ except $\mathscr{A}^4_2(2,2)$ are thin subsets in $\mathscr{A}^n_p$ suggesting that, taking error in data measurements into account, the only robust choice of (n_i) would be, in this case, $(n_i) = (2,2)$.

Quite generally, suppose

$$n = \ell p + r \tag{4.7a}$$

and

$$n_i = \ell + 1 \text{ for } r \text{ of the } n_i\text{'s} = \ell \text{ for } p-r \text{ of the } n_i\text{'s} \tag{4.7b}$$

Proposition 4.1. If μ is a probability measure on $\tilde{\mathscr{A}}^n_p$

which is absolutely continuous with respect to Lebesque measure, then

$$(F,G,H) \in \tilde{\mathscr{A}}^n_p(n_i), \text{ where } (n_i) \text{ satisfies (4.7a-b)}$$

with probability one.

Proof. (See, e.g., [5] or [16].) The assertion follows from the claim that $\tilde{\mathscr{A}}^n_p(n_i)$ contains an open subset of $\mathbb{R}^{n^2+2np}$ just in case $(4.6)^p$ holds. This claim itself follows since to say $\tilde{\mathscr{A}}^n_p(n_i)$ contains an open subset is to say

$$\dim \tilde{\mathscr{A}}^n_p(n_i) = n^2 + 2np \qquad (4.8)$$

Comparing (4.8) with Brockett's formula (4.5) yields the claim by elementary combinatorics.

Q.E.D.

It would seem, then, that convergence of an identification algorithm, to $(A_j^0, D_j^0) \in \mathscr{A}^n_p$, based on estimating the (n_i) is only possible provided

$$(A_j^0, D_j^0) \in \mathscr{A}^n_p(n_i), \text{ where } (n_i) \text{ satisfies (4.7b)} \qquad (4.9)$$

The constraint (4.9) arises in the recent algorithm derived by E.J. Hannan and J.J. Rissanen [16], which is known to apply to a wider class of systems, e.g. ARMAX, and to a broader class of driving noise. Hannan and Rissanen actually work on the subspace

$$\Sigma^n_p = \{(A_j, D_j) \text{ realizable by an observable } \underline{\text{and controllable}} \; (F,K,H)\}$$

of $\mathscr{A}^n_p$ and the corresponding neighborhood $\Sigma^n_p(n_i)$ and constraint

$$(A_j^0, D_j^0) \in \Sigma^n_p(n_i); n_1 = \ldots n_r = \ell+1, n_{r+1} = \ldots = n_p = \ell \qquad (4.9)'$$

According to Proposition 4.3, the constraints (4.9) or (4.9)' are satisfied with probability 1 for any absolutely continuous probability measure. Nevertheless, it is still of importance to ask:

What can one do about relaxing the constraints (4.9) or (4.9)'?

In particular, does there exist a globally convergent identification algorithm on $\mathscr{A}^n_p$ or on $\sum^n_p$? If not, how many charts - such as the "generic neighborhoods" (4.9) or (4.9)' - on which there exist convergent algorithms are needed to cover $\mathscr{A}^n_p$ or $\sum^n_p$.

One choice of such a family of charts [16] is given by

$$\mathscr{A}^n_p[n_i]=\{(A_j,D_j) \text{ realizable by } (F,K,H) \text{ satisfying } (h^t_1,\ldots,(F^t)^{n_1-1}h^t_1,\ldots,h^t_p,\ldots,(F^t)^{n_p-1}h^t_p) \text{ is linearly independent}\}.$$

This differs from the assertion that (n_i) are the observability indices of (F,H) in that no requirements on the depenence, for example, of $(F^t)^{n_1}h^t_1$ on its predecessors are made. In particular, $\mathscr{A}^p_n[n_i]\simeq\mathbb{R}^{2np}$ for all (n_i). Indeed $\mathscr{A}^n_p(n_i) = \mathscr{A}^n_p[n_i]$ if, and only if, (4.7b) is satisfied.

However, there are $\binom{n+m-1}{m-1}$ such charts (see [16]) on $\mathscr{A}^n_p$. Thus, for $n = 25$, $m = 5$ one requires approximately 10^8 charts. Conjecturely, this number far exceeds the number of charts which are actually necessary. A crude corollary of the main results in Delchamps' thesis gives a lower bound, which can be vastly improved in particular cases:

<u>Theorem 4.4. [12]</u>. One needs at least p charts to cover $\sum^n_p$.

As a byproduct, Delchamps also draws inferences concerning global convergence of flows on $\sum^n_p$:

<u>Corollary 4.5. [12]</u>. There does not exist any objective function on $\sum^n_p$ if $p > 1$.

In fact a slightly stronger statement in [12] implies that there does not exist any globally and locally convergent deterministic flows on $\sum^n_p$. Combining this with Ljung's "O.D.E. method" gives further credence to the "Euclidean Ansatz" for global convergence.

Theorem 3.7-Corollary 4.5 of course apply to the cases where the flow is defined on $Rat(p,q) \times M$ where M is any space of measurements, e.g. $\mathbb{R}^N$.

Remark 4.6. $\mathscr{A}_p^n$ (and thus $\sum_p^n$) can be given the structure of a differentiable manifold in such a way that the $\mathscr{A}_p^n(n_i)$ are submanifolds and that the function

$$\pi : \tilde{\mathscr{A}}_p^n \to \mathscr{A}_p^n , \quad \text{defined via}$$

$$\pi(F,K,H) = (A_j(F,K,H), D_j(F,K,H))$$

is differentiable. This latter fact was first envisioned by Kalman in [22] and pursued later within broader investigations independently by Hazewinkel-Kalman [18] and Byrnes-Hurt [9]. The connectivity properties of $\sum_p^n$ are discussed for $p = 1$ in Brockett [4] and for $p > 1$ in Glover [13]. The first correct proof in the literature that $\sum_p^n$ is a manifold was given by Clark in [11], while shortly thereafter proofs that $\mathscr{A}_p^n$ is a manifold (implying Clark's Theorem) were given in the 1976 NASA-AMES Conference Proceedings by Byrnes [6] and by Hazewinkel [17]. These manifolds are also discussed rather explicitly in Hannan [15]. The most detailed information on topology of $\sum_p^n$ can be found in Delchamps' thesis [12], which uses Morse Theory and a differential geometric interpretation of the positive definite solution of the Riccati equation. Complementing this, the recent thesis of Helmke [19] contains the deepest investigation of the topology of $\mathscr{A}_p^n$, starting from the "cell decomposition" (4.6). In particular, these theses contain enough information to deduce the nonexistence of globally convergent flows as well as the separate issue of the nonexistence of globally defined continuous canonical forms (a priori, algebraic), which was part of most of the previous investigations. Although this issue certainly does not impact identification in ARMA parameters, it is interesting as a theorem in realization theory and also as the starting point of the beautiful construction of the "universal family" of systems by Hazewinkel, see [17] or [18]. Finally, the topology of the space of observable (or, dually, controllable) pairs (F,K) with fixed observability indices, up to permutation, of which $\mathscr{A}_p^n(n_i)$ is an open chart is studied in [5], which contains the dimension formula used in Proposition 4.3.

5. BIBLIOGRAPHY

[1] G. Alengrin, "Estimation of Stochastic Parameters for ARMA Models by Fast Filtering Algorithms," in Nonlinear Stochastic Problems (R.S. Bucy, J.F. Moura, eds.), D. Reidel, Dordrecht, 1983. (this volume pp. 65-78)

[2] K.J. Åström and T. Bohlin, "Numerical Identification of Linear Dynamical Systems from Normal Operating Records," in Theory of Self-Adaptive Control Systems (P.H. Hammond, ed.), Plenum: New York, 1966.

[3] K.J. Åström and T. Söderström, "Uniqueness of the Maximum Likelihood Estimates of the Parameters of an ARMA Model," IEEE Trans. Aut. Control 19 (1974) 769-773.

[4] R.W. Brockett, "Some Geometric Questions in the Theory of Linear Systems," IEEE Trans. Aut. Control 21 (1976) 449-455.

[5] R.W. Brockett, "The Geometry of the Set of Controllable Linear Systems," Research Reports of the Aut. Control Lab, Fac. of Eng., Nagoya Univ., Vol. 24, July 1977.

[6] C.I. Byrnes, "The Moduli Space for Linear Dynamical Systems," in Geometric Control Theory (C. Martin, R. Hermann, eds.), Math. Sci. Press, Brookline, 1977.

[7] C.I. Byrnes, "A Brief Tutorial on Calculus on Manifolds," in Nonlinear Stochastic Problems (R.S. Bucy, J.F. Moura, eds.), D. Reidel, Dordrecht, 1983. (this volume pp. 123-150)

[8] C.I. Byrnes and T.E. Duncan, "On Certain Topological Invariants Arising in System Theory," in New Directions in Applied Mathematics (P.J. Hilton, G.S. Young, eds.), Springer-Verlag, N.Y., 1981.

[9] C.I. Byrnes and N.E. Hurt, "On the Moduli of Linear Dynamical Systems," Advances in Math. Suppl. Series 4 (1978) 83-122; also in Modern Mathematical System Theory, MIR Press, Moscow, 1978 (in Russian).

[10] C.I. Byrnes and P.S. Krishnaprasad, "On the Euclidean Ansatz for Stable and Miniphase Systems," (to appear).

[11] J.M.C. Clark, "The Consistent Selection of Local Coordinates in Linear System Identification," in Proc. of the J.A.C.C., Purdue, 1976.

[12] D. Delchamps, "The Geometry of Spaces of Linear Systems with an Application to the Identification Problem," Ph.D. Thesis, Harvard Univ., 1982.

[13] K. Glover, "Some Geometrical Properties of Linear Systems with Implications in Identification," in Proc. of the I.F.A.C. Congress, Boston, 1975.

[14] R.P. Guidorzi, "Canonical Structures in the Identification of Multivariable Systems," Automatica 11 (1975) 361-374.

[15] E.J. Hannan, "System Identification," in Stochastic Systems: The Mathematics of Filtering and Identification and Applications (M. Hazewinkel and J.C. Willems, eds.), D. Reidel, Dordrecht, 1981.

[16] E.J. Hannan, The Statistical Theory of Linear Systems, M.I.T. Course Notes, Spring 1982.

[17] M. Hazewinkel, "Moduli and Canonical Forms for Linear Dynamical Systems III: The Algebraic-Geometric Case," in Geometric Control Theory (C. Martin, R. Hermann, eds.), Math. Sci. Press, Brookline, 1977.

[18] M. Hazewinkel and R.E. Kalman, "On Invariants, Canonical Forms, and Moduli for Linear Constant, Finite-Dimensional Dynamical Systems," Lecture Notes Econ.-Math. System Theory 131 (1976) 48-60, Springer-Verlag, N.Y.

[19] U. Helmke, "A Cell Decomposition for the Orbit Space of the Similarity Action," (to appear).

[20] C. Hermite, "Sur les Nombres des Racines d'une Équation Algébrique Comprises entre des Limites Données," J. Reine Angew. Math 52 (1856) 39-51.

[21] A. Hurwitz, "Über die Bedingungen unter Welchen eine Gleichung nur Wurzeln mit Negativen Reelen Theilen Besitzt," Math. Ann. 46 (1895) 273-284.

[22] R.E. Kalman, "Global Structure of Classes of Linear Dynamical Systems," preprint of a lecture given at NATO-ASI on Geometric and Algebraic Methods for Nonlinear Systems, London, 1971.

[23] P.S. Krishnaprasad, "Geometry of Minimal Systems and the Identification Problem," Ph.D. Thesis, Harvard, 1977.

[24] L. Ljung, "On Positive Real Functions and the Convergence of Some Recursive Schemes," IEEE Trans. Aut. Control 22 (1977) 539-551.

[25] L. Ljung, "Analysis of Recursive Stochastic Algorithms," IEEE Trans. Aut. Control 22 (1977) 551-575.

[26] J.W. Milnor, "Differential Topology," in Lectures in Modern Mathematics Vol. II (T.L. Saaty, ed.), Wiley, N.Y., 1964.

[27] J.W. Milnor, Morse Theory, Princeton Univ. Press, Princeton, 1963.

[28] E. Mosca and G. Zappa, "Overparameterization, Positive Realness, and Multistep Minimum Variance Adaptive Regulators," in Nonlinear Stochastic Problems (R.S. Bucy, J.F. Moura, eds.), D. Reidel, Dordrecht, 1983. (this volume pp. 205-215)

[29] E. Spanier, Algebraic Topology, McGraw-Hill, N.Y., 1966.

[30] E. Tse and H.L. Weinert, "Structure Determination and Parameter Identification for Multivariable Stochastic Linear Systems," IEEE Trans. Aut. Control 20 (1975) 603-612.

[31] F.W. Wilson, Jr., "The Structure of the Level Surfaces of a Lyapunov Function," J. of Diff. Eqns. 3 (1967) 323-329.

MULTIVARIABLE ADAPTIVE REGULATORS BASED ON MULTISTEP QUADRATIC COST FUNCTIONALS°

Edoardo Mosca
ISIS
Università di Firenze
Via S.Marta 3, 50139 Firenze, Italy

ABSTRACT - An innovative approach to adaptive regulation of stochastic multivariable linear plants is described. Its main feature consists of synthesizing a feedback-gain matrix according to the minimization of a quadratic cost functional defined over a control horizon of arbitrary length via an implicit identification algorithm. Indeed, all that is required is to recursively carry out an LS estimation of a multistep prediction model whose parameters almost directly yield the desired feedback. The resulting algorithm, referred to by the acronym MUSMAR, encompasses the basic LS+MV self-tuning regulator as well as adaptive regulators strictly related to steady-state LQG theory. In the latter version, the MUSMAR, inheriting the intrinsic robusteness of LQG design, performs almost optimally even when the plant is nonminimum-phase or does not satisfy a positive realness condition.

1. INTRODUCTION

The output regulation problem, where an output vector of a given plant is to be reduced to zero from an arbitrary initial state, is known to be the basic control problem. This problem arises, for instance, when a plant is to be operated at a desired constant set

---

°This work was supported in part by the Italian CNR under contracts 81.00876.07 and 81.01570.97.

R. S. Bucy and J. M. F. Moura (eds.), Nonlinear Stochastic Problems, 187–204.

point, and hence deviations from the set point are to be regulated to zero. Moreover, suppression of disturbances and tracking control problems can be suitably reformulated as output regulation problems. If the plant is linear and time-invariant and its parameters are known, standard methods can be used to design linear regulators making the overall closed-loop system stable.

If the plant is unknown, adaptive regulators, adjusting the feedback coefficients according to measured input and output variables, make the overall regulated system nonlinear [1], [2]. This significantly complicates the analysis and raises fundamental questions concerning convergence and stability of adaptively regulated systems. Although some major steps have been recently made with the establishment of convergence [3] and global stability [4]-[8] for a number of adaptive control algorithms, so far all adaptive control schemes that are known to enjoy convergence or be globally stable impose specific requirements on the plant parameters. Consequently, they cannot be applied to arbitrary unknown stochastic linear plants. Indeed, major limitations concern: i) nonminimum-phase plants [9]; ii) plant noise not fulfilling a positive realness condition [3]; iii) uncertainty on the plant i/o transport delay [10]. The extensions of such adaptation schemes to multivariable plants [11]-[13] present further limitations. Regulators sharing such limitations all operates according to direct schemes in that feedback parameters are directly estimated on the basis of a single step prediction model of the plant.

Indirect adaptive schemes, proposed to overcome limitations inherent to direct adaptive regulators, have recently received attention [14]. They imply greater computational complexity because of the need of using adaptive state observers in the feedback loop. Further, unlike direct schemes, indirect adaptive regulators require the presence of a sufficiently exciting external input.

In the approach discussed hereafter an attempt is made to retain the advantages of computational simplicity of direct schemes without loosing the greater flexibility and potentially higher performance of indirect schemes. This is obtained by adjusting the feedback parameters according to a multistep quadratic-cost functional of plant future i/o samples, whereas computational simplicity is retained by using LS (Least-Squares) estimates of a multistep prediction model of the closed-loop system. This circumvents the need of explicitly constructing an adaptive state observer. This approach gives rise to the MUSMAR (MUltiStep Multivariable Adaptive Regulator) algorithm that was introduced in [15], and, under semplificative assumptions, analyzed in [16]. An application of the

MUSMAR algorithm in adaptive prediction problems was considered in [17], and in [18] a study on MUSMAR computational complexity was reported.

The main goal of this work is to discuss the MUSMAR approach to adaptive regulation in an organic way.

In Sect.2 the basic MUSMAR optimization problem (BAMOP) is formulated and its relation with steady-state LQG regulation theory is underlined. In Sect.3 BAMOP is solved in a way that directly enlights the role in adaptive regulation of LS parameter estimates of plant multistep prediction models, and some MUSMAR desirable properties are proved. In Sect.4, simulation results are presented exhibiting MUSMAR performance improvements over the LS+MV self-tuning regulator. In Sect.5 a local convergence analysis is carried out, while in Sect.6 some conclusive remarks are made.

2. THE BASIC MUSMAR OPTIMIZATION PROBLEM

This section describes what hereafter will be referred to as the basic MUSMAR optimization problem. It will be shown that the corresponding solution may lead, if suitably iterated, to conveniently obtain the optimal steady-state LQG regulator for an unknown multivariable plant.

Let us consider a *plant* with inputs $u(k) \in R^m$ and outputs $y(k) \in R^p$ whose dynamic behavior can be described by a discrete time Gauss-Markov system in its innovations-representation form [9],[19]

$$\begin{aligned} x(k+1) &= Ax(k) + Bu(k) + Ke(k) \\ y(k) &= Cx(k) + e(k) \end{aligned} \tag{1}$$

where: $x(k) \in R^n$; $e(k)$ is the innovations process of $y(k)$; A, B, K, C, are matrices of consistent dimensions. It is assumed that representation (1) is minimal, i.e. $(A,[B \mid K])$ reachable and (A,C) observable, and that (A,B) is stabilizable. Occasionally, other possible representations, e.g. stochastic vector ARMAX equations, equivalent to (1) from an external behavior point of view, will also be conveniently used hereafter. Most appropriately, (1) or any other equivalent representation should be regarded as an analytic rule laying behind the plant accessible variables, viz. u(k) and y(k). Associated with the plant a *quadratic-cost functional* is considered

$$J(t,t+T) := (T+1)^{-1} E\{ \sum_{k=t}^{t+T} \| y(k+1) \|^2_{Q_y} + \| u(k) \|^2_{Q_u} \} \tag{2}$$

where: T=0,1,2,...; E denotes expectation; $Q_y = Q'_y \geq 0$; $Q_u = Q'_u \geq 0$; $\| \gamma \|^2_Q = \gamma' Q\gamma$, the prime denoting transpose. It is assumed that the *admissible control strategy* amounts to selecting inputs u(k) according to a linear nonanticipative feedback law

$$u(k) = F(k)s(k) \qquad s(k) \in [y^k, u^{k-1}] \tag{3}$$

where: $\gamma^k = \{\gamma(k), \gamma(k-1), \ldots\}$; and $[\gamma^k]$ denotes the subspace of all random variables linearly generated by the ones belonging to γ^k; $s \in [\gamma^k]$ is a shorthand notation denoting that all the components of the random vector s belong to $[\gamma^k]$.

Thus, the *basic* MUSMAR *optimization problem* (BAMOP) can be formulated as follows. Given the sequence of feedback-gain matrices

$$F^{t+T}_{t+1} := \{F(t+1),\ F(t+2),\ \ldots\ ,\ F(t+T)\} \tag{4}$$

find a feedback-gain matrix L(t) such that u(t) = L(t)s(t) minimizes the quadratic-cost functional

$$J(t, t+T | F^{t+T}_{t+1}) \tag{5}$$

for the plant (1).

It is advisable to consider in some details two possible choices for the vector s(k), which hereafter will be referred to as the *pseudostate*:

$$\begin{aligned} s(k) &:= E[x(k+1) | y^k, u^{k-1}, u(k) = 0] \\ &= Ax(k) + Ke(k) := \chi(k) ; \end{aligned} \tag{6}$$

$$\begin{aligned} s(k) &:= [y'(k), u'(k-1), \ \ldots\ , u'(k-n+1), y'(k-n+1)]' \\ &:= \sigma(k). \end{aligned} \tag{7}$$

Although $\chi(k)$ is seldom available in practical instances, the choice (6) is of interest in that related inputs $u(k) = F(k)\chi(k)$ have the same structure as the optimal LQG inputs. The pseudostate (7) is an attractive one since plant i/o records are available in practical applications. Another reason for considering the choice (7) and which has not been enough underlined in the control literature, is that, as can be easily shown, any constant state-feedback control $u(k) = F\chi(k)$ can also be expressed as $u(k) = \bar{F}\sigma(k)$ for suitable $\bar{F}$. Therefore, if $\bar{F}_o$ corresponds to the optimal steady-state LQG feedback-gain matrix F_o, inputs $u(k) = \bar{F}_o\sigma(k)$ yield the optimal plant performance.

Some further considerations are made on the BAMOP relevance in control. Let us assume in (2) T=0.

Thus, F_{t+1}^{t+T} is empty, and $J(t,t+T|F_{t+1}^{t+T}) = J(t,t) = E\{\| y(t+1)\|_{Q_y}^2 + \| u(t)\|_{Q_u}^2 \}$. Therefore, in this case BAMOP coincides with:

either a multivariable minimum-variance (MV) regulation problem with a weight on the control effort if a plant representation is a priori known; or, as will be seen, a multivariable least-squares (LS) plus MV (LS+MV) regulation problem [1],[10],[11],[13] if a plant representation is not a priori known. In the latter case, the equivalence with an LS+MV regulation problem is obtained by adopting a certainty equivalence control.

Further insight into the problem that has been posed, is gained by considering the other extreme situation, viz. in (2) an ideally semiinfinite regulation interval T= ∞. In addition, let us assume that the plant representation (1) is known, $s(t) = \chi(t)$, and a sequence of constant feedback-gain matrices is used

$$F_{t+1}^{\infty} = \{F_1, F_1, \dots , F_1, \dots\} := \vec{F}_1 \qquad (8)$$

such that $(A+BF_1A)$ is a stability matrix. Next, let us denote by F_2 the solution obtained by minimizing $J(t,\infty|\vec{F}_1)$ for the system (1), by F_3 the one obtained by minimizing $J(t,\infty|\vec{F}_2)$ for (1) and so on. Then one can prove [16] the following

<u>Proposition</u> 1 - Let $\{F_i\}_{i=1}^{\infty}$ be the sequence generated according to the above BAMOP iterations. Then

$$\lim_{i\to\infty} F_i = F_o \qquad (9)$$

where F_o is the *optimal steady-state* LQG *feedback-gain matrix* for the problem (1)-(3). More specifically, the sequence $\{F_i\}_{i=1}^{\infty}$ coincides with the one generated by the discrete-time version of Newton-Raphson iterations [20],[21] for computing F_o. ∎

3. BAMOP SOLUTION AND THE MUSMAR ALGORITHM

Sect.2 establishes that BAMOP is strictly related to LQG regulation theory. In particular, according to Proposition 1, one can hope to obtain a good approximation to the optimal solution of the steady-state LQG regulation problem by setting in BAMOP T large enough and F(t+T+1) = L(t). The main goal of this section is to find a BAMOP solution in a form suitable for a priori unknown plants.

Let us denote by z(t,t+T) the vector of the plant *future i/o joint process*

$$z'(t,t+T) = [y'(t+T+1)\ u'(t+T)\ \dots\ y'(t+1)\ u'(t)] \quad (10)$$

Thus, (2) can be rewritten as follows

$$J(t,t+T|F_{t+1}^{t+T}) = E\{\| z(t,t+T)\|^2_{Q_{T+1}} \} /(T+1) \quad (11)$$

where

$$Q_j = \text{block-diag}\ \{Q, Q, \dots , Q\} \quad \text{(j-times)}$$

with

$$Q = \begin{bmatrix} Q_y & 0 \\ 0 & Q_u \end{bmatrix} \quad (12)$$

The results of the following Lemma, whose proof can be found in the Appendix, yield at once a BAMOP solution.

Lemma - Let the regression model of z over u and s be available

$$z = \Theta u + \Psi s + v \quad (13)$$

where $v \perp [u'\ s']'$. Then any matrix L such that $u = Ls$ minimizes $E\{\| z \|^2_Q\}$ is a solution of the LS problem

$$\min_L \| \Theta L + \Psi \|^2_Q \quad (14)$$

where $\| M \|^2_Q = Tr(M'QM)$. ∎

Theorem 1 - All BAMOP solutions can be obtained by: first, introducing a *multistep predictive model* for the future joint i/o process based on the present input and pseudostate

$$[y'(t+T+1)\ u'(t+T)\ \dots\ y'(t+1)\ u'(t)]' =$$

$$\Theta(F_{t+1}^{t+T})u(t) + \Psi(F_{t+1}^{t+T})s(t) + v(t) \quad (15)$$

with

$$v(t) \perp [u'(t)\ s'(t)] \quad ; \quad (16)$$

and, second, setting L(t) equal to any matrix L that solves the LS problem

$$\min_L \| \Theta(F_{t+1}^{t+T})L + \Psi(F_{t+1}^{t+T}) \|^2_{Q_{T+1}} \quad (17)$$

∎

Eqs. (15) and (17) make up one of the key results of the present paper. If a plant stochastic representation such as (1) is known and $s(t) = \chi(t)$, the matrices Θ and Ψ in the multistep predictive model (15) can be computed. Further if $T=\infty$, $F_{t+1}^{\infty} = \vec{F}_i$, F_{i+1} denotes the solution L of (17), and (15) and (16) are iterated, according to Proposition 1 the corresponding sequence $\{F_i\}_{i=1}^{\infty}$ of BAMOP solutions coincides, under the condition that $A + BF_1A$ is a stability matrix, with the one generated by the Newton-Raphson iterations for computing F_o. In this case BAMOP iterations make up just an alternative algorithm for implicitly solving the algebraic Riccati equation associated with the steady-state LQG problem (1) - (2) and computing F_o. Under such circumstances, BAMOP iterations would not be of great practical interest due to the computational burden required to get $\Theta(\vec{F}_{i+1})$ and $\Psi(\vec{F}_{i+1})$ at each iteration step.

On the contrary BAMOP iterations become more interesting as far as their use in practical adaptive regulation algorithms is concerned. In fact, if the plant to be regulated in unknown, one may try to adopt a *certainty equivalence control*. By this we mean that the matrices Θ and Ψ in (17) are replaced by their LS estimates $\hat{\Theta}(t+T+1)$ and $\hat{\Psi}(t+T+1)$ based on all the observed i/o data up to time t+T+1. It is to be underlined that, whatever the pseudostate s(t) might be, the LS estimates of Θ and Ψ must be referred to the multistep predictive model (15). This model is certainly related to more conventional plant stochastic models, e.g. (1) or vector difference stochastic equations such as

$$y(t+1) = B_1u(t) + \Sigma\sigma(t) + \Omega\varepsilon(t) + e(t+1) \tag{18}$$

where

$$\varepsilon'(t) = [e'(t)e'(t-1) \; \dots \; e'(t-n+1)]$$

and Ω is a diagonal p×p matrix. On the other hand, if in BAMOP we set T=0 and $s(t)=\sigma(t)$, and consequently (15) becomes

$$y(t+1) = \Theta u(t) + \Psi\sigma(t) + v(t) \tag{19}$$

consistent estimates of the matrices Θ and Ψ corresponding to a constant feedback will not coincide, unless $\Omega=0$, with the matrices B_1 and Σ in the model (18). Actually, correct identification of the predictive model (19) instead of (18) is what makes the feedback-gain matrix updating (17) trivial from a computational point of view. This fact has become well known among control engineers thanks to the work mainly carried out by the Swedish School of adaptive control. The basic idea of Åström and Wittenmark's LS+MV self-tuning regulator is to rewrite the process model in such a way that the feedback-gain matrix updating step is trivial. In the light of the results of this paper, one may regard the LS+MV as a prototype of adaptive control algorithms of this type, whereas BAMOP iterations can be interpreted as a generalization of the same basic concept to more general adaptive regulation problems.

At this point we are ready to introduce the MUSMAR *algorithm*:

1 - Given all i/o data up to time t+T+1, i.e. $\{y^{t+T+1}, u^{t+T}\}$, compute the LS estimates $\hat{\Theta}(t+T+1)$ and $\hat{\Psi}(t+T+1)$ of the matrices Θ and Ψ in the multistep prediction model

$$[y'(k+T+1)\ u'(k+T)\ \dots\ y'(k+1)\ u'(k)]' = \\ = \Theta u(k) + \Psi\sigma(k) + v(k) \qquad k=t,\ t-1,\ \dots \tag{20}$$

2 - Solve the LS problem

$$\min_{L(t)} \|\hat{\Theta}(t+T+1)L(t) + \hat{\Psi}(t+T+1)\|^2_{Q_{T+1}} \tag{21}$$

3 - Set

$$F(t+T+1) = L(t) \tag{22}$$

and feed the plant at the time t+T+1 by

$$u(t+T+1) = L(t)\sigma(t+T+1) + n(t+T+1) \tag{23}$$

where $n(k)$ is a low-intensity zero-mean white noise independent of the plant, superimposed to the feedback control component so as to make the LS estimation problem of Θ and Ψ nonsingular.

Once again, it is to be remarked that in the MUSMAR algoritm by a proper choice of model structure the regulator parameters are updated almost directly and that cumbersome feedback-gain matrix calculations are avoided. Algorithms of this type are called [22] algorithms based on *implicit identification* of a process model in contrast to adaptive control algorithms based on explicit identification. While algorithms relying on stochastic control theory and based on explicit identification, e.g. the ones carrying out Maximum-Likelihood identification and updating the feedback-gain matrix according to the steady-state solution of a Riccati equation, have been considered for a long time, corresponding implicit algorithms make up still an open problem. We notice that the MUSMAR algorithm is one of the first answers to the latter problem.

Next theorem, whose proof is given in the Appendix, deals with the consistency property of the LS estimates $\hat{\Theta}$ and $\hat{\Psi}$, under the assumption of a constant feedback law.

<u>Theorem</u> 2 - Let the inputs $u(k)$ of the unknown plant (1) be given by

$$u(k) = Fs(k) + n(k) \quad , \quad k = t,\ t+1,\ \dots,\ t+N+1$$

where: $n(k)$ is a stationary zero-mean white noise with intensity $V = E[n(k)\ n'(k)] > 0$, and F makes the closed loop system asymptotically stable, i.e., the plant i/o joint process is stationary. Thus, the LS estimates $\hat{\Theta}(F)$ and $\hat{\Psi}(F)$ of respectively $\Theta(F)$ and $\Psi(F)$ in the multistep prediction model (15) are weakly consistent, i.e., for $N \to \infty$,

$$[\hat{\Theta}(\vec{F})\ \hat{\Psi}(\vec{F})] \xrightarrow{\text{prob}} [\Theta(\vec{F})\ \Psi(\vec{F})]\ . \qquad \blacksquare$$

Next theorem, whose proof is given in the Appendix shows that the BAMOP iterations using the pseudostate $\sigma(t)$ instead of $\chi(t)$, admit F_o as an equilibrium point. This is obviously, in addition to the result of Theorem 2, one of the properties that induced us to test by simulations the MUSMAR behavior in more complex cases, and to undertake the convergence analysis of Sect.5.

Theorem 3 - Let us assume that the BAMOP iterations are carried out for $T=\infty$, $F^{\infty}_{t+1} = \vec{F}_i$ and $s(t)=\sigma(t)$. Let $[\hat{\Theta}(\vec{F}_i)\ \hat{\Psi}(\vec{F}_i)] = [\Theta(\vec{F}_i)\ \Psi(\vec{F}_i)]$. Then, the optimal steady-state LQG feedback-gain matrix F_o is an equilibrium point of the BAMOP iterations, viz.

$$\vec{F}_i = \vec{F}_o \Longrightarrow F_{i+1} = F_o\ . \qquad \blacksquare$$

4. SIMULATION RESULTS

This section presents results of simulations related to plants for which standard adaptive regulation algorithms fail to converge.

In all the figures: μ indicates the forgetting factor used in the LS estimation algorithm for Θ and Ψ, σ_n^2 denotes the intensity of the additive white noise component of the plant input; $\bar{J}$ equals the time-average of the square of the plant output; the arrows indicate the optimal values of the feedback components.

Example 1 - *Adaptive MV control of a nonpositive-real plant.*

The following plant was considered:

$$y(t+1) - 1.6\,y(t) + 0.75\,y(t-1) = u(t) + u(t-1) + 0.9\,u(t-2) + e(t+1) + 1.5\,e(t) + 0.75\,e(t-1) \qquad (24)$$

The MV (Minimum-Variance: $Q_u=0$; $Q_y=1$) controller for the above plant is given by the feedback-law $u(t) = F\sigma(t)$, $F = [\,f_1=-3.1\ \ f_2=-1\ \ f_3=0\ \ f_4=-0.9]$, $\sigma(t) = [y(t)\ u(t-1)\ y(t-1)\ u(t-2)]'$. System (24), first considered in [23] is a pathological one, in that the associated polynomial

$$C(q^{-1}) = 1 + 1.5\,q^{-1} + 0.75\,q^{-2}$$

in not positive-real and, therefore, does not satisfy a condition for the convergence of the LS+MV algorithm [3].

Fig.1a shows the time evolution of the feedback coefficients obtained by applying to the plant (24) the LS+MV algorithm, which coincides with the MUSMAR algorithm with T=0. As soon as the LS+MV feedback coefficients come close to their optimal values, they are thrown away, in full agreement with the analysis carried out in [23]. No convergence is achieved and the closed-loop behavior is rather far from the optimal one. Notice that, since the plant considered is output controllable in a single step, for any T the optimal feedback-gain matrix coincides with the MV one. However, Fig.1b indicates that the MUSMAR algorithm with T=2 achieves a much more satisfactory performance. In fact, the corresponding feedback coefficients exhibit much smoother time-evolutions and are closer to the optimal theoretical values. Finally, notice the substantial decrease (-3dB) of the $\bar{J}$ value that takes place by passing from T=0 to T=2. A performance comparable to the latter can be already achieved for T=1.

Example 2 - *Minimum Variance control of a non-minimum phase plant.* The following non-minimum phase plant [10] was simulated

$$y(t) - 0.95\, y(t-1) = u(t-2) + 2\, u(t-3) + e(t) - 0.7\, e(t-1)$$

whose optimal MV controller that makes the overall closed-loop system stable, is given by [24] $u(t) = F\sigma(t)$, $F = [-0.117 \;\; -0.75 \;\; 0 \;\; -0.246]$, $\sigma(t) = [y(t), u(t-1), y(t-1), u(t-2)]'$

In the MUSMAR algorithm, the general cost-functional (2) was adopted with $Q_y=1$ and Q_u-values equal to 0 and 10^{-4}. For all these values, if T was kept equal to 0, or 1 (notice the time-delay in the plant), computer overflows were experimented after a few steps. On the other hand, Figs.2 indicates that the MUSMAR regulated plant for T=2 and T=4 performs satisfactorily if a non zero but otherwise small value, e.g. 10^{-4}, is assigned to Q_u.

When a zero weight was imposed on the input variance, the MUSMAR regulator though making the closed-loop system stable did not tune itself close enough on the optimal feedback-gain coefficients.

Examples 1 and 2, as well as many other simulations [25], lend them selves to draw the following remarks. First, by increasing the regulation horizon T and setting $Q_u>0$, the MUSMAR tends to tune itself on the corresponding optimal LQG regulator. Under such circumstances, the MUSMAR achieves the optimal regulator performance even with nonminimum-phase plants. Second, an increase of T makes the MUSMAR algorithm stable around the MV equilibrium point even if the plant does not have a positive-real $C(q^{-1})$ polynomial. Finally, in all simulations the MUSMAR, with suitable design parameters, exhibited the self-tuning property in that it tuned itself on the corresponding optimal steady-state LQG regulator, and no different equilibrium point was experimented.

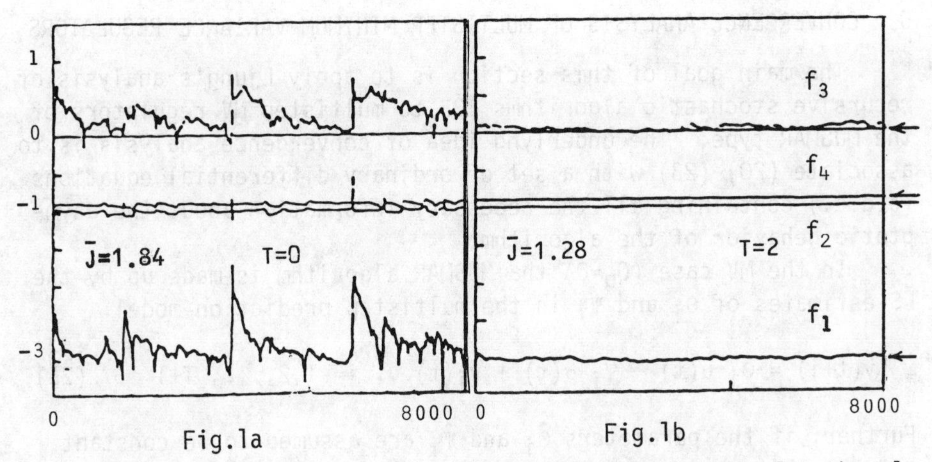

Figs. 1a,1b - Behavior of the four feedback components for the plant of Ex.1 regulated by the MUSMAR (μ=0.995; $\sigma_n^2=10^{-2}$).

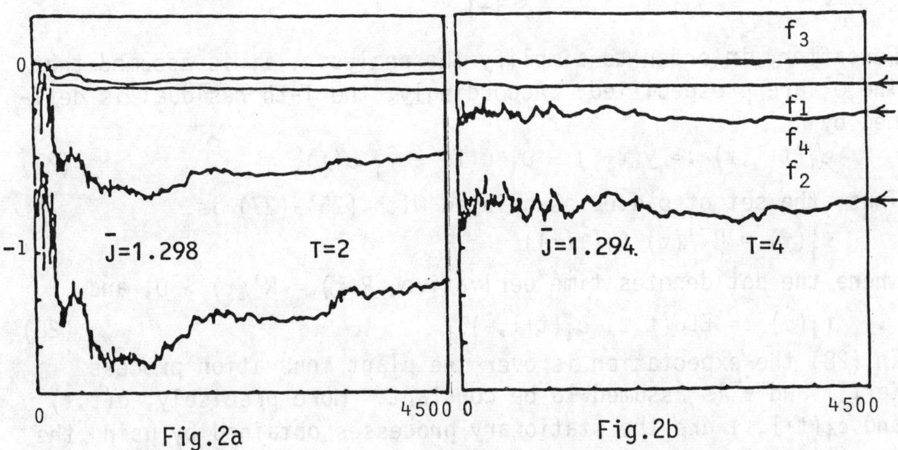

Figs. 2a,2b - Behavior of the four feedback components for the plant of Ex.2 regulated by the MUSMAR (μ=1; $\sigma_n^2=10^{-2}$).

While such an overall satisfactory performance is not unexpected in view of the strict relationship existing between the MUSMAR algorithm and the optimal LQG steady-state regulator, at a first glance the MUSMAR stabilizing action on the MV equilibrium point for nonpositive-real plants looks some what surprising. Next section clarifies the latter aspect of the MUSMAR behavior through an application of Ljung's analysis of recursive stochastic algorithms to multistep MV regulators.

5. CONVERGENCE ANALYSIS OF MULTISTEP MINIMUM VARIANCE REGULATORS

The main goal of this section is to apply Ljung's analysis of recursive stochastic algorithms [3] to multistep MV regulators of the MUSMAR type. The underlyng idea of convergence analysis is to associate (20)-(23) with a set of ordinary differential equations (o.d.e.) containing all the necessary information about the asymptotic behavior of the algorithm.

In the MV case ($Q_u=0$) tha MUSMAR algorithm is made up by the LS estimates of Θ_i and Ψ_i in the multistep prediction model

$$y(t+i) = \Theta_i\, u(t) + \Psi_i\, \sigma(t) + v_i(t) \quad , \; i= 1,2,\dots, T+1 \qquad (25)$$

Further, if the parameters $\hat{\Theta}_i$ and $\hat{\Psi}_i$ are assumed to be constant and the white noise component $n(t)$ negligeable, the plant input is given by

$$u(t) = - \left(\sum_{i=1}^{T+1} \hat{\Theta}_i^2 \right)^{-1} \left(\sum_{i=1}^{T+1} \hat{\Theta}_i\, \hat{\Psi}_i \right) \sigma(t) := F\, \sigma(t) \qquad (26)$$

Hereafter, in order to simplify the analysis, it is assumed that the $\hat{\Theta}_i$ are prespecified. Accordingly, the i-th residual is defined by

$$\varepsilon_i(t+1,\hat{\Psi}) := y(t+i) - \hat{\Theta}_i\, u(t) - \hat{\Psi}_i\, \sigma(t) \qquad (27)$$

Thus, the set of o.d.e. associated with (25)-(27) is

$$\dot{\hat{\Psi}}_i'(\tau) = R^{-1}(\tau)\, f_i(\hat{\Psi}(\tau))$$

where the dot denotes time derivative, $R(\tau) = R'(\tau) > 0$, and

$$f_i(\hat{\Psi}) := E[\sigma(t,\hat{\Psi})\, \varepsilon_i(t+1,\hat{\Psi})] \qquad (28)$$

In (28) the expectation is over the plant innovation process $\{e(t)\}$ and $\hat{\Psi}$ is assumed to be constant. More precisely, $\sigma(t,\hat{\Psi})$ and $\varepsilon_i(t+1,\hat{\Psi})$ are the stationary processes obtained by using the input (26) for the plant to be regulated under the assumption that overall closed-loop system be asymptotically stable. It is not difficult to obtain the o.d.e. for the feedback-gain vector F' by using (26)-(28)

$$\dot{F}'(\tau) = - \left(\sum_{i=1}^{T+1} \hat{\Theta}_i^2 \right)^{-1} \left(\sum_{i=1}^{T+1} \hat{\Theta}_i\, \dot{\hat{\Psi}}_i'(\tau) \right) \qquad (29)$$

$$= R^{-1}(\tau)(b_1/\hat{\Theta}_1) \left(\sum_{i=1}^{T+1} \gamma_i^2 \right)^{-1} E[\sigma(\tau,\hat{\Psi}) H(q^{-1}) \sigma'(\tau,\hat{\Psi})](F_{MV}' - F'(\tau))$$

where:

$$H(q^{-1}) := \left\{ \sum_{i=1}^{T+1} \gamma_i q^{i-1} \right\} / C(q^{-1}) \quad ; \tag{30}$$

$$\gamma_i := \hat{\theta}_i / \hat{\theta}_1 \quad ;$$

$C(q^{-1})$ is the polynomial appearing in the ARMAX model of the plant

$$A(q^{-1})y(t) = B(q^{-1})u(t-1) + C(q^{-1})e(t) \quad ; \tag{31}$$

b_1 is the coefficient of $u(t-1)$ in (31); and F_{MV} is the corresponding MV feedback-gain vector.

A fundamental result of [3] rephrased for the present context is that only stable, stationary points of (29) are possible convergence points of the multistep MV regulator.

We see that $F = F_{MV}$ is a stationary point of (29). Further, if $H(q^{-1})$ is strictly positive real, $F = F_{MV}$ is a stable stationary point of (29). In the MUSMAR algorithm for $F = F_{MV}$, $\hat{\theta}_1 = b_1$, the γ_i in (30) turn ont to be the samples of the impulse response associated with the transfer function $C^{-1}(q^{-1})$

$$\frac{1}{C(q^{-1})} = \sum_{i=1}^{\infty} \gamma_i q^{-i+1} \tag{32}$$

Therefore for T large enough

$$H(q^{-1}) \cong \frac{1}{C(q)C(q^{-1})} \tag{33}$$

which is always strictly positive real. This result explains, in full agreement with the simulations of Example 1 in Sect.4, why an increase of T tends to make the MUSMAR algorithm stable around the MV equilibrium point, whatever the $C(q^{-1})$ plant polynomial might be.

6. CONCLUSIONS

The MUSMAR algorithm appears to be strictly related to LQG theory. One of its features is that it can be implemented by solving two LS problems: the first one consists of identifying a closed-loop multistep prediction model; the second one updates the feedback-gain matrix. This circumvents the drawback of requiring the solution of an associated algebraic Riccati equation at each regulation step and an explicit identification of the plant.

Simulation results show that the MUSMAR algorithm yields a satisfactory and definitely better performance over more standard adaptive regulators when the plant to be regulated does not satisfy conditions of minimum-phase, positive-realmess, and in the presence of uncertainty on the plant i/o transport delay. Since, when the plant is time-varying, these conditions cannot be a priori guaranteed, the MUSMAR approach can provide a robust solution to the adaptive regulation problem of unknown plants.

In the multivatiable case many examples can be considered for which convenzional multivariable adaptive regulators are inadequate, whereas the MUSMAR can yield a satisfactory performance.

Further studies should be carried out in order to analyze the MUSMAR behavior similarly to what has been recently done for other classes of adaptive regulators.

APPENDIX

Proof of Lemma - Under the stated assumption we have

$$E[\| z \|_Q^2] = E\{E[\| z \|_Q^2 | s]\} = E[\| \Theta u + \Psi s \|_Q^2] + Tr(QE[vv'])$$

where we have taken into account that $u \varepsilon [s]$ and $v \perp [s]$. Since $E[vv']$ does not depend on L, the quantity to be minimized is

$$E[\| (\Theta L + \Psi)s \|_Q^2] = Tr[(\Theta L + \Psi)'Q(\Theta L + \Psi)S] \tag{34}$$

where $S = E[ss']$. In turn, this can be replaced by $\| \Theta L + \Psi \|_Q^2$, in that every L minimizing $\| \Theta L + \Psi \|_Q^2$, minimizes also (34). The converse is true only if S is positive definite.

Proof of Theorem 2 - Let z(k) denote the j-th component of the observed vector on the L.H.S. of (20). Thus

$$z(k) = \rho' g(k) + v_j(k) = g'(k)\rho + v_j(k)$$

where: ρ' is the j-th row of $[\Theta(\vec{F})\ \Psi(\vec{F})]$; $g(k) = [u'(k)s'(k)]'$; and $v_j(k)$ is the j-th component of v(k). The whole set of N equations in matrix form is

$$Z = G\rho + V$$

with obvious identification for Z, G and V. The LS estimate $\hat{\rho}$ of ρ based on N observations satisfies the equation

$$\frac{G'G}{N}\hat{\rho} = \frac{G'Z}{N} \quad .$$

But $G'Z = G'G\rho + G'V$. Now

$$\frac{G'V}{N} = \frac{1}{N}\begin{vmatrix} \sum_{k=t}^{t+N-1} u(k)v_j(k) \\ \sum_{k=t}^{t+N-1} s(k)v_j(k) \end{vmatrix} \xrightarrow{\text{prob.}} \begin{vmatrix} E[u(k)v_j(k)] \\ E[s(k)v_j(k)] \end{vmatrix} = 0$$

where the last equality holds true by virtue of (16). Summing up, as $N\to\infty$ $\frac{G'G}{N}\hat{\rho} \xrightarrow{\text{prob.}} \frac{G'G}{N}\rho$, which means that any LS estimate $\hat{\rho}$ is asymptotically equivalent to the desired ρ. Obviously, uniqueness follows the asymptotic nonsingularity of $\frac{G'G}{N}$ which, in turn, holds true iff $E[s(k)s'(k)] > 0$.

Proof of Theorem 3 - Since $\vec{F} = F_o$, according to the comments after (7) one has

$$u_o(k) = F_o\,\sigma_o(k) = \Xi_o\,\chi_o(t), \qquad \forall k \in Z$$

where $\sigma_o(k)$ and $\chi_o(k)$ are the stationary processes obtained by using the optimal input $u_o(k)$. In addition, being by assumption $[\hat{\Theta}(F_o)\;\hat{\Psi}(F_o)] = [\Theta(F_o)\;\Psi(F_o)]$, the next BAMOP iteration amounts to solving the following minimization problem for any T ($L = F_{i+1}$)

$$\min_L \{(T+1)^{-1}E[\|x(t+1)\|^2_{P(\Xi_o)} + \|L\,\sigma_o(t)\|^2_{Q_u}]\}$$

$$x(t+1) = \chi_o(t) + BL\,\sigma_o(t)$$

where the quadratic term in $x(t+1)$ stands for the cost

$$J(t+1,t+T) + \|y(t+1)\|^2_{Q_y}$$

of the deterministic version of the LQG problem (1),(2) under the assumption that $u_o(k) = \Xi_o\,\chi_o(t)$.

At this point it is convenient to use the identity

$$E[-] = E\{E[-|\sigma_o(t)]\} \; .$$

$$E[\| x(t+1)\|^2_{P(\Xi_0)} + \| L\sigma_0(t)\|^2_{Q_u} \mid \sigma_0(t)] =$$

$$= E[\| x(t+1)\|^2_{P(\Xi_0)} \mid \sigma_0(t)] + \| L\sigma_0(t)\|^2_{Q_u}$$

Further,

$$E[\| x\|^2_P \mid \sigma_0] = \| \hat{x}\|^2_P + \text{Trace } (PR_{\tilde{x}})$$

where:

$$\hat{x}(t+1) = E[x(t+1)\mid\sigma_0(t)]$$

$$= \hat{x}_0(t) + BL\ \sigma_0(t);$$

$$\tilde{x}(t+1) = x(t+1) - \hat{x}(t+1); \text{ and}$$

$$R_{\tilde{x}} = E[\tilde{x}(t+1)\ \tilde{x}'(t+1)].$$

does not depend on L. Now

$$\hat{X}_0(t) = \Gamma_0\sigma_0(t)$$

with

$$\Gamma_0 R_{\sigma_0} = R_{X_0\sigma_0} = E[X_0(t)\sigma_0'(t)] .$$

Thus, the quantity to be minimized is

$$\min_L\{(T+1)^{-1}E[\sigma_0'(t)[(\Gamma_0+BL)'P(\Xi_0)(\Gamma_0+BL) + L'Q_uL]\sigma_0(t)]\}$$

For $T \to \infty$, this yields

$$L = - (Q_u + B'P(\Xi_0)B)^{-1}B'P(\Xi_0)\Gamma_0$$

$$= \Xi_0\Gamma_0$$

where the latter equality follows by the definition of Ξ_0 which for $T=\infty$ is a stationary point of the Newton-Raphson iterations. Thus

$$L\ \sigma_0(t) = \Xi_0\Gamma_0\sigma_0(t)$$

$$= \Xi_0 E[X_0(t)\mid\sigma_0(t)] \qquad [\Gamma_0\sigma_0(t) = \hat{X}_0(t)]$$

$$= E[F_0\sigma_0(t)\mid\sigma_0(t)] \qquad [\Xi_0 X_0(t) = F_0\sigma_0(t)]$$

$$= F_0\sigma_0(t).$$

REFERENCES

[1] K.J.Åström, and B.Wittenmark, *On self-tuning regulator*, Automatica, 9, pp.185-199, 1973.

[2] I.D.Landau, *Adaptive control. The model reference approach.* Dekker, New York, 1979.

[3] L.Ljung, *On positive real transfer function and the convergence of some recursive schemes.* IEEE Trans. Automat. Contr., 22, pp.539-551, Aug.1977.

[4] A.S.Morse, *Global stability of parameter adaptive control systems*, IEEE Trans. Automat. Contr., 25, pp.433-439, June 1980.

[5] K.S.Narendra, Y.H.Lin, and L.S.Valavani, *Stable adaptive controller design - Part II: Proof of Stability*, IEEE Trans. Automat. Contr., 25, pp.440-448, June 1980.

[6] G.C.Goodwin, P.J.Ramadge, and P.E.Caines, *Discrete-time multivariable adaptive control*, IEEE Trans. Automat. Contr., 25, pp.449-456, June 1980.

[7] G.C.Goodwin, K.S.Sin, and K.K.Saluja, *Stochastic adaptive control and prediction - The general delay-colored noise case*, IEEE Trans. Automat. Contr., 25, pp.946-950, Oct.1980.

[8] G.C.Goodwin, C.R.Johnson, and K.S.Sin, *Global convergence for adaptive one-step-ahead optimal controllers based on input matching*, IEEE Trans. Automat. Contr., 26, pp.1269-1273, Dec.1981.

[9] K.J.Åström, *Introduction to stochastic control theory*, Academin-Press, New York, 1970.

[10] K.J.Åström, U.Borisson, L.Ljung, and B.Witternmark, *Theory and applications of self-tuning regulators*, Automatica, 13, pp. 457-476, 1977.

[11] U.Borisson, *Self-tuning regulators for a class of multivariable systems*, Preprints 4th IFAC Symp. Identification and System Parameter Estimation, pp.420-429, Tbilisi, 1976.

[12] L.Keviczky, J.Hetthessy, M.Hilger, and J.Kolostory, *Self-tuning adaptive control of cement raw material blending*, Automatica, 14, pp.525-532, 1978.

[13] N.H.Koivo, A *multivariable self-tuning controller*, Automatica, 16, pp.351-356, 1980.

[14] G.Kreisselmeier, *On adaptive state regulation*, IEEE Trans. Automat. Contr., 27, pp.3-17, Feb. 1982.

[15] G.Menga, and E.Mosca, *MUSMAR: multivariable adaptive regulators based on multistep cost functionals*, Advances in Control, D.G. Lainiotis and N.S.Tzannes (Eds.), pp.334-341, D.Reidel Publishing Company, Dordrecht, 1980.

[16] E.Mosca, and G.Zappa, *MUSMAR: basic convergence and consistency properties*, Lecture Notes in Control, 28, pp.189-199, Springer-Verlag, 1980.

[17] C.Manfredi, E.Mosca, and G.Zappa, *Multivariable adaptive prediction by MUSMAR Techniques*, Proc. 4th IASTED Symp. on Measurement and Control, pp.161-164, Cairo, Sept. 1981.

[18] E.Mosca, and G.Zappa, *A robust adaptive control algorithm for microcomputer based implementation*, 2nd IASTED Int. Symp. on Ident. and Contr. Davos, March 1982.

[19] E.Mosca, and G.Zappa, *Consistency conditions for the asymptotic innovations representation and an equivalent inverse regulation ptoblem*, IEEE Trans. Automat. Contr., 24, pp.501-503, June 1979.

[20] D.L.Kleinman, *On an iterative technique for Riccati equation computation*, IEEE Trans. Autom. Contr., 13, pp.114-115, Feb.1968.

[21] G.A.Hewer, *An iterative technique for the computation of the steady state gains for the discrete time optimal regulator*, IEEE Trans. Automat. Contr., 16, pp.382-384, Aug. 1971.

[22] K.J.Åström, *Self-tuning regulators - Design principles and applications*, in Applied Adaptive Control, Narendra and Monopoli (Eds.), Academic Press, 1980.

[23] L.Ljung, and B.Wittenmark, *Asymptotic properties of self-tuning regulators*, Rept. 7404, Lund Inst. of Technology, Lund, Sweden, 1974.

[24] V.Kucera, *Discrete linear control - The polynomial equation approach*, J.Wiley & Sons, Chichester, 1979.

[25] E.Mosca, G.Zappa, L.Menichetti, and C.Manfredi, *The MUSMAR approach to regulating unknown multivariable plants I: Simulation studies on nonpositive-real plants*, ISIS TR-4/81, Univ. of Florence, Florence, Italy.

OVERPARAMETERIZATION, POSITIVE REALNESS AND MULTISTEP MINIMUM VARIANCE ADAPTIVE REGULATORS°

Edoardo Mosca and Giovanni Zappa

ISIS
Università di Firenze
Via S. Marta 3, 50139 Firenze, Italy

ABSTRACT - The convergence properties of a new class of adaptive regulators based on multistep quadratic cost functionalsare analyzed emplcyingthe O.D.E. method introduced by L.Lyung. It is shown that, even if the plant does not satisfy a positive realness condition, local convergence can be achieved by extending the time horizon of the controller. Moreover a comparison is carried out with the class of adaptive minimum variance regulators which overcome the positive realness condition by overparameterization of the plant model.

1. INTRODUCTION

In the last few years considerable attention has been devoted in the literature to convergence properties of recursive stochastic algorithms. Despite the different mathematical tools employed in the analysis (e.g.: hyperstability theory [1], martingale convergence theorems [2], stability of ordinary differential equations [3]), in most cases necessary and/or sufficient conditions for convergence are expressed in terms of the positive realness of an appropriate transfer function depending on the statistics of the plant noise.

°This work was supported in part by the Italian CNR under contracts 81.00876.07 and 81.01570.07.

R. S. Bucy and J. M. F. Moura (eds.), Nonlinear Stochastic Problems, 205–215.

In this context, positive realness can be considered as a measure of the distance between the noise acting on the plant and the white noise, since, if the noise is "white" enough, convergence is guaranteed. Obviously, if the noise spectral density is known, then by a suitable prefiltering of input and output processes, we can limit ourselves to consider only plants with white disturbances. Conversely, if the noise spectral density is unknown, we can modify the recursive algorithm by including noise model parameters in the overall parameter vector to be estimated. However this leads to recursive algorithms much more complicated, and, in most cases, of difficult analysis.

Recently in [5] it has been pointed out that for the LS+MV regulator the positive realness condition can be removed by simply overparameterizing the model of the plant to be controlled, and hence by increasing the regulator complexity. Intuitively, extending the number of parameters to be estimated and hence the number of input-output pairs included in the "state" of the model, reduces noise influence. The aim of this paper is to outline the connections existing for minimum variance (MV) adaptive regulators between the removal of positive realness condition by overparameterization and the convergence properties od a class of adaptive regulators (MUSMAR), recently introduced in the literature (see [6] for a comprehensive survey). These regulators, based on a multistep quadratic cost functional, require the identification of several predictive models, one for each sample of the output variables appearing in the cost functional. Therefore also in this case the plant is overparameterized since the number of estimated parameters exceeds the minimum number of parameters necessary to compute the feedback gain vector. Following Ljung's technique [3],[4], a convergence analysis is carried out on the basis of the stability properties of an ordinary differential equation (O.D.E.). Despite some simplifying assumptions introduced so as to keep the algorithms of interest amenable to analysis, the results achieved turn out to be in good agreement with simulation experiments. In any event, it is believed that they provide a guideline to explain the major effects produced by adaptive regulation algorithms based on multistep cost functionals.

The paper is organized as follows. In Sect. 2 theorems concerning the convergence of overparmeterized LS+MV regulators are reported. In Sect. 3 multistep MV adaptive regulators are analyzed and a theorem on local convergence is proved. Finally, in Sect. 4, the two approaches are compared on the basis of simulation results and the corresponding numerical complexity is considered. Some concluding remarks end the paper.

2. LOCAL CONVERGENCE OF LS+MV REGULATORS

Hereafter we shall consider the MV adaptive control problem of the following SISO unknown plant:

$$A(q^{-1})y(t) = B(q^{-1})u(t) + C(q^{-1})e(t) \tag{1}$$

where

$$A(q^{-1}) = 1 + a_1 q^{-1} + \dots + a_{\bar{n}} q^{-\bar{n}}$$
$$B(q^{-1}) = b_1 q^{-1} + \dots + b_{\bar{n}} q^{-\bar{n}}$$
$$C(q^{-1}) = 1 + c_1 q^{-1} + \dots + c_{\bar{n}} q^{-\bar{n}}$$

$y(t)$ is the output, $u(t)$ the input and $e(t)$ is a stationary sequence of independent random variables such that all moments exist. Moreover it is assumed that the polynomials A,B and C are relatively prime and that B and C are stable, that is they have all their zeroes outsidethe unit circle.

The LS+MV regulator is based on the following one-step ahead predictive model for the output [7]:

$$y(t+1) = \Theta\, u(t) + \Psi\sigma(t) + v(t+1) \tag{2}$$

where

$$\sigma(t)' = [-y(t) \dots -y(t-\bar{n}+1),\; u(t-1) \dots u(t-\bar{n}+1)] \tag{3}$$

and $v(t+1)$ is a prediction residual uncorrelated to $\sigma(t)$ and $u(t)$. While Θ is a priori fixed, the vector Ψ is recursively estimated via a recursive least-square (LS) algorithm

$$\Psi'(t+1) = \Psi'(t) + \frac{1}{t+1}\,\frac{R(t)^{-1}}{1+\frac{1}{t+1}[\sigma' R(t)^{-1}\sigma(t) - 1]}\,\sigma(t)y(t+1) \tag{4a}$$

$$R(t+1) = R(t) + \frac{1}{t+1}[\sigma(t+1)\sigma(t+1)' - R(t)] \tag{4b}$$

and the control $u(t)$ is chosen as

$$u(t) = - \frac{\Psi(t)}{\Theta} \sigma(t) \tag{5}$$

The o.D.E. method, introduced in [3] , gives conditions for convergence. In particular the parameter vector

$$\Psi_{MV} \triangleq \frac{\Theta}{b_1} [a_1-c_1, \ldots ,a_{\bar{n}}-c_{\bar{n}}, b_2, \ldots ,b_{\bar{n}}]$$

corresponding to a MV control law, is the unique stationary point of the O.D.E. associated to the recursive scheme (1),(4),(5), and consequently the only possible convergence point, if the transfer function $H(q^{-1}) = 1 / C(q^{-1})$ is strictly positive real (s.p.r.).

Conversely, if the i/o representation (1) is not minimal, or equivalently, if the predictive model (2) has order $n > \bar{n}$, then all the stationary points of the O.D.E. constitute a hypersurface S in the parameter space. S is parameterized by a vector $\gamma \triangleq [\gamma_1, \ldots , \gamma_m]'$, $m = n-\bar{n}$, with the constraint that the polynomial $C_\gamma(z) \triangleq 1 + \sum_{i=1}^{m} \gamma_i z^i$ is stable. Moreover it can be proved that if $\Psi(t)$ converges and m is such that there exists a stable C_γ that makes the transfer function $1 / C(q^{-1})C_\gamma(q^{-1})$ s.p.r., then the LS+MV adaptive regulator asymptotically performs as the MV regulator, i.e. $y(t) = e(t)$. Now in [5] it was proved that for every stable polynomial C of order n, there exists an integer m such that $1 / C(q^{-1})C_\gamma(q^{-1})$ is s.p.r. for some polynomial C_γ of order greater or equal to m. Therefore it was concluded that the LS+MV regulator locally converges to the MV controller, no matter what $C(q^{-1})$ might be, provided that the predictive model is sufficiently overparameterized ($n \geq \bar{n}+m$). Notice that, in this case, convergence means that the variance of the output of the controlled plant asymptotically equals that of $e(t)$.

3. MULTISTEP MINIMUM VARIANCE REGULATORS

Given the plant described by (1), multistep MV regulators of MUSMAR type are based on several predictive models

$$y(t+i) = \Theta_i u(t) + \Psi_i \sigma(t) + v_i(t+i) \quad , \quad i = 1, \dots , T+1 \qquad (6)$$

where $\sigma(t)$ is given by (3), $v_i(t+i)$ are prediction residuals uncorrelated to the observation vector $\eta(t) \triangleq [u(t),\sigma(t)']'$, and T is a parameter which denotes the time-horizon of the controller. The parameters Θ_i and Ψ_i are now both recursively estimated. Denoting by $\Phi_i(t) \triangleq [\Theta_i(t),\Psi_i(t)]$ the estimates based upon observations up to time t+i, we have the following updating recursions:

$$\Phi_i(t+1)' = \Phi_i(t)' + \frac{1}{t+1} \frac{R(t)^{-1}}{1+ \frac{1}{t+1} [\eta(t)'R(t)^{-1}\eta(t) - 1]} \eta(t)\varepsilon_i(t+i) \qquad (7a)$$

$$\varepsilon_i(t+i) = y(t+i) - \Phi_i(t)\, \eta(t) \qquad (7b)$$

$$R(t+1) = R(t) + \frac{1}{t+1}[\eta(t+1)\, \eta(t+1)' - R(t)] \quad . \qquad (7c)$$

The feedback part of the control is given by

$$u_F(t) = -[\sum_{i=1}^{T+1} \Theta_i(t)^2]^{-1}[\sum_{i=1}^{T+1} \Theta_i(t)\Psi_i(t)]\sigma(t) \triangleq F(t)\sigma(t) \qquad (8)$$

Moreover, in order to make the LS identification problem non singular in the steady state, a low-intensity zero-mean white noise n(t) is superimposed to $u_F(t)$, so that

$$u(t) = u_F(t) + n(t) \qquad (9)$$

Notice that the presence of a white noise component in the control is a realistic assumption since, in many practical situations, the control actuators introduce additional low-intensity noise into the plant.

Due to the white excitation on the control signal (9), the O.D.E. associated to (1),(7),(8), has a stationary point (Φ°,R°) corresponding to

$$F^\circ = F_{MV} \quad , \quad \Theta_i^\circ = b_1 \gamma_i \qquad (10)$$

where γ_i are the samples of the impulse response associated to the transfer function $1/C(q^{-1})$

$$1/C(q^{-1}) = \sum_{i=1}^{\infty} \gamma_i \, q^{-i+1} \quad . \tag{11}$$

Let us now investigate the nature of the MV stationary solution of the O.D.E. In order to perform this analysis, we shall assume that in a neighbourhood of (Φ°, R°), the Θ_i parameters are almost constant. This assumption allows us to simplify the differential equation involving the time evolution of the feedback-gain vector $F(\tau)$, which becomes [6] :

$$\dot{F}(\tau)' = -\left[\sum_{i=1}^{T+1} \Theta_i^{\circ 2}\right]^{-1} \left[\sum_{i=1}^{T+1} \Theta_i^\circ \, \dot{\Psi}_i(\tau)'\right] \tag{12a}$$

$$= R_\sigma(\tau)^{-1} \left[\sum_{i=1}^{T+1} \gamma_i^2\right]^{-1} E[\sigma(t,\Psi)H(q^{-1})\sigma(t,\Psi)'][F'_{MV} - F(\tau)']$$

$$\dot{R}_\sigma(\tau) = E[\sigma(t)\sigma(t)'] - R_\sigma(\tau) \tag{12b}$$

where

$$H(q^{-1}) = \sum_{i=1}^{T+1} \gamma_i q^{i-1} \; / \; C(q^{-1}) \tag{13}$$

dot denotes derivative with respect to τ , and expectation is with respect to the distribution of e(t), assuming that all stochastic processes appearing within expectation brackets are stationary. From Ljung results [4], F_{MV} is a possible convergence point for (12a) only if $H(q^{-1})$ is s.p.r.. In order to test the real positiveness of H, we introduce the following

<u>Lemma</u> - Let D(q) be a n-th degree stable monic polynomial and let

$$1 \; / \; D(q) = \sum_{i=1}^{\infty} \delta_i(D) q^i \tag{14}$$

Then

$$|\delta_i(D)| \le \frac{(n+i+1)!}{(n-1)!.i!} \rho(D)^{-i} \tag{15}$$

where $\rho(D)$ denotes the minimum modulus of the zeroes of D.

Proof. First assume that $D(q) = (1 - \frac{q}{\rho})^n$, $\rho > 1$. Then, from Newton's binomial expansion

$$|\delta_i(D)| = \frac{(n+i+1)!}{i!\,(n-1)!} \rho^{-i}$$

Since, for any other polynomial $\bar{D}(q)$ such that $\rho(\bar{D}) = \rho(D)$, we have $|\delta_i(\bar{D})| \le |\delta_i(D)|$, i=1,2, ... , the lemma is proved. ∎

Now we can prove the following

Theorem - Given the plant representation (1), then the transfer function (13) appearing in the O.D.E. associated to (7),(8),(9), is s.p.r., if the time horizon T is such that

$$\rho(C)^{-T} \le \frac{\rho(C)^{-n-2}}{n \sum_i |c_i|} \cdot \frac{(T+2-n)!\,(n-1)!}{(T+3)!} \tag{16}$$

Proof.

$$H(q^{-1}) = \frac{\sum_{i=1}^{T+1} \gamma_i q^{i-1}}{C(q^{-1})} = \frac{(\sum_{i=1}^{T+1} \gamma_i q^{i-1})\, C(q)}{C(q^{-1})\, C(q)} =$$

$$= \frac{1 - q^T \sum_{i=1}^{n} \sum_{j=i}^{n} [c_j \gamma_{T+1+i-j}]\, q^i}{C(q^{-1})\, C(q)}$$

where the last equality follows from (11). Since the function $C(e^{i\omega})\, C(e^{-i\omega})$ is real and non negative for $\omega \in [-\pi,\pi]$, $H(q^{-1})$ is s.p.r. if

$$|\sum_{i=1}^{n} \sum_{j=i}^{n} c_j \gamma_{T+1+i-j}| \le \sum_{i=1}^{n} \sum_{j=1}^{n} |c_j||\gamma_{T+1+i-j}| < 1 \tag{17}$$

Now, from (16), taking into account (15), (17) follows and the theorem is proved. ∎

Notice that, since $\frac{(T+2-n)!(n-1)!}{(T+3)!} < 1$ and, by assumption $\rho(C) > 1$, (16) can be always satisfied by a time-horizon T sufficiently high. Therefore the previous analysis can be considered as a first rational explanation of simulation experiments [6] : if the time-horizon T of MUSMAR MV regulator is increased, then the time-evolution of the coefficients of the gain vector F becomes smoother, till approximate convergence is achieved.

4. SIMULATION RESULTS

In order to compare the two algorithms discussed in the previous sections, we shall present simulation results concerning the MV adaptive control of the following well known "pathological" system

$$y(t) - 1.6y(t-1) + 0.75y(t-2) = u(t) + u(t-1) + 0.9u(t-2)$$
$$+ e(t) + 1.5e(t-1) + 0.75e(t-2)$$

The polynomial $C(q^{-1})$ associated to this system has a negative real part for $q = \exp\{ i\omega\}$, $\omega \in [1.78 , 2.48]$. The real part of the transfer function $H(q^{-1}) = (\sum_{i=1}^{T+1} \gamma_i q^{i-1}) / C(q^{-1})$ has been plotted for T = 0,2,4. in Fig. 1. Notice that the behavior of this function agrees with the results of simulation experiments reported in [6] . As T increases, the real part of $H(q^{-1})$ becomes positive and the coefficients of the corresponding adaptive feedback vector F(t) exihibitsa smoother time-evolution. Conversely the behavior of overparameterized LS+MV regulator is reported in Fig. 2, which shows the time-evolution of the coefficients f_i, $i = 1, .. ,6$, of the feedback vector F(t), $u(t) = F(t)\sigma(t)$, $\sigma(t) = [y(t), y(t-1), y(t-2), u(t-1), u(t-2), u(t-3)]'$. From a comparison with the standard LS+MV regulator, we notice that the gain of u(t) in the feedback vector, that is f_4, has now a smoother behavior, due to the presence of the additional gains f_3 anf f_6.

On the other hand, overparameterization induces a lack of identifiability for the feedback coefficients as can be deduced from the fact that f_2, f_4 , f_5 vary in almost the same way.

Finally we would underline that MUSMAR numerical complexity does not substantially exceed that of the standard LS+MV regulator, despite that in the MUSMAR several predictive models are identified. This is due essentially to the fact that the recursion (7c), appearing in the LS recursive algorithm, is the same for all predictive models. Notice that (7c) is computationally heavier than (7a) and (7b). In Tab. 1 are reported [8] the number of sums and products required in a Square Root implementation of the MUSMAR algorithm. It is apparent that, in order to limit the numerical complexity, it is better to increase the number of predictive models (T>0), than the dimension of each predictive model ($n>\bar{n}$).

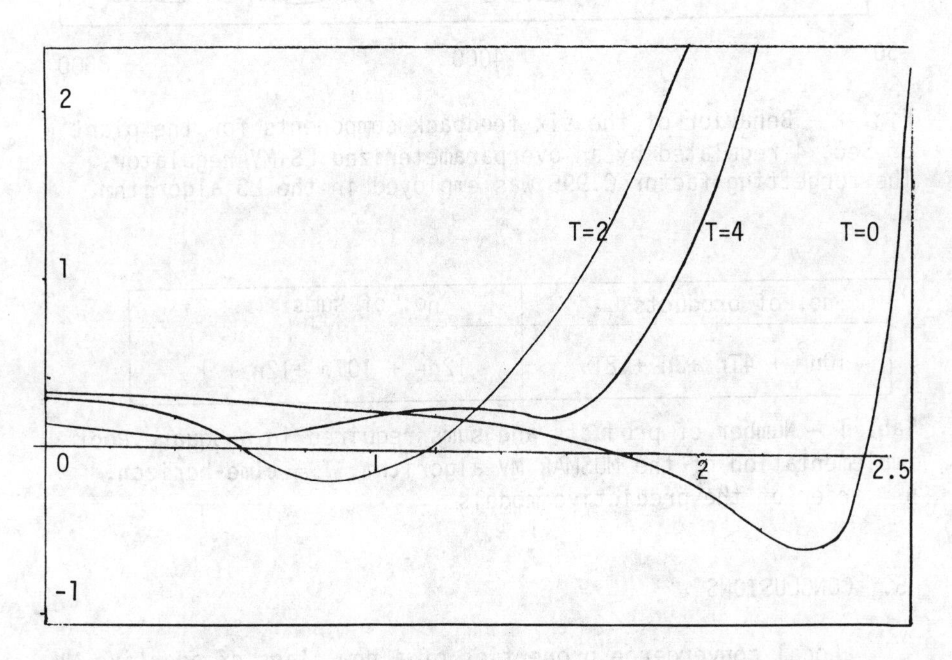

Fig. 1 - Real part of the function $H(e^{-i\omega})$,$\omega \in [0 , 2.5]$ for T = 0, 2, 4.

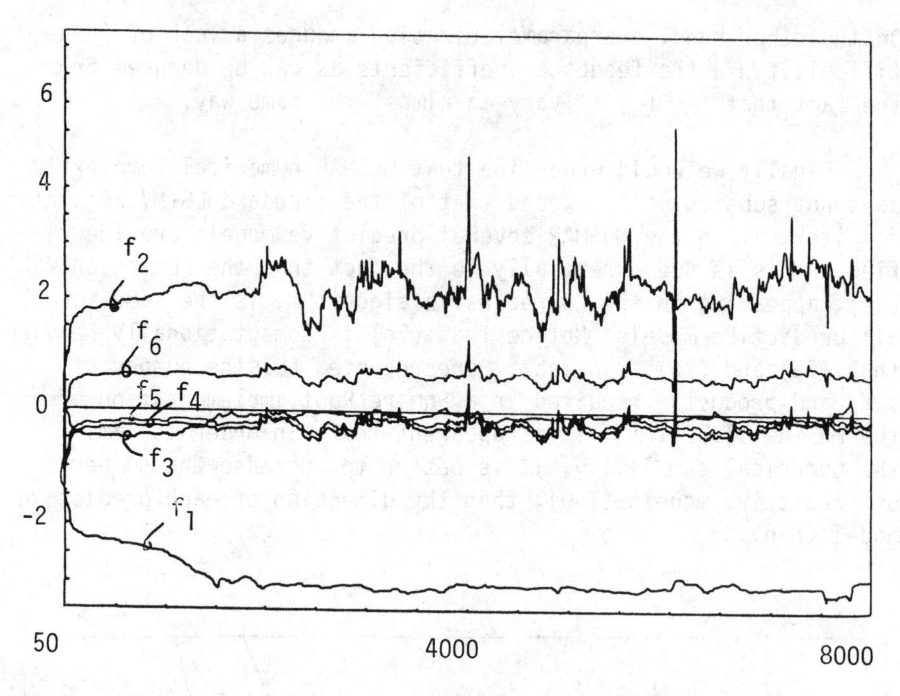

Fig. 2 - Behavior of the six feedback components for the plant of Sec. 4 regulated by an overparameterized LS+MV regulator. The forgetting factor 0.995 was employed in the LS algorithm.

no. of products	no. of sums
$10n^2 + 4Tn + 3n + 2T$	$12n^2 + 10Tn + 12n + 1$

Tab. 1 - Number of products and sums required in a Square Root implementation of the MUSMAR MV algorithm. T = time-horizon, n = order of the predictive models.

5. CONCLUSIONS

Local convergence properties of a new class of adaptive MV regulators (MUSMAR) based on multistep cost-functionals were analyzed. It turns out that, extending the time-horizon of the cost-functional, convergence can be achieved around the MV controller even if the plant does not satisfy a positive realness condition. Within this respect MUSMAR behaves like an overparameterized LS+MV regulator where convergence is guaranteed by overparameterizing the predictive model of the plant, since, in both cases, the influence of the noise acting on the plant is reduced by increasing the number of parameters to be identified. However, the MUSMAR approach has the advantage of requiring less parameters in the feedback gain vector, while simulation experiments have outlined a lack of identifiability for the coefficients of the overparameterized LS+MV regulator. The numerical complexity of a Square Root implementation of the MUSMAR algorithm was reported, which turns out to be comparable with that of the standard LS+MV algorithm.

REFERENCES

[1] L. Dugard and I.D. Landau, *Stochastic model reference adaptive controllers*, Proc. 19th IEEE-CDC, Albunquerque, Dec. 1980.

[2] V. Solo, *The convergence of AML*, IEEE Trans. Autom. Control, 24, pp. 958-963, Dec. 1979.

[3] L. Ljung, *Analysis of recursive stochastic algorithms*, IEEE Trans. Autom. Control, 22, pp. 552-575, Aug. 1977.

[4] L. Lyung, *On positive real transfer function and the convergence of some recursive schemes*, IEEE Trans. Autom. Control, 22, pp. 539-551, Aug. 1977.

[5] S. Shah and G.F. Franklin, *On the removal of the positive real condition by overparameterization*, Proc. 20th IEEE-CDC, San Diego, Dec. 1981.

[6] E. Mosca, *Multivariable adaptive regulators based on multistep quadratic cost functionals*, Proc. NATO ASI on Nonlinear Stochastic Problems, Armacao de Pera, Portugal, May 1982.

[7] K.J. Åström and B. Wittenmark, *On self-tuning regulators*, Automatica, 13, pp. 457-476, 1973.

[8] E. Mosca and G. Zappa, A *robust adaptive control algorithm for microcomputer based implementation*, 2nd IASTED Int. Symp. on Ident. and Control, Davos, March 1982.

ADAPTIVE RECEDING HORIZON CONTROLLERS FOR DISCRETE STOCHASTIC SYSTEMS

Engin Yaz and Yorgo Istefanopulos
Boğaziçi University
Electrical Engineering Department
Istanbul, Turkey

ABSTRACT

An adaptive suboptimal control algorithm is proposed for discrete-time stochastic systems with constant but unknown parameters. The linear control law is certainty-equivalent in the sense that it is linear in the estimates of the states and that the feedback gain matrix is calculated using the estimates of the unknown parameters. In this work the control scheme is separated into an adaptive estimator which simultaneously estimates the states and identifies the parameters of the sytem, and a certainty-equivalent controller which makes use of the state and parameter estimates as if they were the true values. For the estimation stage the adaptive state estimator of Ljung (1) is employed and for the control stage the receding horizon concept of Thomas and Barraud (2) is made use of, which is shown to have the desirable property of making the closed-loop stochastic system asymptotically stable.

1. INTRODUCTION

This paper deals with the problem of adaptive regulation of linear discrete-time systems driven by additive white Gaussian noise sequences. The optimization is posed as the minimization of a quadratic form involving the control variables, subject to the system dynamics and also subject to an equality constraint on the final state. The observations are linear functions of the states with additive white Gaussian noise. Some of

R. S. Bucy and J. M. F. Moura (eds.), Nonlinear Stochastic Problems, 217–228.

the parameters in the system and measurement equations are unknown. The system dynamics and measurement equations are given by

$$x(k+1) = F(\Theta)x(k) + Bu(k) + Ky(k) \tag{1}$$

$$y(k) = \Theta Hx(k) + v(k) \tag{2}$$

where $x(k)$ is the nx1 state vector at the kth time instant, $u(k)$ is the mx1 deterministic input vector, $y(k)$ is the corresponding output vector of dimensions px1 and $v(k)$ is the noise sequence whose statistics are known. The noise sequence is assumed to be zero mean white Gaussian with covariance $E[v(k)v^T(j)] = Q\delta_{kj}$. In this formulation the pxs matrix Θ contains all the unknown parameters in the model. Therefore the system matrix $F(\Theta)$ and the output matrix ΘH are completely specified if the parameter matrix Θ is known. Furthermore in this formulation a particular parametrization, suggested by Ljung (1), is adopted and the system matrix is assumed to be in the form $F(\Theta) = F + G\Theta H$.

The problem is to obtain the control sequence $u(k)$ for $k=0,1,\ldots,N-1$ which minimizes

$$E\left[\sum_{k=0}^{N-1} u^T(k)Ru(k)\right] \tag{3}$$

subject to the system dynamics of Eq. 1 and also subject to the constraint

$$E[x(N)] = 0 \tag{4}$$

where R is a positive definite matrix and N is the predetermined horizon length.

For the overall controller structure to be implementable, we require the control at the k'th stage to be a function of the information state $\{Y_k, U_{k-1}\}$ where $Y_k = \{y(0), y(1), \ldots, y(k)\}$ and $U_{k-1} = \{u(0), u(1), \ldots u(k-1)\}$. If the parameter matrix Θ is known and if constant feedback gains are to be used for the ease of implementation, the results are the direct extention of Thomas' receding horizon controller to the stochastic case as demonstrated by Yaz (3). But with unknown parameters, the controller structure must be improved to include adaptation to the parameter identification process. The configuration of the controller to be employed is shown in Figure 1.

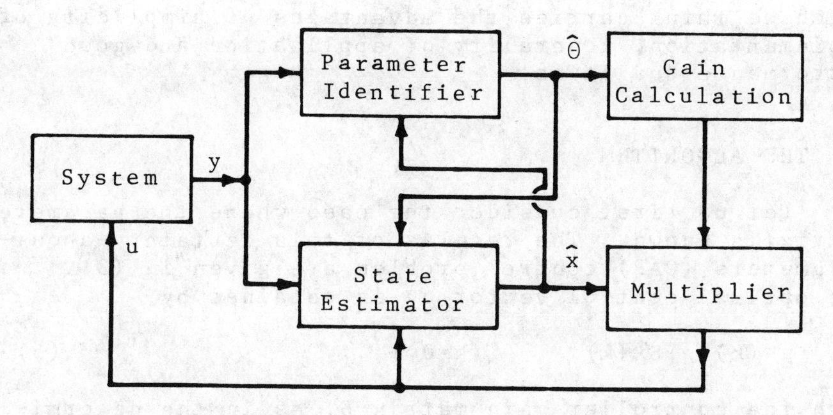

FIGURE 1. Adaptive Stochastic Controller Structure.

As can be noticed from the figure, the controller does not take into account the uncertainty associated with the identification of the parameters, but accepts the parameter estimates as if they were the true values of the parameters; that is, in the terms defined by Wittenmark (4) the controller is not "cautious" but simply "certainty-equivalent".

Similar controller configurations have been implemented so far with different realizations for the constituent subsystems. The parameter adaptive self organizing controller of Saridis (5), for example, is realized with a first order stochastic approximation algorithm for parameter identification, a Kalman filter for state estimation, whereas the control gains are computed by either the steady state dynamic programming equations or by "one-step ahead" approximations. Alag and Kaufman (6), have designed a compensator identifier, a Kalman state estimator and a model-following control law making use of a single-stage performance index. Kreisselmeier (7) also suggests a similar controller configuration where the feedback gains are computed based on the current estimates of the parameters. Cao (8) has used the "per interval controller" of Saridis in conjunction with a first-order stochastic approximation type parameter estimator and a steady state Kalman filter for state estimation. But the bias in the parameter estimates led to control inconsistencies.

In the sequel a new suboptimal control algorithm is suggested for the same controller configuration, which we think carries the advantages of simplicity of implementation, generality of application and good performance qualities.

2. THE ALGORITHM

Let us first consider the case where the parameter matrix is known. The results of this certainty-about-parameters (CAP) control problem are given in (3). The optimal control vector u* is obtained by

$$u^*(k) = S\hat{x}(k) \quad , \quad k=0,1,\ldots \tag{5}$$

with the controller gain matrix S, as in the deterministic case, being calculated as

$$S = -R^{-1}B^T \left[W(o)F^T\right]^{-1} \tag{6}$$

where W(o) is the zeroeth index solution of the backward iteration

$$W(i) = F^{-1}\left[W(i+1) + BR^{-1}B^T\right] F^{-T} \; ; \; W(N)=0 \tag{7}$$

provided that i) F is nonsingular, ii) the system is controllable and iii) $N \geq n-m+1$.
The state vector estimate is obtained by

$$\hat{x}(k+1) = F\hat{x}(k) + Bu(k) + Ky(k) \tag{8}$$

or equivalently, upon substitution of Eq. 5 into Eq. 8, by

$$\hat{x}(k+1) = (F+BS)\hat{x}(k) + Ky(k) \tag{9}$$

When there exist some unknown parameters either in the system equation or in the measurement equation the state vector can be augmented to include the unknown parameters and the augmented vector can be estimated. However, this procedure would lead to a nonlinear control scheme. To avoid this nonlinear control problem a certainty-equivalence is imposed both with respect to the state estimates as well as the parameter estimates ensuring ease of implementation and realizability of the control scheme for high order systems.

As a result of this enforced certainty-equivalence the stochastic adaptive control algorithm becomes

$$u(k) = \hat{S}(k)\hat{x}(k) = -R^{-1}B^T[\hat{W}(k,o)\hat{F}^T(k)]^{-1}\hat{x}(k) \quad (10)$$

where $\hat{F}(k) = F + G\hat{\Theta}(k)H$ and $\hat{W}(k,o)$ is the zeroeth index solution of the backward iteration.

$$\hat{W}(k,i) = \hat{F}^{-1}(k)[\hat{W}(k,i+1) + BR^{-1}B^T]\hat{F}^{-T}(k); \quad \hat{W}(k,N) = 0 \quad (11)$$

If the chosen horizon length N is large, then doubling algorithms may be employed as demonstrated by Yaz (3), thus avoiding matrix inversion at every step.

Notice that the controller gain in Eq. 10 depends on $\hat{F}(k)$ which changes as the parameter estimates are changed at every step along with parameter identification.

The estimates for the states and the parameters are computed by

$$\hat{x}(k+1) = [\hat{F}(k) + B\hat{S}(k)]\hat{x}(k) + Ky(k) \quad (12)$$

$$\hat{\Theta}^T(k) = \hat{\Theta}^T(k-1) + [\gamma(k)/r(k)]H\hat{x}(k)[y(k) - \hat{\Theta}(k-1)H\hat{x}(k)| \quad (13)$$

$$r(k) = r(k-1) + \gamma(k)[||Hx(k)||^2 - r(k-1) + \delta] \quad (14)$$

In this algorithm $\gamma(k)$ is an arbitrary scalar gain sequence which satisfies Dvoretzky's conditions (9). The arbitrary constant δ is a small positive term used to prevent $r(k)$ from taking on the null value. It can also be noticed that the parameter identifier is of a stochastic approximation type with $r(k) = \text{Trace}P(k)$, where $P(k)$ is the uncertainty matrix associated with the parameter identification.

The algorithm is started with an arbitrarily assigned $\hat{\Theta}(o)$ for any given $x(o)$.

3. CONVERGENCE CONDITIONS

The convergence of the proposed algorithm is closely related with the conditions of identifiability for systems operating under feedback. These conditions have been established by Söderström et.al (10) for multivariable systems in a feedback loop. Here an attempt is made to demonstrate that the proposed algorithm meets the conditions for identifiability.

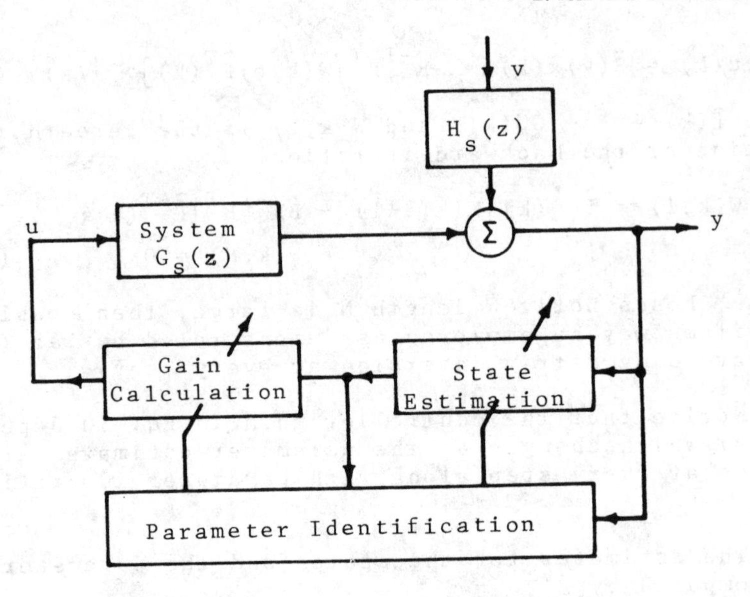

FIGURE 2. System Operating Under Feedback.

Figure 2 depicts the configuration of the adaptive stochastic controller together with the system whose closed-loop identifiability will be examined.

Following the joint input/output identification method which is used to model the dynamics from v to y and from y to u, where y and u are regarded as stochastic processes, the following transfer functions can be derived:

Substituting Eq. 2 into Eq. 1 we get

$$x(k+1) = \left[F + (G+K)\Theta H\right]x(k) + Bu(k) + Kv(k)$$
$$= \Phi x(k) + Bu(k) + Kv(k) \qquad (15)$$

and in conjunction with Eq. 2, the transfer matrix H_s from v to y is found to be

$$H_s(z) = \Theta H(zI - \Phi)^{-1} K + I \qquad (16)$$

Considering the estimator of Eq. 12, the transfer matrix from y to $\hat{x}$ is obtained as $\left[zI - (\hat{F} + B\hat{S})\right]^{-1} K$ and thus the transfer matrix from y to u becomes

$$F_i(z) = \hat{S}\left[zI - (\hat{F} + B\hat{S})\right]^{-1} K \qquad (17)$$

Finally using Eq. 15 again we find the transfer matrix from u to x to be

$$G_s(z) = \Theta H(zI - \Phi)^{-1} B \quad (18)$$

The subscript i in the feedback transfer matrix of Eq. 17 denotes the different values of F(z) due to tuning by the parameter identifier. That is, the feedback law shifts between r different cases (r>1) where each case is to be used in a nonnegligible part of the total control horizon.

Examination of the transfer matrices given above will show compliance with the identifiability conditions of Söderström et. al (10). One can observe from Eq. 17 that the feedback loop contains the necessary time delay for closed loop identifiability. Also, it is assumed that the parameter estimates do not change drastically, and that they constitute a set of pseudo-stationary points of operation each taking sufficiently long duration to ensure an r-shift in feedback laws, where r is the smallest integer such that $r>1+m/p$. If these weak conditions are satisfied then the only remaining condition for identifiability is the asymptotic stability of the closed loop system as required in the theorem in (10). The asymptotic stability is, in turn, secured by the identifiability. That is, in the limit, if the true values of the parameters are known then the proposed stochastic receding horizon controller is sufficient to render the system asymptotically stable (3). This means that in this case asymptotic stability implies, and is in turn implied by, the identifiability of the closed loop system.

4. SIMULATION RESULTS

Several systems have been simulated and the proposed algorithm converged in all cases. Some of the simulated systems and the achieved results are reported below:

<u>System 1</u> $x(k+1) = (.5 + 1.5\theta)x(k) + u(k) + 1.5v(k)$

$$y(k) = \theta x(k) + v(k) \text{ with } v(k):N(0,.25);\ \theta=1$$

Clearly this first system is unstable but controllable.

System_2_ $$x(k+1) = (-.7 + 1.5\theta)x(k) + u(k) + 1.5v(k)$$

$$y(k) = \theta x(k) + v(k) \quad \text{with } v(k):N(0,.25);\theta=1$$

This system is the first standard example reported in Söderström et al (11).

System_3_ $$x(k+1) = (.83 + .15\theta)x(k) + .1u(k) + .15v(k)$$

$$y(k) = \theta x(k) + v(k) \text{ with } v(k):N(0,.25);\ \theta=1$$

This system is obtained from the same continuous time system that System 2 was obtained, but with one tenth the sampling period.

System_4_

$$x(k+1) = \begin{bmatrix} a & b \\ 0 & 0 \end{bmatrix} x(k) + \begin{bmatrix} 0 \\ 1 \end{bmatrix} u(k) + \begin{bmatrix} 1 \\ 0 \end{bmatrix} v(k)$$

$$y(k) = \begin{bmatrix} a & b \end{bmatrix} x(k) + v(k)$$

with $v(k):N(0,1)$ and $a = .95$, $b = -.05$

This system represents the dynamics of the pharmacodynamical application reported by Koivo (12), who studied the infusion rate of a drug for blood pressure regulation. In his paper Koivo used a minimum variance regulator not penalizing the cost of input energy to comply with microprocessor requirements. The proposed algorithm penalizes the input energy, meaning that it restricts the infusion rate of the drug while regulating the blood pressure.

All systems considered above are obtained from an ARMAX model of the type

$$y(k) = ay(k-1) + bu(k-1) + v(k) + cv(k-1)$$

with the following numerical values:

System 1) $a = 2$, $b = 1$, $c = -.5$
System 2) $a = .8$, $b = 1$, $c = .7$
System 3) $a = .98$, $b = .1$, $c = -.83$
System 4) $a = .95$, $b = -.05$, $c = 0$

The simulation of these systems exhibited the following general properties of the proposed algorithm:

1) Effective Regulation to Zero: System 1 and System 2

were controlled by the proposed adaptive receding horizon scheme. In 50 iterations the following results were obtained for the average value of the estimate $(1/50) \sum_{k=1}^{50} \hat{x}(k)$

System 1 with 1-step ahead controller: .0056
System 1 with 4-step ahead controller: .0109
System 2 with 1-step ahead controller: .0316

2) Regulation to Nonzero Set Points: In System 4 the desired final state was chosen to be $(-50,0)^T$, the first number physically representing a 50mmHg reduction in blood pressure, and the second meaning that the infusion of drug must be stopped gradually. For this case, a two-step controller was chosen, which is the minimum possible for a second order system with a single input. In one iteration the values of $x(1)=(-50.82, -3.169\times10^{-5})^T$ and $\hat{\mathbf{x}}(1)=(-50.00,-3169\times10^{-5})^T$ were obtained, which are quite close to $(-50,0)^T$.

3) Success of Adaptation: Jacobs et. al (13) suggests that the incremental costs after many stages of operation can be used as a measure to evaluate the asymptotic properties of controllers. Small incremental costs as compared with incremental costs of the certainty-about-parameters case would suggest successful adaption. The performance for System 1 of the 1-step ahead adaptive controller was compared with that of the 1-step ahead CAP controller using the incremental cost $\hat{x}(1000) + u^2(999)$. These costs differ from each other by 2×10^{-4}, which means very good adaptation.

4) Tunability by the Choice of Horizon Length: Better performance was observed with fewer-step-ahead controller. The table below gives a comparison of the performance of 1-step-ahead and 4-step-ahead controllers for System 1, at k=3000.

	Trace P	$\lvert\hat{\theta}-\theta\rvert$	$\lvert\hat{x}-x\rvert$	$1/k \sum_{i=1}^{k} \hat{x}(i)$	$1/k \sum_{i=1}^{k-1} u^2(i)$
1-step ahead	.5571	.0080	.0000	.0207	2.718
4-step ahead	.7446	.0028	.0000	.0415	2.564

This behavior is attributed to the lazy character of the 4-step ahead controller since it is given the information that it has 4-step before it to achieve the control.

5) Tunability by the Choice of Sampling Period: System 2 and System 3 are obtained from the same continuous time system. However, System 3 has a sampling period equal to one tenth of that used for System 2. The accuracy of control as measured by $(1/k) \sum_{i=1}^{k} \hat{x}(i)$ is 0.023 for System 2 and 0.0065 for System 3, with k=3000, which indicates that in general shortening the sampling interval results in a better controller. The original continuous time system is

$$\dot{x}(t) = -.2x(t) + .8u(t) + 1.5v(t)$$

$$y(t) = x(t) + v(t)$$

and a first order approximation has been used for discretization.

5. DISCUSSION AND CONCLUSIONS

As can be noticed from the above presentation, the algorithm has several distinct features to be emphasized:

1) Due to the inherent property of receding horizon controllers, the designer does not have to choose any state penalization matrix whose choice of relative magnitude with respect to the input penalization matrix is more or less a trial and error procedure, see Athans (14). For single input systems, one does not even have to choose the input penalization constant, because it eventually cancels in the calculations. However, the necessary tradeoff between the input energy and the system behavior quality is always conserved.

2) The choice of horizon length adds a flexibility to design. The number N is bounded below by N_0=dim(x) - rank(u) + 1, see Thomas (2). Stronger control is associated with less number of steps before the controller. If the choice of N is large, one can use the doubling algorithm as applied to this case by Yaz (3).

3) Even though this exposition includes multidimensional systems with multi-parameter-uncertainty, no use is made of matrix update equations for state and parameter estimation uncertainties whose presence constitute most of the computational burden of other methods.

4) The chosen parametrization is general enough to contain the ARMAX model

$$y(k) + A_1 y(k-1) + \ldots A_n y(k-n_a)$$
$$= B_1 u(k-1) + \ldots B_m u(k-n_b)$$
$$+ v(k) + C_1 v(k-1) + \ldots + C_p v(k-n_c)$$

with

$$x(k) = \left[y^T(k-1) \ldots y^T(k-n_a)\, u^T(k-1) \ldots u^T(k-n_b)\, v^T(k-1) \ldots v^T(k-n_c) \right]^T$$

and proper choices of coefficient matrices and vectors.

5) All the above simulations and the convergence analysis is done without resort to an xternal perturbation signal. Presence of such a "sufficiently exciting" signal is considered to be a must by many authors, such as (1), (5), (7), etc. Our simulations verified that the identifier sufficiently tunes the parameters of the controller to provide the necessary shifts in control law, needed for closed loop identifiability.

An adaptive controller for linear stochastic systems with parameter uncertainty has been introduced. The controller is certainty equivalent in both the parameters and the states in the sense that is uses the parameter estimates which depend on the state estimates instead of the true parameters in the controller gain calculation and makes use of the state estimates depending on the parameter estimates instead of the true state in the feedback law. Another way to pose the situation is that the identifier tunes the parameters of both the constant deterministic feedback gain and the state estimator.

The proposed control algorithm represents an improvement over the self-tuning regulators which do not penalize the energy spent in control, thus achieving their aim by using in some cases unacceptably high energy. The proposed controller is also an extension of the popular controllers using linear quadratic cost criteria, but which consider only one-step-ahead effects. This algorithm gives the designer the possibility of penalizing the amount of energy spent and the flexibility of tuning with different horizon lengths. This controller with the enforced certainty equivalence with respect to both the state estimates and the parameter estimates is by far simpler to implement as compared with the control law using the on-line solution of the matrix Riccati equation.

REFERENCES

1. L. Ljung, *Convergence of an Adaptive Filter Algorithm*, Int. J. Cont., 1978, No. 5, pp. 673-679.
2. Y. Thomas, A. Barraud, *Commande Optimal a Horizon Fuyant*, Rev RAIRD, Apr. 1974, pp. 146-150.
3. E. Yaz, PhD Thesis, Dept. of Electrical Engineering Boğaziçi University, Turkey, 1982.
4. B. Wittenmark, *Stochastic Adaptive Control Methods: A Survey*, Int. J. Cont., 1975, No. 9, pp. 705-730.
5. G.N. Saridis, R.N. Lobbia, *Parameter Identification and Control of Linear Discrete-Time Systems*, IEEE Trans. Aut. Cont. No. 1, Feb. 1972.
6. G. Alag and H. Kaufman, IEEE Trans. Auto. Cont. No. 5, October 1977.
7. G. Kreisselmeir, IEEE Trans. Aut. Cont. No. 4, August 1980.
8. Cao, *A Simple Adaptive Concept for the Control of an Industrial Robot*, Proc. of Ruhr Symposium on Adaptive Systems, March 1980.
9. A. Dvoretzky, *On Stochastic Approximation*, Proc. 3rd Berkeley Symp. Mathematics Statistics and Probability, 1965, pp. 35-55.
10. T. Söderström, L. Ljung and I. Gustavsson, *Identifiability Conditions for Linear Multivariable Systems Operating under Feedback*, IEEE Trans. Aut. Con. Dec. 1976.
11. T. Söderström, L. Ljung and I. Gustavsson, *A Theoretical Analysis of Recursive Identification Methods*, Automatica, Vol. 14, pp. 231-244.
12. A.J. Koivo, *Microprocessor-Based Controller for Pharmacodynamical Applications*, IEEE Trans. Aut. Cont. No. 5, October 1981.
13. O.L.R. Jacobs, P. Saratchandran, *Comparison of Adaptive Controllers*, Automatica, Vol. 16, pp. 89-97.
14. M. Athans, *The Role and Use of the Stochastic Linear-Quadratic-Gaussian Problem in Control Systems Design*, IEEE Trans. Aut. Cont., No. 6, December 1971.

AN EXPLICIT SOLUTION TO A PROBLEM IN NONLINEAR STOCHASTIC CONTROL INVOLVING THE WIENER PROCESS

L. A. Shepp

Bell Laboratories
Murray Hill, NJ 07974

Abstract. This is a brief outline of a lecture to be given at a Nato Advanced Study Institute on Nonlinear Stochastic Problems, May 16-28, 1982. The problem to be treated, taken from a paper with V. E. Benes and H. S. Witsenhausen, is to find $\min E \int_0^\infty e^{-\alpha t} |x_o+w_t-\xi_t|^2 dt$ where w is a Wiener process and the min is taken over all nondeterministic processes ξ_t with $\xi_o = 0$ and finite total variation $\int |d\xi| \leq y_o$, where $x_o, y_o \geq 0$, $\alpha > 0$ are given. In presenting the explicit solution our purpose will be to discuss the techniques and concepts involved in the formulation and solution of the problem and the use of the Wiener process as model.

1. A control problem with explict solution

In a recent paper [1], with V. E. Benes and H. S. Witsenhausen, we found the explicit solution to several new problems in stochastic control, among them the "finite fuel follower" problem of optimally tracking a standard Wiener process $x_0 + w_t$ started at x_o by a nonanticipating process ξ_t having $\xi_o = 0$ and total variation (fuel) $\int_0^\infty |d\xi| \leq y_o$ so as to minimize the expected discounted square error $E \int_0^\infty e^{-\alpha t} |x_o+w_t-\xi_t|^2 dt$ where $y_0 \geq 0$, $\alpha > 0$, and $-\infty < x_o < \infty$ are given. Here y_o represents the total fuel available for control, α represents the discount rate for future errors, and x_o the initial error in tracking.

Most problems of the above type are solved by first writing the Bellman equation and then applying numerical techniques. For the above specific problem however, thanks to the special properties of the *homogeneity* of w_t, the *exponential* discounting, and the *squared* error criterion, we give the optimal tracking control ξ explicitly. This special problem is discussed here, not for its practical utility of course, since it is one-dimensional and overly specialized, but to illustrate the use of Wiener process techniques. It is also felt that these explicit results would be useful as a numerical check on accuracy and debugging for a computer program designed to treat similar problems of a more practical type which do not admit explicit solution.

R. S. Bucy and J. M. F. Moura (eds.), Nonlinear Stochastic Problems, 229–230.

The explicit solution we found in [1] is given in terms of the two-dimensional process (x_t, y_t), $x_t = x_o + w_t - \xi_t$, $y_t = y_o - \int_o^t |d\xi_s|$ so that x_t is the (signed) tracking error and y_t is the remaining fuel at any time $t \geq 0$. The optimal tracking control ξ_t is such that (x_t, y_t) remains in the region $C = \{(x,y) : 0 \leq y \leq f(x)\}$ where f is the function uniquely defined by the properties: $f(x) = f(-x); f(0) = \infty$; $f'(x/\sqrt{2\alpha}) = (x^{-1} \tanh x - 1)^{-1}$, $0 < x < u$; $f(x) = 0$, $u/\sqrt{2\alpha} < x < \infty$, where $u \approx 1.199$ is the root of $u \tanh u = 1$, and $f(x)$ is continuous at $x = u/\sqrt{2\alpha}$. The function f is graphed for $x > 0$ for $\alpha = \frac{1}{2}$ below. The control $d\xi_t = 0$ unless $t = 0$ or (x_t, y_t) is on the boundary of C, $y_t = f(x_t)$. In the interior of C, $dy_t = 0$ and no fuel is used. On the boundary of C, fuel is applied in order to keep (x_t, y_t) inside C. As C flares out with decreasing fuel, a larger error is needed to make us want to use it. The optimal expected error $V(x_o, y_o, \alpha)$ is given in [1] explicitly.

The method of solution is to set up a Bellman equation, guess its solution from a heuristic principle of smooth fit which has been discussed earlier for optimal stopping problems [2], and verify the guessed solution by a martingale technique.

[1] V. E. Benes, L. A. Shepp, H. S. Witsenhausen, Some solvable stochastic control problems, Stochastics (4), 1980, 39-83.

[2] L. A. Shepp, Explicit solutions to some problems of optimal stopping, Ann. Math. Stat. (40), 1969, 993-1010.

We use this opportunity to correct some misprints in [1]. On p. 43, line 9 of Problem 3 replace $u\sqrt{2\alpha}$ by $u/\sqrt{2\alpha}$ and $f'(x)$ by $f'(x/\sqrt{2\alpha})$. In the next line replace $u\sqrt{2\alpha}$ by u. In p. 43 l.-4 replace $u\sqrt{2\alpha}$ by $u/\sqrt{2\alpha}$. In the 3rd line of the first display on p. 58 replace what is there, $(\tanh x\sqrt{2\alpha} - x\sqrt{2\alpha})^{-1}$ by $(\tanh x\sqrt{2\alpha}/((x\sqrt{2\alpha})-1))^{-1}$. In p. 58 l.11 replace $\alpha = 1$ by $\alpha = 1/2$.

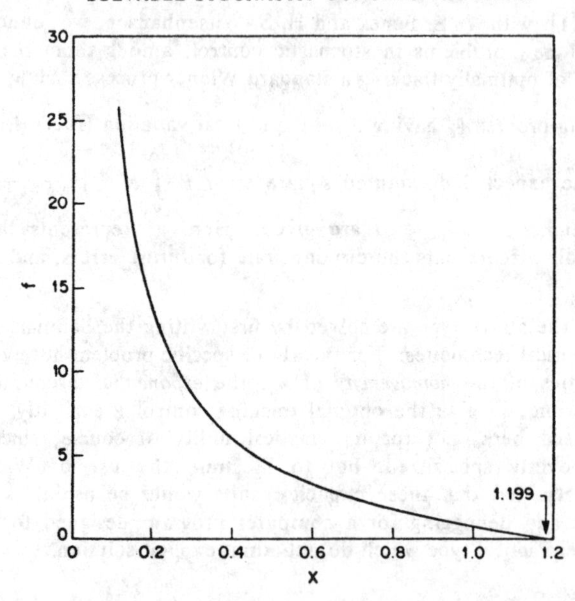

CHAPTER VI

OPTIMAL CONTROL

ON OPTIMAL CONTROL FOR A CLASS OF PARTIALLY-OBSERVED SYSTEMS

K. Helmes

Institut für Angewandte Mathematik der Universität Bonn

1. INTRODUCTION

This note is intended to popularize a class of partially-observed control systems which, like the LQG-problem (cf. [5]), can be solved explicitly. The problem is to steer a 'linear' system to a hyperplane in fixed time using bounded controls and having only partial information about the state available. While state- and observation-process evolve exactly as in the LQG-problem it differs from that one in that (1) the performance index is a quite different one and (2), more important, in that 'hard constraints' are put on the controls. In 1980 Beneš and Karatzas [1] have analyzed the one-dimensional problem and they have shown that the optimal control using partial observations is just $u(t) = -\text{sign}(s(t)\cdot \hat{x}(t))$ where $\hat{x}(t)$ is the conditional mean of the state given the observations up to time t and s(t) is a deterministic function depending on the given data in a prescribed way (cf. also [8]). Interestingly, although this is a non-linear problem it exhibits the 'certainty-equivalence' principle - the optimal control is obtained by estimating the state and then using this estimate as though it were the true state. Recently Christopeit and Helmes [3] derived the analogous result for the multidimensional problem. A cornerstone of their analysis is an existence result on weak solutions to certain stochastic differential equations with degenerate diffusions which has been derived by Christopeit [4] using Skorokhod's imbedding technique. Here we shall give another proof of this result which is based on the martingale formulation of multidimensional diffusions (cf. [9]); proofs of all other statements will because of space limitation at most be sketched; for details we refer to [3].

R. S. Bucy and J. M. F. Moura (eds.), Nonlinear Stochastic Problems, 233–241.

2. PROBLEM FORMULATION

A generic element of the class of partially-observed systems which we consider is described by a *'linear' state equation*

$$dX_t = A(t)X_t dt + B(t)u_t dt + C(t)dW_t^{(1)}, \quad X(0) = X_o , \tag{1}$$

a *'linear' observation equation*

$$dY_t = F(t)X_t dt + G(t)dW_t^{(2)}, \quad Y(0) = 0, \tag{2}$$

and the *performance index*

$$J(u) := E[k(<\zeta_T, X_T>)]. \tag{3}$$

Here $(W_t^{(1)}, W_t^{(2)})$ denotes an (n×k)-dimensional (standard) Wiener process defined on some probability space (Ω, F, P). It is assumed that this Brownian motion is independent of the *Gaussian* random vector X_o. The matrices A(t), B(t), C(t), F(t) and G(t) are assumed to be of size n×n, n×m, n×n, k×n and k×k, respectively, and continuous in t, with C(t)C'(t) and G(t)G'(t) being uniformly positive definite. The vector ζ_T is a fixed element in $\mathbb{R}^n$ and the functional $k: \mathbb{R} \to \mathbb{R}^+$ which, at time $T > 0$, evaluates the distance of the state of the system to the hyperplane $\{x | <\zeta_T, x> = 0\}$ satisfies the conditions

$$\begin{array}{ll} \text{(i)} & k \text{ is symmetric around zero, } k(0) = 0, \\ \text{(ii)} & k \text{ is monotone increasing on } \{x \geq 0\}, \\ \text{(iii)} & k \text{ satisfies an exponential growth condition, i.e.} \\ & |k(x)| = O(\exp[\alpha |x|]) \text{ for some } \alpha > 0. \end{array} \tag{4}$$

The objective will be to minimize the 'cost' (3) choosing a control u in the set of *admissible* controls U. To formulate admissibility and in order to give a precise formulation of the control problem, let us start with the (n×k)-dimensional Brownian motion $(W_t^{(1)}, W_t^{(2)})$ and let $Z_t = (X_t, Y_t)$ denote the solution to

$$dX_t = A(t)X_t dt + C(t)dW_t^{(1)}, \tag{5}$$

$$dY_t = F(t)X_t dt + G(t)dW_t^{(2)}, \quad 0 \leq t \leq T, \tag{6}$$

with initial condition $Z_o = (X_o, 0)$. Let $G_t = \sigma(Y_s, s \leq t)$ be the σ-algebra measuring the past of the observation process up to time t. The class of admissible controls will then be defined by

$$U = \{u = (u_t) | u \text{ is a measurable } G_t\text{-adapted process on } (\Omega, F, P) \text{ with values in } [-1,1]^m\}.$$

Corresponding to an admissible control $u \in U$, we shall define a (weak) solution to (5), (6) via Girsanov's measure transformation technique. To this end, consider the probability measure P^u defined by

$$dP^u = \phi_o^T(u)dP$$

where

$$\phi_o^T(u) = \exp\left[\int_o^T [C(t)^{-1}B(t)u_t]'dW_t^{(1)} - \frac{1}{2}\int_o^T [C(t)^{-1}B(t)u_t]^2 dt\right].$$

By Girsanov's theorem, (Z_t) is a solution to (1), (2) under the measure P^u with (W_t^1, W_t^2) replaced by a different Brownian motion (W_t^u).

The problem posed in the introduction can now be stated as follows (E^u denotes expectation with respect to P^u):

(*P*) Choose an admissible control $u \varepsilon U$ so as to minimize
$J(u) := E^u[k(<\zeta_T, X_T>)]$. (7)

Remark 1. For a somewhat different formulation of the partially-observed control model and for its relationship to problem (*P*) cf. [3].

3. THE OPTIMAL CONTROL

The model (1)-(3) is a highly nonlinear control and filtering problem. We shall briefly describe how it can be solved. The starting point of the analysis is the observation that (Z_t) is conditionally Gaussian under P^u (cf. [7],ch.11). Hence problem (*P*) splits into a filtering problem: to estimate (under P^u) X_t given $(Y_s)_{s\leq t}$, and a completely-observed control problem $(\hat{P})$ (see below) the state to be controlled being the conditional mean $\hat{X}_t^u := E^u[X_t|G_t]$. This process satisfies the equation

$$d\hat{X}_t^u = A(t)\hat{X}_t^u dt + B(t)u_t dt + K(t)d\upsilon_t^u,$$

$$\hat{X}_0^u = E^u[X_0], \tag{8}$$

where $K(t) = R(t)F'(t)[G(t)G'(t)]^{-1}$ denotes the 'gain matrix', $R(t)$ the covariance matrix of the error $(X_t - \hat{X}_t^u)$ and (υ_t^u) the innovation process which is given by

$$d\upsilon_t^u = dY_t - F(t)\hat{X}_t^u dt = G(t)d\hat{w}_t^u, \quad \upsilon_o^u = 0,$$

with a k-dimensional Brownian motion $(\hat{w}_t^u)$ (under P^u) adapted to G_t. The error covariance $R(t)$ is independent of u and can be determined by the well-known Riccati equation associated with (5), (6) (see [5]). Let $\hat{k}$ denote the convolution of the functional k with the normal density having mean value zero and variance

$r(t):=\zeta_T'R(T)\zeta_T$.

The 'separated' control problem $(\hat{P})$ reads as follows:

$(\hat{P})$ *Minimize*

$$\hat{J}(u):= E^u\left[k(<\zeta_T,\xi_T^u>)\right]$$

over U subject to

$$d\xi_t^u = A(t)\xi_t^u dt + B(t)u_t dt + K(t)G(t)d\hat{w}_t^u ,$$

$$\xi_o^u = E^u[X_o]. \tag{9}$$

From the very definition of $\hat{k}$, the fact that $\hat{X}_t^u$ given G_t is conditionally Gaussian and that solutions to equation (9) are strongly unique it follows:

Lemma 1. $\inf_{u\varepsilon U}\{J(u)\} = \inf_{u\varepsilon U}\{\hat{J}(u)\}$.

Problem $(\hat{P})$ can be solved if we assume a certain property relating the matrices A,C,F and G to hold. To formulate this condition let H(t,s) denote the transition matrix associated with A, i.e.

$$\frac{\partial}{\partial t}H(t,s) = A(t)H(t,s),\ H(t,t)=Id;$$

put

$$\zeta_t := H'(T,t)\zeta_T \text{ and } b(t): = B'(t)\zeta_t.$$

From now on we shall assume the following condition to hold:

$$\sigma^2(t): = \zeta_t'K(t)G(t)G'(t)K'(t)\zeta_t > 0 \text{ for all } 0\leq t\leq T. \tag{10}$$

Hypothesis (10) ensures G_t to contain 'information' on $<\zeta_t,X_t>$ and thus allows one to relate problem $(\hat{P})$ to a specific one-dimensional control problem $(\hat{P}^{(1)})$, viz.

$(\hat{P}^{(1)})$ *Minimize*

$$\hat{J}^{(1)}(v): = E[\hat{k}(x_T)]$$

within the class of all (F_t)-adapted processes v *taking values in the interval*

$$I(t) := \left[-\sum_{i=1}^{m} |b_i(t)|,\ \sum_{i=1}^{m} |b_i(t)|\right],$$

v *being defined on some probability space*(Ω,F,P) *which carries a Wiener process* (β_t,F_t) *and* (x_t) *being defined by*

$$dx_t = v_t dt + \sigma(t)d\beta_t\ ,\quad 0\le t\le T\ ,$$

$$x_o = x.$$

Problem $(\hat{P}^{(1)})$ has been analyzed by several authors employing a variety of techniques (cf. [2] and the literature cited therein). Since $\hat{k}$ inherits the properties (i)-(iii) (see (4)) from k an optimal control for $(\hat{P}^1)$ is specified by the Markov strategy

$$v^*(t,x) := -\sum_{i=1}^{m} |b_i(t)|\,\mathrm{sign}(x). \tag{11}$$

The most difficult part of our analysis is related to the question of how an optimal control for $(\hat{P})$ can be derived from the one-dimensional result. To this end, define the Markov strategy, $\xi\in\mathbb{R}^n$,

$$u_i^*(t,\xi) := -\mathrm{sign}(b_i(t)<\zeta_t,\xi>),\qquad i=1,\ldots,m.$$

Let (w_t) be any k-dimensional Brownian motion under some probability measure P. Then by Theorem 3 stated in the following section the stochastic differential equation

$$d\xi_t = [A(t)-K(t)F(t)]\xi_t dt + B(t)\hat{u}^*(t,\xi_t)dt + K(t)G(t)dw_t,$$

$$\xi_o = E[X_o], \tag{12}$$

has a unique strong solution (ξ_t) which may be represented as a nonanticipative functional of the paths $\int_o^{\cdot} G(s)dw_s$,

$$\xi_t = \Xi(t,\int_o^{\cdot} G(s)dw_s),$$

where Ξ is a measurable process on $C([0,1];\mathbb{R}^n)$ adapted to the natural filtration. Put

$$u_t^* = \hat{u}^*(t,\Xi(t,Y)), \tag{13}$$

where Y is the process given by (5), then $u^*\in U$. Define $P^*=P^{u^*}$ and $w^*=\hat{w}^*$ as in Section 3, and let $\hat{x}_t^* = \hat{x}_t^{u^*} = E^*[x_t|G_t]$. Based on Lemma 1 and the construction just given optimality of u^* can be shown.

<u>Theorem 1.</u> *The control* u^* *(see (13)) is optimal for (P) and*

admits a feedback representation in terms of the estimated state x* *which is given by*

$$u_t^* = \hat{u}^*(t,\hat{x}_t^*).$$

4. EXISTENCE OF STRONG SOLUTIONS

In this section we shall prove existence and uniqueness of strong solutions to stochastic differential equations like (12) only assuming condition (10) to be fulfilled. It is well known, see [9], ch.8, or [6], ch.4, that the existence of a unique strong solution is implied by the existence of a (unique) weak solution and pathwise uniqueness. If K(t)G(t) is assumed to be nondegenerate, the existence of a (unique) weak solution to (12) is an immediate consequence of Girsanov's theorem. If the drift term of Eq. (12) were 'smooth' Itô's construction would readily give a strong solution, irrespective of whether or not K(t)G(t) is nondegenerate. Since the drift term (in (12)) is a not too irregular function which depends on a nondegenerate Itô process (see (18) below) both methods can be combined to construct a weak solution.

Throughout this paragraph we shall adopt the following notation: L(t) denotes any continuous n×n-matrix valued function and A, B, C, F, G, ζ_t, σ^2, b and $\hat{u}^*$ the quantities which were defined in previous sections. Put $\varepsilon(t):=K(t)G(t)$, $v_N(\alpha):=(I_{(-\infty,-1/N)} - N\alpha I_{[-1/N,1/N]} - I_{[1/N,\infty)})(\alpha)$, $\alpha\in\mathbb{R}$, $N\in\mathbb{N}$, where I_Γ denotes the indicator function of a set Γ, $u_N^i(t,x):=\mathrm{sign}(b_i(t)v_N(<\zeta_t,x>)$, $x\in\mathbb{R}^n$, $1\leq i\leq m$, $\varphi(t,x):=(A(t)-L(t))x+B(t)\hat{u}^*(t,x)$, $\gamma(t,x):=-<\zeta_t,L(t)x>+<b(t),\hat{u}^*(t,x)>$, $e(t):=\varepsilon\varepsilon'(t)$ and φ_N, γ_N respectively, is defined exactly as φ is, γ respectively, only $\hat{u}^*$ is replaced by u_N.

Since φ contains a linear term and thus is unbounded, the following construction had to be made locally. But in order not to overburden the presentation we refrain from spelling out all the details and tacitly assume all unbounded terms to be truncated whenever necessary; see [9], ch. 10, for detailed information. Let us denote by Θ the space $C([0,\infty),\mathbb{R}^n)$ equipped with the Borel σ-algebra M and the natural filtration $M_t=\sigma(\theta(s),s\leq t)$, $t\geq 0$, $\theta\in\Theta$.

For every $N\in\mathbb{N}$, (e,φ_N) is a pair of Lipschitz continuous (in x) functions which satisfy a linear growth condition. Hence, for every $x\in\mathbb{R}^n$ there is a unique probability measure P_x^N such that

$$P_x^N[\theta(0)=x] = 1$$

and

$$f(\theta(t)) - \int_0^t L_s^N f(\theta(s))ds$$

is a P_x^N-martingale for all $f \in \mathcal{D} := C_o^\infty(\mathbb{R}^n)$ where

$$\mathcal{L}_s^N := \frac{1}{2}\Sigma_{i,j=1}^n e_{ij}(s)\frac{\partial^2}{\partial x_i \partial x_j} + \Sigma_{j=1}^n \varphi_N^j(s,x)\frac{\partial}{\partial x_j} . \tag{16}$$

The probability measure P_x^N is called the 'diffusion with coefficients e and φ_N starting from x at time 0' or, equvalently, a solution to the martingale problem for $\mathcal{L}_t^N$ (or e and φ_N) starting from (0,x).

Theorem 2. There is a (unique) solution to the martingale problem for e and φ starting from (0,x).

Proof. The tightness criterion given in [9], p. 39, implies $\{P_x^N\}$ is precompact (φ_N is assumed to be truncated). Thus, all that we need to show is that if $(P^{N'})$ is a subsequence of (P^N) which converges to P, then P solves the martingale problem for e and φ starting from (0,x). From now on we shall also denote the subsequence by (P^N). Clearly $P[\theta(0)=1]=1$. To complete the proof, let $0 \leq s < t$ and $\psi : \Theta \to \mathbb{R}$ be bounded, continuous and M_t-measurable. Given $f \in \mathcal{D}$, we must show that (E, E^N respectively, denotes expectation with respect to P, P^N respectively)

$$E\left[\psi[f(\theta(t)) - f(\theta(s)) - {}_s^t\!\int \mathcal{L}_r f(\theta(r))dr]\right] = 0$$

where $\mathcal{L}$ denotes the differential operator (16) with coefficients e and φ. Since

$$E^N\left[\psi[f(\theta(t)) - f(\theta(s)) - {}_s^t\!\int \mathcal{L}_r^N f(\theta(r))dr]\right] = 0$$

for all N and since e(r), (A(r)-L(r))x are continuous functions in (r,x), it is clear that we need only show that

$$E^N\left[\psi[{}_s^t\!\int \nabla f \cdot B(r)u_N(r,\theta(r))dr]\right] \underset{N\to\infty}{\to} E\left[\psi[{}_s^t\!\int \nabla f \cdot B(r)\hat{u}^*(r,\theta(r))dr]\right]$$

where ∇f stands for the gradient of f and $\nabla f \cdot B(r)\hat{u}^*(r,\theta(r))$ denotes the scalar product ot the vectors $\nabla f(\theta(r))$ and $B(r)\hat{u}^*(r,\theta(r))$. For any $m \in \mathbb{N}$ fixed

$$E^N\left[\psi[{}_s^t\!\int \nabla f \cdot B(r)u_m(r,\theta(r))dr]\right] \underset{N\to\infty}{\to} E\left[\psi[{}_s^t\!\int \nabla f \cdot B(r)u_m(r,\theta(r))dr]\right].$$

It is therefore enough for us to show that

$$E\left[\psi[{}_s^t\!\int \nabla f \cdot B(r)\hat{u}^*(r,\theta(r))dr]\right] \underset{m\to\infty}{\to} E\left[\psi[{}_s^t\!\int \nabla f B(r)u_m(r,\theta(r))dr]\right]$$

and that

$$\overline{\lim_{m\to\infty}}\ \overline{\lim_{N\to\infty}} \left|E^N\left[\psi[{}_s^t\!\int \nabla f \cdot B(r)[u_N - u_m](r,\theta(r))dr]\right]\right| = 0 .$$

To prove the first of these note that

$$E\left[\psi[{}_s^t\int\nabla f\cdot B(r)[\hat{u}^*-u_m](r,\theta(r))dr]\right] \leq C\, {}_s^t\int P\left[z_r\in\left(\frac{-1}{M},\frac{1}{M}\right)\right]dr \qquad (17)$$

where $z_r:=\langle\zeta_r,\theta(r)\rangle$ and C a constant which depends on f, B and ψ. Since (10) holds z_t is a nondegenerate Itô process (for every P^N) which satisfies the equation

$$dz_t = \gamma_N(t,\theta(t))dt + \sigma(t)d\beta^N(t), \quad (P^N\text{-a.e.}) \qquad (18)$$

where (β_t^N) is some Brownian motion with respect to P^N. Moreover, the distribution of z_r under P^N tends to the distribution of z_r under P. Since

$$\sup_N \left\{P^N\left[z_r\in\left(\frac{-1}{M},\frac{1}{M}\right)\right]\right\} \leq \hat{C}\,\frac{1}{M}$$

the right hand side of (17) tends to zero. To prove the second, observe that

$$\begin{aligned}&\overline{\lim_{m\to\infty}}\ \overline{\lim_{N\to\infty}} \left|E^N\left[\psi[{}_s^t\int\nabla f\cdot B(r)[u_N-u_m](r,\theta(r))dr]\right]\right| \\ &\quad\leq C\ \overline{\lim_{m\to\infty}}\ \sup_{N\geq 1}\ {}_s^t\int P^N\left[z_r\in\left(\frac{-1}{M},\frac{1}{M}\right)\right]dr \\ &\quad= 0\end{aligned}$$

again by the same argument. □

Remark 2. Theorem 2 treats the case of a specific stochastic differential equation; but the proof of the theorem goes through in the case of more general equations provided the coefficients satisfy conditions which are natural generalizations of those satisfied by e and φ; cf. [4] for a possible extension.

Put $L(t) = K(t)F(t)$; Theorem 2 together with pathwise uniqueness of solutions to Eq. (12) (proved in the Appendix of [3]) implies our final result:

Theorem 3. The stochastic differential equation (12) has a unique strong solution.

REFERENCES

[1] BENEŠ, V.E. and KARATZAS, I.:1980, *Examples of optimal control for partially observable systems: Comparison, classical and martingale methods*, Stochastics 5, pp. 43-64.

[2] CHRISTOPEIT, N. and HELMES, K.: 1982, *On Beneš' bang-bang control problem*, Appl. Math. Optim., to appear.

[3] CHRISTOPEIT, N. and HELMES, K: 1982, *Optimal control for a class of partially observable systems*, Stochastics, to appear.

[4] CHRISTOPEIT, N.: 1982, *On the existence of weak solutions to stochastic differential equations, preprint.*

[5] FLEMING, W. and RISHEL, R.: 1975, *Deterministic and Stochastic Optimal Control,* Springer-Verlag, New York.

[6] IKEDA, N. and WATANABE, S.: 1981, *Stochastic Differential Equations and Diffusion Processes,* North-Holland Publ., Amsterdam.

[7] LIPTSER, R.S. and SHIRYAYEV, A.N.: 1977, *Statistics of Random Processes, vols. 1, 2,* Springer-Verlag, New York.

[8] RUZICKA, J.: 1977, *On the separation principle with bounded coefficients,* Appl. Math. Optim. 3, pp. 243-261.

[9] STROOCK, D.W. and VARADHAN, S.R.S.: 1979, *Multidimensional Diffusion Processes,* Springer-Verlag, New York.

OPTIMAL STOCHASTIC CONTROL OF LINEAR SYSTEMS WITH STATE AND CONTROL DEPENDENT NOISE: EFFICIENT COMPUTATIONAL ALGORITHMS

F.C. Incertis

IBM Madrid Scientific Center

In this paper two new numerically efficient algorithms to solve the state-and control-dependent noise linear-quadratic optimal control problem are proposed. The formulation extends the (Hoskins-Walton, 1978) method for square-root matrix computations and some recent results (Incertis, 1982) for deterministic algebraic Riccati equations resolution. Implementation requires as basic operations only additions, products and inversions in the field of the positive definite matrices being the approach quite general since it can be applied to solve optimal control problems associated to processes with multiple independent noises in the control and state vectors.

1. INTRODUCTION

Consider the stochastic linear system formally described by the stochastic Itô equation (Kleinman, 1969, McLane, 1971)

$$dx(t) = Ax(t)dt + Bu(t)dt + Fx(t)d\alpha + Gu(t)d\beta + dw \tag{1}$$

and

$$y(t) = Cx(t) \tag{2}$$

where A, F are constant n x n matrices, B, G are constant n x m matrices; C is an r x n constant matrix; F and G have full rank. The pair <A,B> is stabilizable and the pair <A,C> is completely detectable. Moreorver $w(t)$, $\alpha(t)$, $\beta(t)$ are independent Wiener processes with autocorrelations

$$E\{[w(t) - w(\tau)][w(t) - w(\tau)]^T\} = W|t-\tau| \tag{3}$$

R. S. Bucy and J. M. F. Moura (eds.), Nonlinear Stochastic Problems, 243–253.

$$E\{[\alpha(t)-\alpha(\tau)]^2\}=\sigma|t-\tau| \quad (4)$$

$$E\{[\beta(t)-\beta(\tau)]^2\}=\rho|t-\tau| \quad (5)$$

Now, if we define the steady-state cost

$$J=E\{y^T(t)y(t)+u^T(t)Ru(t)\} \quad (6)$$

where $R=R^T>0$, it is well known (McLane, 1971) that the optimal linear feedback control that minimizes J is given by

$$u_0(t)=-(R+\rho G^TKG)^{-1}B^TKx(t) \quad (7)$$

where K is the unique positive definite solution (when exists) of the "stochastic algebraic Riccati equation" (SARE)

$$L+A^TK+KA+\sigma F^TKF-KB(R+\rho G^TKG)^{-1}B^TK=0 \quad (8)$$

where $L=L^T=C^TC\geq 0$. Aside from the terms σF^TKF and ρG^TKG this equation is commonly known as the "deterministic" algebraic Riccati equation (ARE).

Several numerical techniques have been investigated in the literature for solving (8); most of them are natural extensions of the (Kleinman, 1968) iterative method for deterministic ARE problems. To summarize this computationally efficient method (Kleinman, 1976) let us consider equation (8) and the following iterative procedure:
Algorithm: If a solution $K>0$ of the SARE (8) exists then $K=\lim K_i$, where $K_i>0$, $i=1,2,\ldots$ is the unique (positive definite) solution of the ARE

$$A^TK_i+K_iA-K_iB(R+\rho G^TK_{i-1}G)^{-1}B^TK_i+\sigma F^TK_{i-1}F+L=0 \quad (9)$$

where $K_0=0$. If a solution $K>0$ does not exists then, K_i diverges. Hence, the technique advocated by Kleinman consists in solving the SARE as a limiting sequence of solutions of conventional ARE and, by using the author's method, the solution of each ARE is obtained as a limiting sequence of solutions of linear matrix Liapunov type equations, (see Kleinman, 1968). The proof of the algorithm convergence is founded on the fact that the ARE solution is a monotone function (in the positive definite matrix norm sense) of the additive term $\sigma F^TK_{i-1}F$ being also monotone with respect to $R_i = R+\rho G^TK_{i-1}G$; hence, if K exists sequence K_i; $i=1,2,\ldots$ is bounded above and then we have convergence.

By using this procedure, jointly with an iterative algorithm to solve deterministic ARE for each K_i, it is advisable to initialize this algorithm by utilizing the last computed K_{i-1}. By using this strategy a drastic reduction in the number of linear equations needed to compute K_i from K_{i-1} has been reported (see Kleinman, 1976).

In this paper some previous results on square-root matrix computations (Hoskins and Walton, 1978) and on deterministic ARE computations (Incertis and Martínez, 1977; Incertis, 1981,82) are generalized to the stochastic case and two new numerical methods for the solution of SARE are presented. Application of both algorithms requires computation of a sequence of positive definite matrices by means of an iterative procedure that makes use of matrix additions, products and inversions. The results are particularly useful when the system matrix A is a near-diagonal or diagonal dominant matrix, this is the case when dealing with many spatially quantized distributed parameter systems. For this class of systems the proposed numerical algorithms will converge in a few program rounds.

Another advantage of the proposed approach is due to the fact that all computations are performed in the field of the positive definite matrices. Thus, it is not required to resort to complex number arithmetics as occurs in most today available efficient methods (Laub, 1979; Larin et al., 1981).

The remainder of this paper is organized as follows: Section 2 presents a reformulation of the SARE problem and some previous results and theorems which are useful to the development of new computational algorithms. Section 3 is devoted to the presentation of the two new algorithms and the corresponding convergence conditions. In Section 4 an analysis of the computational cost and storage requirements of the main algorithm is performed and comparisons with today available efficient methods are given. In Section 5 an example is solved to show the behaviour of Algorithm 2. Finally, Section 6 summarizes the main contributions and features of the algorithms in this paper.

2. MAIN RESULTS

In what follows we shall assume that matrix C is n x n and invertible, hence $L = C^T C$ will be a positive definite symmetric matrix. Under fulfillment of the existence and uniqueness conditions the unique positive definite solution K of (8) must be computed. Now, under those assumptions, K^{-1} is also unique and positive definite. The pre-and post-multiplication of (8) by K^{-1} then gives

$$K^{-1}LK^{-1} + K^{-1}A^T + AK^{-1} + \sigma K^{-1}F^T KFK^{-1} - B(R+\rho G^T KG)^{-1}B^T = 0 \quad (10)$$

The pre-multiplication of this equation by matrix L leads to

$$LK^{-1}LK^{-1} + LK^{-1}A^T + LAK^{-1} + \sigma LK^{-1}F^T KFK^{-1} - LB(R+\rho G^T KG)^{-1}B^T = 0 \quad (11)$$

which may be factored as

$$(LK^{-1}+A)(LK^{-1}+A^T)-(A-LAL^{-1})LK^{-1}=LB(R+\rho G^T KG)^{-1}B^T+AA^T-\\ -\sigma LK^{-1}F^T KFK^{-1} \quad (12)$$

Now, introducing the notation

$$X = LK^{-1} + A^T \quad (13)$$

after an straightforward computation problem (8) becomes to find a matrix X such that

$$X^2+(LAL^{-1}-A^T)X = LAL^{-1}A^T+LB[R+\rho G^T(X-A^T)^{-1}LG]^{-1}B^T-\\ -\sigma(X-A^T)F^T(X-A^T)^{-1}LFL^{-1}(X-A^T) \quad (14.1)$$

$$X - A^T > 0 \quad (14.2)$$

where condition (14.2) is equivalent to impose matrix K to be positive definite. Moreover, let by definition

$$2D = LAL^{-1} - A^T \quad (15.1)$$

$$T(X) = LAL^{-1}A^T + LB[R+\rho G^T(X-A^T)^{-1}LG]^{-1}B^T -\\ -\sigma(X-A^T)^{-1}F^T(X-A^T)^{-1}LFL^{-1}(X-A^T) \quad (15.2)$$

By utilizing this notation equation (14.1) can be written more compactly as

$$X^2 + 2DX = T(X) \quad (16)$$

Thus, the problem becomes to find a solution X of equation (16) and verifying condition (14.2). In order to formulate new algorithms for the stochastic optimal control problem some previous results, (Incertis, 1981), for deterministic algebraic Riccati equations are required.

2.1. Deterministic Matrix Riccati Equations

In the deterministic case F and G are null matrices, hence equation (8) takes the form

$$L + A^T K + KA - KBR^{-1}B^T K = 0 \quad (17)$$

Particularizing (16) to this case, we can write

$$X^2 + 2DX = T \quad (18)$$

where

$$T = LAL^{-1}A^T + LBR^{-1}B^T \tag{19}$$

Furthermore, consider transformation

$$H = X + D \tag{20}$$

in (18). A straightforward computation then gives

$$H^2 + DH - HD = M \tag{21}$$

where

$$M = T + D^2 = LAL^{-1}A^T + LBR^{-1}B^T + D^2 \tag{22}$$

Now, if M is a positive definite matrix, then a positive definite square root matrix $M^{1/2}$ exists and is unique. For instance, if M is semisimple it may always be diagonalized in the form

$$M = P\Lambda P^{-1} \tag{23}$$

where $\Lambda > 0$ is strictly diagonal. The square root matrix $M^{1/2}$ is then given by

$$M^{1/2} = P\Lambda^{1/2}P^{-1} > 0 \tag{24}$$

From the precedent formulation the following fundamental theorem is demonstrated (Incertis, 1981):
Theorem 1: If $[M^{1/2}, D] \triangleq M^{1/2}D - DM^{1/2} = 0$, then the unique solution of (18) which fulfills the positive definiteness condition (14.2) is given by

$$X = M^{1/2} - D \tag{25}$$

Notice that since (13) the solutions of (17) and (18) can be related by the following theorem, which is stated here without proof:
Theorem 2: If $\langle A,B\rangle$ is stabilizable, $\langle A,C\rangle$ is detectable, $L = C^TC$ is positive definite and a solution X of (18) is obtained for which condition (14.2) is fulfilled, then the unique positive definite solution of the ARE (17) is given by

$$K = (X - A^T)^{-1}L \tag{26}$$

In order to apply this theoretical result to general ARE problems for which commutativity condition $[M^{1/2}, D] = 0$ is not identically satisfied, the following numerical algorithm has been formulated (see Incertis, 1982) as an extension of the Hoskins-Walton method for square-root matrix computations: ARE Algorithm: If a solution X of (18) exists then $X = \lim X_i$, where $X_0 = T$ and

$$X_{i+1} = -D + 1/2(X_i + TX_i^{-1}) \tag{27}$$

In order to use this algorithm it is required that X_i; i=0,1,2,... to be a convergent sequence of positive definite matrices. The following necessary and sufficient condition for the convergence of this iterative procedure is given in (Incertis, 1982).

$$||T^{1/2} - D|| > 1/\sqrt{3}||T^{1/2}|| > 0 \tag{28}$$

We shall note at this point that this algorithm can be also initialized by utilizing any good a priori estimate of the solution, when available, such as $X_0 = T^{1/2}$, or some positive definite matrix X_0 for which $||X_0^{1/2} - D|| > 0$. In the following section the previous theoretical and algorithmical results on deterministic Riccati equations are generalized and applied to solve the more complex and difficult SARE problem.

3. SARE ALGORITHMS

Two main algorithms are discussed in this section: The first algorithm is founded on the monotonic convergence behaviour of deterministic ARE sequences jointly with the above given computational scheme to solve ARE. The second algorithm will be developed as a direct iteration procedure derived from equation (16).

Algorithm 1: Solution of (8) is obtained as $K = \lim K_i$ where $K_i > 0$; $i = 1,2,...$ is the unique solution of the quadratic equation

$$A^T K_i + K_i A - K_i A - K_i B R_i^{-1} B^T K_i + L_i = 0 \tag{29}$$

where, recursively

$$L_i = L + \sigma F^T K_{i-1} F \tag{30.1}$$

$$R_i^{-1} = (R + \rho G^T K_{i-1} G)^{-1} \tag{30.2}$$

and $K_0 = 0$. Moreover, for the i-th main iteration step the following imbedded iterative procedure is applied to determine solution K_i of (29).

Inner iterations: For every $i = 1,2,...$

i) Compute auxilliar coefficient matrices

$$D_i = 1/2(L_i A L_i^{-1} - A^T) \tag{31.1}$$

$$T_i = L_i(A L_i^{-1} A^T + B R_i^{-1} B^T) \tag{31.2}$$

ii) Starting from the initial value

$$X_{i,0} = L_{i-1} K_{i-1}^{-1} + A^T \; ; \; X_{0,0} = T_0 \tag{32}$$

compute, recursively

$$X_{i,j+1} = -D_i + 1/2(X_{i,j} + T_i X_{i,j}^{-1}) \quad (33)$$

for $j = 1,2,\ldots$ and stop iterations when, for some $j = N$ $\|X_{i,N+1} - X_{i,N}\| < \varepsilon_0$, where ε_0 is the admissible error bound

iii) Let $\hat{X}_i = X_{i,N+1}$ be the last computed matrix in ii), by using this value compute K_i by formula

$$K_i = (\hat{X}_i - A^T)^{-1} L_i \quad (34)$$

Outer iterations on the sequence K_i; $i = 0,1,2,\ldots$ will use the stopping criteria $\|K_{N+1} - K_N\| < \varepsilon_1$, for some $i = N$ and some admissible tolerance ε_1. Observe that, since (32) the last computed value K_{i-1} during the (i-1)-th outer iteration is used to obtain an initial guess for the i-th inner iteration step. Experience with this algorithm shows that, like in the Klainman's method, only one to four inner iterations are required to get K_i from K_{i-1} for $i > 3$.

Algorithm 2: Solution of (8) is obtained as

$$K = (\hat{X} - A^T)^{-1} L \quad (35)$$

where $\hat{X}$ is the solution of (16), computed recursively as follows. For every $i = 0,1,2,\ldots$ and choosing $X_0=0$:

i) Compute the auxilliar matrix

$$T(X_i) = LAL^{-1}A^T + LB[R+\rho G^T(X_i-A^T)^{-1}LG]^{-1}B^T - \\ -\sigma(X_i-A^T)F^T(X_i-A^T)^{-1}LFL^{-1}(X_i-A^T) \quad (36)$$

ii) Iterate, by using recursive formula

$$X_{i+1} = D + 1/2(X_i + T(X_i)X_i^{-1}) \quad (37)$$

then, do $i = i+1$ and go to step i).

As in the above Algorithm 1, the stopping criteria will be $\|X_{N+1} - X_N\| < \varepsilon$, for some i=N and some given tolerance ε. A necessary condition for convergence of this algorithm is that $T(X_i) > 0$, for all i=1,2,..., however, no monotone convergence properties can be given for this case. Computational experience with this procedure shows that it can be advantageous over Algorithm 1 when the states and controls noise levels are small (i.e. $\|\sigma F\|$ and $\|\rho G\| \ll 1$). For more general problems Algorithm 1 is preferable since it assures monotonic convergence towards solution if the existence conditions are fulfilled (i.e. if the stochastic system is controlable and observable).

We lay stress here on the computational advantages of those algorithms mainly due to the fact that the only operations involved in implementation are matrix additions, products and inversions in the field of positive definite matrices, (see equations 30.1 - 34). Furthermore the set of transformations applied on the former SARE (8) to get polynomial like form (16) enables us to obtain additional computational advantages when system matrix is a band diagonal matrix and the stochastic components $||F||$ and $||G||$ are small: Under those conditions $T^{1/2}$, with T as given by (19), gives a good first approximation of the solution of (16) and then Algorithms 1 and 2 may converge in a few program rounds. This is the case of many distributed parameter systems where because of finite elements quantization each state variable is functionally linked with a few neighbour state variables, thus leading to the typical band diagonal structure of the system matrix. To improve convergence of iteration equations (27), (33) and (37) those equations can be substituted by the more general procedure

$$X_{i+1} = -2/\mu D + 1/\mu((\mu-1)X_i + TX_i^{-1}), \quad \mu > 2 \tag{38}$$

and, on the computer implementation, μ is estimated by means of a special subroutine in order to assure convergence and to minimize the number of iterations.

4. OPERATIONS COUNTS AND STORAGE

We shall give an approximate operation count for the solution of n-th order SARE of the form (8) by utilizing Algorithm 1. Each operation is assumed to be equivalent to compute $a+(b\times c)$ where a, b, c are floating-point numbers.

It is almost impossible to give an accurate operation count for the proposed algorithm mainly due to the fact that the number of inner and outer iterations strongly depends on problem structure, numerical values of coefficient matrices and noise covariances σ and ρ. We shall give only a ballpark $0(n^3)$ figure for the entire process. The main steps are

	Operations
a) Outer iterations	Operations
i) Computations of L_i and R_i^{-1}	$N_0(5n^3)$
b) Inner iterations	Operations
i) Computation of D_i, T_i	$7n^3$
ii) Computation of $\hat{X}_i$	$N_i(7/3\,n^3)$
iii) Computation of K_i	$7/3\,n^3$

N_i represents the number of steps required to compute K_i from K_{i-1}

for a given number of figures using iteration procedure (31.1-34); N_0 represents the total number of outer iterations required to obtain K. Thus we have a ballpark estimate of about $((40+7N_i)/3)n^3 N_0$ operations for the entire process. By solving a large number of test problems it can be estimated that the total number of inner iterations tends to be approximately equal to three times the total number of outer iterations, which gives an average estimate of $\bar{N}_i=3$. Comparisons with other methods, like the Kleinman's method and the Schur method (Laub, 1979) can now be made on the basis of this average number of inner iterations which is shared by most methods.

By utilizing the most today efficient Hessenberg-Schur method (Golub et al., 1979) the solution of one linear matrix Liapunov equation requires about $20\,n^3$ operations: Hence, the total cost by using Kleinman's method to solve SARE is rougly given by $65\,n^3 N_0$ operations.

By utilizing the well known and efficient Schur method to solve deterministic ARE the computational cost is about $75\ n^3$ operations per equation . Thus, to solve SARE by using this inner procedure the total computational cost is roughly given by $75\ n^3 N_0$ operations.

Finally, by utilizing Algorithm 1 given in this paper the total computational cost is about $20\ n^3 N_0$ operations. Thus we see that the proposed method significatively reduces the computational cost for a large class of SARE problems, as compared with the most efficient today available algorithms.

With respect to storage considerations the algorithm requires $6\ n^2$ storage locations, less than a helf of the required storage by using the "eigenvector methods" (Larin et al., Laub, 1979) which are limited to solve matrix equations on the order of 100 or less in many common computing environments. Additionally, implementation of the proposed algorithms in this paper only requires standard routines for matrix inversion, addition and product which are inmediately available in all computer libraries.

Application of Algorithm 2 must be restricted to SARE computations for which $||\sigma F||$ and $||\rho G||$ are small to assure convergence. When these conditions are satisfied the computational results presented in (Incertis, 1982) enable us to determine approximate bounds for operations counts and required storage.

Computational experience with both algorithms show effectiveness for low or moderate values of $||\sigma F||$ and $||\rho G||$. As the stochastic components increase computational burden also increases, numerical stability decreases and the solution norm $||K||$ tends to be unbounded for very high noise components. We should remark that the rerults in this paper can be generalized to the case of several (independent) noises α_i, with convariances σ_i, and β_j with covariances

ρ_j by replacing stochastic components in (8) by the general terms

$$\sum_i \sigma_i F_i^T K F_i \quad \text{and} \quad \sum_i \rho_j G_j^T K G$$

To end this section it is worth to remark that the proposed approach can be extended to the case of a singular matrix L by means of perturbation theory or by an extension of the formulation given in (Incertis, 1982) for ARE problems.

5. AN EXAMPLE

The following example of a system with state dependent noise and zero control noise illustrates de usefulness of Algorithm 2 and the unstable behaviour of the solutions for increasing values of σ. Let us consider a system of the form (1)-(6) where the system and performance matrices are given by

$$A = \begin{bmatrix} 0 & 1 \\ 0 & 0 \end{bmatrix}; \; B = \begin{bmatrix} 0 \\ 1 \end{bmatrix}; \; F = \begin{bmatrix} 0.5 & 0.1 \\ 0.2 & 0.5 \end{bmatrix}; \; C = I; \; R = 1; \; G = 0$$

By utilizing Algorithm 2 to solve this problem equation (36) adopts the particular form

$$T(X_i) = T - \sigma P_i F^T P_i^{-1} L F L^{-1} P_i \; ; \quad P_i = X_i - A^T$$

where

$$T = LAL^{-1}A^T + LBR^{-1}B^T = I \text{ and } L = I$$

Also, matrix D in (37) is given by

$$D = 1/2\,(LAL^{-1} - A^T) = \begin{bmatrix} 0 & 0.5 \\ -0.5 & 0 \end{bmatrix}$$

The following table shows the computational effort versus the covariance noise parameter σ for Algorithm 2.

Noise Covariance	Number of iterations	Max. Element in K
0	10	1.73205
0.25	12	1.84523
0.5	13	1.96614
0.75	13	2.09412
1	14	2.22838
2	20	2.83521
3	35	3.54571
4	136	4.35715
5	> 500	-

The results show the effectiveness of Algorithm 2 for small or intermediate values of σ. As σ increases toward values for which the stochastic system cannot be stabilized by feedback, the computational burden increases and the SARE solution norm grows proving that state-dependent noise calls for vigorous control (large gains).

6. CONCLUDING REMARKS

We have discussed in considerable detail some new theoretical results and extensions leading to new algorithms for solving matrix equations arising in stochastic system analysis and optimizations. A number of numerical issues have been addressed and the detailed formulation of the algorithms has been given. The method performs reliably on large dimensional systems with a band diagonal structure in the system matrix, is in practice more rapid than actual available methods, is more robust and saves a large amount of computer memory.

REFERENCES

(1) Golub, G.H. et al.: 1979, IEEE Trans. Automat. Contr., AC-24, pp.909-913.
(2) Hoskins, W.D. and Walton, D.J.: 1978, IEEE Trans. Automat. Contr., AC-23, pp. 494-495.
(3) Incertis, F.C. and Martínez, J.M.: 1977, IEEE. Automat. Contr., AC-22, pp. 128-129.
(4) Incertis, F.C.: 1981, IEEE Trans. Automat. Contr., AC-26, pp. 768-770.
(5) Incertis, F.C.:1982, Procc. Second IASTED Symposia on Modelling, Identification and Control, Davos, Switzerland.
(6) Kleinman, D.L.: 1968, IEEE Trans. Automat. Contr. AC-13, pp. 114-115.
(7) Kleinman, D.L.:1969, IEEE Trans. Automat. Contr. AC-14, pp. 673-677.
(8) Kleinman, D.L.:1976, IEEE Trans. Automat Contr. AC-21, pp. 419-420.
(9) Larin, V.B., Aliev, F.A., Bordug, B.A.: 1981, Report 81-41, Institute of Math., Ukranian Academy of Science, URSS, Kiev, (personal communication).
(10) Laub, A.J.:1979, IEEE Trans. Automat. Contr., AC-24, pp. 913-921.
(11) McLane, P.J.: 1971, IEEE Trans. Automat. Contr. AC-16, pp. 793-798.
(12) Wonham, W.M.: 1967, SIAM J. Contr., 5, pp.486-500.

CHAPTER VII

NONLINEAR FILTERING

JOINT INFORMATION AND DEMODULATION [1]

Richard S. Bucy

University of Southern California, Los Angeles

For some time the theories of Information and Control have been derived under differing assumptions which has lead to theories with little common ground except on the philosophical level. However recently new interest in uniting these areas has been sparked by the use of Information theory to produce lower bounds for the Nonlinear filtering problem. In this paper we will examine the phase demodulation problem in the context of these theories.

I. ,INTRODUCTION

We consider the following problem a signal process is integrated Brownian motion $X_1(t)$ and noisy observations $(z_i(t), i=1,2)$ are made of the signal as

$$\begin{aligned} dz_1 &= \cos x_1(t)\,dt + dv_1 \\ dz_2 &= \sin x_1(t)\,dt + dv_2 \end{aligned} \qquad (1.0)$$

with 1.0 denoting stochastic differential equations, where V_i are independent Brownian motions which are also independent of the signal process and satisfy

$$E(V_i(t)V_j(\tau)) = r \min(t,\tau)\,\delta_{ij} \qquad i,j=1,2$$

The signal process can be thought of as the solution of the stochastic differential equations

$$\begin{aligned} dx_1 &= x_2\,dt \\ dx_2 &= d\beta \end{aligned} \qquad (1.1)$$

(1) This research was supported in part by a Nato research grant 1557.

R. S. Bucy and J. M. F. Moura (eds.), Nonlinear Stochastic Problems, 257–265.

with β Brownian, $(E(\beta(t)\beta(\tau)) = q \min(t,\tau))$ and the process is independent of the initial conditions on (1.1).

The nonlinear filtering problem consists of finding the conditional density of $x_1(t), x_2(t)$ given $\mathcal{O}_t$ the minimum sigma-field induced by the observation $z_i(s)$ $i=1,2$ $s \leq t$. For phase demodulation one is interested in estimating the phase and a measure of performance could be the expected cyclic loss where if ϵ is the error the cyclic loss is $1-\cos\epsilon$

A way of understanding the cyclic loss and its estimate

$$x^*(t) = \tan^{-1} \frac{\widehat{\sin}_t}{\widehat{\cos}_t} \tag{1.2}$$

with $\widehat{\sin}_t = E \sin x_1(t)|\mathcal{O}_t$ and $\widehat{\cos}_t = E \cos x_1(t)|\mathcal{O}_t$ is the following; choose $x^*(t)$ so that $e^{ix^*(t)}$ is the minimum mean square error estimator of $e^{ix_1(t)}$ then $E(\epsilon(t)\overline{\epsilon}(t)) = E(e^{ix_1(t)} - e^{ix^*(t)})(e^{-ix_1(t)} - e^{-ix^*(t)}) = 2(1 - E(\cos x_1(t) - x_1^*(t)))$ where $\epsilon(t) = e^{ix_1(t)} - e^{ix^*(t)}$ (1.3) more details on how (1.2) can be shown to minimize (1.3) can be found in [12] and [13] In particular, defining $D(t)$ as

$$\begin{aligned} \widehat{\sin}_t &= D(t)\sin(x^*(t)) \\ \widehat{\cos}_t &= D(t)\cos(x^*(t)) \end{aligned} \tag{1.4}$$

easy calculations give

$$E\,\epsilon^o(t)\overline{\epsilon^o(t)} = 2(1-D(t)) \tag{1.5}$$

$$\dot{I}(e^{ix_1(\cdot)}; z_1(\cdot), z_2(\cdot)) = \frac{1-D(t)D(t)}{2r} \tag{1.6}$$

where $\dot{I}$ is the steady state information rate for the indicated process time histories, see [4].

II. UPPER BOUNDS FOR PHASE ESTIMATION

We will consider linear gaussian filtering problem:

$$dz = y_t\,dt + dv_t \tag{2.1}$$

where V_t is a complex gaussian white noise process of zero mean and $E V_t \overline{V_t} = 2rt$ while y_t is an independent complex gaussian process of mean $\exp(-\frac{1}{2}R(t,0))$ and satisfying

$$E\, y_t \overline{y_\tau} = \exp(-\tfrac{1}{2}(R(t,t) + R(\tau,\tau) - 2R(t,\tau))) \tag{2.2}$$

For the problem (2.1) to have wide sense error, see [13] which is an upper bound on the optimal filtering error associated with (1.1) and (1.0), y_t should have the same moments up to second order as $e^{ix_1(t)}$, or $R(t,\tau)$

should satisfy

$$R(t,\tau)=\begin{cases} q(\frac{\tau^2 t}{2}-\frac{1}{6}\tau^3) & t>\tau \\ q(\frac{t^2\tau}{2}-\frac{1}{6}t^3) & \tau>t \end{cases}$$

$$B(t,\tau)=\begin{cases} q(\frac{t^3}{3}-\tau^2 t+\frac{2}{3}\tau^3) & t>\tau \\ q(\frac{\tau^3}{3}-t^2\tau+\frac{2}{3}t^3) & \tau>t \end{cases} \quad (2.3)$$

where $B(t,\tau)=R(t,t)+R(\tau,\tau)-2R(t,\tau)$
now with $\phi(t,\tau)=\exp\{-\frac{1}{2}B(t,\tau)\}$ the Wiener Hopf equation for the minimum mean square linear estimate weighting function $W(t,\tau)$ for the estimate $\hat{y}_t$ (i.e. $\hat{y}_t=\int_0^t W(t,\tau)\,dz_\tau$), is

$$\phi(t,\tau)=\int_0^t W(t,\sigma)\,\phi(\sigma,\tau)\,d\sigma+2rW(t,\tau) \quad (2.4)$$

for $0\leq\tau\leq t$
Further it is easily seen that the mean square error satisfies

$$E(y_t-\hat{y}_t)\overline{(y_t-\hat{y}_t)}=2rW(t,t)=1-\int_0^t W(t,\sigma)\,\phi(\sigma,t)\,d\sigma \quad (2.5)$$

— denotes complex conjugate, but

$$\phi(\tau,t)=\exp\{-\tfrac{qt}{2}(\tau-t)^2-\tfrac{q}{3}(\tau-t)^3\}$$

and for large t and $\tau<t$

$$2rW(t,t)\cong 1-\tfrac{1}{2}\sqrt{\tfrac{2\pi}{qt}}\,W(t,t)+O(\tfrac{1}{t^{1/2}}) \quad (2.6)$$

as $W(t,\tau)$ is bounded. However yields

$$E(y_t-\hat{y}_t)\overline{(y_t-\hat{y}_t)}\to 1 \text{ as } t\to\infty$$

so that

$$2(1-D)\leq 1 \quad \text{as } t\to\infty \quad (2.7)$$

Note that in [5] e^{ix_t} is studied to model Brain waves.

<u>Remark</u>

For the case of a Brownian phase process

$$dx_t=d\beta_t \quad (2.8)$$

$$dz_t=e^{ix_t}\,dt+dv_t$$

an analogous analysis yields the upper bound on the cyclic error as

$$qr\left(\sqrt{1+\tfrac{2}{qr}}-1\right) \quad (2.9)$$

since in this case $y_t=\exp(ix_t)$ has covariance $\exp(-\frac{q}{2}|t-\tau|)$, where q and $2r$ are the spectral parameters of $\beta.$ and $v.$ respectively. This bound is an upper bound on the steady state cyclic error of the nonlinear filter. It is interesting that (2.8) is in general smaller than the cyclic error of the Phase lock loop which is $2\left(1-\frac{I_1(\alpha)}{I_0(\alpha)}\right)$ where $\frac{1}{\alpha}=\sqrt{2qr}$ see [12], I_0 and I_1 are modified Bessel functions).

Another method for generating an upper bound is to determine the cyclic loss of the classical phase lock loop which for the problem (1.0) and(1.1) , is given by

$$\begin{aligned} dx^{**} &= y^{**}dt + g_1(\cos x^{**}dz_2 - \sin x^{**}dz_1) \\ dy^{**} &= g_2(\cos x^{**}dz_2 - \sin x^{**}dz_1) \end{aligned} \qquad (2.10)$$

with $g_1=\sqrt{2}\,(q/r)^{1/4}$, $g_2=(q/r)^{1/2}$. The cyclic loss associated with (2.10) has no simple analytic expression as in the case of Brownian phase and simulation must be resorted to. In the sequel it will be convenient to think of the discrete version of(2.10) x_n^{**}, y_n^{**} and $\Delta_n=\cos x^{**}dz_2-\sin x^{**}dz_1$ and $\Delta_n^{\perp}=\cos x^{**}dz_1+\sin x^{**}dz_2$ in order to provide arguments to validate certain information theoretic equalities.

III. JOINT INFORMATION

In [1] , it was shown that for a linear gaussian problem the joint information between the signal at time t, x_t and the observation history up to was given by

$$I(x_t; z^t) = \tfrac{1}{2}\ln\{\det(Ex(t)x'(t)P^{-1}(t))\} \qquad (3.1)$$

where $P(t)$ is the matrix solution of the Riccati associated with the linear filtering problem and of course represents the error covariance matrix of the optimum filter. Further it was shown that

$$I(x_t;\hat{x}_t) = I(x_t;z^t) \qquad (3.2)$$

where $\hat{x}_t$ is the conditional mean. In his thesis Glados, see [14],gave conditions which extended (3.2) to the general nonlinear filtering case. For the phase demodulation problem we will assume that

$$I(e^{ix_1(t)}; z_1^t, z_2^t) = I(e^{ix_1(t)}; \widehat{\sin}_t, \widehat{\cos}_t) \qquad (3.3)$$

In our simulation of the nonlinear cyclic filtering problem we actually compute the conditional density of x_1 modulo 2π given z_1^t, z_2^t, $p(\tilde{x}_1(t)/\mathcal{Q}_t)$ and since

$$I(e^{ix_1(t)}; z_1^t, z_2^t) = H(e^{ix_1(t)}) - E\,p(\tilde{x}_1(t)/\mathcal{Q}_t) \qquad (3.4)$$

we can by Monte Carlo methods measure (3.4).
The result appears in the sixth column of Table 1 for various values of $R=\sqrt{2}\,q^{1/4}r^{3/4}$ the output signal to noise ratio, which appear in column 1 of Table 1 in decibels. Now analogously by computing the solution of a <u>second</u> nonlinear filtering problem that of finding the conditional density of $x_1(t)$ modulo 2π given the <u>in-phase</u> phase lock loop observations $\{$ i.e. $d\nu_t = \cos x_t^{**} dz_2(t) - \sin x^{**}(t)\, dz_1(t)\}$ $P(\tilde{x}_1(t)|\nu^t)$ we can find

$$I(e^{ix_1(t)}; \nu^t) = I(e^{ix_1(t)}; x^{**t}, y^{**t}) \quad (3.5)$$

since it is easy to see that the $\nu.$ history determines and is determined by the $x.^{**}, y.^{**}$ history. The joint information between the phase lock loop estimate history and the sensor $e^{ix_1(t)}$ is given in Table 1, column 7. Column 3, 4 and 5, of Table 1 give the mean square errors of the cyclic estimate based on z_1^t and z_2^t, the phase lock loop, the cyclic estimate based on ν^t respectively. The second column of Table 1 gives the discrete time step which is chosen to be ten samples per time constant of the phase lock loop, so that the discrete problem is a close approximation to the continuous problem. The statistics in Table 1 were based on 30,000 sample paths of length 130 time steps.

In figure 1 the joint informations for the non-linear filter and for the phase lock loop are compared. Notice that in reality the phase lock loop discards the out of phase measurements and suffers a consequent loss of information and error performance.

In [6], the joint information is used to provide lower bounds on the optimal filter mean square error performance.

IV. CONCLUSIONS

In order to evaluate the joint information between a signal and a set of observations or functionals of these observations the expected value of the entropy of the conditional distribution of the signal given the observations must be computed. Consequently any optimal nonlinear filter realization program can with very little extra coding find this joint information. In this paper we have measured the effect on the information and mean square error of neglecting the out of phase component of the observations, via Monte

Carlo simulation.

Acknowledgements

The extensive Monte Carlo runs reported here were possible because of equipment provided by Airforce office of Scientific Research Electronics Division, specifically an AP120 B array processor, Graduate students Faramesh Ghovanlou and James Hiroshiga contributed to this paper.

TABLE I

1	2	3	4	5	6	7
+ 2	.6818	1.9400	2.447	2.3264	.15283	.07139
+ 1	.6314	1.6470	2.2093	2.1026	.22049	.10071
0	.5848	1.38896	2.0013	1.9039	.30208	.07962
- 1	.5416	1.20538	1.7829	1.69948	.40809	.20414
- 3	.4645	.69624	1.1486	1.1023	.66871	.42829
- 4	.4302	.53478	.86258	.82078	.81802	.60347

1 - Noise to signal ratio, R in db.
2 - Delta, the time step.
3 - Mean square error, nonlinear filter in radians^2.
4 - Mean square error, phase lock loop in radians^2.
5 - Mean square error "in phase" nonlinear filter in radians^2.
6 - Nonlinear filter average information in nats.
7 - Phase lock loop average information in nats.

FIGURE 1

2 Dimensional Nonlinear Filter

30,000 Monte Carlo Runs

Average Joint Information

—— Joint Information Nonlinear filter
- - - Joint Information Phase lock loop

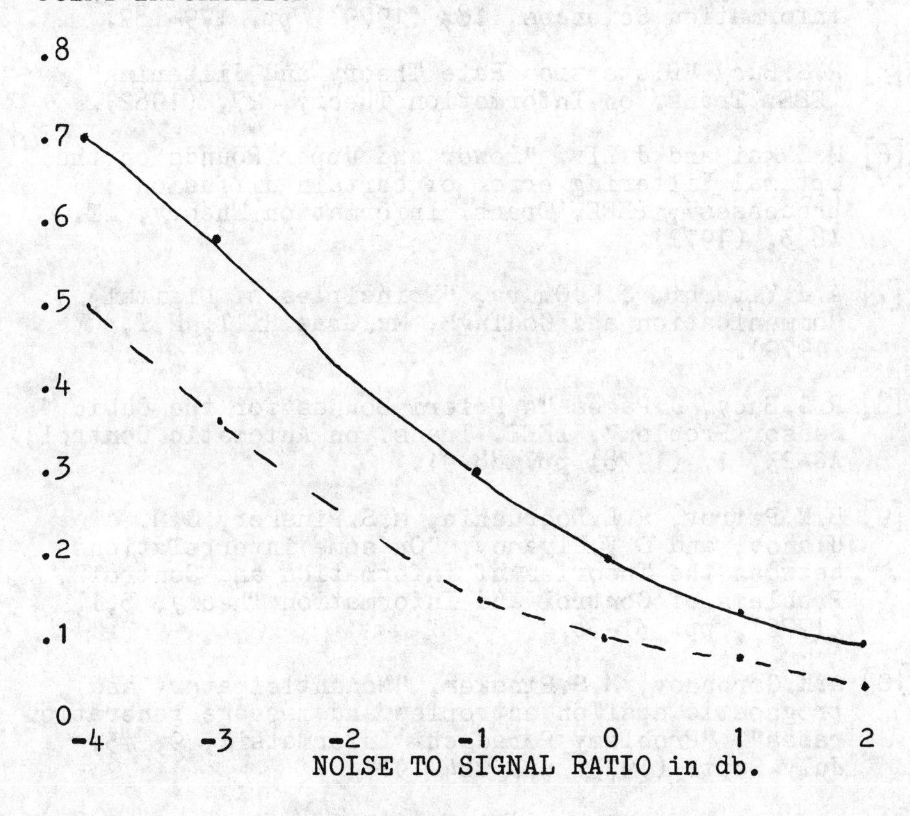

REFERENCES

[1] R.S.Bucy, P.D.Joseph "Filtering for Stochastic Processes with Applications to Guidance", Interscience(1968).

[2] I.M.Gelfand, A.M.Yaglom "Calculation of the Amount of Information About one Random Function Contained in Another such Function",Translations of A.M.S. 41, (1959). pp. 199-246.

[3] M.S.Pinsker "The Quantity of Information about a Gaussian Randon Stationary Process contained in a second process connected with it in a Stationary manner". Dokl. Akad. Nauk. SSSR (N.S.) 99 (1954), pp. 213-216.

[4] R.S.Bucy "Filtering and Information", Journal of Information Sciences, 18, (1979), pp. 179-187.

[5] R.S.Bucy "Distortion Rate Theory and Filtering", IEEE. Trans. on Information Theory, 27, (1982).

[6] M.Zakai and J.Ziv. "Lower and Upper Rounds on the Optimal Filtering error of certain diffusion processes", IEEE. Trans. Information Theory, IT. 18,3, (1972).

[7] A.J.Viterbi, J.K.Omura, "Principles of Digital Communication and Coding", Mc.Graw Hill, N.Y, (1979).

[8] R.S.Bucy, J.Pages "A Priori Bounds for the Cubic Sensor Problem", IEEE. Trans. on Automatic Control, AC-23, 1, (1978) pp. 88-91.

[9] B.N.Petrov, R.L.Dobrushin, M.S.Pinsker, G.M. Ulanov, and S.V.Ulyanov, "On some interrelations between the Theories of Information and Control", Problems of Control and Information Theory, 5,1, (1976), pp. 31-38.

[10] A.K.Gorbunov, M.S.Pinsker, "Nonanticipatory and prognostic epsilon entropies and message generation rates", "Problemy Peradachi Informatsii, 9, #3, July-Sept. (1973) pp. 184-191.

[11] __________, "Prognostic epsilon entropy of a Gaussian message and a Gausian source", Ibid. 10, #2, April-June, (1974), pp. 93-109.

[12] R.S.Bucy, A.J.Mallinckrodt "An Optimal Phase Demodulator", Stochastics, (1973), Vol. i, pp. 3-23.

[13] J.L.Doob. "Stochastic Processes", Wiley, N.Y., (1953).

[14] J.Glados, Thesis, M.I.T. Electrical Engineering Department, (1978).

[15] N.Wiener, "Nonlinear Problems in Random Theory", Wiley, N.Y. (1958).

[16] J.Glados, "A Rate distortion theory lower bound on desired function filtering error", IEEE Trans. on Information Theory, IT-27, (1981), pp.366-368.

GENERALIZED FINITE-DIMENSIONAL FILTERS IN DISCRETE TIME

G.B. Di Masi, W.J. Runggaldier* and B. Barozzi*

LADSEB-CNR and Istituto di Elettrotecnica, Università di Padova, Italy
*Seminario Matematico, Università di Padova, Italy

Nonlinear filtering problems in discrete-time are studied using the Bayesian approach. The concept of a generalized finite-dimensional filter is introduced and it is shown that under suitable assumptions such a filter exists and can be implemented in a recursive way. The resulting technique is successfully applied to various situations.

1. INTRODUCTION

For t=0,1,... consider a partially observable process (x_t,y_t), $x_t \in X \subset R$, $y_t \in Y \subset R$, defined as follows.
The unobservable component x_t satisfies

$$x_{t+1} = T(x_t) + v_{t+1} \quad ; \quad x_o = v_o \tag{1}$$

where $T:R \rightarrow R$ is some measurable mapping and v_t is a sequence of independent random variables (r.v.'s) with generalized probability density function (g.p.d.f.) $p(v_t)=g(v_t;\psi_t) \in G$ where G denotes a given family

$$G = \left\{ g(.;\psi) \mid \psi \in \Phi \subset R^p \right\} \tag{2}$$

The observable component y_t is such that $y_o=0$ and, given x_{t-1}, y_t is independent of the entire process $\{v_t\}$ and of $y^{t-1} := \{y_o,\ldots,y_{t-1}\}$. The distribution of $\{y_t\}$ is characterized by the family

$$\left\{ p(y_t \mid x_{t-1}),\ t=1,2,\ldots \right\} \tag{3}$$

The filtering problem then consists in constructing the family of

R. S. Bucy and J. M. F. Moura (eds.), Nonlinear Stochastic Problems, 267–277.

conditional g.p.d.f.'s

$$\{ p(x_t|y^t) \ , \ t = 1,2,\dots \} \tag{4}$$

The solution to the filtering problem will be given here in terms of a generalized finite-dimensional filter (Def.1 below) in the sense that $p(x_t|y^t)$ is, for all t, representable as a linear combination of distributions belonging to a given parametrized family, where the number of terms in the combination may possibly vary with time. This filter generalizes that in [9], [6], [7], which does not apply to various models where the support of the disturbances in (1) is not all of R (most non-Gaussian disturbances) and nonlinear models with Gaussian disturbances. The filter considered here is also more general than the one in [1] , [8], where combinations of only Gaussian densities are considered, thereby obtaining for most of the above models only approximate solutions. Using our generalized filter concept, we are able to derive a technique to obtain exact recursive solutions for various linear models with non-Gaussian disturbances, as well as for one non-linear model with Gaussian disturbances. The latter can be used in more general situations to obtain approximate solutions with bounds on the quality of the approximation [5].

We shall use the Bayesian approach to filtering, which for model (1)-(3) provides the recursive formula [2]

$$p(x_t|y^t) \propto \int p(x_t|x_{t-1}) p(y_t|x_{t-1}) p(x_{t-1}|y^{t-1}) dx_{t-1} \ , \tag{5}$$

where the symbol $\propto$ denotes proportionality. Formula (5) can be shown to be equivalent to the Zakai-equation of the martingale approach [4].

2. DEFINITIONS AND MAIN RESULTS

Let G denote a given family of g.p.d.f.'s of the form

$$G = \{ g(.;\psi) \mid \psi \in \Psi \subset R^p \} . \tag{6}$$

Associated to G consider for $m \in N$ the parameter set

$$\Theta_m := \{ \vartheta = (c_1,\dots,c_m;\psi_1,\dots,\psi_m) \mid c_i \in R, \sum_{i=1}^m c_i = 1; \psi_i \in \Psi \} \tag{7}$$

and the family of g.p.d.f.'s

$$F_m := \{ f(.;\vartheta) = \sum_{i=1}^m c_i g(.;\psi_i) \mid f(.;\vartheta) \text{ is a g.p.d.f.,}$$

$$\vartheta = (c_1,\dots,c_m;\psi_1,\dots,\psi_m) \in \Theta_m ; \ g(.;\psi_i) \in G \} \tag{8}$$

Notice that $\Theta_1=\Psi$, $F_1=G$. Letting

$$\Theta := \bigcup_{m=1}^{\infty} \Theta_m \tag{9}$$

define

$$d(.) : \Theta \to N \tag{10}$$

as the function that associates to $\vartheta=(c_1,\ldots,c_m; \psi_1,\ldots,\psi_m)$ the value $d(\vartheta)=m$. Finally, let

$$F := \bigcup_{m=1}^{\infty} F_m \tag{11}$$

<u>Definition 1.</u> A partially observable process (x_t,y_t), $x_t \in X \subset R$, $y_t \in Y \subset R$, admits a <u>generalized finite dimensional filter</u>, if there exists a family G of g.p.d.f.'s and a sequence of measurable mappings

$$\Phi_t : \Theta \times Y \to \Theta \tag{12}$$

such that $p(x_o)=f(x_o; \vartheta_o) \in F$ and for $t=1,2,\ldots,$

$$p(x_{t-1}|y^{t-1})=f(x_{t-1}; \vartheta_{t-1}) \in F \Rightarrow p(x_t|y^t)=f(x_t;\Phi_t(\vartheta_{t-1},y_t)) \in F \tag{13}$$

<u>Definition 2.</u> A generalized finite dimensional filter is called an <u>m_t-filter</u>, where m_t is a sequence in N, if $d(\Phi_t(\vartheta_{t-1},y_t))=m_t$, independent of y^t for all t.

Notice that from a computational point of view, essentially only m_t-filters can be considered.

Some of the following definitions are generalizations of those in [10].

<u>Definition 3</u>. A class of g.p.d.f.'s G as in (6) is said to be <u>weakly stable under addition</u> if there exists a mapping

$$\alpha : \Psi \times \Psi \to \Theta \tag{14}$$

such that for all pairs of independent r.v.'s X_1,X_2 with g.p.d.f.'s $g(x_1; \psi_1)$, $g(x_2; \psi_2) \in G$ the r.v. $Z=X_1+X_2$ has g.p.d.f.

$$f(z; \alpha(\psi_1, \psi_2)) \in F$$

If $d(\alpha(\psi_1, \psi_2)) \equiv 1$ we simply say that G is <u>stable under addition</u>.

Remark 1. Recalling that the g.p.d.f. of a sum of two independent r.v.'s is the convolution of the corresponding g.p.d.f.'s, weak stability under addition also implies that there exists a mapping

$$\alpha^* : \Theta \times \Theta \to \Theta \tag{15}$$

such that for all pairs of independent r.v.'s X_1, X_2 with g.p.d.f.'s $f(x_1; \vartheta_1), f(x_2; \vartheta_2) \in F$, the r.v. $Z=X_1+X_2$ has g.p.d.f.

$$f(z; \alpha^*(\vartheta_1, \vartheta_2)) \in F$$

Notice that weak stability under addition allows a convolution integral to be computed without actually performing the integration, thus reducing an infinite-dimensional problem to a finite-dimensional one.

Definition 4. Let $T : R \to R$ be a measurable mapping. A class of g.p.d.f.'s G as in (6) is said to be weakly stable under T, if there exists a mapping

$$\tau : \Theta \to \Theta \tag{16}$$

such that, if X has g.p.d.f. $f(x; \vartheta) \in F$, the r.v. $Z=T(X)$ has g.p.d.f.

$$f(z; \tau(\vartheta)) \in F$$

A particular case of Def.4. arises for $T(x)=a.x$ $(a \in A \subset R)$, in which case we say that G is (weakly) stable under multiplication by a constant $a \in A \subset R$.

Definition 5. A class of g.p.d.f.'s G as in (6) is said to be weakly closed under multiplication, if there exists a mapping

$$\mu : \Psi \times \Psi \to \Theta \tag{17}$$

such that, if $g(x; \psi_1)$, $g(x; \psi_2) \in G$, then $g(x; \psi_1).g(x; \psi_2) \propto$ $\propto f(x; \mu(\psi_1, \psi_2)) \in F$. If $d(\mu(\psi_1, \psi_2)) \equiv 1$, we simply say that G is closed under multiplication.

Remark 2. Weak closedness under multiplication also implies that there exists a mapping

$$\mu^* : \Theta \times \Theta \to \Theta \tag{18}$$

such that, if $f(x; \vartheta_1)$, $f(x; \vartheta_2) \in F$, then $f(x; \vartheta_1).f(x; \vartheta_2) \propto$ $\propto f(x; \mu^*(\vartheta_1, \vartheta_2)) \in F$.

Theorem 1. Let a partially observable process (x_t, y_t) of the form (1)-(3) be given. Under the assumptions:
A.1. There exists a mapping $\gamma : Y \to \Theta$ such that for all t

$$p(y_t \mid x_{t-1}) \propto f(x_{t-1}; \gamma(y_t)) \in F, \tag{19}$$

A.2. G is weakly stable under addition (Def.3.),
A.3. G is weakly stable under T (Def.4.),
A.4. G is weakly closed under multiplication (Def.6.),
there exists a generalized finite dimensional filter (Def.1.) and

$$\Phi_t(\vartheta_{t-1}, y_t) = \alpha^*(\tau(\mu^*(\gamma(y_t), \vartheta_{t-1})), \psi_t) \tag{20}$$

where $\alpha^*, \tau, \mu^*, \gamma$ are given in (15), (16), (18) and (19) respectively and $\psi_t \in \Psi$ is such that $p(v_t)=g(v_t; \psi_t)$.

Proof. From (1) we have $p(x_o)=g(x_o; \psi_o) \in G \subset F$. Assume now $p(x_{t-1} \mid y^{t-1}) = f(x_{t-1}; \vartheta_{t-1}) \in F$. From the Bayes formula (5), using (1), A.1. and A.4. one then has

$$p(x_t|y^t) \propto \int p(x_t|x_{t-1})p(y_t|x_{t-1})p(x_{t-1}|y^{t-1})dx_{t-1} \propto$$

$$\propto \int g(x_t - T(x_{t-1}); \psi_t)\ f(x_{t-1}; \gamma(y_t))f(x_{t-1}; \vartheta_{t-1})dx_{t-1} \propto$$

$$\propto \int g(x_t - T(x_{t-1}); \psi_t) f(x_{t-1}; \mu^*(\gamma(y_t), \vartheta_{t-1}))dx_{t-1} \tag{21}$$

It now remains to compute the integral in (21); to this effect notice that (1) implies (p(.) denotes a generic density)

$$p(x_t) = \int p(x_t|x_{t-1})p(x_{t-1})dx_{t-1} =$$

$$= \int g(x_t - T(x_{t-1}); \psi_t)p(x_{t-1})dx_{t-1} \tag{22}$$

so that the integral in (21) represents the distribution of x_t when

$$p(x_{t-1}) = f(x_{t-1}; \mu^*(\gamma(y_t), \vartheta_{t-1})) \in F \tag{23}$$

Observe now that from (1), using the independence of v_t and $T(x_{t-1})$ as well as A.3., A.2. and Remark 1., it follows that

$$p(x_t|y^t) = f(x_t; \alpha^*(\tau(\mu^*(\gamma(y_t), \vartheta_{t-1})), \psi_t) \in F \tag{24}$$

thereby completing the proof.

Remark 3. Assumptions A.1. and A.4. together imply that, in a weak sense, G is a conjugate family of distributions for $\{p(y_t \mid x_{t-1}),\ t=1,2,\dots\}$.

In order for the filter in (20) to be an m_t-filter (Def.2.), one has to make sure that $d(\alpha^*(\tau(\mu^*(\gamma(y_t),\vartheta_{t-1})),\psi_t)$ does not depend on y^t for all t. A-priori conditions that guarantee this property in general are overly restrictive, so that one has to take into account the particular structure of a given problem. This is done for various situations in the next Section, where we obtain conditions that are easily verifiable and where the resulting m_t-filter assumes a tractable form.

3. APPLICATIONS

3.1. Linear models with Gamma-type disturbances.

For r, $s>0$, denote by $g(x;r,s)$ the Gamma density with parameter (r,s),namely

$$g(x;r,s) = \frac{s^r}{\Gamma(r)}\, x^{r-1}\, e^{-sx}\, I_{(0,\,+\infty)}(x) \tag{25}$$

and let G be the class of such densities with integer r parameter

$$G = \{ g(x;\psi) \mid \psi=(r,s),\ r\in Z^+,\ s>0 \}. \tag{26}$$

Consider a partially observable process (x_t, y_t) of the form (1)-(3) with $T(x)=ax$, $a>0$, and $p(v_t)=g(v_t;r_t,s_t)$. We shall show here that for several choices of $p(y_t \mid x_{t-1})$ it is possible to satisfy the requirements of Theorem 1 and moreover to obtain an m_t-dimensional filter.

By direct calculation it is possible to show that the class G is closed under multiplication and furthermore stable under multiplication by a positive real constant. The following Proposition shows that G is also weakly stable under addition. The proof, requiring essentially only formal manipulations, can be found in [3].

Proposition. Let X and X' be independent r.v.'s with densities $g(x;r,s)$ and $g(x';r',s')$ respectively; then defining $Z=X+X'$,
i) if $s=s'$, the density of Z is $g(z;r+r',s)$,
ii) if $s\neq s'$, the density is

$$\sum_{k=1}^{r} c_k g(z;k,s) + \sum_{k=1}^{r'} c'_k g(z;k,s'),$$

where

$$c_k = \binom{r+r'-k-1}{r'-1} \left(\frac{s'}{s'-s}\right)^{r'} \left(\frac{s}{s-s'}\right)^{r-k},$$

$$c'_k = \binom{r+r'-k-1}{r-1} \left(\frac{s}{s-s'}\right)^{r} \left(\frac{s'}{s'-s}\right)^{r'-k}$$

The following classes of g.p.d.f.'s will be considered as possible choices for $p(y_t|x_{t-1}=x)$:

C.1. (Gamma) $p(y_t|x)=\frac{x^q}{\Gamma(q)} y_t^{q-1} \exp(-xy_t)I_{(0,+\infty)}(y_t)\cdot$

$$\cdot I_{(0,+\infty)}(x), \quad q\in N$$

C.2. (Beta) $p(y_t|x)=xy_t^{x-1} I_{(0,1)}(y_t) I_{(0,+\infty)}(x)$

C.3. (Pareto) $p(y_t|x)=xw^x y_t^{x-1} I_{(w,+\infty)}(y_t)I_{(0,+\infty)}(x)$, $w>0$

C.4. (Laplace) $p(y_t|x)=\frac{1}{2}\frac{x^q}{\Gamma(q)} |y_t|^{q-1}\exp(-x|y_t|)I_{(0,+\infty)}(x), q\in N$

C.5. (Poisson) $p(y_t|x)=\frac{1}{y_t!} e^{-x}x^{y_t} I_N(y_t)I_{(0,+\infty)}(x)$

It is easily seen that for these classes, assumption A.1. of Theorem 1 is satisfied with

$$\gamma(y_t):=(\gamma_t^1, \gamma_t^2) = \begin{cases} (q+1,\ y_t) & \text{for C.1.} \\ (2,-\lg y_t) & \text{for C.2.} \\ (2,-\lg(w/y_t)) & \text{for C.3.} \\ (q+1,|y_t|) & \text{for C.4.} \\ (y_t+1,\ 1) & \text{for C.5.} \end{cases}$$

Therefore, for these classes of models, Theorem 1 provides a generalized finite-dimensional filters. Under additional conditions, these filters are actually m_t-filters.
More precisely, for $p(y_t|x_{t-1})$ as given in C.5., under the additional condition $s_t+1=a\cdot s_{t+1}$, $t=0,1,\ldots,$(20) provides an m_t-filter with $m_t\equiv 1$, that is given by

$$p(x_t|y^t) = g(x_t;\ \sum_{k=0}^{t} (r_k+y_k),\ s_t) \tag{27}$$

This result is also obtained in [10] using a different approach. Furthermore, if $p(y_t|x_{t-1})$ is as in C.1. through C.4., under the additional condition $as_{t+1}-s_t \leqslant 0$, $t=0,1,\dots$, (20) provides an m_t-dimensional filter given by

$$p(x_t|y^t) = \sum_{k=0}^{t} \sum_{h=1}^{A_k^t} D_{kh}^t g(x_t;h,B_k^t) ,$$

with $\quad m_t = \sum_{k=0}^{t} A_k^t ,$

where

$$A_k^t = A_k^{t-1} + \gamma_t^1 - 1 , \quad t=1,2,\dots; \ k=0,1,\dots,t-1,$$

$$A_0^0 = r_0, \ A_t^t = r_t \text{ (implying that in fact } m_t \text{ does not depend on } y^t),$$

$$B_k^t = (B_k^{t-1} + \gamma_t^2)/b \quad t=1,2,\dots; \ k=0,1,\dots,t-1,$$

$$B_0^0 = s_0, \ B_t^t = s_t ,$$

$$D_{kh}^t = P_{kh}^t/N^t , \ N^t = \sum_{k=0}^{t} \sum_{h=1}^{A_k^t} P_{kh}^t$$

with

$$P_{kh}^t = \sum_{i=1 \vee h-\gamma_t^1+1}^{A_k^{t-1}} F_h(i+\gamma_t^1-1, r_t; B_k^t, s_t) D_{ki}^{t-1} K_{ki}^{t-1} ,$$

$$t=2,3,\dots; \ k=0,1,\dots, t-1; \ h=1,2,\dots,A_k^t$$

$$P_{th}^t = \sum_{i=0}^{t-1} \sum_{j=1}^{A_i^{t-1}} F_h(\gamma_t^1, j+\gamma_t^1-1; s_t, B_i^t) D_{ij}^{t-1} K_{ij}^{t-1} ,$$

$$t=2,3,\dots; \ h=1,2,\dots,r_t$$

$$P_{0h}^1 = F_h(A_0^1, r_1; B_0^1, s_1)$$

$$P_{1h}^1 = F_h(r_1, A_0^1; s_1, B_0^1) ,$$

where

$$F_h(u,u';v,v') := \binom{u+u'-h-1}{u'-1}\left(\frac{v'}{v'-v}\right)^{u'}\left(\frac{v}{v-v'}\right)^{u-h}, \quad u,u' \in N$$

$$K_{ij}^t = G(j, \gamma_t^1; B_i^t, \gamma_t^2), \; t=1,2,\ldots; \; i=0,1,\ldots,t; \; j=1,2,\ldots,A_i^t$$

$$G(u,u',v,v') := v^u (v')^{u'} (v+v')^{1-u-u'} \left[(u+u'-1)B(u,u')\right]^{-1},$$

where B is the beta function.
Further m_t-filters for linear models with non-Gaussian disturbances are examined in [3].

2.2. A nonlinear model with Gaussian disturbances

Denote by $g(x;m,\sigma^2)$ the normal density with mean m and variance σ^2 and let

$$G = \left\{ g(x;m,\sigma^2) \,\middle|\, m, \sigma^2 \in R, \sigma^2 \geqslant 0 \right\},$$

Consider a partially observable process (x_t, y_t) of the form (1)-(3) with $p(v_t)=g(v_t; m_t, \sigma_t^2)$, $p(y_t|x_{t-1})=g(y_t;x_{t-1},s_t^2)$, $s_t^2>0$ and $T(x)$ a step-function of the form

$$T(x) = \sum_{k=1}^{n} b_k I_k(x), \tag{28}$$

where $b_k \in R(k=1,\ldots,n)$ and I_k is the indicator function of the interval $(a_k, a_{k+1}]$ with $a_k \in R(k=2,..,n-1)$ and $-\infty=a_1 < a_2 \; \ldots <$ $a_n = +\infty$. We shall show that, using Theorem 1, it is possible to obtain for this model an m_t-filter with $m_t \equiv n$.

By direct calculation it is easily seen that G is closed under multiplication. Furthermore, observing that T maps every r.v. into a discrete r.v., whose distribution can be interpreted as a convex combination of normal g.p.d.f.'s, it follows that G is weakly stable under T. Finally, notice that G is stable under addition and that $p(y_t|x_{t-1})$ satisfies assumption A.1. in Theorem 1 with

$$\gamma(y_t) = (y_t, s_t^2)$$

Applying Theorem 1, it is possible [5] to obtain a filter of constant dimension n, given by

$$p(x_t|y^t) = \sum_{k=1}^{n} c_{tk} g(x_t; m_t + b_k, \sigma_t^2), \quad t=1,2,\ldots, \tag{29}$$

where

$$c_{tk} = \frac{1}{N_t} \sum_{h=1}^{n} c_{(t-1)h} K_{(t-1)h} \left[\mathcal{N}\left(\frac{a_{k+1} - M_{th}}{W_t}\right) - \mathcal{N}\left(\frac{a_k - M_{th}}{W_t}\right) \right] ,$$

$$t = 1,2,\ldots;\ k=1,2,\ldots,n$$

$$c_{01} = 1;\ c_{0k} = 0,\ k = 2,\ldots,n\ .$$

$$K_{th} = g(m_t + b_h;\ y_{t+1},\ \sigma_t^2 + s_t^2),\quad t=1,2,\ldots;\ h=1,\ldots,n$$

$$K_{01} = 1;\ K_{0h} = 0\ ,\ h = 2,\ldots,\ n\ .$$

$$M_{th} = \frac{(m_{t+1} + b_h)\, s_{t-1}^2 + y_t\, \sigma_{t-1}^2}{s_{t-1}^2 + \sigma_{t-1}^2}\ ,\ t=2,3,\ldots;\ h=1,\ldots,n$$

$$M_{11} = M_1\ ;\ M_{1h} = 0\ ,\ h=2,\ldots,n$$

$$W_t^2 = \frac{s_{t-1}^2 \cdot \sigma_{t-1}^2}{s_{t-1}^2 + \sigma_{t-1}^2}$$

$$N_t = \sum_{h=1}^{n} c_{(t-1)h}\ K_{(t-1)h}\ ,$$

and $\mathcal{N}(x)$ is the standard normal distribution function. This filter can also be used for the approximation of more general nonlinear filtering problems [5] .

REFERENCES

1. D.L. Alspach and H.W. Sorenson, "Nonlinear Bayesian estimation using Gaussian sum approximations". IEEE Trans. AC 17, pp.439-448 (1972).

2. M. Aoki, "Optimization of stochastic systems". Academic Press (1967).

3. B. Barozzi, "Filtri a tempo discreto di dimensione finita generalizzati". Thesis, Università di Padova (1982).

4. G.B. Di Masi and W.J. Runggaldier, "On measure transformations for combined filtering and parameter estimation in discrete time". Systems & Control Letters (to appear).

5. G.B. Di Masi and W.J. Runggaldier, "Approximations and bounds for discrete-time nonlinear filtering". Preprint (1982).

6. G. Sawitzki, "Exact filtering in exponential families: discrete time", in "Stochastic control theory and stochastic differential systems" (M. Kohlmann and W. Vogel eds.). L.N. in Contr. and Inf. Sci. 16, Springer-Verlag (1979).

7. G. Sawitzki, "Finite-dimensional filter systems in discrete time". Stochastics 5, pp.107-114 (1981).

8. H.W. Sorenson and D.L. Alspach, "Recursive Bayesian estimation using Gaussian sums". Automatica 7, pp.465-479 (1971).

9. J.H. Van Schuppen, "Stochastic filtering theory: a discussion of concepts, methods and results", in "Stochastic control theory and stochastic differential systems" (M. Kohlmann and W. Vogel eds.). L.N. in Contr. and Inf. Sci. 16, Springer Verlag (1979).

10. J.H. Van Schuppen, "A study of estimation and filtering by the Bayesian method", SSM-Report 7607, Washington University, St. Louis (1976).

NONLINEAR FILTERING EQUATION FOR HILBERT SPACE VALUED PROCESSES

H. KOREZLIOGLU

Ecole Nationale Supérieure des Télécommunications
Département Systèmes et Communications
46, rue Barrault - 75634 PARIS CEDEX 13 - FRANCE

RESUME

The integration of operator-valued processes with respect to a Hilbert space valued Brownian motion is shown to be equivalent to the integration of Hilbert-Schmidt operator valued processes with respect to a cylindrical Brownian motion. As a consequence, a more general filtering equation than those already known is obtained.

INTRODUCTION

The construction of the stochastic integrals of linear operator valued processes with respect to Hilbert space-valued square-integrable martingales was given by Métivier and Pistone in [8] and this type of stochastic integrals was used for the representation of martingales ([9]), for the elaboration of Girsanov theorems([7], [10]) and for the derivation of filtering equations in the field of Hilbert space-valued processes ([6], [10]). In [3] and [4] applications were considered for the filtering of two parameter processes. The necessity of putting into one-to-one correspondence Hilbertian stochastic integrals and one sided two-parameter stochastic integrals inspired the present work.

After a paragraph of notations and introductory definitions, in Paragraph 2, stochastic integration is considered. By means of the stochastic integral of processes whose values are linear functionals on a real separable Hilbert space H, with respect to a H-valued Brownian motion W, a cylindrical Brownian motion $\widetilde{W}$ on a certain Hilbert space $\widetilde{H}$ of linear functionals on H is obtained. The definition of the stochastic integral of Hilbert-Schmidt operator-valued processes on $\widetilde{H}$ is then given according to [5], [7], [12]. It

R. S. Bucy and J. M. F. Moura (eds.), Nonlinear Stochastic Problems, 279–289.

is shown here that there is a one-to-one correspondence between this type of stochastic integrals and the stochastic integrals, in the sense of Métivier-Pistone [8], of operator-valued processes on H with respect to W. Stochastic integrals defined here depend on an arbitrary factorization DD^* of the covariance operator Q of W, whereas in [8] their construction is based on the particular factorization for which $D=D^*=Q^{1/2}$. There is an isometric correspondence between the two cases. The construction based on a different factorization than in [8] was again necessitated by the applications considered in [3] and [4].

In Paragraph 3, the nonlinear filtering equation for Hilbert space-valued processes is given where the martingale term is expressed as a stochastic integral with respect to the cylindrical Brownian motion corresponding to the innovation process. As the Girsanov theorem obtained here is more general than the one used in [6] and [10], the filtering equation derived here extends those obtained in these two works.

1. NOTATIONS AND PRELIMINARY DEFINITIONS

All the Hilbert spaces considered here are real. H, K and G represent separable Hilbert spaces. Scalar products and norms are denoted by $(.|.)$ and $\|.\|$, respectively. In order to precise, if necessary, the space on which they are defined they will be indexed by the symbol representing the space.

L(H,K) is the space of bounded linear operators from H into K with the uniform norm $\|.\|_{HK}$, $L^1(H,K)$ the space of nuclear operators with the trace norm $\|.\|_{H1K}$ and $L^2(H,K)$ the space of Hilbert-Schmidt operators with the Hilbert-Schmidt norm $\|.\|_{H2K}$. $H\hat{\otimes}_1K$ (resp. $H\hat{\otimes}_2K$) is the projective (resp. Hilbertian) tensor product of H with K. For notational conveniences, $H\hat{\otimes}_1K$ (resp. $H\hat{\otimes}_2K$) will be identified with $L^1(H,K)$ (resp. $L^2(H,K)$) under the isometry which puts into one-to-one correspondence $h\otimes k$ with $(.|h)_H k$ for $h\in H$, $k\in K$.

The transpose of a linear operator U is denoted by U^*, its domain by Dom U and range by Rg U.
The closure of a set S in denoted by $\overline{S}$.
Every Hilbert space will be identified with its topological dual.
We shall often write h(x) instead of (x|h).

For the terminology of stochastic analysis used here we refer to [1] and [7].
All random variables and processes are supposed to be defined on a complete probability space $(\Omega,\underline{\underline{A}},P)$ with a filtration $\underline{\underline{F}}=(\underline{\underline{F}}_t,$ $t\in[0,T])$ satisfying the usual completeness and right continuity conditions. We take $\underline{\underline{A}}=\underline{\underline{F}}_T$. Unless the contrary is mentioned, adapted processes, predictable processes, martingales, etc. are with

respect to $\underline{\underline{F}}$. Although almost all the results presented here are valid for processes indexed by $\mathbb{R}_+$, for brevity only processes indexed by $[0,T]$ will be considered. $\mathcal{B}$ represents the Borel field of $[0,T]$ and $\mathcal{P}$ the σ-algebra of $\underline{\underline{F}}$-predictable elements of $\mathcal{B} \otimes \underline{\underline{F}}_T$. $L^2_H([0,T] \times \Omega, \mathcal{B} \otimes \underline{\underline{F}}_T, dt \times dP)$ (resp. $L^2_H([0,T] \times \Omega, \mathcal{P}, dt \times dP)$) is denoted by $L^2_H(\mathcal{B} \otimes \underline{\underline{F}}_T)$ (resp. $L^2_H(\mathcal{P})$).

For two square integrable martingales M and N, with values in H and K, respectively, there is a unique $H \hat{\otimes}_1 K$-valued predictable process denoted by $\langle M,N \rangle$ such that $M_o \otimes N_o = \langle M,N \rangle_o$, $E \int_o^T \| d\langle M,N \rangle \|_{H1K} < \infty$ and $M \otimes N - \langle M,N \rangle$ is a $H \hat{\otimes}_1 K$-valued martingale. When M=N (for H=K), $\langle M,N \rangle$ is called the increasing process of M.

A H-valued Brownian motion $W=(W_t)$ is a continuous square-integrable martingale such that $W_o = o$ and $\langle W,W \rangle = tQ$ where Q is a nonnegative element of $H \hat{\otimes}_1 H$, called the covariance operator of W.

Throughout this paper W will be a H-valued Brownian motion with covariance operator Q.

$\{q_n, n \in \mathbb{N}\}$ represents the nonincreasing sequence of positive eigenvalues of Q, each eigenvalue being repeated as many times as its multiplicity, and $\{e_n, n \in \mathbb{N}\}$ is the corresponding sequence of eigenvectors. Operator Q has then the following representation

$$(1.1) \quad Q = \sum_{n=o}^{\infty} q_n e_n \otimes e_n$$

D represents an operator that realizes the factorization :

$$(1.2) \quad Q = DD^\star$$

Such an operator D is always a Hilbert-Schmidt operator and it has the following representation :

$$(1.3) \quad D = \sum_{n=o}^{\infty} q_n^{1/2} (D^\star q_n^{-1/2} e_n) \otimes e_n \in H \hat{\otimes}_2 H$$

Let $\mathcal{L}^2(H,K,D)$ be the space of (not necessarily continuous) linear operators A from H into K such that Dom $A \supset$ Rg D and $AD \in L^2(H,K)$. The bilinear form $(A|B) = \mathrm{Tr}\,[(BD)(AD)^\star]$ defines a scalar product with respect to which the above space $\mathcal{L}^2$ is complete. Consequently, for $A \in \mathcal{L}^2$, $\|AD\|_{H2K}$ defines a seminorm and the quotient of $\mathcal{L}^2$ by the equivalence relation : $A \sim B$ iff $\|(A-B)D\|_{H2K} = o$, is a Hilbert space. With an abuse of notation we shall still denote by $\mathcal{L}^2(H,K,D)$ the quotient space and not distinguish operators with their equivalence classes. We denote by $\widetilde{L}^2(H,K,D)$ (or shortly by $\widetilde{L}^2$) the smallest sub-Hilbert space of $\mathcal{L}^2(H,K,D)$ containing $L(H,K)$.

The space $\widetilde{L}^2(H,\mathbb{R},D)$ of functionals on H is denoted by $\widetilde{H}$. H, identified with its topological dual, is dense in $\widetilde{H}$. If a continuous functional $g^\star$ on H is represented by g then $g^\star D$ is represented by

$D^\star g$. Therefore,

$$(1.4) \quad (f|g)_{\widetilde{H}} = (D^\star f|D^\star g)_H \quad \text{for } f, g \in H$$

So, $D^\star$ defines an isometry from H into H which extends to $\widetilde{H}$. We denote by I this isometry of $\widetilde{H}$ into H. Notice that $I\widetilde{H} = \overline{D^\star H}$.

If $h \in \widetilde{H}$ then there is a sequence $\{h_n, n \in \mathbb{N}\}$ in H converging to h in the topology of $\widetilde{H}$. Considered as a linear functional on H, h can be evaluated by $h(x) = \lim_{n\to\infty} (h_n|x)_H$ on the subset of H on which the limit exists. This subset contains Rg D and $h(Dx) = (x|Ih)_H$ defines a continuous functional on H.

According to (1.4) it is seen that $\{\widetilde{e}_n = q_n^{-1/2} e_n, n \in \mathbb{N}\}$ is an orthonormal basis of $\widetilde{H}$.

PROPOSITION 1.1 : The spaces $\widetilde{L}^2(H,K,D)$ and $L^2(\widetilde{H},K)$ are isometric.

Proof : Let $\widetilde{A} \in L^2(\widetilde{H},K)$ be defined by

$$(1.5) \quad \widetilde{A} = \sum_{n=0}^{\infty} a_n \widetilde{h}_n \otimes k_n \in \widetilde{H} \hat{\otimes}_2 K$$

where (a_n) is a square-summable sequence of positive numbers, $(\widetilde{h}_n)$ and (k_n) are orthonormal sequences in $\widetilde{H}$ and K, respectively. Let A be defined by

$$(1.6) \quad Ax = \sum_{n=0}^{\infty} a_n \widetilde{h}_n(x) k_n$$

Approximating by operators of finite rank, it can be proved that $A \in \widetilde{L}^2(H,K,D)$. Since $\|AD\|^2_{H2K} = \sum_n a_n^2 = \|\widetilde{A}\|^2_{\widetilde{H}2K}$ it is seen that the above correspondence $\widetilde{A} \to A$ defines an isometry I_1 of $L^2(\widetilde{H},K)$ into $\widetilde{L}^2(H,K,D)$.

Now, let $B \in L(H,K)$ be such that

$$(1.7) \quad BD = \sum_{n=0}^{\infty} b_n h_n \otimes k_n \in H \hat{\otimes}_2 K$$

where (h_n) and (k_n) are orthonormal sequences. We have $D^\star B^\star k_n/b_n = h_n \in D^\star H$. Let then $\widetilde{h}_n$ be an element of H such that $D^\star\widetilde{h}_n = h_n$ and define

$$(1.8) \quad \widetilde{B} = \sum_{n=0}^{\infty} b_n \widetilde{h}_n \otimes k_n \in \widetilde{H} \hat{\otimes}_2 K$$

We have $\|\widetilde{B}\|^2_{\widetilde{H}2K} = \|BD\|^2_{H2K} = \|B\|^2_{\widetilde{L}^2} = \sum_n b_n^2$

Therefore, the correspondence $B \to \widetilde{B}$ extends to an isometry I_2 of $\widetilde{L}^2$ into $L^2(\widetilde{H},K)$.

Consider the operator $\overline{B} = I_1 I_2 B \in \widetilde{L}^2(H,K,D)$. We have $\|\overline{B}-B\|_{\widetilde{L}^2} = 0$. This implies that I_1I_2 reduces to the identity operator on $\widetilde{L}^2(H,K,D)$. ∎

The above isometry of $\widetilde{L}^2(H,K,D)$ onto $L^2(\widetilde{H},K)$, will be denoted by J and for $A \in \widetilde{L}^2(H,K,D)$ we shall also write $\widetilde{A}$ instead of JA.

2. STOCHASTIC INTEGRATION

We put $\Lambda^2(H,K,D,\mathcal{P}) = L^2_{\widetilde{L}^2}(\mathcal{P})$ where $\widetilde{L}^2 = \widetilde{L}^2(H,K,D)$ and $\widetilde{\Lambda}^2(\widetilde{H},K,\mathcal{P}) = L^2_{L^2}(\mathcal{P})$ where $L^2 = L^2(\widetilde{H},K)$

We consider here the stochastic integration, in the sense of Métivier-Pistone [8], of processes in $\Lambda^2(H,K,D,\mathcal{P})$ with respect to W, deduce from this a cylindrical Brownian motion $\widetilde{W}$ on $\widetilde{H}$ and show that the above mentioned stochastic integration is equivalent to the integration of processes in $\widetilde{\Lambda}^2(\widetilde{H},K,\mathcal{P})$ with respect to $\widetilde{W}$.

The isometry J of $\widetilde{L}^2(H,K,D)$ onto $L^2(\widetilde{H},K)$ induces an isometry of $\Lambda^2(H,K,D,\mathcal{P})$ onto $\widetilde{\Lambda}^2(\widetilde{H},K,\mathcal{P})$ that we denote again by J.

A process $X \in \Lambda^2(H,K,D,\mathcal{P})$ is said to be elementary if it is of the form :

$$(2.1) \quad X_t(\omega) = \sum_{k=o}^{n} A_k \, 1_{F_k}(\omega) \, 1_{]t_k,\, t_{k+1}]}(t)$$

where $A_k \in L(H,K)$, $o \leqslant t_o < t_1 < \ldots < t_n$ and $F_k \in \underline{\underline{F}}_{t_k}$. The space of elementary processes is dense in $\Lambda^2(H,K,D,\mathcal{P})$. If X is an elementary process given by (2.1), then its stochastic integral with respect to W is given by

$$(2.2) \quad \int_o^T X_t \, dW_t := \sum_{k=o}^{n} 1_{F_k} A_k (W_{t_{k+1}} - W_{t_k}).$$

This correspondence $X \to \int X \, dW$ defines an isometry from the pre-Hilbert space of elementary processes into $L^2_K(\Omega, \underline{\underline{F}}_T, P)$.

For an arbitrary $X \in \Lambda^2(H,K,D,\mathcal{P})$, the integral $\int_o^T X \, dW$ is defined by an extension of this isometry to the whole space $\Lambda^2(H,K,D,\mathcal{P})$.

For $X \in \Lambda^2(H,K,D,\mathcal{P})$, the process X.W defined by $\int_o^t X_s dW_s$ is a square-integrable K-valued continuous martingale. On the other hand if $Y \in \Lambda^2(H,G,D,\mathcal{P})$ then

$$(2.3) \quad \langle X.W,\, Y.W \rangle_t = \int_o^t (Y_s D)(X_s D)^{\star} ds$$

Although slightly different in its presentation the above integration is the one developed in [7] and [8] with $D = Q^{1/2}$.

For $K = \mathbb{R}$, $\Lambda^2(H,K,D,\mathcal{P})$ reduces to $L^2_{\widetilde{H}}(\mathcal{P})$ and of course, if X is an element in this space then X.W is a square-integrable real martingale.

Suppose now $h \in \widetilde{H}$ and consider the process $(\widetilde{W}_t(h))$ defined by the integration of the constant process h, i.e.

(2.4) $\widetilde{W}_t(h) := \int_o^t h\, dW_s$

This random variable can also be calculated as follows. Let $\{h_n, n \in \mathbb{N}\} \subset H$ converge to h in $\widetilde{H}$. Then $W_t(h_n) = (h_n|W_t)_H$ converges in $L^2(\Omega, \underline{\underline{F}}_T, P)$ and the limite random variable coincides with $\widetilde{W}_t(h)$. The process $(\widetilde{W}_t(h))$ has the following properties.

(i) $(\|h\|_{\widetilde{H}}^{-1} \widetilde{W}_t(h))$ is a real Brownian motion

(ii) For each t, $h \to \widetilde{W}_t(h)$ is a continuous linear mapping of $\widetilde{H}$ into $L^2(\Omega, \underline{\underline{F}}_t, P)$ such that $\|\widetilde{W}_t(h)\|^2 = t\|h\|^2$.

Therefore, $\{\widetilde{W}_t(h), t \times h \in [0,T] \times \widetilde{H}\}$ is a standard cylindrical Brownian motion on $\widetilde{H}$ (cf. [7]). Stochastic integration with respect to a cylindrical Brownian motion was considered in [5], [7], [12]. We just give its definition here.

For $X \in \Lambda^2(H, R, D, \mathcal{P}) = L^2_{\widetilde{H}}(\mathcal{P})$ we put

(2.5) $\int_o^T X_t\, d\widetilde{W}_t := \int_o^T X_t\, dW_t$

We have $X_t = \sum_{n=o}^{\infty} x_{n,t}\, \widetilde{e}_n$

where $(\widetilde{e}_n)$ is the orthonormal basis of $\widetilde{H}$ considered in paragraph 1 and $x_{n,.} = (X.|\widetilde{e}_n)_{\widetilde{H}} \in L^2_{\mathbb{R}}(\mathcal{P})$

Let us put

(2.6) $W_{n,t} = \widetilde{W}_t(\widetilde{e}_n) = q_n^{-1/2}(e_n|W_t)_H$

$\{W_n, n \in N\}$ is a sequence of mutually independent Brownian motions. We then have

(2.7) $\int_o^T X_t\, d\widetilde{W}_t = \sum_{n=o}^{\infty} \int_o^T x_{n,t}\, dW_{n,t}$

where the series converges in $L^2(\Omega, \underline{\underline{F}}_T, P)$.

Starting from this last equality, it is easy to define the stochastic integral of processes in $\widetilde{\Lambda}^2(\widetilde{H},K,\mathcal{P})$. If X is such a process then it can be represented as

(2.8) $X_t = \sum_{(m,n)\in\mathbb{N}^2} x_{m,n,t}\, \widetilde{e}_m \otimes k_n$

where (k_n) is an orthonormal basis of K and $x_{m,n} \in L^2_{\mathbb{R}}(\mathcal{P})$. Then

(2.9) $\int_o^T X_t\, d\widetilde{W}_t = \sum_{(m,n)\in\mathbb{N}^2} (\int_o^T x_{m,n,t}\, dW_{m,t})k_n$,

where the series converges in $L^2_K(\Omega, \underline{\underline{F}}_T, P)$.

We denote by $X.\widetilde{W}$ the process defined by $\int_o^t X\, d\widetilde{W}$. It is elementary to prove that if $Y \in \widetilde{\Lambda}^2(\widetilde{H},G,\mathcal{P})$ then

(2.10) $\langle X.\widetilde{W}, Y.\widetilde{W} \rangle_t = \int_o^t Y_s X_s^{\star}\, ds$

We see that the integration of elements of $\widetilde{\Lambda}^2(\widetilde{H},K,\mathcal{P})$ with respect to $\widetilde{W}$ as defined by (2.9) reduces to the integration of real valued processes with respect to real valued Brownian motions.

The relation between the stochastic integrals of the two types we considered above is given by the following theorem.

THEOREM 2.1 : For $X \in \Lambda^2(H,K,\mathcal{P})$ or, equivalently, $\widetilde{X} = JX \in \widetilde{\Lambda}^2(\widetilde{H},K,\mathcal{P})$ we have

$$(2.11) \qquad \int_0^T X_t \, dW_t = \int_0^T \widetilde{X}_t \, d\widetilde{W}_t$$

Proof : Consider the operators B, $\widetilde{B}$ and $\overline{B}$ introduced in the proof of Proposition 1.1. We have $BW_t = \overline{B}W_t$. But we also have by definition

$$(\widetilde{B}.\widetilde{W})_t = \sum_{n=o}^{\infty} b_n \widetilde{W}_t(\widetilde{h}_n)k_n$$

$$= \sum_{n=o}^{\infty} b_n \left(\int_0^t \widetilde{h}_n \, dW_s\right)k_n = \overline{B}\, W_t = B\, W_t$$

This equality extends to elementary processes in $\Lambda^2(H,K,D,\mathcal{P})$ and hence to the whole space to give equality (2.11).■

REMARKS :

(i) The set of random variables which are defined by stochastic integrals $(X.W)_T = (\widetilde{X}.\widetilde{W})_T$ does not depend on the factorization $DD^\star$ of Q, because the integral of elementary processes is independent on this factorization and various spaces Λ^2 corresponding to various factorizations are isometric to each other.

(ii) In the definition of the stochastic integral with respect to a Brownian motion one can replace predictable processes by measurable and adapted processes and reach the same conclusions.

One of the main motivations of the preceding paragraph is the improvement of the Girsanov Theorem. We present it below.

THEOREM 2.2 : Let $\widetilde{\ell} \in \Lambda^2(H, R, D,\mathcal{P}) = L^2_{\widetilde{H}}(\mathcal{P})$ be given and put

$$(2.12) \qquad h = D(I\widetilde{\ell}) \in L^2_H(\mathcal{P})$$

$$(2.13) \qquad Y_t = \int_0^t h_s \, ds + W_t$$

$$(2.14) \qquad Z_t = \exp\left(-\int_0^t \widetilde{\ell}_s \, dW_s - \frac{1}{2}\int_0^t \|\widetilde{\ell}_s\|^2_{\widetilde{H}} \, ds\right)$$

If $E(Z_T) = 1$, then $dQ = Z_T\, dP$ is a probability measure on $\underline{\underline{F}}_T$ and Y is a Brownian motion under Q (with respect to the filtration $\underline{\underline{F}}$).

A theorem of this form, but with $D = Q^{1/2}$, was also elaborated in [11]. In the Girsanov theorem as in [7], used in [6] and [10] for the derivation of the filtering equation, h was taken to be an element of $QH = D(D^{\star}H)$, whereas here it is an element of $D(\overline{D^{\star}H})$. This improvement was made necessary for the applications to two-parameter processes considered in [3] and [4].

In the next paragraph we give the filtering equation taking into account the above version of the Girsanov Theorem.

3. FILTERING EQUATION

The "state and observation" model presented below extends to Hilbert space valued processes the model of [2] in the finite dimensional case.
The state process is a K-valued process given by

$$(3.1) \quad X_t = X_o + \int_o^t f_s \, ds + \int_o^t g_s \, dB_s$$

and the observation process is a H-valued process given by Equation (2.13).

$X_o \in L^2_K(\Omega, \underline{\underline{F}}_T, P)$, B is a G-valued Brownian motion, and X_o is independent of B and W.

For all the objects related to W, we use the notations of the preceding paragraphs. Q_B denotes the covariance operator of B.

The filtration $\underline{\underline{F}}$ considered here is the smallest filtration to which B and W are adapted and such that $\sigma(X_o) \subset \underline{\underline{F}}_o$.

It is supposed that B and W are both Brownian motions with respect to $\underline{\underline{F}}$. In this case, there exists a predictable $L^1(H,G)$-valued process C with bounded norm such that $\langle W,B\rangle_t = \int_o^t C_s \, ds$.

g is a weakly predictable process with values in L(G,K).
($\forall$ u $\in$ G, k $\in$ K, $(g_t u|k)_K$ is a predictable process). We suppose $E \int_o^T \|g_t\|^4_{GK} \, dt < \infty$.

f is a K-valued $\underline{\underline{F}}$-progressive process such that $E \int_o^T \|f_t\|^4 \, dt < \infty$.
We suppose that h and $\tilde{\ell}$ are as in Theorem 2.2 and that $E \int_o^T \|\tilde{\ell}_t\|^4_H \, dt < \infty$.

In order to be able to express the filtering equation we shall need the following lemmas 3.1 and 3.2.

$\underline{\underline{G}}$ denotes the smallest filtration to which Y is adapted.

LEMMA 3.1 : Let L be a separable Hilbert space and let $Z \in L^2_L(\mathcal{B} \otimes \underline{\underline{F}}_T)$ be adapted to $\underline{\underline{F}}$. Then there is a process $\hat{Z}$, in the same space, adapted to $\underline{\underline{G}}$ such that $\hat{Z}_t = E(Z_t/\underline{\underline{G}}_t)$a.s. In case Z is $\underline{\underline{F}}$-predictable then $\hat{Z}$ can be chosen to be $\underline{\underline{G}}$-predictable.

The proof is based on the classical arguments.

LEMMA 3.2 : Let M be the martingale g.B in (3.1) then there is a process $\psi \in \Lambda^2(H,K,D,\mathcal{P})$ such that

(3.2) $g\,C = \psi\,Q$
(3.3) $M = g.B = \psi.W + M^{\perp}$

where $M^{\perp}$ is a square-integrable martingale orthogonal to W.

Proof : Representation (3.3) is proved in [9]. We have

$$< W, g.B>_t = \int_0^t g_s\, C_s\, ds = < W,\ \psi.W > = \int_0^t \psi_s\, Q\, ds\ ,$$

from which (3.2) is deduced. ∎
Here is now the filtering equation.

THEOREM 3.3 : (i) The innovation process ν defined by

(3.4) $\nu_t = Y_t - \int_0^t \hat{h}_s\, ds$

is a H-valued $\underline{\underline{G}}$-Brownian motion with covariance operator Q.

(ii) Let $\tilde{\nu}$ be the corresponding cylindrical Brownian motion on $\tilde{H}$. Then we have

$$(3.5)\quad \hat{X}_t = E(X_o) + \int_o^t \hat{f}_s\, ds + \int_o^t (\widehat{\tilde{\ell}_s \otimes X_s} - \hat{\tilde{\ell}}_s \otimes \hat{X}_s + \hat{\tilde{\psi}}_s)\, d\tilde{\nu}_s$$

where $\tilde{\psi} = J\psi \in \tilde{\Lambda}^2(\tilde{H},K,\mathcal{P})$ and ψ is given in Lemma 3.2.

Proof : The proof is similar to those of [2] and [10], respectively in the finite dimensional case and the infinite dimensional case.
By an application of the Ito formula we have,

$$E\,[\exp(i \int_u^t f_s\, d\,\nu_s(h))/\underline{\underline{G}}_u\,] = \exp[-\tfrac{1}{2}\,\|h\|^2_{\tilde{H}} \int_u^t f_s^2\, ds]$$

where f is a nonrandom measurable bounded function and $h \in \tilde{H}$. Then we conclude (i).
Consider the process N defined by

(3.6) $N_t = \hat{X}_t - E(X_o) - \int_o^t \hat{f}_s\, ds$

It is proved as in [2] that N is a square-integrable martingale. At this point we need the following representation lemma.

LEMMA 3.4 : Let N be any square-integrable $\underline{\underline{G}}$-martingale with values in K. Suppose $N_o = o$. Then there is an element $U \in \Lambda^2(H,K,D,\rho)$ such that $N = U.\nu$ or equivalently $N = \widetilde{U}.\widetilde{N}$.

The proof of this lemma is similar to that of [10] in the linear case. But the exponential process used for the change of probability is given here by (2.14).

We only have to prove that the martingale N of (3.6) is represented by the stochastic integral in the right-hand side of the filtering equation (3.5). Following the lines of [10] we find

$$< \nu, N >_t = \int_o^t (\widehat{h_s \otimes X_s} - \hat{h}_s \otimes \hat{X}_s + \widehat{g_s C_s})ds$$

From this the derived representation of N is obtained.∎

REFERENCES

[1] C. DELLACHERIE, P-A MEYER : Probabilités et Potentiel, Tome 2, Hermann, Paris, 1980.

[2] M. FUJISAKI, G. KALLIANPUR, H. KUNITA : "Stochastic differential equations for the nonlinear filtering problem", Osaka J. Math., 9,pp.19-40,1972.

[3] H. KOREZLIOGLU : "Two-parameter Gaussian Markov processes and their recursive linear filtering", Ann.Scient. l'Univ. Clermont, n°67,pp.69-93,1979.

[4] H. KOREZLIOGLU, G. MAZZIOTTO, J. SZPIRGLAS :"2-D filtering via infinite dimensional filtering". Int. Symp. Math. Theory of Networks and Systems, Aug. 5-7, 1981, Santa Monica.

[5] D. LEPINGLE, J-Y. OUVRARD :"Martingales browniennes hilbertiennes", C.R. Acad. Sc. Paris, t. 276, Série A, pp. 1225-1228, 1973.

[6] C. MARTIAS :"Filtrage non-linéaire dans des espaces de Hilbert réels et séparables", Ann. Scient. l'Univ. Clermont-Ferrand II, n°69,pp.87-113,1981.

[7] M. METIVIER, J. PELLAUMAIL : Stochastic Integration, Academic Press, New York, 1980.

[8] M. METIVIER, G. PISTONE :"Une formule d'isométrie pour l'intégrale stochastique hilbertienne et équations d'évolution linéaire stochastiques", Z. Wahrschein. verw. Gebiete 33,pp.1-18,1975.

[9] J-Y. OUVRARD :"Représentation des martingales vectorielles de carré intégrable à valeurs dans des espaces de Hilbert séparables", Z. Wahrschein. verw. Gebiete 33,pp.195-208,1975.

[10] J-Y. OUVRARD :"Martingale projection and linear filtering in Hilbert spaces", SIAM J. Control and Optimization, Vol.16, n°6, pp.912-937, 1978.

[11] A. PENOEL :"Représentation d'un processus gaussien équivalent à un mouvement brownien en dimension infinie". Thèse de Troisième Cycle,Université Paris VI,1977.

[12] M. YOR :"Existence et unicité de diffusion à valeurs dans un espace de Hilbert", Ann. Inst. Henri Poincaré, Vol.X, n°1, pp.55-88, 1974.

OPTIMAL ORTHOGONAL EXPANSION FOR ESTIMATION I: SIGNAL IN WHITE GAUSSIAN NOISE

James Ting-Ho Lo and Sze-Kui Ng

Department of Mathematics, University of Maryland
Baltimore County, Catonsville, MD 21228

ABSTRACT

The purpose of the paper is to present a systematic method to develop an approximate *recursive* estimator which is optimal for the given structure and approaches the best estimate, when the order of approximation increases.

The minimal variance estimate is projected onto the Hilbert subspace of all Fourier-Hermite (FH) series, driven by the observations, with the same given index set. The projection results in a system of linear algebraic equations for the FH coefficients, the parameters of the desired approximate estimator.

The estimator consists of finitely many Wiener integrals of the observations and a memoryless nonlinear postprocessor. The postprocessor is an arithmatic combination of the Hermite polynomials evaluated at the Wiener integrals.

If a complete orthonormal set (CONS) of step functions is chosen for the FH series, the approximate estimator uses only the observations at discrete times. A discrete-time recursive algorithm is given in which the CONS of Rademacher functions is employed.

Acknowledgement

This work was supported in part by the Air Force Office of Scientific Research, Air Force Systems Command, USAF, under Grant AFOSR-80-0241.

R. S. Bucy and J. M. F. Moura (eds.), Nonlinear Stochastic Problems, 291–309.

I. INTRODUCTION

It was perhaps first observed by Norbert Wiener [1]-[2] that a functional is the mathematical equivalent of a system. This observation made the functional series techniques available for the analysis and synthesis of systems. In the 1980's, V. Volterra studied the functional series named after him as a generalization of Taylor series [3]. The limitations of the Volterra series such as the difficulties in convergence and kernel determination, some of which are unsurprisingly shared by the Taylor series, motivated Wiener to develop orthogonal functional series. The orthogonal set of functionals discovered by Wiener are now called the Wiener G-functionals or the multiple Wiener integrals. Based on Wiener's work on stochastic integrals, Cameron and Martin [4] proposed a new complete orthogonal set of functionals, which are simple arithmatic combinations of Hermite polynomials evaluated at Wiener integrals of the input. They named these functionals the Fourier-Hermite functionals.

A minimum-variance (MV) filter is a system. We know also that when the signal and sensor dynamics are nonlinear, the MV filter is usually infinite-dimensional. From the above functional viewpoint, a natural approach to approximating a MV filter is to use the functional series expansion. In fact, this approach was first taken by Katzenelson and Gould [5], published in 1962. They studied the problem, in their own words,: "Given an input (signal plus noise) which is a sample function from a stationary, ergodic, random process and using the mean-square error criterion, what is the optimum filter consisting of a finite number of Volterra kernels to filter the signal from the noise?" An "n^{th} order Wiener-Hopf equation" was obtained which characterizes the Volterra kernels of their optimum n^{th} order filter.

The importance of the work is in the sense that it points out a systematic way to derive finite-dimensional approximate filters, which are optimum for the given structure and approach the MV filter as the order of approximation increases. Unfortunately the paper was largely ignored in the late 1960's and early 1970's. This is probably due to the following limitations of their results:

(1) The observation process is assumed to be stationary. The specification of the nth order filter requires the first 2n autocorrelation of the observation process and n cross-correlations between the signal and the observation process.
(2) The nth order filters are not recursive.
(3) The nth order Wiener-Hopf equation is not easy to solve, although an iterative method to solve it is given.

The functional series approach was taken up again by Mitter

and Ocone [6]-[8] in the context of the modern nonlinear filtering theory [11]-[13], in which the signal is received through a nonlinear function in additive white Gaussian noise. Instead of the Volterra functionals used by Katzenelson and Gould, they used the multiple Wiener integrals. Their main result is also an "n^{th} order Wiener-Hopf equation" characterizing the integrands of the multiple integrals. While they succeeded in removing part of the above first limitation, their Wiener-Hopf equation is still hard to solve and their best n^{th} order filter is still not recursive.

In this paper, we shall show that all the above three limitations can be eliminated by the use of the Fourier-Hermite functionals developed by Cameron and Martin [4] and a Carleman linearization technique developed in [16].

Let $\{\phi_i\}$ be a complete orthonormal set (CONS) of functions $[0,t]$. A finite FH series driven by the observation $y^t := \{y_\tau, 0 \leq \tau \leq t\}$ is

$$b(t,0) + \sum_{k=1}^{n} \sum_{\xi \in \underset{\sim}{\xi}_k} b(t,k{:}\xi) G_k(t,\phi_{\xi_1} \ldots \phi_{\xi_k}, y)$$

where G_k denotes the multiple Wiener integral of $\phi_{\xi_1} \ldots \phi_{\xi_k}$ wrt y [1], [4], [9] and $\underset{\sim}{\xi}_k$ is a finite set of k-tuples $\xi = [\xi_1,\ldots,\xi_k]$ of nonnegative integers. The multiple Wiener integrals in the above series are called FH functionals.

The FH functionals are orthonormal (therefore Fourier) with respect to the Wiener measure. This property enables us to project the minimum variance estimate $\hat{\emptyset}(x)$ of a measurable functional $\emptyset$ of the signal process x, such that $E\emptyset^2(x) < \infty$, directly onto the Hilbert subspace $H_n(t,\xi)$ of all finite FH series with the same index set $\underset{\sim}{\xi} = \underset{\sim}{\xi}_1 \cup \ldots \cup \underset{\sim}{\xi}_k$. A typical element in $H_n(t,\underset{\sim}{\xi})$ is the FH series displayed above. The direct projection for a preassigned index set does not need the "n^{th} order Wiener-Hopf equation". In fact, it produces a system of linear algebraic equations for the FH coefficients $b(t,k;\underset{\sim}{\xi})$!

A most remarkable property of an FH functional is that it is the product of finitely many Hermite polynomials (therefore Hermite). The argument of each Hermite polynomial is simply a Wiener integral $\int_0^t \phi_{\xi_i}(t,s_i)dy(s_i)$! Such a structure for similar systems is discussed in detail by Schetzen [2, pp. 369-404].

By expanding $b(t,k;\xi)$ into finite Fourier series on the time interval $[0,T]$, over which the estimator is applied, we obtain a further and simpler approximate estimator. The Fourier expansion is carried out by minimizing a weighted cumulative estimation error over $[0,T]$.

The choice of the CONS $\{\phi_i\}$ depends on three factors: (1) the estimation error and the total number of terms in the FH series; (2) the recursive calculation of the Wiener integrals; (3) the determining of $b(t,k;\xi)$. Among all the CONS; the Rademacher functions seem to be the most promising, especially for recursive estimation. Using Rademacher functions reduces the Wiener integrals to additions of the observations at discrete times and multiplications of the orthonormal functions by real constants. The dependence on only the observations at discrete times makes Rademacher functions especially suitable for the sample-and-hold type of measurement devices and, obviously, for digital implementation of the estimator.

This paper is abstracted from the report [16, III], where the multivariable case is also treated.

II. PRELIMINARIES

Consider the scalar observation process that is described by the following equation in the sense of Ito:

$$dy = h(x,t)dt + r_t^{1/2}dw \tag{1}$$

where w is a standard Wiener process and the signal process x is a stochastic process independent of w [11]-[13]. It is assumed that the measures μ_y and μ_w induced by y and w respectively on the measurable space $(C[0,t],\mathscr{B})$ of the continuous functions and the associated Borel field are equivalent and, almost surely with respect to μ_y,

$$\frac{d\mu_y}{d\mu_w}(y) = E_{\mu_x}(\theta^t(x,y)) > 0$$

$$\theta^t := \theta^t(x,y) := \exp\left[\int_0^t h_\tau r_\tau^{-1} dy_\tau - \frac{1}{2}\int_0^t h_\tau^2 r_\tau^{-1} d\tau\right]$$

$$h_t := h(x_t,t)$$

Furthermore the conditional expectation $\hat{\phi}(x) := E(\phi(x)|y^t)$ of a measurable functional $\phi(x)$ of x, such that $E\phi^2(x) < \infty$, given the measurement $y^t := \{y_\tau, 0 \le \tau \le t\}$ has the following representation [11]-[13]:

$$\hat{\emptyset}(x) = \sigma_t(\emptyset)/\sigma_t(1)$$

$$\sigma_t(\emptyset) := (\sigma_t(\emptyset))(y) := E_{\mu_x}(\emptyset(x)\theta^t(x,y))$$

where 1 is the constant function with value one everywhere.

Throughout the paper, we will adopt the following symbols:

$$\int^k (\ )d^k w := \int_0^t \int_0^{s_1} \cdots \int_0^{s_{k-1}} (\ )dw_{s_k} \cdots dw_{s_1}$$

$$⨍^k (\ )d^k \sigma := \int_0^t \cdots \int_0^t (\ )d\sigma_k \cdots d\sigma_1$$

$$F(t,k,\xi,y) := G_k(t,\phi_{\xi_1} \cdots \phi_{\xi_k},y)$$

Here the first denotes the iterated stochastic integral.

We note that θ^t satisfies

$$d\theta^t = h_t r_t^{-1} \theta^t dy_t \tag{2}$$

Hence by the Peano-Baker procedure, it has the following expansion in iterated integrals:

$$\theta^t = \sum_{k=0}^{n} \int^k (h_{s_1} r_{s_1}^{-1}) \ldots (h_{s_k} r_{s_k}^{-1}) d^k y + \int^{n+1} (h_{s_1} r_{s_1}^{-1}) \ldots (h_{s_{n+1}} r_{s_{n+1}}^{-1}) \theta^{s_{n+1}} d^{n+1} y \tag{3}$$

Let $\emptyset(x)$ be a measurable functional of x. Then we have the following finite Volterra series representation of $\sigma_t(\emptyset)$:

$$\sigma_t(\emptyset) = \sum_{k=0}^{n} \int^k E[(h_{s_1} r_{s_1}^{-1}) \ldots (h_{s_k} r_{s_k}^{-1}) \emptyset(x)] d^k y + \int^{n+1} E_{\mu_x}[(h_{s_1} r_{s_1}^{-1}) \ldots (h_{s_{n+1}} r_{s_{n+1}}^{-1}) \emptyset(x) \theta^{s_{n+1}}] d^{n+1} y \tag{4}$$

If $\emptyset(x) = x_\tau$, $0 \leq \tau \leq t$, we have the filtering or the smoothing problem according to $\tau = t$ or $\tau < t$, respectively. We note that computing the Volterra kernels in the above equation is a difficult task. A straightforward and effective method to calculate the kernels, when the signal x satisfies an Ito differential equation, is provided in [16] and thus removes the correlation requirements in the first limitation stated in the Introduction.

Let a and b be L^2 random fields with n-dimensional and m-dimensional parameters, respectively. Then the product of two iterated stochastic integrals has a finite Volterra series expansion as follows:

$$[\int^n a(s_1,\ldots,s_n)d^n w][\int^m b(s_1,\ldots,s_m)d^m w]$$

$$= \sum_{k=0}^{\min(n,m)} \frac{1}{k!} \int^{n+m-2k} [\int^k \sum_{\pi} \bar{a}(\sigma_1,\ldots,\sigma_k,s_{\pi(1)},\ldots,s_{\pi(n-k)})\bar{b}(\sigma_1,\ldots,\sigma_k,s_{\pi(n-k+1)},\ldots,s_{\pi(n+m-2k)})d^k\sigma]d^{n+m-2k}w \tag{5}$$

$$\bar{a}(s_1,\ldots,s_n) := a(s_{\tau(1)},\ldots,s_{\tau(n)}) \tag{6}$$

τ := a permutation of $1,\ldots,n$ such that

$s_{\tau(1)} \geq s_{\tau(2)} \geq \ldots \geq s_{\tau(n)}$.

$d^k\sigma := d\sigma_k \ldots d\sigma_1$

where the summation $\sum_{\pi}$ is taken over all combinations π of $n+m-2k$ elements $\{1,\ldots,n+m-2k\}$ taken $n-k$ at a time.

The above product-to-sum formula (5) is a consequence of Ito's rule. According to Mitter and Ocone [6], it was first derived by T. Hida. The proof given in [6]-[8] uses only Ito's rule and induction. A new proof by the authors [16] illustrates clearly its relationship with the product-to-sum formula for the Hermite polynomials.

III. FOURIER-HERMITE SERIES

Let $\{\phi_i, i = 1,2,\ldots\}$ be a CONS on $[0,t]$ and consider the multiple Wiener integral,

$$F(t,k,\xi,w) := G_k(t,\phi_{\xi_1}\ldots\phi_{\xi_k}, w) \tag{7}$$

where ξ_i is a nonnegative integer for $i = 1,\ldots,k$ and $\xi_1 \leq \ldots \leq \xi_k$. The set of all such integrals, which are orthonormal, was called the Fourier-Hermite (FH) set by Cameron and Martin [4]. It was shown in [4] that if $A(w^t)$ is any functional for which $E[A^2(w^t)] < \infty$, then it can be approximated arbitrarily close by a FH series in the form

$$A(w^t) \simeq \sum_{k=0}^{n} \sum_{\xi\in\underset{\sim}{\xi}_k} a(k;\xi)F(t,k,\xi,w) \tag{8}$$

$a(k;\xi) := E[A(w^t)F(t,k,\xi,w)]$

where ξ_k is a finite subset of $\{\xi = [\xi_1,\ldots,\xi_k] \mid \xi_1 \leq \ldots \leq \xi_k,\ \xi_i$ is a nonnegative integer for $i = 1,\ldots,k\}$.

It is necessary to impose the restriction $\xi_1 \leq \ldots \leq \xi_k$ for each index vector, ξ, in order to have orthonormal HF functionals associated with different index vectors. For instance, $G(t,\phi_1\phi_2,w)$ and $G(t,\phi_2\phi_1,w)$ are identical FH functionals.

We will need the following terminologies:

(i) The integral (7) is called the ξ-th FH functional;
(ii) The term $a(k;\xi)F(t,k,\xi,w)$ is called the ξ-th component;
(iii) The coefficient $a(k;\xi)$ is called the ξ-th FH coefficient of $A(w^t)$;
(iv) The set $\xi := \xi_1 \cup \ldots \cup \xi_n$ is called the index set of the finite FH series on the right side of (8).

Suppose that $A(w^t)$ is an arbitrary multiple Wiener integral $G_k(t,a,w)$. Its ξ-th FH coefficient $a(k,\xi)$ can easily be determined from (8) by calculating the inner product of G_k and $F(t,k,\xi,w)$:

$$a(k,\xi) = E(G_k(t,a,w)F(t,k,\xi,w)) = \int^k a(s_1,\ldots,s_k)\sum_\tau \phi_{\tau(\xi_1)}(s_1)\ldots\phi_{\tau(\xi_k)}(s_k)d^k s \tag{9}$$

Now if $A(w^t)$ is a finite Volterra series

$$A(w^t) := \sum_{k=0}^{n} G_k(t,a_k,w) + 0(n+1)$$

where $0(n+1)$ is orthogonal to any multiple Wiener integral of order less than or equal to n, then the ξ-th FH coefficient is also (9), since two multiple Wiener integrals of different orders are orthogonal to each other.

By (4), we have

$$\sigma_t(\emptyset) = \alpha(t,0) + \sum_{k=1}^{n}\int^k \alpha_k(t,s_1,\ldots,s_k)d^k y + \int^{n+1} r_{n+1}(t,s_1,\ldots,s_{n+1},y^{s_{n+1}})d^{n+1}y$$

$$\sigma_t(1) = \beta(t,0) + \sum_{k+1}^{2n}\int^k \beta_k(t,s_1,\ldots,s_k)d^k y + \int^{2n+1} r_{2n+1}(t,s_1,\ldots,s_{2n+1},y^{2n+1})d^{2n+1}y$$

Given index sets $\underset{\sim}{\xi} = \underset{\sim}{\xi}_1 \cup \ldots \cup \underset{\sim}{\xi}_n$ and $\underset{\sim}{\eta} = \underset{\sim}{\eta}_1 \cup \ldots \cup \underset{\sim}{\eta}_{2n}$, the finite FH series representations of $\sigma_t(\emptyset)$ and $\sigma_t(1)$ can be written as

$$\sigma_t(\emptyset) = \alpha(t,0) + \sum_{k=1}^{n} \sum_{\xi \in \underset{\sim}{\xi}_k} \alpha(t,k;\xi) F(t,k,\xi,y) + R_1(y^t, \underset{\sim}{\xi}) \tag{10}$$

$$\sigma_t(1) = \beta(t,0) + \sum_{k=1}^{2n} \sum_{\eta \in \underset{\sim}{\eta}_k} \beta(t,k;\eta) F(t,k,\eta,y) + R_2(y^t, \underset{\sim}{\eta}) \tag{11}$$

$$\alpha(t,k;\xi) := \frac{1}{k!} \int^k \bar{\alpha}_k(t,s_1,\ldots,s_k) \sum_\tau \phi_{\tau(\xi_1)}(s_1) \ldots \phi_{\tau(\xi_k)}(s_k) d^k s$$

$$:= \int^k \bar{\alpha}_k(t,s_1,\ldots,s_k) \prod_{j=1}^{k} \phi_{\xi_j}(s_j) d^k s$$

$$\beta(t,k;\eta) := \int \bar{\beta}_k(t,s_1,\ldots,s_k) \prod_{j=1}^{k} \phi_{\eta_j}(s_j) d^k s$$

Here we have used the identity

$\int^k a_k d^k y = \frac{1}{k!} G_k(t,\bar{a},y)$ and (9).

We stress that (10) and (11) are orthogonal expansions w.r.t. μ_w. Not μ_y! For instance,

$$E(R_1(y^t, \underset{\sim}{\xi}) F(t,k,\xi,y)) \neq 0$$

$$E(R_1(w^t, \underset{\sim}{\xi}) F(t,k,\xi,w)) = 0 \quad \forall\ \xi \in \underset{\sim}{\xi}_k, \quad k = 1,\ldots,n \tag{12}$$

IV. PROJECTION FOR FOURIER-HERMITE SERIES REPRESENTATION

Consider the set of all the FH series indexed by $\underset{\sim}{\xi}$ and denote it by $H_n(t,\underset{\sim}{\xi})$:

$$H_n(t,\underset{\sim}{\xi}) := \left\{ \sum_{k=0}^{n} \sum_{\xi \in \underset{\sim}{\xi}_k} a(t,k;\xi) F(t,k,\xi,y) \,\middle|\, a(t,k;\xi) \text{ are real numbers} \right\}$$

We will call an element of $H_n(t,\underset{\sim}{\xi})$ the $\underset{\sim}{\xi}$-optimal estimate of $\emptyset(x)$ and denote it by $\underset{\sim}{\xi}_t(\emptyset) := \underset{\sim}{\xi}_t(\emptyset(x))$, if

$$E(\underset{\sim}{\xi}_t(\emptyset(x)) - \emptyset(x))^2 \leq E(z - \emptyset(x))^2, \quad \forall z \in H_n(t,\xi)$$

As $\sum_{\xi \in \underset{\sim}{\xi}_k} a(t,k;\xi) \prod_{j=1}^{k} \phi_{\xi_j}(t,s_j) \in L^2([0,t]^k)$, $k = 1,\ldots,n$, we see

that $H_n(t,\underset{\sim}{\xi})$ is a closed Hilbert subspace of $H_n(t)$:

$$H_n(t,\underset{\sim}{\xi}) \subset H_n(t) := \left\{ \sum_{k=0}^{n} \int^{k} a_k(t,s_1,\ldots,s_k)d^k y, \right.$$

$$\left. a_k \in L^2([0,t]^k),\ k = 1,\ldots,n \right\}$$

Let the norm of the Hilbert space H of all finite variance random variables z be denoted by $||z|| := (Ez^2)^{1/2}$.

It follows immediately that
(i) $\underset{\sim}{\xi}_t(\emptyset)$ is unique.

(ii) An element z_n of $H_n(t,\underset{\sim}{\xi})$ is $\underset{\sim}{\xi}$-optimal, if

$E[(\emptyset - z_n)z] = 0,\ \forall z \in H_n(t,\underset{\sim}{\xi})$, i.e., z_n is the projection of $\emptyset(x)$ onto $H_n(t,\underset{\sim}{\xi})$.

(iii) If $\underset{\sim}{\xi} \subset \underset{\sim}{\eta}$, then $||\emptyset - \underset{\sim}{\eta}_t(\emptyset)||^2 = ||\emptyset - \underset{\sim}{\xi}_t(\emptyset)||^2 - ||\underset{\sim}{\xi}_t(\emptyset) - \underset{\sim}{\eta}_t(\emptyset)||^2$.

The projection theorem, Theorem 4.1, to be developed provides a simple method of projecting $\hat{\emptyset}(x)$ onto $H_n(t,\underset{\sim}{\xi})$. First, we prove a lemma.

Lemma 4.1: Consider the Fourier-Hermite series representation (10) of $\sigma_t(\emptyset)$. For every

$z(y) \in H_n(t,\underset{\sim}{\xi})$,

$$E[(\emptyset(x)(\sigma_t(1))(w) - \sum_{k=0}^{n} \sum_{\xi \in \underset{\sim}{\xi}_k} \alpha(t,k;\xi)F(t,k,\xi,w))z(w)] = 0 \quad (13)$$

Proof: LHS of (13)

$$= E[(\emptyset(x)(\sigma_t(1))(w) - \sum_{k=0}^{n} \sum_{\xi \in \underset{\sim}{\xi}_k} \alpha(t,k;\xi)F(t,k,\xi,w) - R(w^t,\underset{\sim}{\xi}))z(w)]$$

$$= E[(\emptyset(x)(\sigma_t(1))(w) - (\sigma_t(\emptyset))(w))z(w)]$$

$$= E[(\frac{\emptyset(x)(\sigma_t(1))(w) - (\sigma_t(\emptyset))(w)}{(\sigma_t(1))(w)})z(w)\frac{d\mu_y}{d\mu_w}(w)]$$

$$= E[(\emptyset(x) - \hat{\emptyset}(x))z(y)] = 0$$

where the first equality follows from (12).

Theorem 4.1: The functional $z_n(y) \in H_n(t,\underset{\sim}{\xi})$ is the $\underset{\sim}{\xi}$-optimal estimate of

$\emptyset(x)$ if and only if $D(y) := z_n(y)\sigma_t(1)$

$$- \sum_{k=0}^{n} \sum_{\xi \in \underset{\sim}{\xi}_k} \alpha(t,k;\xi)F(t,k,\xi,y) \tag{14}$$

is orthogonal to $H_n(t,\underset{\sim}{\xi})$, w.r.t. μ_w, i.e. for every $z(y) \in H_n(t,\xi)$,

$$E(D(w)z(w) = 0 \tag{15}$$

Proof: (if): For every $z(y) \in H_n(t,\underset{\sim}{\xi})$,

$$E[(\emptyset(x) - z_n(y))z(y)] = E[(\emptyset(x)(\sigma_t(1))(w) - z_n(w)(\sigma_t(1))(w))z(w)]$$

$$= E[(\emptyset(x)(\sigma_t(1))(w) - \sum_{k=0}^{n} \sum_{\xi \in \underset{\sim}{\xi}_k} \alpha(t,k;\xi)F(t,k,\xi,w) - D(w))z(w)]$$

$$= 0$$

The second equality follows from (14) and the third is a consequence of Lemma 3.1 and (15).

(only if): For every $z(y) \in H_n(t,\underset{\sim}{\xi})$,

$$E(D(w)z(w)) = E[z_n(w)(\sigma_t(1))(w) - \sum_{k=0}^{n} \sum_{\xi \in \underset{\sim}{\xi}_k} \alpha(t,k;\xi)F(t,k,\xi,w))z(w)]$$

$$= E[(z_n(w)\sigma_t(1)(w) - \emptyset(x)(\sigma_t(1))(w))z(w)] = E[(z_n(y) - \emptyset(x))z(y)]$$

$$= 0$$

The second equality results from Lemma 3.1.

V. THE $\underset{\sim}{\xi}$-OPTIMAL ESTIMATION

Given an index set $\underset{\sim}{\xi}$, let the $\underset{\sim}{\xi}$-optimal estimate $\underset{\sim}{\xi}_t(\emptyset(x))$ be written as

$$\underset{\sim}{\xi}_t(\emptyset) = b(t,0) + \sum_{k=1}^{n} \sum_{\xi \in \underset{\sim}{\xi}_k} b(t,k;\xi)F(t,k,\xi,y) \tag{16}$$

where $b(t,k,\xi)$, for $\xi \in \underset{\sim}{\xi}_k$ and $k = 1,\ldots,n$, are unknown functions of t to be determined.

Recall that the Fourier-Hermite (FH) series (10) and (11) of $\sigma_t(\emptyset)$ and $\sigma_t(1)$ are available for any given index sets $\underset{\sim}{\xi}$ and $\underset{\sim}{\eta}$. We will in this section, apply Theorem 4.1 to project $\hat{\emptyset}(x) = \sigma_t(\emptyset)/\sigma_t(1)$ onto $H_n(t,\underset{\sim}{\xi})$. The projection can be obtained by considering all the FH components of $\sigma_t(1)$ that affect each such

ζ-component of the product $\xi_t(\emptyset)\sigma_t(1)$ that $\zeta \in \xi$. The index set for all those FH components of $\sigma_t(1)$ will be denoted η. Its determination necessitates the following lemmas:

Lemma 5.1: The product of two FH functionals can be expressed as follows:

$$F(t,k_1,\xi,y)F(t,k_2,\eta,y) = \sum_{k=0}^{\min(k_1,k_2)} \frac{1}{k!}\int^{k_1+k_2-2k} \sum_{p_1}\sum_{p_2} \delta(p_1(\xi_1), p_2(\eta_1))\ldots\delta(p_1(\xi_k),p_2(\eta_k))\sum_\pi \phi_{p_1(\xi_{k+1})}(s_{\pi(1)})\ldots\phi_{p_1(\xi_{k_1})}(s_{\pi(k_1-k)})$$

$$\cdot\phi_{p_2(\eta_{k+1})}(s_{\pi(k_1-k+1)})\ldots\phi_{p_2(\eta_{k_2})}(s_{\pi(k_1+k_2-2k)})d^{k_1+k_2-2k}y \qquad (17)$$

$$\delta(i,j) := \begin{cases} 1 \text{ for } i = j \\ 0 \text{ for } i \neq j \end{cases}$$

Proof: By the product-to-sum formula (5), we have

$$\text{LHS} = \left(\int^{k_1}\sum_{p_1}\prod_{j=1}^{k_1}\phi_{p_1(\xi_j)}(s_j)d^{k_1}y\right)\left(\int^{k_2}\sum_{p_2}\prod_{j=1}^{k_2}\phi_{p_2(\eta_j)}(s_j)d^{k_2}y\right)$$

$$= \sum_{k=0}^{\min(k_1,k_2)}\frac{1}{k!}\int^{k_1+k_2-2k}\int^{k}[\sum_\pi\sum_{p_1}\sum_{p_2}\phi_{p_1(\xi_1)}(\sigma_1)\ldots\phi_{p_1(\xi_1)}(\sigma_k)\phi_{p_1(\xi_{k+1})}$$

$$\cdot(s_{\pi(1)})\ldots\phi_{p_1(\xi_{k_1})}(s_{\pi(k_1-k)})\phi_{p_2(\eta_1)}(\sigma_1)\ldots\phi_{p_2(\eta_k)}(\sigma_k)\phi_{p_2(\eta_{k+1})}$$

$$\cdot(s_{\pi(k_1-k+1)})\ldots\phi_{p_2(\eta_{k_2})}(s_{\pi(k_1+k_2-2k)})]d^k\sigma d^{k_1+k_2-2k}y = \text{RHS}$$

From this expression, we see in order that the product (17) have a ζ-component, the vector η must contain all the components in ξ and ζ other than those that ξ and ζ have in common. Here repeated components are regarded as independent one's. For instance, if $\xi = [1,2,2,3]$ and $\zeta = [2,3,4]$, then η must contain $\{1,2,4\}$. Furthermore, the vector η may contain any number of the common components of ξ and ζ, but they have to be included in duplicate. In the above example, the common components are $\{2,3\}$ and the set of all η's that make (17) have a ζ-component is $\{[1,2,4], [1,2,2,2,4], [1,2,3,3,4], [1,2,2,2,3,3,4]\}$.

There is a bijective relationship between index vectors and products of ϕ's, namely $[i_1,\ldots,i_k] \leftrightarrow \phi_{i_1}\ldots\phi_{i_k}$, where ϕ's are regarded as independent variables and $i_1 \leq \ldots \leq i_k$. Let $G := GCD(\phi_{\xi_1}\ldots\phi_{\xi_{k_1}}, \phi_{\zeta_1}\ldots\phi_{\zeta_{k_3}})$, $F := \phi_{\xi_1}\ldots\phi_{\xi_{k_1}}/G$, and $H := \phi_{\zeta_1}\ldots\phi_{\zeta_{3}}/G$. If the power of G is ℓ, then there are 2^{ℓ} different factors (including 1) of it. Taking the square of each factor and multiplying it to F•H yields the set $\underset{\sim}{\eta}(\xi,\zeta)$ of all the η that make (17) have a ζ-component.

When $\eta \in \underset{\sim}{\eta}(\xi,\zeta)$ we see that the ζ-component of (17) $= C(\xi,\eta,\zeta)$ $\cdot F(t,k(\zeta),\zeta,y)/[(k(\xi) + k(\eta) - k(\zeta))/2]!$ (18)

where $k(\xi)$, $k(\eta)$, and $k(\zeta)$ are the dimension of ξ, η, and ζ, respectively, and the constant $C(\xi,\eta,\zeta)$ results from the permutations p_1 and p_2 in (17). To see how $C(\xi,\eta,\zeta)$ can be determined, let us first consider an example: $\xi = [1,2_1,2_2,3]$, $\eta = [1,2_3,2_4]$, and $\zeta = [2,2,3]$. The subscripts of 2 are used to distinguish different permutations. Simple substitution yields that

$$C(\xi,\eta,\zeta) = \delta(1,1)\delta(2_1,2_3) + \delta(2_1,2_3)\delta(1,1) + \delta(1,1)\delta(2_1,2_4)$$

$$+ \delta(2_1,2_4)\delta(1,1) + \delta(1,1)\delta(2_2,2_3) + \delta(2_2,2_3)\delta(1,1) + \delta(1,1)\delta(2_2,2_4)$$

$$+ \delta(2_2,2_4)\delta(1,1) = (\text{\# ways that } \{1,2\} \text{ can be drawn from } \xi)$$

$$\cdot (\text{\# ways that } \{1,2\} \text{ can be drawn from } \eta)\cdot(\text{\# permutations of } \{1,2_1\} \text{ or } \{1,2_2\})\cdot(\text{\# ways that } \{1,2_3\} \text{ or } \{1,2_4\} \text{ can match a permutation of } \{1,2_1\}) = 2 \cdot 2 \cdot 2 \cdot 1 = 8$$

Recall the bijective relationship $[i_1,\ldots,i_k] \leftrightarrow \phi_{i_1}\ldots\phi_{i_k}$. Denoted by γ the index vector associated with $(\phi_{\xi_1}\ldots\phi_{\xi_{k(\xi)}}\cdot\phi_{\eta_1}\ldots\phi_{\eta_{k(\eta)}}/\phi_{\zeta_1}\ldots\phi_{\zeta_{k(\zeta)}})^{1/2}$. A general formula for $C(\xi,\eta,\zeta)$ is then the following:

$C(\xi,\eta,\zeta) =$ (# ways that the components of γ can be drawn from those of ξ) • (# ways that the components of γ can be drawn from those of η) • ((# components of γ)!) • ((# components of γ)!/(# permutations of the components of γ with subscripts removed)).

Remark: Denote the index set of the RHS of (17) by $\underset{\sim}{\zeta}(\xi,\eta)$. Some reflection shows that $\underset{\sim}{\zeta}(\xi,\eta)$ can be constructed from ξ and η in exactly the same way as $\underset{\sim}{\eta}(\xi,\eta)$ is constructed from ξ and η. Combining this observation and (18), Lemma 5.1 can be converted into a product-to-sum formula for FH functionals:

$$F(t,k(\xi),\xi,y)F(t,k(\eta),\eta,y)$$

$$= \sum_{k=|k(\xi)-k(\eta)|}^{k(\xi)+k(\eta)} \sum_{\zeta \in \underset{\sim}{\zeta}_k(\xi,\eta)} \frac{C(\xi,\eta,\zeta)}{[(k(\xi)+k(\eta)-k)/2]!} \cdot F(t,k,\zeta,y) \quad (19)$$

where $\underset{\sim}{\zeta}_k(\xi,\eta)$ is the collection of all k-tuples in $\underset{\sim}{\zeta}(\xi,\eta)$. We note that $\underset{\sim}{\zeta}_k(\xi,\eta)$ is empty when $(k(\xi) + k(\eta) - k)/2$ is not an integer.

Lemma 5.2: A product-to-sum formula for the finite FH series is

$$\left(\sum_{k=0}^{n} \sum_{\xi \in \underset{\sim}{\xi}_k} \gamma_1(t,k;\xi)F(t,k,\xi,y)\right)\left(\sum_{k=0}^{n} \sum_{\eta \in \underset{\sim}{\eta}_k} \gamma_2(t,k;\eta)F(t,k,\eta,y)\right)$$

$$= \sum_{\zeta \in \underset{\sim}{\zeta}(\underset{\sim}{\xi},\underset{\sim}{\eta})} \left[\sum_{k=0}^{n} \sum_{\xi \in \underset{\sim}{\xi}_k} \sum_{\eta \in \underset{\sim}{\eta} \cap \underset{\sim}{\eta}(\xi,\zeta)} \frac{C(\xi,\eta,\zeta)\gamma_1(t,k;\xi)\gamma_2(t,k(\eta),\eta)}{[(k + k(\eta) - k(\zeta))/2]!}\right]$$

$$\cdot F(t,k(\zeta),\zeta,y) \quad (20)$$

$$\underset{\sim}{\zeta}(\underset{\sim}{\xi},\underset{\sim}{\eta}) := \bigcup_{\xi \in \underset{\sim}{\xi}} \bigcup_{\eta \in \underset{\sim}{\eta}} \underset{\sim}{\zeta}(\xi,\eta)$$

Now we consider the product

$$\left(\sum_{k=0}^{n} \sum_{\xi \in \underset{\sim}{\xi}_k} b(t,k;\xi)F(t,k,\xi,y)\right)\left(\sum_{k=0}^{2n} \sum_{\eta \in \underset{\sim}{\eta}_k} \beta(t,k;\eta)F(t,k,\eta,y)\right)$$

Applying Lemma 5.2 and setting the ζ-components of $\hat{\emptyset}(x_t,\xi)\sigma_t(1)$ and $\sigma_t(\emptyset)$ equal, we obtain

$$\left[\sum_{k=0}^{n} \sum_{\xi \in \underset{\sim}{\xi}_k} \sum_{\eta \in \underset{\sim}{\eta}(\xi,\zeta)} \frac{C(\xi,\eta,\zeta)b(t,k,\xi)\beta(t,k(\eta),\eta)}{((k + k(\eta) - k(\zeta))/2)!}\right]F(t,k(\zeta),\zeta,y)$$

$$= \alpha(t,k(\zeta),\zeta)F(t,k(\zeta),\zeta,y)$$

Hence for every $\zeta \in \underset{\sim}{\xi}_k$, $k = 0,\ldots,\eta$,

$$\sum_{k=0}^{n} \sum_{\xi \in \underset{\sim}{\xi}_k} \left(\sum_{\eta \in \underset{\sim}{\eta}(\xi,\zeta)} \frac{C(\xi,\eta,\zeta)\beta(t,k(\eta),\eta)}{[(k+k(\eta) - k(\zeta))/2]!}\right) b(t,k,\xi) = \alpha(t,k(\zeta),\zeta) \quad (21)$$

We note that these form an algebraic system of simultaneous linear equations, in which the number of unknowns $b(t,k,\xi)$ is the same as that of the equations.

VI. EXPRESSION IN HERMITE POLYNOMIALS

Consider a FH functional,

$$G_k(t, \prod_{j=1}^{k} \phi_{\xi_j}, y) := F(t,k,\xi,y)$$

where not all ξ_j are different. In this section, we will express it as a product of the Hermite polynomials of the Wiener integrals $\int_0^t \phi_{\xi_j}(s)dy_s$.

Let us define the symbol $\phi^{(i)}$ by

$$G_k(t,\phi^{(i)},y) := G_k(t, \prod_{j=1}^{i} \phi(s_j),y) \tag{22}$$

In other words, $\phi^{(i)}$ denotes that a function ϕ appears i times in the integrand of a multiple Wiener integral. In view of Lemma 5.1, an immediate consequence of the product-to-sum formula (5) is

$$G_{i+j}(t,\phi_\ell^{(i)}\phi_k^{(j)},y) = G_i(t,\phi_\ell^{(i)},y)G_j(t,\phi_k^{(j)},y)$$

If ϕ_{λ_i} appears k_i times in $\prod_{j=1}^{k} \phi_{\xi_j}$ for i = 1,...,q, then we write

$$\prod_{j=1}^{k} \phi_{\xi_j} = \prod_{i=1}^{q} \phi_{\lambda_i}^{(k_i)}$$

Hence we have

$$G_k(t, \prod_{j=1}^{k} \phi_{\xi_j},y) = \prod_{i=1}^{q} G_{k_i}(t,\phi_{\lambda_i}^{(k_i)},y)$$

A well-known property of the multiple Wiener integral [14, p. 37] is

$$G_{k_i}(t,\phi_{\lambda_i}^{(k_i)},y) = H_{k_i}(\int_0^t \phi_{\lambda_i} dy)$$

Hence

$$F(t,k,\xi,y) = G_k(t, \prod_{j=1}^{k} \phi_{\xi_j},y) = \prod_{i=1}^{q} H_{k_i}(\int_0^t \phi_{\lambda_i} dy) \tag{23}$$

Thus we have reduced the $\underset{\sim}{\xi}$-optimal estimation (16) to the calculation of the Wiener integrals, $\int_0^t \phi_{\lambda_i} dy$. The $\underset{\sim}{\xi}$-optimal estimate $\underset{\sim}{\xi}_t(\emptyset)$ can be obtained by processing these Wiener integrals through a memoryless nonlinear system consisting essentially of finite many Hermite polynomials H_{k_i}. The synthesis of memoryless nonlinear systems that are similar to these was discussed in great detail in Chapters 17 and 18 in [2, pp. 369-404]. The main questions that remain are how to choose the orthonormal functions $\{\phi_i\}$ and how to calculate the above Wiener integral recursively. Of course, these two questions are highly correlated.

VII. FINITE FOURIER EXPANSION OF $b(t,k;\xi)$

It was shown in Section V that the function $b(t,k;\xi)$ can easily be determined at each t by solving a linear system of simultaneous equations (21). Although the solution of (21) can be carried out at a large (but finite) number of time points before the implementation of the estimator, the storage of all $b(t,k;\xi)$ is obviously difficult especially when the time interval $[0,T]$ over which the estimator will be applied is large.

Therefore we will, in this section, expand $b(t,k;\xi)$ in a finite Fourier series. By choosing an appropriate set of orthonormal functions on $[0,T]$, which can easily be stored or reproduced, the above difficulty can be removed.

The expansion can be obtained by minimizing the cumulative error $\rho(\underset{\sim}{\xi}_t(\emptyset), \hat{\hat{\emptyset}})$ over the constant coefficients of $b_j(k,\xi)$, where

$$\rho(\underset{\sim}{\xi}_t(\emptyset), \hat{\hat{\emptyset}}) := \int_0^t E(\underset{\sim}{\xi}_t(\emptyset) - \hat{\hat{\emptyset}})^2 \nu_t \, dt$$

$$\hat{\hat{\emptyset}}(x) := \sum_{k=0}^{n} \sum_{\xi \in \underset{\sim}{\xi}_k} b'(t,k;\xi) F(t,k,\xi,y)$$

$$b'(t,k,\xi) :=: \sum_{j=0}^{j(k,\xi)} b_j(k,\xi)\psi_j(t) \tag{24}$$

Here $\{\psi_j, j = 1,2,..\}$ on $[0,T]$ is a CONS on $[0,T]$ w.r.t. the weight function ν.

We will now set about to calculate $\rho(\xi_t(\emptyset), \hat{\emptyset})$. Applying Lemma 5.2, we obtain

$$(\xi_t(\emptyset) - \hat{\hat{\emptyset}})^2 = \sum_{\zeta \in \underset{\sim}{\zeta}(\underset{\sim}{\xi},\underset{\sim}{\xi})} f(t,k(\zeta);\zeta) F(t,k(\zeta),\zeta,y) \tag{25}$$

$$f(t,k(\zeta);\zeta) := \sum_{k=0}^{n} \sum_{\xi \in \underset{\sim}{\xi}} \sum_{\eta \in \underset{\sim}{\xi} \cap \underset{\sim}{\eta}(\xi,\zeta)} \frac{C(\xi,\eta,\zeta)\gamma(t,k;\xi)\gamma(t,k(\eta),\eta)}{[(k + k(\eta) - k(\zeta))/2]!}$$

$$\gamma(t,k;\xi) := b(t,k;\xi) - \sum_{j=0}^{j(k,\xi)} b_j(k,\xi)\psi_j(t)$$

Since the FH functionals are orthonormal and the Radon-Nikodym derivative $d\mu_y/d\mu_w$ is $\sigma_t(1)$, it is easy to see from (25) and (11) that

$$E(\xi_t(\emptyset) - \hat{\hat{\emptyset}})^2 = E((\xi_t(\emptyset))(y^t) - (\hat{\hat{\emptyset}})(y^t))^2$$

$$= E[((\xi_t(\emptyset))(w^t) - (\hat{\hat{\emptyset}})(w^t))^2(\sigma_t(1))(w)]$$

$$= \sum_{\zeta \in \underset{\sim}{\zeta}(\underset{\sim}{\xi},\underset{\sim}{\xi})} \beta(t,k(\zeta);\zeta)f(t,k(\zeta);\zeta)$$

Expressing $\rho(\xi_t(\emptyset), \hat{\hat{\emptyset}})$ as a quadratic polynomial in the unknown constants $b_j(k,\xi)$, we obtain

$$\rho(\xi_t(\emptyset), \hat{\hat{\emptyset}}) = \sum_G B_0 + \sum_G \sum_{j=0}^{j(k,\xi)} B_1 b_j(k,\xi) + \sum_G \sum_{j=0}^{j(k(\eta),\eta)}$$

$$\cdot B_2 b_j(k(\eta),\eta) + \sum_G \sum_{j_1=0}^{j(k,\xi)} \sum_{j_2=0}^{j(k(\eta),\eta)} B_3 b_{j_1}(k,\xi)$$

$$\cdot b_{j_2}(k(\eta),\eta); \quad \sum_G := \sum_{\zeta \in \underset{\sim}{\zeta}(\underset{\sim}{\xi},\underset{\sim}{\xi})} \sum_{k=0} \sum_{\xi \in \underset{\sim}{\xi}_k} \sum_{\eta \in \underset{\sim}{\xi} \cap \underset{\sim}{\eta}(\xi,\zeta)} \tag{26}$$

$$Q := \frac{C(\xi,\eta,\zeta)\beta(t,k(\zeta);\zeta)}{[(k + k(\eta) - k(\zeta))/2]!}; \quad B_0 := \int_0^T Qb^2(t,k;\eta)\nu_t dt;$$

$$B_1 := -\int_0^T Qb(t,k(\eta);\eta)\psi_j(t)\nu_t dt; \quad B_2 := -\int_0^T Qb(t,k;\xi)\psi_j(t)\nu_t dt;$$

$$B_3 := \int_0^T Q\psi_{j_1}(t)\psi_{j_2}(t)\nu_t dt$$

By minimizing the above quadratic function over b_j, we get a finite Fourier expansion (24) of $b(t,k;\xi)$.

VIII. RECURSIVE ESTIMATION USING RADEMACHER FUNCTIONS

Rademacher functions [15] are orthonormal step functions and they have a feature very important for recursive algorithm. It is that every Rademacher function becomes a longer Rademacher function (i.e. with longer time domain) after being concatenated with one more period of the same function and being renormalized.

To illustrate how to concatenate Rademacher functions for recursive computation, let us consider a simple example. Suppose that the sampling period is Δ and we want to use Rademacher functions of periods 2Δ, 4Δ, and 8Δ. If the present time is 8Δ, then we have the three Radamacher function depicted in the first panel of Fig. 1. They are normalized for clearer illustration.

At $t = 8\Delta$, only three numbers, the values of three Wiener integrals are used and need to be stored. At $t = 9\Delta$, a 1Δ bar is obtained. Since it cannot be connected to any of the existing Rademacher functions to form longer ones, it joins them as an orthonormal member in a larger orthonormal set of functions. Four numbers are now used and stored. At $t = 10\Delta$, another 1Δ bar is obtained. Together with the first 1Δ bar, it is used to elongate the Rademacher function of period Δ and to form a 2Δ bar, a useful building block for the Rademacher functions of periods 2Δ. Still 4 numbers are used and stored. Similarly, at $t = 11\Delta$, a 1Δ bar joins as an orthonormal member and 5 numbers are used and stored. At $t = 12\Delta$, both the Rademacher functions of periods Δ and 2Δ are elongated, and a 4Δ bar joins as an orthonormal member. There are four numbers to be used and stored. Continuing in this fashion, at $t = 13\Delta$, 14Δ, 15Δ, 16Δ, we have 5, 5, 6, 3 numbers, respectively, to be used and stored. This completes one cycle for the Rademacher functions of periods Δ, 2Δ, and 3Δ. We stress that normalization should be done every time a Rademacher function is elongated or a bar function is used as an orthonormal member.

The above procedure can be repeated for as many cycles as needed. The number of orthonormal functions, which is the same as the number of numbers used and stored, remains to be 3, 4, 4, 5, 5, 6, 3 in each cycle. The calculation of the Wiener integrals is thus carried out recursively, and it involves only simple addition and subtraction of the observation y at $t = \Delta$, $2\Delta, \ldots$. Normalization of functions involves only multiplication by real constants.

The choice of the Rademacher functions to be used in an estimator depends very much on the frequency band width of the observation function $h(x_t, t)$. More research is required to understand this dependency.

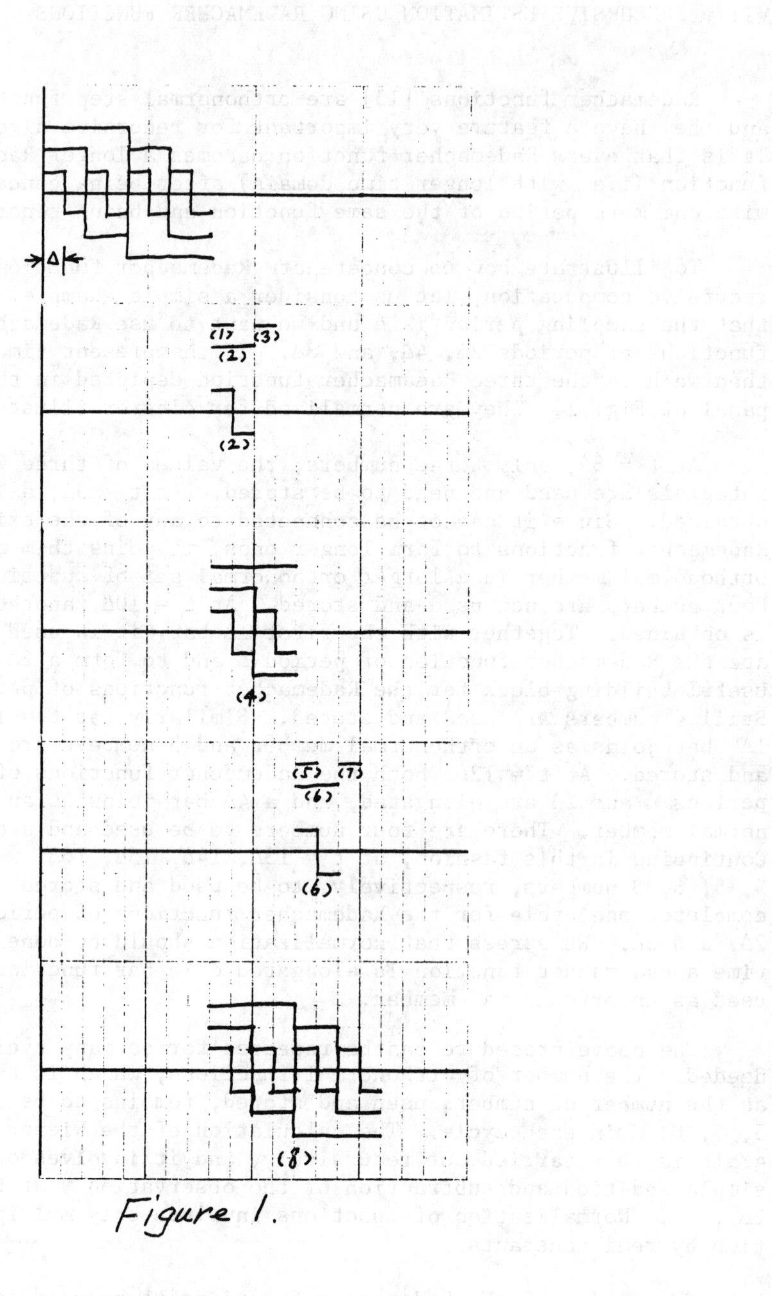

Figure 1.

REFERENCES

[1] N. Wiener, *Nonlinear Problems in Random Theory*, The Technology Press, M.I.T. and J. Wiley & Sons Inc., N. Y., 1958.
[2] M. Schetzen, *The Volterra and Wiener Theories of Nonlinear Systems*, J. Wiley & Sons, Inc., N. Y., 1980.
[3] V. Volterra, *Theory of Functionals and of Integral and Integro-Differential Equations*, Dover Publications, Inc., N. Y., 1959.
[4] R. M. Cameron and W. T. Martin, "The Orthogonal Development of Nonlinear Functionals in Series of Fourier-Hermite Functionals", *Ann. of Math.* 48, 1947, 358-389.
[5] J. Katzenelson and L. A. Gould, "The Design of Nonlinear Filters and Control Systems, Part I", *Information and Control* 5, pp. 108-143, 1962.
[6] S. K. Mitter and D. Ocone, "Multiple Integral Expansions for Nonlinear Filtering", *Proceedings of the 18th IEEE Conference on Decision and Control*, Ft. Lauderdale, Florida, 1979.
[7] D. Ocone, *Topics in Nonlinear Filtering Theory,* Ph.D. Thesis, Dept. of Math., M.I.T., 1980.
[8] D. Ocone, "Multiple Integral Expansions for Nonlinear Filtering", to appear in *Stochastics*.
[9] K. Ito, "Multiple Wiener Integrals", *J. Math. Soc. Japan*, Vol. 3 (1951), pp. 157-169.
[10] M. Schetzen, *The Volterra and Wiener Theories of Nonlinear Systems*, J. Wiley & Sons, Inc., 1980.
[11] R. S. Bucy and P. D. Joseph, *Filtering for Stochastic Processes with Applications to Guidance*, Interscience, N. Y., 1968.
[12] G. Kallianpur, *Stochastic Filtering Theory*, Springer-Verlag, N. Y., 1980.
[13] R. S. Liptser and A. N. Shiryayev, *Stochastics of Random Processes I: General Theory*, Springer-Verlag, N. Y., 1977.
[14] H. P. McKean, Jr., *Stochastic Integrals*, Academic Press, N. Y. and London, 1969.
[15] H. Rademacher, "Einge Satze Uber Reihen von Allgemen Orthogonal Function", *Ann. Math.*, Vol. 87, (1922), 712-738.
[16] J. T.-H. Lo and S. K. Ng, "Finite Order Nonlinear Filtering I-III", Mathematics Research Reports Nos. 81-8, 9 & 13, Dept. of Math., UMBC, Aug. 1981, Aug. 1981, and Sept. 1981.

OPTIMAL ORTHOGONAL EXPANSION FOR ESTIMATION II: SIGNAL IN COUNTING OBSERVATIONS

James Ting-Ho Lo and Sze-Kui Ng

Department of Mathematics, University of Maryland
Baltimore County, Baltimore, Maryland 21228

ABSTRACT

The purpose of the paper is to show that the approach developed in the preceding part can also be used to construct approximate estimators for the signal that modulates the rate of a counting process. The estimators are optimal for the given structure and approach the best estimate, when the approximation order increases.

A product-to-sum formula for the multiple Wiener integrals, with respect to the counting process martingale, is derived. By using the formula, the minimal variance estimate is projected onto the Hilbert subspace of all Fourier-Charlier (FC) series, driven by the counting observations, with the same given index set. The projection results in a system of linear algebraic equations for the FC coefficients, the parameters of the desired approximate estimator.

The estimator consists of finitely many Wiener integrals of the counting observations and a nonlinear postprocessor. The nonlinear postprocessor, however, is not memoryless. More work is required here.

I. INTRODUCTION

The estimation for signals from counting observations was formulated and studied by Galchuk, Snyder, Frost, Rubin, Bermaud, Davis, Van Schuppen, Boel, Varaiya, Wong, and Segall, Kailath, etc. in [1] - [12]. It was observed that the problem can be for-

R. S. Bucy and J. M. F. Moura (eds.), Nonlinear Stochastic Problems, 311–338.

mulated in a way very similar to that for signals in additive white Gaussian noise (AWGN). The analogy was brought out very strikingly by the martingale theory approach in [7] - [12].

In this paper, the analogy is pursued further between the AWGN and the counting observation cases. The optimal orthogonal expansion method presented in the preceding Part for the AWGN problem will now be developed for the counting observation problem.

As in the AWGN case, the minimal variance estimate is projected onto the Hilbert subspace of all the orthogonal series of a certain type, driven by the counting observations, with the same given index set. The projection results also in a system of linear algebraic equations for the series coefficients, the parameters of the desired approximate estimator.

The orthogonal series to be used are analogous to the Fourier-Hermite series and are called Fourier-Charlier (FC) series. Each FC functional can also be realized by finite many Wiener integrals w.r.t. the observations followed by a nonlinear postprocessor. Unfortunately, this postprocessor is not memoryless. Before a better method is found, a good way to carry out the computation required by the postprocessor is perhaps the iterative method found by Segall and Kailath [13, p. 292].

The above projection also calls for a product-to-sum formula for the iterated integrals w.r.t. the martingale process associated with the observations, a counting process in the present case. Although a major portion of this paper is devoted to the development of the formula, which is circuitous and different from that for the AWGN case, the resemblance between the two formulas is indeed striking. This, in part. lead the authors to believe that a similar formula exists for the iterated integrals w.r.t. any martingale process.

II. PRELIMINARIES

In this section, we shall establish probability spaces and state some fundamental results for counting processes.

Let $(\Omega,\mathcal{F},P)$ be a complete probability space and let $(\mathcal{F}_t)$, $t \geq 0$ be a nondecreasing family of right continuous sub-σ-algebras of $\mathcal{F}$ augmented by sets of zero probability. A random process (ξ_t), $t \geq 0$ is said to be adapted to $(\mathcal{F}_t)$ and denoted by $(\xi_t,\mathcal{F}_t)$, $t \geq 0$, if for any $t \in T$ the random variable ξ_t is $\mathcal{F}_t$-measurable.

We call a process $(\lambda_t, \mathcal{F}_t)$, $t \geq 0$ predictable if it is measurable w.r.t. the smallest σ-algebra on $R^+ \times \Omega$ generated by all the random processes that are adapted to $(\mathcal{F}_t)$ and have left-continuous sample paths.

A counting process $N = (N_t, \mathcal{F}_t)$, $t \geq 0$ is a random process with sample paths that are piecewise constant with positive unit jumps, (P—a.s.) right continuous, and have limits from the left. For each counting process N, let $\tau_n(N) = \inf\{s \geq 0 | N_s = n\}$ and $\tau_\infty = \lim_{n\to\infty} \tau_n$. The Markov time $\tau_n(N)$ is called the n-th jump time of N.

By the Doob-Meyer decomposition theorem, it can be shown that there exists a unique right-continuous, predictable, increasing process $\tilde{A} = (\tilde{A}_t, \mathcal{F}_t)$ such that $(N_t - \tilde{A}_t, \mathcal{F}_t)$ is a τ_∞-local martingale on $(\mathcal{F}_t)$, i.e., for each n, $(N_{t\wedge\tau_n} - \tilde{A}_{t\wedge\tau_n}, \mathcal{F}_t)$, $t < \tau_\infty$, is a uniformly integrable martingale. The process $\tilde{A}$ is called the compensator (or the integrated rate) of the counting process N.

Consider a probability space $(\mathbb{N}, \mathcal{B}, \mu)$ where $\mathbb{N}$ is the space of piecewise constant and right continuous functions $y = (y_t)$, $t \geq 0$ with positive unit jumps and $y_0 = 0$; and $\mathcal{B}$ is the σ-algebra of Borel sets of $\mathbb{N}$. Let $\mathcal{B}_t$, $t \geq 0$ be the sub-σ-algebra of $\mathcal{B}$ generated by Borel sets of elements in $\mathbb{N}$ restricted to $[0,t]$. Let μ_1 and μ_2 be two probability measures on $(\mathbb{N}, \mathcal{B})$. Then w.r.t. the measure μ_i, the random process $(y_t, \mathcal{B}_t)$ on $(\mathbb{N}, \mathcal{B}, \mu_i)$ is a counting process, which we shall denote by $N_i = (y_t, \mathcal{B}_t, \mu_i)$. the compensator of N_i will be denoted by $A_i = (A_{it}, \mathcal{B}_t, \mu_i)$, for $i = 1$ and 2.

The proofs of the following two theorems can be found on p. 315 and p. 258 in Liptser and Shiryayev [14], respectively.

Theorem 2.1. Suppose that the sample paths of A_2 are continuous. Then a necessary and sufficient condition for $\mu_1 \ll \mu_2$ is that there exists a nonnegative predictable process $\dot{\lambda} = (\dot{\lambda}_t, \mathcal{B}_t)$ such that the following conditions are satisfied:

$$A_{1t}(y) = \int_0^t \dot{\lambda}_s(y) dA_{2s}(y), \quad \text{for } \mu\text{—a.s. } y \in \mathbb{N} \tag{1}$$

$$\int_0^{\tau_\infty(y)} \left[1 - \sqrt{\dot{\lambda}_s(y)}\right]^2 dA_{2s}(y) < \infty, \quad \text{for } \mu\text{—a.s. } y \in \mathbb{N} \tag{2}$$

Furthermore, the Radon-Nykodim derivative $\Lambda_t(y) := \frac{d\mu_1}{d\mu_2}(t,y)$ satisfies the following equation:

$$\Lambda_t(y) = 1 - \int_0^t \Lambda_{s-}(y)(\dot{\lambda}_s(y) - 1)d(y_s - A_{2s}) \tag{3}$$

Theorem 2.2. Let $A = (A_t, \mathscr{B}_t)$, $t \geq 0$ be a process with continuous and increasing sample paths with $A_0 \equiv 0$. Then A is the compensator of a counting process.

III. A MODEL FOR ESTIMATION FROM COUNTING OBSERVATIONS

A number of models for estimating the signal from randomly modulated counting processes were proposed by many recent papers such as [1] - [12]. Among them the martingale model studied in [7] - [12] provides us with a rather general framework of the estimation problem and brings out nicely the analogy to the additive white Gaussian noise model.

In this section, we shall state a model developed in [11], [14] and impose a few restrictions, which will allow us to avoid some complex technicalities in illustrating our approach. It is believed that some of these restrictions can be removed. The analogy mentioned above will then be used to set up the estimation problem.

We consider a counting observation process $N := (N_t, \mathscr{F}_t)$, $t \geq 0$, of which the compensator $\tilde{A}$, the integrated rate, is a measurable functional of the signal process $x := (x_t, \mathscr{F}_t)$, $t \geq 0$ with sample paths from a measurable space $(X, \mathscr{A})$. It is noted that the observation process N does not enter the functional $\tilde{A}$. Thus no feedback of the observation into the compensator is allowed. This assumption is made only to guarantee that the n-th order Wiener-Hopf equation for the Wiener kernels and the linear system of algebraic equations for the FC coefficients, to be obtained in Sections V and VI respectively, do not involve the observation and can be solved "off-line". The assumption is not necessary for our arguments in deriving those equations.

On the measurable spaces $(X, \mathscr{A})$, $(\mathbb{N}, \mathscr{B})$ and $(X \times \mathbb{N}, \mathscr{A} \times \mathscr{B})$, we define the probability measures: for every $A \in \mathscr{A}$ and $B \in \mathscr{B}$,

$$\mu_x(A) := P(x \in A)$$

$$\mu_N(B) := P(N \in B)$$

$$\mu_{x,N}(A \times B) := P(x \in A, N \in B)$$

We are now ready to state the restrictions as follows:

(I) There exists a regular version of the conditional probability $P(N \in B | \mathscr{F}^x)|_{x=a}$, given the σ-algebra $\mathscr{F}^x$ generated by x, for μ_x-a.s. $a \in X$. The regular version will be denoted by $\mu_N^a(B)$.

(II) For μ_x — a.s. $a \in X$, the compensator $\tilde{A}_t(a)$ is absolutely continuous w.r.t. the measure dA_t for some deterministic, continuous, and nondecreasing function A_t in t. The continuity assumption on A_t is not essential, it is introduced to simplify our discussion. The Radon-Nykodim derivative $d\tilde{A}_t(a)/dA_t$ is positive and will be denoted by $\lambda_t(a)$, i.e.,

$$\tilde{A}_t(a) = \int_0^t \lambda_s(a)dA_s, \text{ for } \mu_x - \text{a.s. } a \in X \tag{4}$$

The process $(\lambda_t(x), \mathcal{F}_t)$ is predictable and

$$\int_0^\infty \left(1 - \sqrt{\lambda_s(a)}\right)^2 dA_s < \infty, \quad \text{for } \mu_x - \text{a.s. } a \in X \tag{5}$$

(III) For μ_x— a.s. $a \in X$, the process $(\tilde{A}_t(a), \mathcal{B}_t)$ is the compensator of the counting process $(y_t, \mathcal{B}_t, \mu_N^a)$.

These restrictions lead immediately to the following consequences:

(1) By Theorem 2.2, the properties of A_s assumed in (II) yield a counting process $(N_{1t}, \mathcal{B}_t, \mu_{N_1})$ with A_s as its compensator.

(2) By Theorem 2.1, it follows from (II) and (III) that for μ_x — a.s. $a \in X$, we have $\mu_N^a << \mu_{N_1}$ and

$$\frac{d\mu_N^a}{d\mu_{N_1}}(t,y) = 1 + \int_0^t \frac{d\mu_N^a}{d\mu_{N_1}}(s^-,y)(\lambda_s(a) - 1)d(y_s - A_s) \tag{6}$$

We shall need the following lemma in the sequel.

Lemma 3.1. With the notation established above, we have $\mu_{x,N} << \mu_x \times \mu_{N_1}$ and

$$\frac{d\mu_{x,N}}{d(\mu_x \times \mu_{N_1})}(t,a,y) = \frac{d\mu_N^a}{d\mu_{N_1}}(t,y), \ \mu_x \times \mu_{N_1} \ - \text{a.s. } (a,y) \tag{7}$$

Furthermore, we have $\mu_N << \mu_{N_1}$ and

$$\frac{d\mu_N}{d\mu_{N_1}}(t,y) = \int_X \frac{d\mu_{x,N}}{d(\mu_x \times \mu_{N_1})}(t,a,y)d\mu_x(a), \ \ \mu_{N_1} - \text{a.s. } y \tag{8}$$

Proof. Let $f(a,y)$ be a bounded $\mathcal{A} \times \mathcal{B}$ measurable functional.

By the restriction (I) and the Fubini theorem,

$$\int_{X\times \mathbb{N}} f(a,y)\,d\mu_{x,N}(a,y)$$
$$= E(E(f(x,N)|\mathcal{F}^x))$$
$$= \int_X E(f(x,N)|\mathcal{F}^x)|_{x=a}\, d\mu_x(a)$$
$$= \int_X \left[\int_{\mathbb{N}} f(a,y)\frac{d\mu_N^a}{d\mu_{N_1}}(y)\,d\mu_{N_1}(y)\right] d\mu_x(a)$$

This proves (7).

For any $B\in\mathcal{B}$, we now have

$$\mu_N(B) = \int_X\int_B d\mu_{x,N}(a,y)$$
$$= \int_X\int_B \frac{d\mu_{x,N}}{d(\mu_x\times\mu_{N_1})}(a,y)\,d(\mu_x\times\mu_{N_1})(a,y)$$
$$= \int_B\left[\int_X \frac{d\mu_{x,N}}{d(\mu_x\times\mu_{N_1})}(a,y)\,d\mu_x(a)\right] d\mu_{N_1}(y)$$

This proves (8).

We are now ready to state a Baysian-type representation theorem for the conditional mean $E(\emptyset(x)|N^t)$ of a measurable functional $\emptyset$ of the signal x given the counting observation $N^t := \{N_s,\ 0 \le s \le t\}$.

Theorem 3.1. The conditional mean $\hat{\emptyset}(x) := E(\emptyset(x)|N^t)$ of a measurable functional $\emptyset$ of x such that $E\emptyset^2(x) < \infty$ has the following representation:

$$\hat{\emptyset}(x) = \sigma_t(\emptyset)/\sigma_t(1)$$
$$\sigma_t(\emptyset) := \int_X \emptyset(a)\Lambda_t(a,y)\,d\mu_x(a)$$
$$\sigma_t(1) := \int_X \Lambda_t(a,y)\,d\mu_x(a) = \frac{d\mu_N}{d\mu_{N_1}}(t,y),\ \mu_{N_1}\text{— a.s. } y$$
$$\Lambda_t(a,y) := \frac{d\mu_{x,N}}{d(\mu_x\times\mu_{N_1})}(t,a,y)$$

Proof. By a known theorem in the probability theory (See Loeve [15, p. 10]), the conditional mean $\hat{\emptyset}$ can be written as

$$\hat{\emptyset}(x) = \frac{E_1(\emptyset(x)\Lambda_t(x,N)|N^t)}{E_1(\Lambda_t(x,N)|N^t)}$$

here the expectation E_1 is taken w.r.t. the measure $\mu_x \times \mu_{N_1}$. The theorem is an immediate consequence of this expression.[1]

Let us obtain iterated integral expansions of $\sigma_t(\emptyset)$ and $\sigma_t(1)$. From Lemma 3.1 and (6), we have the following equation for Λ_t:

$$\Lambda_t(a,y) = 1 + \int_0^t \Lambda_{s-}(a,y)(\lambda_s(a,y) - 1)d(y_s - A_s) \tag{9}$$

Substituting (9) into the right-hand side of itself and repeating the substitution, we get

$$\Lambda_t(a,y) = 1 + \sum_{k=1}^{n}\int^k (\lambda_{s_1} - 1)\dots(\lambda_{s_k} - 1)d(y_{s_k} - A_{s_k})\dots d(y_{s_1} - A_{s_1})$$

$$+ \int^{n+1} (\lambda_{s_1} - 1)\dots(\lambda_{s_{n+1}} - 1)\Lambda_{s_{n+1}} d^{k+1}(y_s - A_s)$$

Thus we obtain the following expressions for $\sigma_t(\emptyset)$ and $\sigma_t(1)$:

$$\sigma_t(\emptyset) = \alpha_0(t) + \sum_{k=1}^{n}\int^k \alpha_k(t,s_1,\dots,s_k)d^k(N_s - A_s) + \int^{n+1} r'_{n+1}d^{n+1}(N_s - A_s) \tag{10}$$

$$\sigma_t(1) = \beta_0(t) + \sum_{k=1}^{2n}\int^k \beta_k(t,s_1,\dots,s_k)d^k(N_s - A_s) + \int^{2n+1} r''_{2n+1}d^{2n+1}(N_s - A_s) \tag{11}$$

$$\alpha_k(t,s_1,\dots,s_k) := E_{\mu_x}[\emptyset(x)(\lambda_{s_1}(x) - 1)\dots(\lambda_{s_k}(x) - 1)]$$

$$r'_{n+1} := E_{\mu_x}[\emptyset(x)(\lambda_{s_1}(x) - 1)\dots(\lambda_{s_{n+1}}(x) - 1)\Lambda_{s_{n+1}}(x,y)]$$

$$\beta_k(t,s_1,\dots,s_k) := E_{\mu_x}[(\lambda_{s_1}(x) - 1)\dots(\lambda_{s_k}(x) - 1)]$$

$$r''_{2n+1} := E_{\mu_x}[(\lambda_{s_1}(x) - 1)\cdots(\lambda_{s_{2n+1}}(x) - 1)\Lambda_{s_{2n+1}}(x,y)]$$

We note that the above expansions are orthogonal w.r.t. μ_{N_1}, but not μ_N, as A is the compensator of N_1, but not N.

IV. A PRODUCT-TO-SUM FORMULA FOR ITERATED INTEGRALS

In this section, we shall derive a product-to-sum formula for the iterated integrals w.r.t. the L_2 martingale of a counting process. Analogous to the case with the additive white Gaussian noise, such a product-to-sum formula plays a central role in the optimal orthogonal series expansion for estimation from counting observations.

Our approach is rather circuitous (A more direct one is desirable.). We shall first set out to "generalize" the basic ingredient $x(x - 1)\ldots(x - n + 1)$ of the Charlier polynomials in a Poisson process and obtain a product-to-sum formula for those "generalized Charlier ingredient". The conversion formulas between a "generalized Charlier ingredient" and an iterated integral will also be obtained. The product-to-sum formula for iterated integrals will then be deduced from that for the "generalized Charlier ingredients".

We assume that the counting process $(N_t, \mathcal{F}_t)$ has a continuous compensator A_t and the associated margingale $Z_t := N_t - A_t$ is square-integrable. For each continuous function f in time, we define

$$Y_n := \int_0^t f_s^n dN_s, \quad n \geq 1$$

Let $F_n(s_1,\ldots,s_n) := f(s_1)\ldots f(s_n)$, $n \geq 1$. We define a polynomial $Y^{(n)}(F_n)$ in Y_i, $i \leq n$ by

$$Y^{(n)}(F_n) := Y^{(n-1)}(F_{n-1})Y_1 + (-1)(n - 1)Y^{(n-2)}(F_{n-2})Y_2 + \ldots$$
$$+ (-1)^{n-2}(n-1)!Y^{(1)}(F_1)Y_{n-1} + (-1)^{n-1}(n-1)!Y_n \quad (12)$$

$$Y^{(1)}(F_1) := Y_1$$

$$Y^{(0)}(F_0) := 1$$

It is noted that when $f \equiv 1$, the polynomials $Y^{(n)}(F_n)$ reduce to the basic ingredient $N_t(N_t - 1)\ldots(N_t - n + 1)$ of the Charlier polynomials in N_t.

Let the n-th iterated integral be denoted by P_n as follows:

$$P_n(f_n) := \int_0^t \int_0^{\bar{s}_1} \cdots \int_0^{\bar{s}_{n-1}} f_n(s_1,\ldots,s_n)dZ_{s_n}\ldots dZ_{s_1}$$

$$=: \int^n f_n d^n Z$$

It is known [13], [16] that $P_n(f_n)$ for different n are orthogonal. The following lemma shows that an iterated integral $P_n(F_n)$ for $F_n(s_1,\ldots,s_n) = f(s_1)\ldots f(s_n)$ has an expression that resembles a Charlier polynomial.

Lemma 4.1. For the symbols F_n, $Y^{(n)}$, and P_n defined above, we have the conversion formula,

$$n!P_n(F_n) = \sum_{k=0}^{n} (-1)^k \binom{n}{k} \left(\int_0^t f_s dA_s\right)^k Y^{(n-k)}(F_{n-k}) \tag{13}$$

Proof. The formula will be proven by induction. It is trivial for n = 1. Assume it holds for all n up to m. Segall and Kailath [13] showed that

$$(m+1)!P_{m+1} = m![P_m\bar{Z}_t + (-1)P_{m-1}[\bar{Z},\bar{Z}]_t + (-1)^2P_{m-2}\sum_{s\le t} f_s^3(\Delta Z_s)^3$$
$$+\ldots+ (-1)^{m-1}P_1\sum_{s\le t} f_s^m(\Delta Z_s)^m + (-1)^m \sum_{s\le t} f_s^{m+1}(\Delta Z_s)^{m+1}] \tag{14}$$

where $\bar{Z}_t := \int_0^t f_s dZ_s$, $P_m := P_m(F_m)$, and $[\bar{Z},\bar{Z}]_t$ is the quadratic variation of the martingale $\bar{Z}_t$. Under the assumption that A_t is continuous, we have

$$[\bar{Z},\bar{Z}]_t = \int_0^t f_s^2 dN_s = Y_2$$

$$\sum_{s\le t} f_s^n(\Delta Z_s)^n = \int_0^t f_s^n dN_s = Y_n, \quad n \ge 3$$

By the induction hypothesis, we rewrite (14) as

$$(m+1)!P_{m+1} = \left[\sum_{i=0}^{m} (-1)^i \binom{m}{i} \bar{A}^i Y^{(m-i)}\right](Y_1 - \bar{A}) + (-1)m\left[\sum_{i=0}^{m-1} (-1)^i \binom{m-1}{i}\right.$$

$$\cdot \bar{A}^i Y^{(m-i-1)}\Big] Y_2 + (-1)^2 m(m-1)\Big[\sum_{i=0}^{m-2} (-1)^i \binom{m-2}{i} \bar{A}^i Y^{(m-i-2)}\Big] Y_3 + \ldots$$

$$+ (-1)^k m(m-1)\ldots(m-k+1)\Big[\sum_{i=0}^{m-k} (-1)^i \binom{m-k}{i} \bar{A}^i Y^{(m-i-k)}\Big] Y_{k+1}$$

$$+\ldots+ (-1)^{m-1} m!\,(Y^{(1)} - \bar{A})Y_m + (-1)^m m!\,Y_{m+1}$$

$$= [Y^{(m)}Y_1 + (-1)mY^{(m-1)}Y_2 + (-1)^2 m(m-1)Y^{(m-2)}Y_3 + \ldots$$

$$+ (-1)^{m-1} m!\,Y^{(1)}Y_m + (-1)^m m!\,Y_{m+1}] - \bar{A}\Big\{Y^{(m)} + [mY^{(m-1)}Y_1$$

$$+ (-1)m\binom{m-1}{1}Y^{(m-2)}Y_2 + (-1)^2 m(m-1)\binom{m-2}{1}Y^{(m-3)}Y_3 + \ldots$$

$$+ (-1)^{m-1} m!\,(1)Y_m]\Big\} + (-1)^2 \bar{A}^2\Big\{\binom{m}{1}Y^{(m-1)} + [\binom{m}{2}Y^{(m-2)}Y_1$$

$$+ (-1)m\binom{m-1}{2}Y^{(m-3)}Y_2 + (-1)^2 m(m-1)\binom{m-2}{2}Y^{(m-4)}Y_3 + \ldots$$

$$+ (-1)^{m-2} m(m-1)\ldots(3)\binom{m-(m-2)}{2}Y_{m-1}]\Big\} +\ldots+ (-1)^{m+1}\bar{A}^{m+1}$$

$$= Y^{(m+1)} + (-1)(m+1)\bar{A}Y^{(m)} + (-1)^2 \frac{(m+1)m}{2} \bar{A}^2 Y^{(m-1)} + \ldots$$

$$+ (-1)^{m+1}\bar{A}^{m+1}$$

Thus (13) holds for $n = m+1$ and this completes the induction proof. □

We shall now set about to generalize Lemma 4.1 for the iterated integral of a symmetric function. First of all, let us have a more general definition of $Y^{(n)}$. We note that iteratively from (12) it is easy to see that

$$Y^{(n)}(F_n) = \sum_p^{(n)} \beta_{n,p} Y_1^{p_1} Y_2^{p_2} \ldots Y_n^{p_n}$$

where the summation $\sum_p^{(n)}$ is taken over all such $p = (p_1,\ldots,p_n)$ that p_i are nonnegative integers and $\sum_{i=1}^{n} ip_i = n$, and $\beta_{n,p}$ are constants independent of f. We note also that $\beta_{n,p} = 1$, for $p = (n, 0,\ldots,0)$.

Hence for each symmetric function $F \in C(R^n)$, it is consistent for us to define $Y^{(n)}(F)$ by

$$Y^{(n)}(F) := \sum_p^{(n)} \beta_{n,p} \int^{|p|} F(\underbrace{s_1,\ldots,s_{p_1}}_{\substack{p_1 \text{ variables} \\ \text{each appearing} \\ \text{once}}}, \underbrace{s_{p_1+1}, s_{p_1+1},\ldots,s_{p_1+p_2}, s_{p_1+p_2}}_{\substack{p_2 \text{ variables each} \\ \text{appearing twice}}},$$

$$\underbrace{\ldots,s_{|p|-p_n+1},\ldots,s_{|p|-p_n+1},\ldots,s_{|p|},\ldots,s_{|p|}}_{\substack{p_n \text{ variables each} \\ \text{appearing n times}}})d^{|p|}N_s$$

$$=: \sum_p^{(n)} \beta_{n,p} \int^{|p|} F[p]d^{|p|}N \tag{15}$$

$$|p| := p_1+\ldots+p_n$$

The equation (13) can now be rewritten as

$$n!P_n(F_n) = \sum_{k=0}^{n} (-1)^k \binom{n}{k} Y^{(n-k)}\left(\int^k F_n(\sigma_1,\ldots,\sigma_k,\cdot)d^kA_\sigma\right) \tag{16}$$

A generalization of Lemma 4.1 thus follows.

Lemma 4.2. For a symmetric function $F \in C(R^n)$, the following conversion formula holds:

$$n!P_n(F) = \sum_{k=0}^{n} (-1)^k \binom{n}{k} Y^{(n-k)}_{\cdot}\left(\int^k F(\sigma_1,\ldots,\sigma_k,\cdot)d^kA_\sigma\right) \tag{17}$$

Proof. Let f be a linear combination $\sum_{i=1}^{n} \alpha_i f_i$ of $f_i \in C(R^1)$ with real coefficients α_i. Applying (16) to $F_n(s_1,\ldots,s_n) := f(s_1)\ldots f(s_n)$ and equating the coefficients of $\alpha_1\ldots\alpha_n$ on the both sides of (16), we obtain (17) for F equal to the symmetrization of the function $(s_1,\ldots,s_n) \mapsto f_1(s_1)\ldots f_n(s_n)$.

It follows by an argument of passing to the limit that (17) holds for every symmetric function $F \in C(R^n)$. □

We note that the continuity assumption in the above lemma can be removed. However, the above version is sufficient for our development in the sequel.

An inverse of the conversion formula (17) can be obtained in a similar way by first inverting (17). It is stated in the following lemma. The proof is omitted.

Lemma 4.3. For a symmetric function $F \in C(R^n)$, the following conversion formula holds:

$$Y^{(n)}(F) = \sum_{k=0}^{n} \binom{n}{k} (n-k)! P_{n-k}\left\{ \int^{k} F(\sigma_1,\ldots,\sigma_k,\cdot) d^k A_\sigma \right\} \tag{18}$$

Before we derive the product-to-sum formulas for P_n and $Y^{(n)}$, we shall now make an observation, in the following lemma, that the product of two iterated stochastic integrals P_n can indeed be expressed as a finite sum of iterated integrals. The integrands of those iterated intergrals in the sum exhibit a special feature, which will help us screen terms in the derivation of the product-to-sum formulas. We shall need the following symbols:

$\bar{a}(s_1,\ldots,s_n) := a(s_{\pi(1)},\ldots,s_{\pi(n)})$, for a permutation π of $(1,\ldots,n)$ such that $s_{\pi(1)} \geq s_{\pi(2)} \geq \cdots \geq s_{\pi(n)}$

$$L_A^2([0,t]^n) := \left\{ a : [0,t]^n \to R \,\middle|\, \int^{n} a^2 d^n A < \infty \right\}.$$

Lemma 4.4. For $a \in L_A^2([0,t]^n)$ and $b \in L_A^2([0,t]^m)$, the product $P_n(a)P_m(b)$ can be expressed as follows:

$$P_n(a)P_m(b) = \sum_{k=0}^{n+m} P_k(c_k) \tag{19}$$

where for each k, c_k is a linear combination of functions of the form,

$$\int \bar{a}(\sigma_1,\ldots,\sigma_\ell,\ldots,s_i,\ldots)\bar{b}(\sigma_1,\ldots,\sigma_\ell,\ldots,s_j,\ldots) d^\ell A_\sigma \tag{20}$$

with the following properties:

(1) The integral is taken over some subset of $[0,t]^\ell$;

(2) The total number of distinct variables s_i and s_j in $\bar{a}$ and $\bar{b}$ is k;

(3) The variables s_i (resp. s_j) occurring in $\bar{a}$ (resp. $\bar{b}$) are distinct, although some s_i in $\bar{a}$ may be repeated in $\bar{b}$.

Proof. The lemma can be proved by induction. For m = n = 1, we have, by the product rule [17, p. 101],

$$\int_0^t a(s)dZ_s \int_0^t b(s)dZ_s = \int_0^t \int_0^{s_1^-} b(s_2)a(s_1)dZ_{s_2}dZ_{s_1}$$

$$+ \int_0^t \int_0^{s_1^-} a(s_2)b(s_1)dZ_{s_2}dZ_{s_1} + \sum_{s \le t} a(s)b(s)(\Delta Z_s)^2$$

$$= \int_0^t \int_0^{s_1^-} (b(s_2)a(s_1) + a(s_2)b(s_1))dZ_{s_2}dZ_{s_1} + \int_0^t a(s)b(s)dZ_s$$

$$+ \int_0^t a(\sigma)b(\sigma)dA_\sigma$$

satisfying (19). Now suppose that (19) holds for all n and m such that $n + m \le M$. Let us consider the case $n + m = M + 1$.

By the product rule [17] again, we have

$$P_n(a)P_m(b) = P_n(\bar{a})P_m(\bar{b})$$

$$= \int_0^t \left[\int_0^{s_1^-} \cdots \int_0^{s_m^-} \bar{b}(s_2,\ldots,s_{m+1})d^mZ_s\right]\left[\int_0^{s_1^-} \cdots \int_0^{s_{n-1}^-} \bar{a}(s_1,\ldots,s_n) \cdot d^{n-1}Z_s\right]dZ_{s_1}$$

$$+ \int_0^t \left[\int_0^{s_1^-} \cdots \int_0^{s_n^-} \bar{a}(s_2,\ldots,s_{n+1})d^nZ_s\right]\left[\int_0^{s_1^-} \cdots \int_0^{s_{m-1}^-} \bar{b}(s_1,\ldots,s_m) \cdot d^{m-1}Z_s\right]dZ_{s_1}$$

$$+ \sum_{s_1 \le t} \left[\int_0^{s_1^-} \cdots \int_0^{s_{n-1}^-} \bar{a}(s_1,\ldots,s_n)d^{n-1}Z_s\right]\left[\int_0^{s_1^-} \cdots \int_0^{s_{m-1}^-} \bar{b}(s_1,\ldots,s_m) \cdot d^{m-1}Z_s\right](\Delta Z_{s_1})^2$$

It is easy to see from the induction hypothesis that the first two terms can be expressed in the form (19). The third term is first written as

$$\int_0^t\left[\int_0^{\bar{s}_1}\cdots\int_0^{\bar{s}_{n-1}}\bar{a}(s_1,\ldots,s_n)d^{n-1}Z_s\right]\left[\int_0^{\bar{s}_1}\cdots\int_0^{\bar{s}_{m-1}}\bar{b}(s_1,\ldots,s_m)d^{m-1}Z_s\right]$$

$$\cdot(dZ_{s_1}+dA_{s_1})$$

$$=:\int_0^t\sum_{k=0}^{n+m-2}P_k'(c_k')dZ_{s_1}+\int_0^t\sum_{k=0}^{n+m-2}P_k'(c_k')dA_{s_1}$$

The prime denotes dependency on s_1. The first term is now in the form (19) with s_1 in $\bar{a}$ repeated in $\bar{b}$. By carefully interchanging the integration order in the second term, we can also express the second term in the form (19). This completes the proof. □

We are now in a position to derive the product-to-sum formulas for $Y^{(n)}$ and P_n.

Theorem 4.1. Given the symmetric functions $a\in C(R^n)$ and $b\in C(R^m)$, we have the product-to-sum formula,

$$Y^{(n)}(a)Y^{(m)}(b)=\sum_{k=0}^{\min(m,n)}\alpha_{nm,k}Y^{(n+m-k)}((a[k]b[k])^{\sim})\tag{21}$$

$$\alpha_{nm,k}:=\frac{n!m!}{(n-k)!(m-k)!k!}\tag{22}$$

where $(a[k]b[k])^{\sim}$ denotes the symmetrization of the function $a[k]b[k]:(s_1,\ldots,s_{n+m-k})\mapsto a(s_1,\ldots,s_k,s_{k+1},\ldots,s_n)b(s_1,\ldots,s_k,s_{n+1},\ldots,s_{n+m-k})$.

Proof. We shall first establish the form (21) and then evaluate $\alpha_{nm,k}$. From (15), it follows

$$Y^{(n)}(a)Y^{(m)}(b)=\left(\sum_p^{(n)}\beta_{n,p}\int^{|p|}a[p]d^{|p|}N\right)\left(\sum_q^{(m)}\beta_{m,q}\int^{|q|}b[q]d^{|q|}N\right)$$

$$=\int^n ad^nN\int^m bd^mN+\sum_{\substack{p,q\\ p\neq(n,0,\ldots,0)\\ \text{or } q\neq(m,0,\ldots,0)}}^{(n)(m)}\beta_{n,p}\beta_{m,q}\int^{|p|}a[p]d^{|p|}N\int^{|q|}b[q]d^{|q|}N$$

$$=:A_1+\sum{}^2A_2$$

$$=Y^{(n+m)}((a[0]b[0]^{\sim})-\sum_{p\neq(n+m,0,\ldots,0)}^{(n+m)}\beta_{n+m,p}\int^{|p|}(a[0]b[0])^{\sim}[p]$$

$$\cdot\, d^{|p|}N + \sum^{2} A_2 = Y^{(n+m)}((a[0]b[0]^{\sim})$$

$$+ \int^{n+m-1}\Big[-\beta_{n+m,(n+m-2,1,0,\ldots,0)}(a[0]b[0])^{\sim}(s_1,s_1,s_2,s_3,\ldots,s_{n+m-1})$$

$$+ \beta_{n,(n-2,1,0,\ldots,0)}\beta_{m,(m,0,\ldots,0)}a(s_1,s_1,s_2,s_3,\ldots,s_{n-1})$$

$$\cdot\, b(s_n,\ldots,s_{n+m-1})$$

$$+ \beta_{n,(n,0,\ldots,0)}\beta_{m,(m-2,1,0,\ldots,0)}b(s_1,s_1,s_2,s_3,\ldots,s_{m-1})$$

$$\cdot\, a(s_m,\ldots,s_{n+m-1})\Big]d^{n+m-1}N_s - \sum_{\substack{p \\ p\neq(n+m,0,\ldots,0) \\ \text{or } (n+m-2,1,0,\ldots,0)}}^{(n+m)} \beta_{n+m,p}\int^{|p|}$$

$$\cdot\, (a[0]b[0])^{\sim}[p]d^{|p|}N + \sum_{\substack{(p,q)\neq((n-2,1,0,\ldots,0),(m,0,\ldots,0)) \\ \text{or } ((n,0,\ldots,0),(m-2,1,0,\ldots,0))}}^{2} A_2$$

$$= Y^{(n+m)}((a[0]b[0])^{\sim}) + \alpha_{nm,1}Y^{(n+m-1)}((a[1]b[1])^{\sim})$$

$$+ Y^{(n+m-1)}(c_1(a(s_1,s_1,s_2,\ldots,s_{n-1})b(s_n,\ldots,s_{n+m-1}))^{\sim} + c_2(b(s_1,s_1,$$

$$s_2,\ldots,s_{m-1})a(s_m,\ldots,s_{m+n-1}))^{\sim}) - \sum_{\substack{p \\ p\neq(n+m-1,0,\ldots,0)}}^{(n+m-1)}$$

$$\beta_{n+m-1,p}\int^{|p|}\Big\{\alpha_{nm,1}(a[1]b[1])^{\sim}$$

$$+ c_1(a(s_1,s_1,s_2,\ldots,s_{n-1})b(s_n,\ldots,s_{n+m-1}))^{\sim}$$

$$+ c_2(b(s_1,s_1,s_2,\ldots,s_{m-1})a(s_m,\ldots,s_{m+n-1}))^{\sim}\Big\}[p]d^{|p|}N_s$$

$$- \sum_{\substack{p \\ p\neq(n+m,0,\ldots 0) \\ \text{or } (n+m-2,1,0,\ldots,0)}}^{(n+m)} \beta_{n+m,p}\int^{|p|}(a[0]b[0]^{\sim}[p]d^{|p|}N$$

$$+ \sum_{\substack{(p,q)\neq((n-2,1,0,\ldots,0),(m,0,\ldots,0)) \\ \text{or } ((n,0,\ldots,0),(m-2,1,0,\ldots,0))}}^{2} A_2$$

Continuing in the above fashion, we can arrive at the expression,

$$Y^{(n)}(a)Y^{(m)}(b) = \sum_{k=0}^{\min(m,n)} \alpha_{nm,k} Y^{(n+m-k)}((a[k]b[k])^{\sim})$$

$$+ \sum_{k=1}^{m+n} Y^{(k)}(f_k) \tag{23}$$

where for each k, the function f_k is a linear combination of functions of the form

$$[a(\ldots,s_i,\ldots)b(\ldots,s_j,\ldots)]^{\sim}$$

Here the total number of distinct variables s_i and s_j in a and b is k, and there is at least one s_i in a or one s_j in b repeating in a or b respectively.

We now wish to show that $\sum_{k=1}^{m+n} Y^{(k)}(f_k) = 0$. By Lemma 4.3, $Y^{(k)}(f_k)$ can be expressed as a linear combination of P_ℓ operating on the functions of the form

$$⨍^r a(\ldots,\zeta_i,\ldots,s_i,\ldots)b(\ldots,\zeta_j,\ldots,s_j,\ldots)d^rA_\zeta \tag{24}$$

where the total number of distinct variables s_i and s_j in a and b is ℓ, and at least one of ζ_i, s_i, ζ_j, s_j repeats in a or b, whichever it appears in.

On the other hand, by Lemma 4.3 and then Lemma 4.4,

$$Y^{(n)}(a)Y^{(m)}(b) = \left[n!\sum_{k=0}^{n}\frac{1}{k!}P_{n-k}\left(⨍^k a(\sigma_1,\ldots,\sigma_k,\cdot)d^kA_\sigma\right)\right]$$

$$\cdot\left[m!\sum_{j=0}^{m}\frac{1}{j!}P_{m-j}\left(⨍^j b(\sigma_1,\ldots,\sigma_j,\cdot)d^jA_\sigma\right)\right]$$

$$= \sum_{k=0}^{n}\sum_{j=0}^{m}\frac{n!m!}{k!j!}\sum_{i=0}^{n+m-k-j} P_i(c_{kji}) \tag{25}$$

where c_{kji} are linear combinations of functions of the form,

$$\int\left[⨍^k a(\sigma_1,\ldots,\sigma_k,\zeta_1,\ldots,\zeta_r,\ldots,s_p,\ldots)d^kA_\sigma ⨍^j b(\eta_1,\ldots,\eta_j,\zeta_1,\ldots,\zeta_r,\ldots,s_q,\ldots)d^jA_\eta\right]d^rA_\zeta \tag{26}$$

Here we note that the total number of distinct variables s_p and s_q is r and none of the arguments of a and b repeat in a or b, whichever it appears in.

Equating (23) and (25) and noting that the equality holds for all functions a and b, we conclude that $\sum_{k=1}^{m+n} Y^{(k)}(f_k) = 0$ and (21) holds.

As (21) is valid for any functions a and b, we shall now evaluate $\alpha_{nm,k}$ by setting a = b = 1. First, observe that

$$Y^{(n)}(1) = N_t^{(n)} := N_t(N_t - 1)\ldots(N_t - n + 1)$$

$$Y^{(m)}(1) = N_t^{(m)} := N_t(N_t - 1)\ldots(N_t - m + 1)$$

In the following the subscript t shall be suppressed for notational simplicity.

We shall prove (22) by induction. For m + 1 = 1, it is trivial. Supposing that (22) holds for $m + n \leq M$, we now show that it holds also for m + n = M + 1. Assume, without loss of generality, $n \geq m$. Simple algebraic manipulation yields

$$N^{(m)}N^{(n+1)} = N^{(m)}N^{(n)}(N - (n + 1) + 1)$$

$$= [N^{(n+m)} + \alpha_{nm,1}N^{(n+m-1)} + \alpha_{nm,2}N^{(n+m-2)} +\ldots+ \alpha_{nm,m}N^{(n)}](N-n)$$

$$= N^{(n+m)}(N - n) + \alpha_{nm,1}N^{(n-m-1)}(N - n) +\ldots+ \alpha_{nm,m}N^{(n)}(N - n)$$

$$= N^{(n+m)}[N - (n + m + 1) + 1 + m] + \alpha_{nm,1}N^{(n+m-1)}$$

$$\cdot [y - (n + m) + 1 + (m - 1)]$$

$$+ \alpha_{nm,2}N^{(n+m-2)}[N - (n + m - 1) + 1 + (m - 2)]$$

$$+\ldots+ \alpha_{nm,m}N^{(n)}[N - (n + 1) + 1]$$

$$= N^{(n+m+1)} + [m + \alpha_{nm,1}]N^{(n+m)} + [(m - 1)\alpha_{nm,1} + \alpha_{nm,2}]N^{(n+m-1)}$$

$$+\ldots+ [(m - k + 1)\alpha_{nm,k-1} + \alpha_{nm,k}]N^{(n+m-k+1)} +\ldots+ [\alpha_{nm,m-1} + \alpha_{nm,m}]N^{(n+1)}$$

Hence we conclude that

$$\alpha_{(n+1)m,k} = (m - k + 1)\alpha_{nm,k-1} + \alpha_{nm,k}$$

$$= \frac{(n + 1)!m!}{(n + 1 - k)!(m - k)!k!}$$

Similarly,

$$\alpha_{n(m+1),k} = \frac{n!(m + 1)!}{(n - k)!(m + 1 - k)!k!}$$

Thus (22) holds for $n + m = M + 1$, completing the proof. ❑

Theorem 4.2. For $a \in L_A^2([0,t]^n)$ and $b \in L_A^2([0,t]^m)$, we have the following product-to-sum formula for iterated stochastic integrals:

$$P_n(a)P_m(b) = \sum_{j=0}^{\min(m,n)} \sum_{i=0}^{j} \frac{1}{i!} P_{n+m-i-j}\Bigg(\int\!\!\!\!-^{i} \sum_{(\pi_1,\pi_2,\pi_3)}^{(j-i,n-j,m-j)} \bar{a}(\sigma_1,\ldots,\sigma_i,$$

$$s_{\pi_1(1)},\ldots,s_{\pi_1(j-i)},s_{\pi_2(1)},\ldots,s_{\pi_2(n-j)})\bar{b}(\sigma_1,\ldots,\sigma_i,s_{\pi_1(1)},\ldots,$$

$$s_{\pi_1(j-i)},s_{\pi_3(1)},\ldots,s_{\pi_3(m-j)})d^iA_\sigma\Bigg) \tag{27}$$

where π_1 is a combination of $j - i$ numbers taken from the set $\{1,\ldots, n + m - i - j\}$; π_2 is a combination of $n - j$ numbers taken from the remaining subset $\{1,\ldots, n + m - i - j\} - \pi_1$; and π_3 is the remaining subset $\{1,\ldots, n + m - i - j\} - \pi_1 - \pi_2$. The summation $\sum_\pi$ is taken over all such combinations (π_1, π_2, π_3).

Proof. Assume first that $a \in C(R^n)$ and $b \in C(R^m)$. By Lemma 4.2, Theorem 4.1, and Lemma 4.3, we have

$$n!P_n(a)m!P_m(b) = m!n!P_n(\bar{a})P_n(\bar{b})$$

$$= \left[\sum_{k=0}^{n} (-1)^k \binom{n}{k} Y^{(n-k)}\left(\int\!\!\!\!-^{k} \bar{a}(\sigma_1,\ldots,\sigma_k,\cdot)d^kA_\sigma\right)\right]\left[\sum_{\ell=0}^{m} (-1)^\ell \binom{m}{\ell} Y^{(m-\ell)}\right.$$

$$\left.\cdot\left(\int\!\!\!\!-^{\ell} \bar{b}(\sigma_1,\ldots,\sigma_\ell,\cdot)d^\ell A_\sigma\right)\right]$$

$$= \sum_{k=0}^{n} \sum_{\ell=0}^{m} (-1)^{k+\ell} \binom{n}{k}\binom{m}{\ell} \sum_{j=1}^{\min(n-k,m-\ell)} \alpha_{(n-k)(m-\ell),j}\, Y^{(n+m-k-\ell-i)}$$

$$\cdot \left\{ \left[\int^{k+\ell} \bar{a}(\sigma_1,\ldots,\sigma_k,s_1,\ldots,s_j,s_{j+1},\ldots,s_{n-k})\bar{b}(\sigma_{k+1},\ldots,\sigma_{k+\ell},s_1,\ldots,s_j,s_{n-k+1},\ldots,s_{n+m-k-\ell-j})d^{k+\ell}A_\sigma \right]^\sim \right\}$$

$$= \sum_{k=0}^{n} \sum_{\ell=0}^{m} (-1)^{k+\ell} \binom{n}{k}\binom{m}{\ell} \sum_{j=1}^{\min(n-k,m-\ell)} \alpha_{(n-k)(m-\ell),j}$$

$$\cdot \sum_{i=0}^{n+m-k-\ell-j} \frac{(n+m-k-\ell-j)!}{i} P_{n+m-k-\ell-j-i} \left\{ \int^{i} [\text{as above}]^\sim (\eta_1,\ldots,\eta_i,\cdot)d^iA_\eta \right\}$$

Let us examine the integrand under $P_{n+m-k-\ell-j-i}$ in the above expression. When $(k,\ell) \neq (0,0)$, there is at least one σ_i in $\bar{a}$ or in $\bar{b}$ which does not match pair with the same σ_i in $\bar{b}$ or respectively in $\bar{a}$. When $(k,\ell) = (0,0)$, if $i > j$, then there is at least such a η_i in $\bar{a}$ or in $\bar{b}$. In view of Lemma 4.4, those terms in the above expression that involve such σ_i and η_i should add up to zero and disappear. Hence,

$$n!P_n(a)m!P_m(b) = \sum_{j=1}^{\min(n,m)} \alpha_{nm,j} \sum_{i=0}^{j} \frac{(n+m-j)!}{i}$$

$$\cdot P_{n+m-j-i}\left\{ \int^{i} \left[\bar{a}(\eta_1,\ldots,\eta_i,s_1,\ldots,s_{j-1},s_{j+i},\ldots,s_n)\bar{b}(\eta_1,\ldots,\eta_i,s_1,\ldots,s_{j-i},s_{n+1},\ldots,s_{n+m-j}) \right]^\sim d^iA_\eta \right\} \tag{28}$$

We recall that the symmetrization $[..]^\sim$ is taken of the function $(s_1,\ldots,s_{m+n-j}) \mapsto a(s_1,\ldots,s_n)\bar{b}(s_1,\ldots,s_j,s_{n+1},\ldots,s_{n+m-j})$.

The iterated integral in (28) thus becomes

$$P_{n+m-j-i}(\text{as above}) = \left[\binom{n+m-j}{j}\binom{n+m-2j}{n-j}\right]^{-1} P_{n+m-j-i}\left\{ \int^{i} \sum_{\pi} \bar{a}(\sigma_1,\ldots,\sigma_i,s_{\pi_1(1)},\ldots,s_{\pi_1(j-i)},\ldots,s_{\pi_2(1)},\ldots,s_{\pi_2(n-j)})\bar{b}(\sigma_1,\ldots,\sigma_i,s_{\pi_1(1)},\ldots,s_{\pi_1(j-i)},s_{\pi_3(1)},\ldots,s_{\pi_3(m-j)}) \right\}$$

Substituting this into (28) yields (27), the desired result.

By an argument of passing to limit, it can be shown that (27) is valid for $a \in L_A^2([0,t]^n)$ and $b \in L_A^2([0,t]^m)$, completing the proof. □

Two alternative expressions of (27) will now be given. As they are consequences of simple algebraic manipulation, their proof is omitted.

Corollary 4.1. The equation (27) in Theorem 4.2 can be replaced with the following

$$P_m(a)P_n(b) = \sum_{j=0}^{2\min(m,n)} \sum_{i=\max(j-\min(m,n),0)}^{[j/2]} \frac{1}{i!} P_{n+m-j}$$

$$\cdot \left(\int^{i} \sum_{(\pi_1,\pi_2,\pi_3)}^{(j-2i,n+i-j,m+i-j)} \bar{a}\,\bar{b}\, d^i A_\sigma \right) \tag{29}$$

$$= \sum_{k=|n-m|}^{n+m} P_k \left(\sum_{\substack{i=n+m-k-2[n+m-k/2] \\ (n+m-k-i)/2 \text{ is an integer}}}^{\min(k-|n-m|,n+m-k)} \right.$$

$$\cdot \left[\left(\frac{n+m-k-i}{2} \right)! \right]^{-1} \int^{(n+m-k-i)/2} \sum_{(\pi_1,\pi_2,\pi_3)}^{(i,(n-m-i+k)/2,(m-n-i+k)/2)}$$

$$\left. \cdot\, \bar{a}\,\bar{b}\, d^{(n+m-k-i)/2} A_\sigma \right) \tag{30}$$

V. OPTIMAL EXPANSION IN ITERATED INTEGRALS FOR ESTIMATION

Consider the set $H_n(t)$ of all the n-th order series of iterated integrals driven by $Z_s := N_s - A_s$ on $[0,t]$:

$$H_n(t) := \left\{ Z \,\middle|\, Z = a_0(t) + \sum_{k=1}^{n} \int^{k} a_k(t,s_1,\ldots,s_k) d^k Z_s,\ a_k \in L_2([0,t]^k) \right\}$$

The set $H_n(t)$ is obviously a subspace of the Hilbert space of the square-integrable random variables.

Given a measurable functional $\phi(x)$ of x such that $E\phi^2(x) < \infty$, the element $\hat{\phi}^{n} \in H_n(t)$ is called the n-th order optimal estimate of $\phi(x)$, if

$E(\hat{\emptyset}^n(x) - \emptyset(x))^2 \leq E(z - \emptyset(x))^2$ for every $z \in H_n(t)$.

The purpose of this section is to derive the n-th order Wiener-Hopf equation that characterizes the kernels, namely the integrands, of $\hat{\emptyset}^n$. The following theorem will enable us to project the Bayesian-type representation of $\hat{\emptyset}^n$ onto $H_n(t)$ by straightforward algebraic manipulation. The proof of the theorem is almost verbatim the same as that [18] of its counter part for the additive white Gaussian noise case and thus omitted.

In the rest of the paper, we assume that $\sigma_t(\emptyset)$ and $\sigma_t(1)$ have been expanded as follows:

$$\sigma_t(\emptyset) = \alpha_0(t) + \sum_{k=1}^{n} \int^k \alpha_k(t,s_1,\ldots,s_k)d^kZ_s + \int^{n+1} r'_{n+1}(t,s_1,\ldots,s_{n+1}, N^{s_{n+1}})d^{n+1}Z_s$$

$$\sigma_t(1) = \beta_0(t) + \sum_{k=1}^{2n} \int^k \beta_k(t,s_1,\ldots,s_k)d^kZ_s + \int^{2n+1} r''_{2n+1}(t,s_1,\ldots,s_{2n+1}, N^{s_{2n+1}})d^{2n+1}Z_s$$

$$Z_s := N_s - A_s$$

Let the optimal n-th order estimate $\hat{\emptyset}^n$ be written as

$$\hat{\emptyset}^n(x) = b_0(t) + \sum_{k=1}^{n} \int^k b_k(t,s_1,\ldots,s_k)d^kZ_s$$

where b_k are the kernels to be determined.

Theorem 5.1. An element $z \in H_n(t)$ is the optimal n-th order estimate $\hat{\emptyset}^n$, if and only if

$$R_{n+1}(N - A) := z(N - A)(\sigma_t(1))(N - A) - \alpha_0(t) - \sum_{k=1}^{n} \int^k \alpha_k d^k(N_s - A_s)$$

is orthogonal to $H_n(t)$ w.r.t. μ_{N_1}, i.e., for every $z_n \in H_n(t)$, $E(R_{n+1}(N_1 - A)z_n(N_1 - A)) = 0$.

We are now ready to derive the n-th order Wiener-Hopf equation for the kernels of $\overset{n}{\hat{\emptyset}}$. By Corollary 4.1 and some algebraic manipulation, we have

$$\overset{n}{\hat{\emptyset}}\left(\sum_{j=0}^{2n}\int^{j}\beta_j d^j (N - A)\right)$$

$$= \sum_{i=0}^{n}\sum_{j=0}^{2n}\sum_{r=0}^{2\min(i,j)}\sum_{k=\max(r-\min(i,j),0)}^{[r/2]}\int^{i+j-r}\frac{1}{k!}\int^{k}\sum_{(\pi_1,\pi_2,\pi_3)}^{(r-2k,i+k-r,j+k-r)}$$

$$\cdot\ \overline{b}_i\overline{\beta}_j d^{i+j-r} Z_s$$

$$= \left[\sum_{m=0}^{n}\sum_{i=0}^{n}\sum_{r=\max(2i-2m,0)}^{2i} + \sum_{m=n+1}^{3n}\sum_{i=0}^{n}\sum_{r=\max(2i-2m,0)}^{\min(2n+i-m,2i)}\right]$$

$$\sum_{k=\max(r-\min(i,m+r-i),0)}^{[r/2]}\int^{m}\frac{1}{k!}\int^{k}\sum_{(\pi_1,\pi_2,\pi_3)}^{(r-2k,i+k-r,m-i+k)}$$

$$\cdot\ \overline{b}_i\overline{\beta}_{m+r-i} d^k A_\sigma d^m Z_s$$

It follows from Theorem 5.1 and the orthogonality property of the iterated integrals w.r.t. μ_{N_1} that

$$\sum_{i=0}^{n}\sum_{r=\max(2i-2m,0)}^{2i}\sum_{k=\max(r-\min(i,m+r-i),0)}^{[r/2]}\frac{1}{k!}\int^{k}\sum_{(\pi_1,\pi_2,\pi_3)}^{(r-2k,i+k-r,m-i+k)}$$

$$\cdot\ \overline{b}_i\overline{\beta}_{m+r-i} d^k A_\sigma = \alpha_m \qquad m = 0,\dots,n \tag{31}$$

This is the desired n-th order Wiener-Hopf equation.

VI. OPTIMAL EXPANSION IN FOURIER-CHARLIER FUNCTIONALS FOR ESTIMATION

Let $\{\phi_i,\ i = 1, 2,\dots\}$ be a complete orthonormal set (CONS) of functions on $[0,t]$ w.r.t. the measure dA_s. Associated with each n-vector $\xi = (\xi_1,\dots,\xi_n)$ of nonnegative integers ξ_i such that $\xi_1 \le \xi_2 \le \cdots \le \xi_n$, there is a symmetric function $[\phi_{\xi_1}(s_1)\dots\phi_{\xi_n}(s_n)]^{\sim}$. The multiple Wiener integral of this function w.r.t. $Z := N_1 - A$ over $[0,t]^n$ will be denoted by

$$F(t,n,\xi,N_1 - A) := n!P_n([\phi_{\xi_1}\ldots\phi_{\xi_n}]^{\sim}) \tag{32}$$

These integrals are orthogonal to one another and will be called the Fourier-Charlier (FC) functionals. It was shown by Ito [19, p. 257] that if $A(N_1^t)$ is a L^2-functional of $\{N_{1s}, 0 \leq s \leq t\}$, then it can be approximated arbitrarily close by a finite FC series as follows:

$$A(N_1^t) \cong \sum_{k=0}^{n} \sum_{\xi \in \underset{\sim}{\xi}_k} a(k,\xi)F(t,k,\xi,N_1 - A)$$

$$a(k,\xi) := E[A(N_1^t)F(t,k,\xi,N_1 - A)]$$

where $\underset{\sim}{\xi}_k$ is a finite subset of $\{\xi = (\xi_1,\ldots,\xi_k) \mid \xi_1 \leq \cdots \leq \xi_k, \xi_i$ are nonnegative integers$\}$.

For any given index sets $\underset{\sim}{\xi} := \underset{\sim}{\xi}_1 \cup \ldots \cup \underset{\sim}{\xi}_n$ and $\underset{\sim}{\eta} := \underset{\sim}{\eta}_1 \cup \ldots \cup \underset{\sim}{\eta}_{2n}$, the functionals $\sigma_t(\emptyset)$ and $\sigma_t(1)$ have the following FC series representations:

$$\sigma_t(\emptyset) = \alpha(t,0) + \sum_{k=1}^{n} \sum_{\xi \in \underset{\sim}{\xi}_k} \alpha(t,k,\xi)F(t,k,\xi,N - A) + R_1(N - A,\underset{\sim}{\xi}) \tag{33}$$

$$\sigma_t(1) = \beta(t,0) + \sum_{k=1}^{2n} \sum_{\eta \in \underset{\sim}{\eta}_k} \beta(t,k,\eta)F(t,k,\eta,N - A) + R_2(N - A,\underset{\sim}{\eta}) \tag{34}$$

where

$$\alpha(t,k,\xi) := \frac{1}{k!}\int^k \bar{\alpha}_k(t,\sigma_1,\ldots,\sigma_k) \sum_{\tau} \phi_{\tau(\xi_1)}(\sigma_1)\ldots\phi_{\tau(\xi_k)}(\sigma_k)d^k A_\sigma$$

$$:= \int^k \bar{\alpha}_k(t,\sigma_1,\ldots,\sigma_k) \prod_{j=1}^{k} \phi_{\xi_j}(\sigma_j)d^k A_\sigma$$

$$\beta(t,k,\eta) := \int^k \bar{\beta}_k(t,\sigma_1,\ldots,\sigma_k) \prod_{j=1}^{k} \phi_{\eta_j}(\sigma_j)d^k A_\sigma$$

τ := a permutation of $\{\xi_1,\ldots,\xi_k\}$

We note that these are orthogonal expansions w.r.t. μ_{N_1}, but not μ_N.

Consider now the set $H_n(t,\underset{\sim}{\xi})$ of all the FC series indexed by $\underset{\sim}{\xi}$:

$$H_n(t,\underset{\sim}{\xi}) := \left\{ \sum_{k=0}^{n} \sum_{\xi \in \underset{\sim}{\xi}_k} a(t,k,\xi)F(t,k,\xi,N - A) \,\middle|\, a(t,k,\xi) \text{ are real numbers} \right\} \tag{35}$$

We will call an element of $H_n(t,\underset{\sim}{\xi})$ the ξ-optimal estimate of $\emptyset(x)$ and denote it by $\underset{\sim}{\xi}(\emptyset) := \xi_t(\emptyset(x))$, if for every $z \in H_n(t,\underset{\sim}{\xi})$,

$$E(\xi_{\tilde{t}}(\emptyset(x)) - \emptyset(x))^2 \leq E(z - \emptyset(x))^2 \tag{36}$$

It is noted that $H_n(t,\underset{\sim}{\xi})$ is a closed subspace of the Hilbert space of L^2-functionals of $\{N_s,\ 0 \leq s \leq t\}$.

The following theorem is analogous to theorem 5.1 and its proof is also omitted.

Theorem 6.1. The functional $z_n(N - A) \in H_n(t,\underset{\sim}{\xi})$ is the ξ-optimal estimate of $\emptyset(x)$ if and only if

$$R(N - A) := z_n(N - A)(\sigma_t(1))(N - A) - \sum_{k=0}^{n} \sum_{\xi \in \underset{\sim}{\xi}_k} \alpha(t,k,\xi)F(t,k,\ ,N - A) \tag{37}$$

is orthogonal to $H_n(t,\underset{\sim}{\xi})$ w.r.t. μ_{N_1}, i.e., for every $z \in H_n(t,\underset{\sim}{\xi})$,

$$E(R(N_1 - A)z(N_1 - A)) = 0 \tag{38}$$

For each index set ξ let the ξ-optimal estimate $\underset{\sim}{\xi}(\emptyset)$ be written as

$$\underset{\sim}{\xi}(\emptyset) := b(t,0) + \sum_{k=1}^{n} \sum_{\xi \in \underset{\sim}{\xi}_k} b(t,k,\xi)F(t,k,\xi,\ N - A) \tag{39}$$

where $b(t,k,\xi)$ are the constants (for a fixed t) to be determined. Using the above theorem, we shall be able to derive a linear system of simultaneous algebraic equations for $b(t,k,\xi)$. We shall need to consider all those FC components of $\sigma_t(1)$ that affect each such ζ-component of the product $\underset{\sim}{\xi}(\emptyset)\sigma_t(1)$ that $\zeta \in \underset{\sim}{\xi}$. This calls for the following lemma, which is an immediate consequence of Theorem 4.2.

Lemma 6.1. The product of two FC functionals can be expressed as follows:

$$F(t,i,\xi,\ N - A)F(t,j,\eta,\ N - A)$$

$$= \sum_{k=0}^{2\min(i,j)} \sum_{k=\max(r-\min(i,j),0)}^{[r/2]} \int^{i+j-r} \left[\frac{1}{k!} \sum_{\tau_1}\sum_{\tau_2} \delta(\tau_1(\xi_1),\tau_2(\eta_1))\right.$$

$$\cdots\delta(\tau_1(\xi_k),\tau_2(\eta_k)) \sum_{(\pi_1,\pi_2,\pi_3)}^{(r-2k,i-r+k,j-r+k)} \phi_{\tau_1(\xi_{k+1})}(s_{\pi_1(1)})$$

$$\cdots\phi_{\tau_1(\xi_{r-2k})}(s_{\pi_1(r-2k)})\phi_{\tau_1(\xi_{r-2k+1})}(s_{\pi_2(1)})\cdots\phi_{\tau_1(\xi_i)}(s_{\pi_2(i-r+k)})$$

$$\cdot\phi_{\tau_2(\eta_{k+1})}(s_{\pi_1(1)})\cdots\phi_{\tau_2(\eta_{r-2k})}(s_{\pi_1(r-2k)})\phi_{\tau_2(\eta_{r-2k+1})}(s_{\pi_2(i-r+k+1)})$$

$$\cdots\phi_{\tau_2(\eta_j)}(s_{\pi_2(i+j-2r+2k)})\Big] d^{i+j-r}[N_s - A_s] \tag{40}$$

$$\delta(i,j) := \begin{cases} 1 & \text{if } i = j \\ 0 & \text{if } i \neq j \end{cases}$$

We note that a typical integrand in (40) is $\phi_{\zeta'}(s_1,\ldots,s_k)$

$$:= \phi_{\xi_{i_1}}(s_1)\phi_{\eta_{j_1}}(s_1)\ldots\phi_{\xi_{i_\ell}}(s_\ell)\phi_{\eta_{j_\ell}}(s_\ell)\phi_{\zeta_{i_{\ell+1}}}(s_{\ell+1})\ldots\phi_{\zeta_{i_k}}(s_k).$$

Here let us use ζ' to denote the index object $[(\xi_{i_1},\eta_{j_1}),\ldots,$ $(\xi_{i_\ell},\eta_{j_\ell}),\zeta_{i_{\ell+1}},\ldots,\zeta_{i_k}]$. We shall call $k = k(\zeta')$ the dimension of ζ' and $\ell = \ell(\zeta')$ the number of pairings in ζ'.

For a fixed index vector ξ (without pairings), let $\eta(\xi,\zeta')$ denote the set of all the index vectors η that make (40) have a ζ'-component. The determination of $\eta(\xi,\zeta')$ can be seen by first considering some examples. If $\xi = [1, 2, 3]$ and $\zeta' = [(1, 4), 2]$, then $\eta(\xi,\zeta') = \{[4, 3], [4, 3, 2, 2]\}$. If $\xi = [1, 2, 3]$ and $\zeta' = [(5, 4)]$, then $\eta(\xi,\zeta')$ is empty. Some reflection shows that the general rule to determine $\eta(\xi,\zeta')$ is the following: $\eta \in \eta(\xi,\zeta')$ if and only if (a). some components of η can match pair with some components of ξ to form all the pairings in ζ'; (b). η contains all the nonmatching uncommon components of ξ and ζ'; (c). η may contain those nonmatching uncommon components of ξ and ζ', but they have to be included in duplicate.

By Lemma 6.1 and some combinatoric consideration, we see that

$$\text{the } \zeta'\text{-component of (40)} = \frac{c(\xi,\eta,\zeta')}{[(k(\xi) + k(\eta) - k(\zeta') - \ell(\zeta'))/2]!}$$

$$\cdot P_{k(\zeta')}(\phi_{\zeta'}) \tag{41}$$

where the coefficient $c(\xi,\eta,\zeta')$ is determined in exactly the same way as that for $c(\xi,\eta,\zeta'')$ in Part I with $\zeta'' := [\xi_{i_1},\eta_{j_1},\ldots,\xi_{i_\ell},$ $\eta_{j_\ell},\zeta_{i_{\ell+1}},\ldots,\zeta_{i_k}]$ (Some permutation of the components may be necessary here to make ζ'' an index vector.): Let γ be the set of the common components of ξ and η excluding the components of ζ'' defined above. Then

$c(\xi,\eta,\zeta') = $ (# ways that the components of γ can be drawn from those of ξ)
 • (# ways that the components of γ can be drawn from those of η)
 • ((# components of γ)!)
 • ((# components of γ)!/(# permutations of the distinguished components of γ)) (42)

When $\ell(\zeta') \neq 0$, the multiple Wiener integral $k(\zeta')!P_{k(\zeta')}(\phi_{\zeta'})$ is in general not a basic FC functional and we need to transform it into a sum (finite or infinite) of FC functionals. To simplify our illustration, we make the assumption: The CONS$\{\phi_i, i = 1, 2,...\}$ has the property that the product of any two elements in it can be expressed as a finite sum of its elements, i.e.,

$$\phi_i(s)\phi_j(s) = \sum_{k \in K(i,j)} \lambda(k,i,j)\phi_k(s) \tag{43}$$

where $K(i,j)$ is a finite set of integers.

Let $\Gamma(\xi)$ denote the set of all the ζ' that make (40) have a ζ-component through (43), and let $\lambda(\zeta',\zeta)$ denote the product string of $\lambda(k,i,j)$ that appears when ζ' is converted into ζ through (43). Then, from (41), for each η, the ζ-component of (40) is

$$\sum_{\zeta' \in \Gamma(\zeta)} \frac{c(\xi,\eta,\zeta')\lambda(\zeta',\zeta)}{[(k(\xi) + k(\eta) - k(\zeta') - \ell(\zeta'))/2]!} F(t,k(\zeta),\zeta,N - A) \tag{44}$$

A product-to-sum formula for finite FC series is readily obtained as follows:

Lemma 6.2. A product-to-sum formula for finite FC series is

$$\left[\sum_{k=0}^{n} \sum_{\xi \in \underset{\sim}{\xi}_k} \gamma_1(t,k,\xi)F(t,k,\xi, N - A)\right]\left[\sum_{k=0}^{m} \sum_{\eta \in \underset{\sim}{\eta}_k} \gamma_2(t,k,\eta)F(t,k,\eta, N - A)\right] = \sum_{\zeta \in \underset{\sim}{\zeta}(\underset{\sim}{\xi},\underset{\sim}{\eta})} \left[\sum_{\zeta' \in \Gamma(\zeta)} \sum_{k=0}^{n} \sum_{\xi \in \underset{\sim}{\xi}_k} \sum_{\eta \in \underset{\sim}{\eta} \cap \underset{\sim}{\eta}(\xi,\zeta')} \cdot \frac{\gamma_1(t,k,\xi)\gamma_2(t,k(\eta),\eta)c(\xi,\eta,\zeta')\lambda(\zeta',\zeta)}{[(k + k(\eta) - k(\zeta') - \ell(\zeta'))/2]!}\right] F(t,k(\zeta),\zeta, N - A)$$

$$\underset{\sim}{\zeta}(\underset{\sim}{\xi},\underset{\sim}{\eta}) := \bigcup_{\xi \in \underset{\sim}{\xi}} \bigcup_{\eta \in \underset{\sim}{\eta}} \underset{\sim}{\zeta}(\xi,\eta) \tag{45}$$

where $\zeta(\xi,\eta)$ denotes the set of all the index vectors ζ that result from the product $F(t,k(\xi),\xi, N - A)F(t,k(\eta),\eta,N - A)$.

Applying Theorem 6.1 and Lemma 6.2, we obtain, for every $\zeta \in \bigcup_{k=0}^{n} \underset{\sim}{\xi}_k$,

$$\sum_{\zeta' \in \Gamma(\zeta)} \sum_{k=0}^{n} \sum_{\xi \in \underset{\sim}{\xi}_k} \sum_{\eta \in \underset{\sim}{\eta}(\xi,\zeta')} \frac{\beta(t,k(\eta),\eta)c(\xi,\eta,\zeta')\gamma(\zeta',\zeta)}{[(k + k(\eta) - k(\zeta) - \ell(\zeta'))/2]!} b(t,k,\xi) = \alpha(t,k(\xi),\zeta) \qquad (46)$$

This is the desired linear system of simultaneous algebraic equations governing the FC coefficients $b(t,k,\xi)$.

REFERENCES

[1] L. I. Galchuck, "Filtering of Markov Processes with Jumps", *Uspekhi Matem. Nauk*, Vol. XXV, 5, 1970, pp. 237-238.

[2] D. L. Snyder, "Filtering and Detection for Doubly Stochastic Poisson Processes", *IEEE Trans. Inform. Theory*, Vol. IT-18, Jan. 1972, pp. 91-102.

[3] __________, "Smoothing for Doubly Stochastic Poisson Processes", *IEEE Trans. Inform. Theory*, Vol. IT-18, September 1972, pp. 558-562.

[4] P. A. Frost, "Estimation and Detection for a Simple Class of Conditionally Independent Increment Processes", in *Proc. IEEE Decision and Control Conf.*, 1971. Also published as lecture notes for Washington Univ. summer course on Current Trends in Automatic Control, St. Louis, Mo., 1970.

[5] __________, "Examples of Linear Solutions to Nonlinear Estimation Problems", in *Proc. 5th Princeton Conf. Information Sciences and Systems*, 1971.

[6] I. Rubin, "Regular Point Processes and Their Detection", *IEEE Trans. Inform. Theory*, Vol. IT-18, September 1972, pp. 547-557.

[7] P. Bermaud, "A Martingale Approach to Point Processes", University of Calif., Berkeley, ERL Memo M345, August 1972.

[8] M. H. A. Davis, "Detection of Signals with Point Processes Observation", *Dept. Computing and Control, Imperial College*, London, England, Publ. 73-8, 1973.

[9] J. H. Van Schppen, *Estimation Theory for Continuous-Time Processes, A Martingale Approach*, Ph.D. Dissertation, Dept. Elec. Eng., University of Calif., Berkeley, September 1973.

[10] R. Boel, P. Varaiya and E. Wong, "Martingales on Jump Processes, II: Applications", *SIAM J. Control*, Vol. 13, 5, 1975, pp. 1022-1061.

[11] A. Segall and T. Kailath, "The Modeling of Randomly Modulated Jump Processes", *IEEE Trans. Inform. Theory*, Vol. IT-21, No. 2, March 1975, pp. 135-143.

[12] A. Segall, M.H.A. Davis and T. Kailath, "Nonlinear Filtering with Counting Observations", *IEEE Trans. Inform. Theory*, Vol. IT-21, No. 2, March 1975, pp. 143-149.

[13] A. Segall and T. Kailath, "Orthogonal Functionals of Independent-Increment Processes", *IEEE Trans. Inform. Theory*, Vol. IT-22, No. 3, May 1976.

[14] R. S. Liptser and A. N. Shiryayev, *Statistics of Random Processes, II: Applications*, Springer-Verlag, New York, 1977.

[15] M. Loève, *Probability Theory II*, 4th Edition, Springer-Verlag, New York, 1978.

[16] P. J. Feinsilver, "Special Functions, Probability Semigroups, and Hamiltonian Flows", *Lecture Notes in Mathematics* No. 696, Springer-Verlag, New York, 1978.

[17] C. Doleans-Dade and P. A. Meyer, "Integrales Stochastiques par Rapport aux Martingales Locales, Seminaire de Probabilities IV", *Lecture Notes in Mathematics* No. 124, Springer-Verlag, New York, 1970, pp. 77-107.

[18] J. T. H. Lo and S. K. Ng, "Finite Order Nonlinear Filtering I: General Theory", *UMBC Mathematics Research Report* No. 10, 1981.

[19] K. Ito, "Spectral type of Shift Transformations of Differential Process with Stationary Increments", *Trans. Amer. Math. Soc.*, Vol. 81, 1956, pp. 253-263.

APPROXIMATIONS FOR NONLINEAR FILTERING[1]

Sanjoy K. Mitter

Department of Electrical Engineering and Computer Science
and Laboratory for Information and Decision Systems
Massachusetts Institute of Technology
Cambridge, Massachusetts 02139

1. INTRODUCTION

The objective of this paper is to show how recent work on nonlinear filtering can give qualitative insight into practical nonlinear filtering and suggest approximation schemes for optimal nonlinear filters.

To simplify the exposition we shall consider filtering problems in which the state and observation processes are scalar stochastic processes. We shall also present formal derivations. For the rigorous derivation of these results and precise hypotheses see the authors' papers FLEMING-MITTER [1982], MITTER [1982], OCONE [1980] and the references cited there.

2. PROBLEM FORMULATION

Consider the nonlinear filtering problem

$$dx(t) = b\Big(x(t)\Big)dt + \sigma\Big(x(t)\Big)dw(t) \tag{2.1}$$

$$dy(t) = h\Big(x(t)\Big)dt + d\tilde{w}(t) \tag{2.2}$$

where $x(t)$ is the state process, $y(t)$ is the observation process and $w,\tilde{w}$ are assumed to be independent standard Wiener processes. It is required to construct an estimate

(1) This research has been supported by the Air Force Office of Scientific Research under Grant No. AF-AFOSR 82-0135.

R. S. Bucy and J. M. F. Moura (eds.), Nonlinear Stochastic Problems, 339–345.

$$\hat{\phi}\left(x(t)\right) = E\left[\phi\left(x(t)\right|\mathscr{F}_t^y\right], \qquad (2.3)$$

where ϕ is some suitable function. In many situations we are required to estimate $x(t)$ itself in which case $\hat{x}(t)$ is just the conditional mean. Later we shall be discussing other estimates such as the conditional mode estimate and the maximum-likelihood estimate which are based on the conditional density of $x(t)$ given $\mathscr{F}_t^y$.

3. BASIC EQUATIONS AND BASIC STRATEGY OF SOLUTION

It is now well known and well-established that working with the conditional density of $x(t)$ given $\mathscr{F}_t^y$ need not be the right approach to the solution of the nonlinear filtering problem. Instead one works with the Zakai equation for the unnormalized conditional density $q(x,t)$ which satisfies

$$dq = A^* q dt + hq dy \qquad t \geq o \quad , \qquad (3.1)$$

where A is the generator of the state process $x(t)$ and * denotes formal adjoint. It can be shown that the conditional density $\rho\ (x,t)$ is given by

$$\rho(x,t) = \frac{q(x,t)}{\int_{\mathbb{R}} q(x,t)\ dx} \quad , \qquad (3.2)$$

and the estimate $\hat{\phi}\left(x(t)\right)$ given by

$$\hat{\phi}\left(x(t)\right) = \frac{\int_{\mathbb{R}} \phi(x) q(x,t) dx}{\int_{\mathbb{R}} q(x,t) dx} \quad . \qquad (3.3)$$

It should be noted that the equation (3.1) is essentially a linear equation and a simpler object to analyze than the Kushner-Stratanovich equation for the unnormalized conditional density.

We shall rewrite (3.1) in Stratanovich form and write it formally as:

$$q_t = (A^* - \frac{1}{2} h^2) q + \dot{y}(t) hq, \ t \geq o \qquad (3.4)$$

where $\cdot$ denotes formal differentiation. Everything we say can be made rigorous, for example, by working with the pathwise filtering equation, which is obtained from $q(x,t)$ by defining $p(x,t)$ as:

$$q(x,t) = \exp\left(y(t)h(x)\right)\ p(x,t) \quad , \qquad (3.5)$$

and noting that $p(x,t)$ satisfies:

$$p_t = (A^y)^* p + \tilde{V}^y p \quad , \tag{3.6}$$

where

$$A^y \phi = A\phi - y(t)\sigma^2(x)h_x(x)\phi_x \tag{3.7}$$

$$\tilde{V}^y(x,t) = -\frac{1}{2}h(x)^2 - y(t)Ah(x) + \frac{1}{2}y(t)^2 h_x(x)\sigma^2(x)h_x(x). \tag{3.8}$$

The pathwise filtering equation is an ordinary partial differential equation and not a stochastic partial differential equation. For a discussion on the pathwise filtering equations see CLARK [1978]. The basic strategy of solution schemes for filtering proceeds by analyzing equation (3.1) to answer questions such as existence of finite-dimensional statistics, invariance of (3.1) under groups of transformations and also by obtaining estimates based on equation (3.1) itself. This will be illustrated in the later sections.

4. THE DUALITY BETWEEN ESTIMATION AND CONTROL

The duality between estimation and control is understood by giving equation (3.4) a stochastic control interpretation. We follow FLEMING-MITTER [1982] in this section. To simplify the exposition let us assume $\sigma \equiv 1$.

$$q(x,t) = \exp\left(-S(x,t)\right) , \tag{4.1}$$

where we are using the fact that q is positive. This transforms equation (3.4) into the Bellmen-Hamilton-Jacobi equation

$$S_t = \frac{1}{2}S_{xx} + H(x,t,S_x) , \quad t \geq o \tag{4.2}$$

$$S(x,o) = S^o(x) = -\log p^o(x) , \tag{4.3}$$

where $p^o(x)$ is the initial density of $x(o)$ assumed to be positive where

$$H(x,t,S_x) = -b(x)S_x - \frac{1}{2}S_x \; S_x \quad + \frac{1}{2}h^2 - \dot{y}(t)h + b_x . \tag{4.4}$$

To get an explicit solution to (3.4) we need to solve (4.2) or equivalently the stochastic control problem

$$\begin{cases} d\xi = b\left(\xi(\tau)\right)d\tau + \underline{u}\left(\xi(\tau),\tau\right)d\tau + dw & o \leq \tau \leq t \\ \xi(o) = x , & \end{cases} \tag{4.5}$$

where the feedback control

$$u(\tau) = \underline{u}\left(\xi(\tau),\tau\right) \tag{4.6}$$

is chosen to minimize

$$J(x,t,\underline{u}) = E_x\left\{\frac{1}{2}\int_0^t \left[\left|u(t-\tau)\right|^2 + b_x(t-\tau) + \frac{1}{2}\left|h(t-\tau)\right.\right.\right. \tag{4.7}$$

$$\left.\left.\left. - \dot{y}(t-\tau)\right|^2\right] d\tau + S^{\circ}\left(\xi(t)\right)\right\}$$

where we have added the harmless term $\frac{1}{2}(\dot{y})^2$.

For the Kalman-Bucy Filtering problem this stochastic control problem turns out to be the linear regulator problem with white Gaussian process noise, quadratic criterion, but perfectly observable. This explains the celebrated duality between filtering and control, first enunciated by Kalman.

The present formulation shows that solving the Zakai equation (3.4) corresponds to solving a nonlinear least-squares problem. Its interest lies in the fact that approximation schemes developed for solving stochastic control problems can be brought to bear on solving the Zakai equation and as we shall show in constructing estimates.

Conversely, perfectly observable, stochastic control problems can be converted by this transformation to the solution of a linear parabolic equation, and if the linear proabolic equation can be explictly solved, this would give an explicit solution to the stochastic control problem.

There are many other possibilities in using these ideas. One possibility is to factor

$$q(x,t) = \eta(x,t)\ L(x,t) \tag{4.8}$$

where $\eta(x,t)$ is a priori density for the x-process and $L(x,t)$ is a Likelihood function (unnormalized). A Bellman-Hamilton-Jacobi equation for $-\log L(x,t)$ can be obtained, which will involve the reverse Markov process corresponding to (4.5). This idea has been investigated by PARDOUX [1981] and BENSOUSSAN [1982] and can also be used to obtain estimators.

Finally, by formulating parameter identification problems as nonlinear filtering problems, we can use these ideas to treat them as nonlinear least-squares problems.

5. THE EXTENDED KALMAN-BUCY FILTER REVISITED

We now want to show how the ideas used in the previous

section can be used to construct filters, and in particular gives us insight into the Extended Kalman-Bucy filter.

As an estimate for $x(t)$, one possible choice is the <u>conditional-mode estimate</u> obtained as

$$\operatorname{Arg\,Max}_x q(x,t) , \tag{5.1}$$

giving rise to the trajectory $\hat{x}(t)$.

This corresponds to

$$\operatorname{Arg\,Min}_x S(x,t) , \tag{5.2}$$

by virtue of (4.1).

By slight abuse of terminology, we call this the Maximum-Likelihood estimate and consider the Likelihood equation

$$S_x(x,t) = 0 . \tag{5.3}$$

For the requisite smoothness properties which would make the sequal rigorous see FLEMING-MITTER [1982].

We obtain an equation for $\hat{x}(t)$ by considering

$$\frac{d.}{dt} \left[S_x\big(x,t\big) \right] = 0$$

along the trajectory $\hat{x}(t)$.

This gives us an equation for $\hat{x}(t)$:

$$\frac{d\hat{x}}{dt} = b\big(\hat{x}(t)\big) + S_{xx}^{-1}\big(\hat{x}(t)\big)\Big[h_x\big(\hat{x}(t)\big)\big(\dot{y}(t) - h(\hat{x}(t))\big) - \frac{1}{2} S_{xxx}\big(\hat{x}(t)\big) - b_{xx}\big(\hat{x}(t)\big)\Big] . \tag{5.4}$$

This derivation requires that S_{xx} be invertible. It is possible to derive differential equations for S_{xx}, S_{xxx} etc. but these couple together and hence we do not get a closed-form solution.

Several remarks are in order. Firstly, in the Kalman-Bucy situation, one can easily show that S is a quadratic function and in that case the process $S_{xx}^{-1}\big(\hat{x}(t)\big)$ turns out to be independent of y and is indeed the error covariance and we recover the Kalman-Bucy filter.

Secondly, if we make the assumption that S_{xx} is invertible at $\hat{x}$, then by the Morse Lemma (MILNOR [1963]), in the neighborhood of the nondegenerate critical point $\hat{x}$, S is a quadratic in a suitable coordinate system. In this coordinate system S_{xxx} is zero and we get the structure of the extended Kaman filter, but with the additional term $-S_{xx}^{-1}\, b_{xx}$ (which is non-zero unless b is linear). Indeed, a possible choice for $S_{xx}^{-1}\left(\hat{x}(t)\right)$ is obtained by solving the Riccati equation

$$\dot{\sigma}(t) = 2b_x\left(\hat{x}(t)\right)\sigma(t) - h_x^2\left(\hat{x}(t)\right)\sigma^2(t) + 1$$

as in the Extended Kalman filter.

The invertibility of S_{xx} is connected with the observability of the nonlinear system and related to "hypoellipticity" concepts for stochastic partial differential equations and the filtering problem.

With the aid of the process $S_{xx}\left(\hat{x}(t)\right)$ one can define the analog of the Fisher-Information matrix. The whole line of enquiry has close connections to the Cramer-Rao lower bound for estimation and its generalizations to the filtering situation by Bobrovsky and Zakai (cf. BOBROVSKY-ZAKAI [1976]), and information-theroetic ideas in nonlinear filtering (GALDOS [1975]).

Finally, the invertibility of S_{xx} is related to the conjugate point phenomenon in the Calculus of Variations.

6. FILTERING WITH SMALL PROCESS NOISE

Consider the situation where

$$dx(t) = b\left(x(t)\right) dt + \sqrt{\varepsilon}\, dw(t) \quad , \tag{6.1}$$

where $\varepsilon > 0$ is a small parameter. This case can be analyzed using the work of Fleming (FLEMING[1971]), provided a regularity condition eliminating conjugate points is imposed. In the limit as $\varepsilon \downarrow 0$, we get a Hamilton-Jacobi equation (as opposed to a Bellman equation) and we get an ordinary optimal control problem parametrized by y.

The details of this as well as a more rigorous discussion of section 5 will appear in a joint paper with Fleming.

REFERENCES

Bensoussan, A.: 1982, Unpublished Manuscript on Nonlinear Filtering.

Bobrovsky, B.Z., Zakai, M.: 1976, A Lower Bound on the Estimation Error for Certain Diffusion Processes, IEEE Trans. on Information Theory, Vol., IT-2, No. 1, pp. 45-51.

Clark, J.M.C.: 1978, The Design of Robust Approximations to the Stochastic Differential Equations of Nonlinear Filtering in Communication Systems and Random Process Theory, ed. J.K. Skwirzynski, Sijthoff and Noordhoff.

Fleming, W.: 1971, Stochastic Control for Small Noise Intensities, SIAM Journal on Control.

Fleming,W.H., Mitter, S.K.: 1982, Optimal Control and Nonlinear Filtering for Nondegenerate Diffusion Processes, to appear Stochastics.

Galdos, J.: 1975, Information and Distortion in Filtering Theory, Doctoral Dissertation, M.I.T.

Milnor, J.: 1963, Morse Theory, Princeton University Press, Princeton, NJ.

Mitter, S.K.: 1982, Lectures on Nonlinear Filtering and Stochastic Control, Proceedings of CIME School on Nonlinear Filtering, and Stochastic Control, July 1981, proceedings to be published by Springer-Verlag.

Ocone, D.: 1980, Topics in Nonlinear Filtering Theory, Doctoral Dissertation, M.I.T.

Pardoux, E.: 1981, Nonlinear Filter as a Likelihood Function, proceedings Decision and Control Conference, San Diego, CA.

PHASE DEMODULATION: A NONLINEAR FILTERING APPROACH*

José M. F. Moura

CAPS, Complexo,Instituto Superior Tecnico
Av. Rovisco Pais, P-1096 Lisboa Codex
PORTUGAL
Ph: (351-1) 572399; (351-1) 884022

Abstract: The paper studies nonlinear filtering techniques. It discusses their application to several classes of phase modulation (PM) problems: cyclic and absolute PM problems. It considers the long time or tracking behavior of the PM demodulators, and their short time or acquisition ability. After motivating examples, the mathematical model is presented. The continuous and the discrete optimal nonlinear filtering solutions are then analysed.The PM problems are described. A brief exposition of the PLL theory follows, with emphasis on a operator type solution of the associated Kolmogoroff equations. Finally, the implementation of the nonlinear demodulator and their performance in the context of absolute PM tracking and acquisition is pursued.

*- This work was partially supported by Nato Scientific Affairs Division under research grant no. 1557/78 and by JNICT under research contract no.118.79.80/II

R. S. Bucy and J. M. F. Moura (eds.), Nonlinear Stochastic Problems, 347–386.

1.INTRODUCTION

The paper studies the application of nonlinear filtering techniques to a class of problems of significant interest in the applications. In doing so, it has the following objectives:

i) To gain insight on methods and actual practical ways of implementing nonlinear filters.

ii) To show what is gained by resorting to nonlinear techniques.

In this process, situations will be distinguished where:

iii) Optimal filters exhibit a significant improvement of performance when compared to linearized structures.

iv) Nonlinear filters are the only possible solution, simply because the linearized receivers cannot improve upon the knowledge available apriori.

Further, the paper will discuss:

v) The tradeoffs and compromises to be made, vis a vis performance requirements versus associated computational effort.

vi) The computational infrastructure needed to devellop and apply the nonlinear machinery.

2.MOTIVATING EXAMPLES

Problem no. 1 (Classical Phase and Frequency Demodulation)

In telecommunications, a message process is to be transmitted through a noisy channel from a transmitter to a possibly distant receiver. For several technical reasons (radiating efficiency, bandwidth, multiplexing, etc.), this is done by first modulating a carrier with this message process (figure 1). The diagram of figure 1 is a simplified one, for example no preemphasis has been assumed.

```
                                        |exp{j[wt+x(t)]}:PM
                                        |
         ------------                   |
x(t)-->| MODULATOR |-->s(t)= <
         ------------                   |
                                        |
                                        |exp{j[wt+∫x(t)dt]}:FM
```

Figure 1: Modulation Diagram

The process x(t) is herein considered to be the state of a system driven by wideband noise (figure 2). The modulated carrier s(t) is transmitted through a noisy channel. The task is to design, in an optimal fashion, the processor that recovers the message x(t) (figure 3).

```
              ---------
u(t)------>| SOURCE |------>x(t)
              ---------
```

Figure 2: Message Process

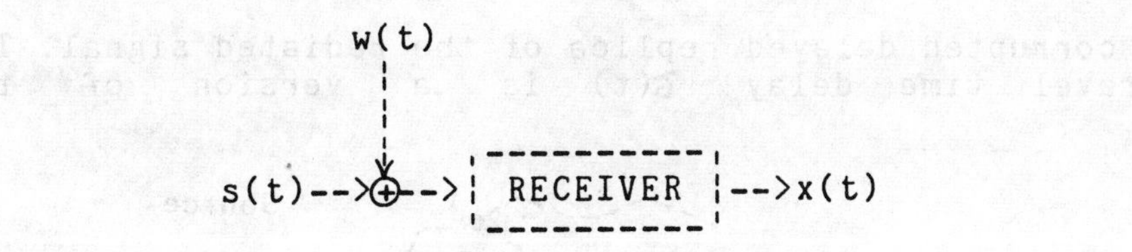

Figure 3: Channel and Receiver

The channel is additive, the disturbance w(t) being wideband. The receiver satisfies a given criterion of quality. The structure of the modulated signal (see figure 1), as well as the additive disturbance characteristic of the channel are the features to emphasize. The state variable framework assumed casts this problem in a context quite different from the one of the traditional communication litterature. References [1] and [2] represent some of the early work on this and more complicated state variable formulations of phase problems. However, these references apply linearizing arguments based on extended Kalman / Bucy filtering techniques to design the demodulator. In many phase modulated (PM) systems, the applications are insensitive to the absolute value of the phase error. What matters is solely a 2π-modulated version of the phase error. This is a local or cyclic phase problem. If the signal is sufficiently strong against the background noise, the error remains small with high probability. The expected local behavior of the linearized PM receiver almost equals the optimal performance. When the noise becomes stronger, the loop degrades significantly its performance with respect to the theoretically optimal level. Even for cyclic PM systems, it is then of interest to further explore this margin of improvement.

Problem no. 2 (Passive Positioning)

In underwater acoustics applications, a noisy platform radiates its own signature (eg. propeller noise), which is sensed by a passive receiver (see figure 4). The received waveform is

$$r(t)=s(t-\tau(t))+w(t), \tag{2.1}$$

a corrupted delayed replica of the radiated signal. The travel time delay $\tau(t)$ is a version of the

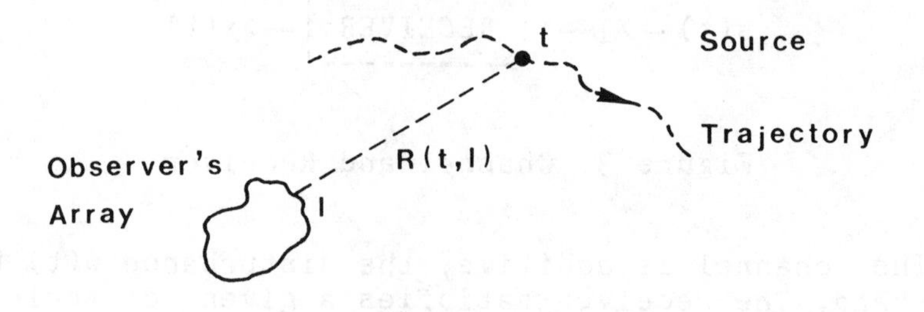

Figure 4: Passive Positioning Geometry

source-receiver separation R(t)

$$\tau(t) = R(t)/c \tag{2.2}$$

where c is the medium propagating speed. For signals which are narrowband, an appropriate model is

$$\begin{aligned} s(t-\tau(t)) &= \exp j w_o (t-\tau(t)) \\ &= \exp j[w_o t - 2\pi R(t)/\lambda] \end{aligned} \tag{2.3}$$

where λ is the wavelenght. Physical platforms have constrained motions (eg. bounded accelerations). By collecting in a state vector the range R(t), the bearing θ(t), and their derivatives up to a certain order, these constraints are modeled by a stochastic differential equation. A state variable framework results. The state vector modulates in phase, through the delay, the emitted signal.Once again, the problem lies on the design of a processor that reconstructs the state vector (the source position and dynamics) from the incoming PM signals, see figure 5. For these applications, it is the absolute error process, and not its modulated 2π-version, that must be kept small. It is an absolute phase problem.

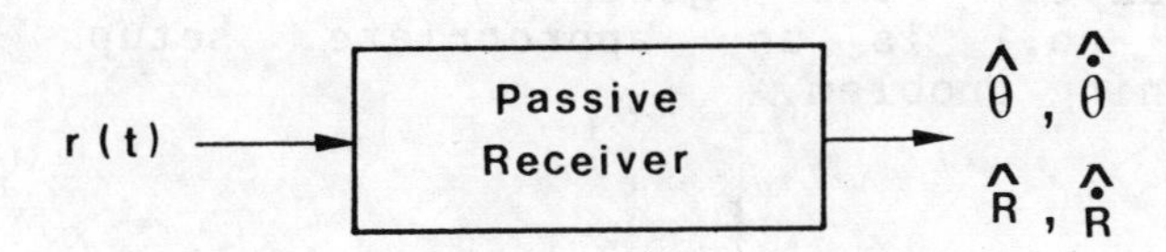

Figure 5: Geometry Demodulation

In passive contexts, one has no knowledge of the travel delay at the time origin, ie. the absolute phase reference $\tau(0)$. If this is irrelevant in the cyclic problem mentioned previously, in the present situation, the demodulation of the relative geometry to be effective requires the overcoming of this lack of knowledge. This is referred to as an acquisition step in the PM application. The absolute PM and the acquisition step are examples where nonlinear techniques provide significant improvements over the linearized algorithms.

Still fitting the passive positioning framework is the problem of tracking a space sonde or a drifting buoy. Besides transmiting to monitoring stations meteorological or oceanographic data, the sonde or the buoy also sends a tracking signal. This may be used to adjust the actual trajectory about a nominal nonlinear path.

It is an interesting point to observe, that the classical problem of underwater acoustics can also be cast in a state variable framework. In this problem, see figure 6, a stationary array of sensors tracks a stationary source. Stationary here means not moving. By considering the array as a curve in space, different parametric descriptions are possible. For example, instead of the time, the array line description may use the arc lenght as a parameter. The geometric constraints, represented in the problem by the line array shape, are then described by a set of stochastic differential equations. This is an appropriate setting for the application of optimal nonlinear filtering techniques, reformulating the classical beam forming theory in a recursive framework. For further comments, see [3].

Under the assumption of narrowband transmitted signals and additive wideband channel noise disturbances, the general communication model of problem no.1 is an appropriate setup for the positioning problem.

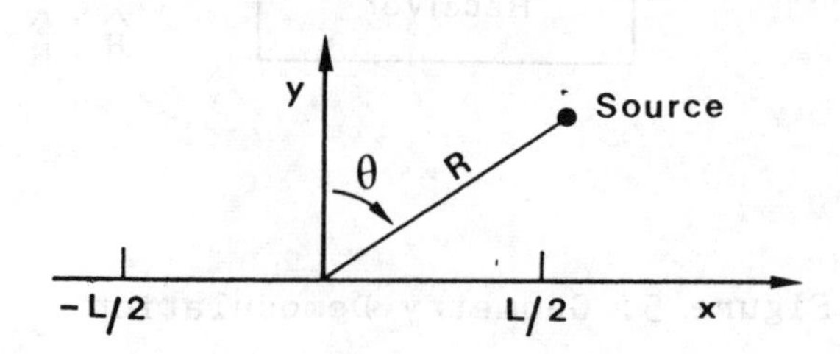

Figure 6: Classical Array Problem

Problem no.3 (Navigation)

Assume, like in the global navigation system OMEGA, or any other navigation system (e.g. satellite based), that beacons (or transponders) emit signals used for location and navigation. A problem which arises from time to time is an overall or global resetting of the navigation system. The framework is dual of the one in problem 2, with the roles of the source and of the receiver reversed, see figure 7.

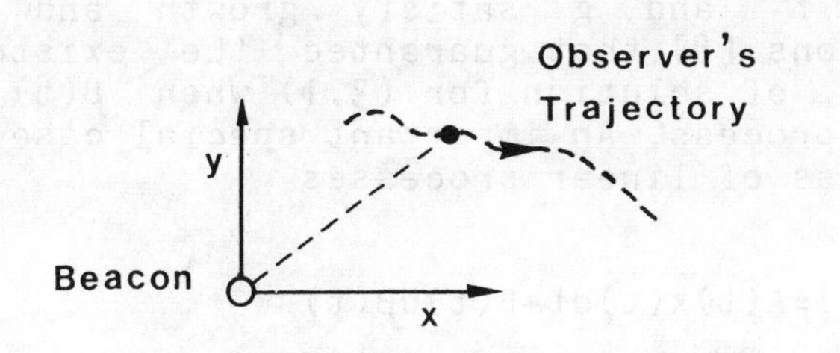

Figure 7: Navigation System

Associated with the navigation system is a chart of hyperbolas defining a grid used to locate the receiver. The receiver starts from a known location (e.g. ship departs at a certain epoch from given harbor). By integrating its dynamics, uncertainty is constrained to the grid mesh. Due to accumulation of errors, or to unexpected system malfunctions, the receiver may loose track of which cell it is in, imposing a need for global resetting of the navigation system. Applying here the same techniques of problem no. 2, an optimal phase demodulation problem may be formulated for the navigation application. For further motivation on problem no. 2 and no. 3 see [4]-[7].

3. MATHEMATICAL MODEL

The examples of section 2 fit a general framework which is presented here. This model is also suitable in many other applications. The terminology of communication theory is kept throughout the section.

Message Process

An appropriate model for the message $x(t)$ describes it as a vector Markov-diffusion process

$$dx(t)=f(x(t),t)dt+g(x(t),t)d\beta(t)$$
$$x(0)=x_o,\ t>0 \tag{3.1}$$

where f and g satisfy growth and measurability conditions [8] that guarantee the existence and the unicity of solution for (3.1) when $\beta(t)$ is a Brownian vector process. An important special case of (3.1) is the class of linear processes

$$dx(t)=A(t)x(t)dt+B(t)d\beta(t)$$
$$x(0)=x_o,\ t>0 \tag{3.2}$$

The state vector $x(t)$ is n-dimensional, ie, $x(t)\in\mathbb{R}^n$, $\beta(t)\in\mathbb{R}^m$. In (3.1), $f\in\mathbb{R}^n$ and $g\in\mathbb{R}^m$. In (3.2), $A(t)$ and $B(t)$ have appropriate dimensions. Equations (3.1) or (3.2) are a state variable dynamical description. They should be interpreted as formal writing, standing for their integral formulations. The underlying theory involves the apparatus of a suitable stochastic calculus. This is provided by the Ito definition of a stochastic integral [8].

Observational Process

The channel assumed is additive. The received signal is then

$$dr(t)=h(x(t),t)dt+dW(t),\ t>0 \tag{3.3}$$

with the nonlinear cyclic sensor

$$h(x(t),t)=[\cos x(t) \mid \sin x(t)]^T. \qquad (3.4)$$

Equations (3.3) and (3.4) model the so called low pass signal. The carrier frequency w_o has been eliminated by an heterodyning stage followed by low pass filtering. Equation (3.3) is also to be interpreted in the sense of Ito calculus. The additive noise W(t) is a Brownian motion which is independent of $\beta(t)$.

4. THE NONLINEAR FILTERING PROBLEM

The nonlinear filter yields from the observations' span a best estimate of the message process x(t). This estimate is constructed from the conditional probability distribution, i.e., the probability distribution of the message x(t) given the sigma algebra R_t generated by the observations in the interval [0,t]

$$R_t = \sigma\text{-}\{r(s),\ 0 \leq s \leq t\}. \qquad (4.1)$$

The conditional probability distribution function will be assumed to admit a probability density function (pdf) (Radon Nykodim derivative)

$$p(t,x \mid R_t) \qquad (4.2)$$

which is measurable and adapted to R_t for each t and x. This pdf is propagated forward in time by the stochastic - integro differential equation

$$dp=L^* pdt+(h-\hat{h})^T pdI. \qquad (4.3)$$

To simplify the notation, the arguments of the several quantities in equation (4.3) have been ommitted, e.g. p stands for the object in equation (4.2), h=h(t,x), etc. This writing will be oftenly used in the sequel. The operator L^* is the adjoint of the (extended) generator associated with the message proces. It is given by

$$L^*(.)=-\sum_i \partial f_i(.)/\partial x_i + \qquad (4.4)$$
$$+(1/2)\sum_{i,j} \partial^2 G_{ij}(.)/\partial x_i dx_j$$

where $G=[G_{ij}]=gg^T$. In equation (4.3),

$$\hat{h}=E[h(t,x)|R_t] \qquad (4.5)$$

is the conditional mean of the sensor vector h with respect to the observational field. The innovations are

$$dI(t)=dr(t)-\hat{h}(t)dt. \qquad (4.6)$$

Equation (4.3) is the Stratonovich - Kushner equation. Loosely speaking, it consists of two parts. The first propagates the pdf according to the dynamics of the model, via the operator L^*. In fact, under general conditions on f and g, the Backward - Kolmogorof equation

$$-dP(t_0,x_0,t,x)=LP(t_0,x_0,t,x) \qquad (4.7)$$

with terminal condition

$$\lim_{t_o \uparrow t} P(t_o,x_o,t,x)=\begin{cases}1 & x_i>x_{oi} \ \forall\, i\\ 0 & \text{otherwise}\end{cases} \qquad (4.8)$$

has a unique solution. The function P is the transition probability function associated with the Markov process

$$P(t_o,x_o,t,A)=Pr\{x(t)\in A|x(t_o)=x_o\} \qquad (4.9)$$

where A is an event. Under the same conditions, P is differentiable with respect to x, and the transition density $\Pi(t_o, x_o, t, x)$ satisfies the Forward - Kolmogoroff or Fokker-Planck equation

$$d\Pi = L^* \Pi dt \tag{4.10}$$

with initial condition

$$\lim_{t \to t_o} \Pi(t_o, x_o, t, x) = \delta(x - x_o). \tag{4.11}$$

To the (extended) generator L of the diffusion Markov process is associated the integral operator

$$\Phi(t,s)(.) = \int_{R^n} (.) P(s, x(s), t, dy) \tag{4.12}$$

which satisfies

$$\Phi(t,s)(.) = I(.) + \int_s^t \Phi(t,u) L_u(.) du. \tag{4.13}$$

The second part of equation (4.3) incorporates the observations via the innovations (4.6). The interpretation of this second part poses additional problems due to its stochastic and integro - differential nature. An alternative solution to the nonlinear filtering problem, studies the so called Zakai equation. It propagates an unnormalized version q(t,x) of the conditional pdf according to

$$dq = L^* q dt + h^T dr(t) q \tag{4.14}$$

with

$$p(t,x|R_t)=q(t,x)\Big/\int_{R^n} q(t,x)dx. \tag{4.15}$$

Equation (4.14) has as driving term the physical observations. Also, the integral term is not present. It has been the basis for recent theoretical work on the continuous time nonlinear filtering problem, vis a vis the so called robust interpretation and pathwise filtering theory, see e.g. the works of [9] - [13]. In particular, reference [14] collects contributions relevant to this topic. The Stratonovich version of (4.9), see for example [15], is

$$dq=L^{*}qdt+h^{T}(dr-(1/2)hdt)q. \tag{4.16}$$

The second term of (4.16) is now suitably interpreted as follows. It is the infinitesimal generator of the multiplicative integral operator

$$\Psi(t,s)(.)=\exp\{\int_s^t h^T(u,x)dr(u)- -(1/2)\int_s^t ||h(u,x)||^2du\} \tag{4.17}$$

The integral operators (4.12) and (4.17) will be put in the perspective of the discrete optimal nonlinear filtering equations of the next section.

5.DISCRETE NONLINEAR FILTERING

The problem of actual implementation of optimal nonlinear filters will be addressed here from a different point of view. As commented in the previous section, equation (4.3) is not suitable for direct implementation. Also, other filtering equations derived having as starting point the Zakai equation, transfer the numerical difficulties to the infinitesimal operators associated with these equations, which become path dependent [11]. Further work is required to show the feasibility of the numerical procedures that implement these structures. This feasibility should be evaluated in terms of the (numerical) stability and expediency of the associated algorithms. In this paper, work will be reported on the actual implementation of nonlinear filters in a digital computer. Other promising technologies may be viable alternatives in the near future, e.g., saw devices for fast implementation of convolutions, or optical technologies [16]. Hybrid digital/analog implementations have also been constructed in the past [17]. The recent general availability of specialized digital hardware, with pipelining and parallelism in its architecture, as the array processors, or the super fast computers, make presently possible to simulate nonlinear filters for certain classes of problems. A comparison of implementations using alternative machines has been carried out in [18]. For further details, see [19].

The filtering problem to be studied will be the discrete optimal nonlinear filtering problem. Instead of time discretization of the continuous version of the optimal nonlinear filter, discretization is first carried out on the model assumed. For dense sampling, the discrete model will be sufficiently near the continuous model. Discretization of equations (3.1) and (3.3) leads to

Dynamics:

$$x(n+1)=f(x(n))+g(x(n))u(n) \qquad (5.1)$$

$$x(0)=x_o$$

Observations:

$$z(n)=h(x(n))+v(n) \qquad (5.2)$$

where f, g, h, are suitable discrete counterparts of their continuous versions. The noise sequences $\{u(n)\}$ and $\{v(n)\}$ are white, independent, Brownian sequences with variance parameters q and r respectively, obtained from the corresponding continuous values q_c and r_c by suitable normalizations

$$q=q_c\Delta \tag{5.3}$$

$$r=r_c/\Delta \tag{5.4}$$

where Δ is the (assumed uniform) sampling interval. The optimal nonlinear discrete filter propagates the conditional probability distribution function (cpdf) F_n by

$$\tilde{P}(n+1,x)=\tilde{F}(n,x)*Q(n,x) \tag{5.5}$$

$$\tilde{F}(n,x)=H(n,x)\tilde{P}(n,x) \tag{5.6}$$

$$F(n,x)=\tilde{F}(n,x)/K(n). \tag{5.7}$$

Equation (5.5) stands for convolution on the state space domain of the unnormalized cpdf $\tilde{F}(n,x)$ with the dynamics kernel $Q(n,x)$. Equation (5.6) represents a pointwise function multiplication of the sensor factor $H(n,x)$ with the unnormalized predicted cpdf $\tilde{P}(n,x)$. Normalization by $K(n)$, guarantees that $F(n,x)$ stays a probability distribution. Equation (5.5) propagates forward in time $\tilde{F}(n,x)$ according to the dynamics (5.1). This parallels the comment on the infinitesimal generator L_* following equation (4.3), see equations (4.12) and (4.13). In section 7, more will be said on the relation between the integral (5.5) and the integral operator of (4.13). Equation (5.6) updates $\tilde{P}(n,x)$ when the new measurement (a posteriori knowledge) is available for processing. The measurement kernel is directly related to the Gaussian statistics of $\{v(n)\}$,

$$H(n)\sim\exp(-||z_n-h||^2/2r) \tag{5.8}$$

or,

$$H(n)\sim\exp(-||h(x)||^2/2r+z_n{}^T h/r). \tag{5.9}$$

Having in mind the correspondence between the continuous and discrete time quantities, one can formally establish the relation between (5.9) and (4.17). Equations (5.5) - (5.7) represent a discrete time iteration for (4.14) and (4.15).

6.PHASE MODULATION PROBLEMS

Depending on the nature of the application under study, one may define several classes of phase modulation (PM) problems. Under the general model of section 3, their most characteristic feature is the common observation vector, the cyclic or periodic sensor $h(x(t))$ given by equation (3.3). To clarify the underlying issues, the error process

$$e(t)=x(t)-\hat{x}(t) \tag{6.1}$$

where $\hat{x}(t)$ is a suitable estimate of the message $x(t)$, is decomposed as

$$e(t)=\tilde{e}(t)+N(t).2\pi \tag{6.2}$$

where the cyclic error component

$$\tilde{e}(t)=e(t)\bmod 2\pi, \tag{6.3}$$

and the integer valued process $N(t)$ accumulates the so called phase slip errors. In numerous problems, the application is insensitive to the cycle slips. They belong to the cyclic or modular PM class. Optimal performance relates to second order statistics of the modulated process $e(t)$. In other problems, cycle slips are the limiting factor. These are the absolute PM problems. The rate of growth of the counting process $N(t)$ is to be reduced. In still other problems,one is interested in resolving, whenever possible, in the fastest way, the initial ambiguity $e(0)$. This is referred to as the acquisition step. In fact, the acquisition is here considered as a state reconstruction at time t, and not a state observability at time t=0. Acquisition problems arise typically in absolute PM contexts like the passive positioning applications studied in section 2. Figure 8 organizes accordingly the PM classes.

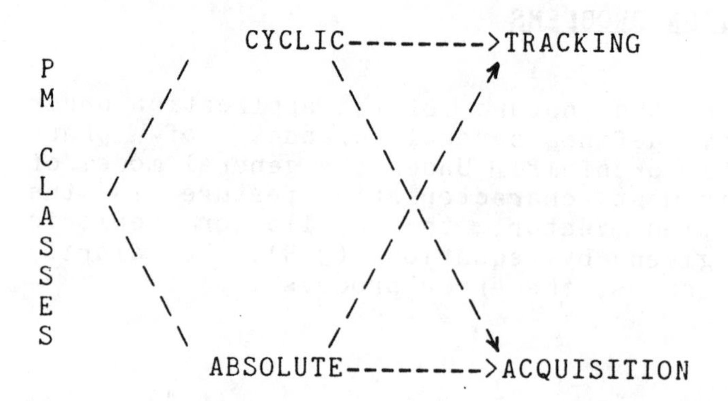

Figure 8: Diagram for Phase Problems

From a temporal horizon point of view, the phase tracking is a long term (steady state type) behavior, while the phase acquisition represents a transient question. To structure the PM study as an optimal problem, a performance criterion must be chosen. This criterion reflects the specifics of the application of interest. In cyclic tracking applications, the following periodic loss function is commonly used

$$L(e)=E\{(1-\cos e)/2\}. \qquad (6.4)$$

In absolute tracking problems, the mean square error loss function is an appropriate choice

$$L(e)=E\|e\|^2. \qquad (6.5)$$

Minimization of the criterion (6.4) leads to the cyclic estimate

$$\hat{x}_{cyc}(t)=ATAN(\widehat{\sin x}/\widehat{\cos x}) \tag{6.6}$$

while (6.5) leads to the conditional mean. The acquisition problem is basically a first passage time problem. Convenient performance measures relate for example to the minimum of the acquisition time, or to the optimization of a probability of a correct or incorrect decision. These loss functions are used to evaluate and compare by simulation the performance of several phase estimates.

7. THE CLASSICAL SOLUTION: THE PHASE LOCKED LOOP

The standard solution to phase modulation (PM) problems is the Phase Locked Loop (PLL). Reasons to consider it herein are the following:

i) The PLL is an ubiquitous circuit in Control and Communications. It is simple, inexpensive and effective. For nonlinear demodulators to have impact in the real world, their complexity and performance should be studied relative to the PLL.

ii) The PLL's behavior is well documented in the litterature. For low dimension PM problems, the theory leads to closed form expressions capable of being analysed.

iii) The PLL provides a benchmark against which one can measure the performance improvements obtained with the nonlinear techniques. The importance of this observation is underlined by the fact that there is no available body of theory for the computation of error bounds for nonlinear filters. On this issue, see for example [20] - [23].

iv) The PLL nonlinear error behavior may be studied via a Fokker - Planck (or Forward - Kolmogoroff) equation, [1], [2], [24], [25]. Techniques to study this equation's solution shed light on some of the difficulties that arise with nonlinear filters.

This section briefly considers the PLL theory from the Forward and Backward Kolmogoroff equations point of view. This study provides insight on the expected form and behavior of the nonlinear filter's solution. The analysis is carried out for the one dimensional phase problem.

The message is

$$dx(t)=ax(t)dt+d\beta(t)$$
$$x(0)=x_o \quad (7.1)$$

where $\{\beta(t)\}$ is a Brownian motion with variance parameter q_c. In the sequel, to obtain meaningful analytical results, the message pole a is assumed to be a nonpositive real. The observations are

$$dr(t)=h(x(t))dt+dv(t) \quad (7.2)$$

where, as before, the low pass periodic sensor

$$h(x(t))=[\cos x \mid \sin x]^T \quad (7.3)$$

and $v(t)=[v_1 \mid v_2]^T$ is a vector Brownian motion with variance matrix $r_c I$, independent of $\beta(t)$. The PLL may be viewed as a (steady state) extended Kalman Bucy (EKB) filter. It is of interest to observe that the associated Riccati equation is not noise dependent. It reaches a steady state solution

$$\gamma=ar_c+\sqrt{(ar_c)^2+q_c r_c}. \quad (7.4)$$

The PLL is given by

$$d\hat{x}(t)=a\hat{x}(t)-k[I^* h(\hat{x})]^T dr(t) \quad (7.5)$$

where the PLL gain is

$$k=\gamma/r_c \quad (7.6)$$

and I^* represents the alternating or sympletic 2-dimensional matrix

$$I^* = \begin{bmatrix} 0 & 1 \\ -1 & 0 \end{bmatrix}. \quad (7.7)$$

The PLL error

$$e_{PLL}(t) = x(t) - \hat{x}(t) \quad (7.8)$$

propagates according to (the subindice PLL is omitted)

$$de(t) = ae(t) - k\sin e(t) + kd\mu(t) + d\beta(t) \quad (7.9)$$

where

$$d\mu(t) = [I^* h(\hat{x}(t))]^T dv(t). \quad (7.10)$$

It can be shown that $\mu(t)$ is a Brownian motion process independent of $\beta(t)$, with variance parameter r_c. By equation (7.9), the process $e(t)$ is a Markov - diffusion. It is characterized by the infinitesimal generator

$$L(.) = [ae - \sin e]\partial(.)/\partial e + (\sigma/2)\partial^2(.)/\partial e^2 \quad (7.11)$$

where the diffusion coefficient

$$\sigma = q_c + k_c^2 r_c. \quad (7.12)$$

The transition probability density function $p(t,e)$ satisfies the Backward Kolmogoroff equation

$$-dp = Lp\,dt, \quad t<0 \quad (7.13)$$

and the Forward Kolmogoroff equation (also known as Fokker - Planck equation)

$$dp=L^{*}p\,dt,\ t>0 \qquad (7.14)$$

where L^{*} is the adjoint operator

$$L^{*}=-\partial\{(ae-\sin e)(.)\}/\partial e+(\ \ /2)\partial^{2}(.)/\partial e^{2}. \qquad (7.15)$$

The boundary conditions for (7.14) are, for the absolute PM problem,

$$p(t,+\infty)=p(t,-\infty)=0. \qquad (7.16)$$

The initial condition (tracking problem) is

$$p(0,e)=\delta(e). \qquad (7.17)$$

When $a<0$, equation (7.15) has a nontrivial steady - state solution

$$p(e)=C\exp[(ae^{2}+2k\cos e)/\gamma] \qquad (7.18)$$

where C is a normalizing constant (so that p(e) integrates to 1). In what follows, this constant is ignored. Adopting the notation

$$N(e)\sim\exp(ae^{2}/\gamma) \qquad (7.19)$$

$$J(e)\sim\exp(2k\cos e/\gamma), \qquad (7.20)$$

the periodic train J(e) can be written as

$$J(e)=\sum_{k=-\infty}^{\infty}\delta(e-k2\pi)*\tilde{J}(e) \qquad (7.21)$$

where

$$\tilde{J}(e)=J(e),\ e[-\pi,\pi]$$
$$=0 \quad \text{elsewhere.} \qquad (7.22)$$

In (7.21), * stands for convolution. The transient solution of (7.14), p(e,t) is now written as the product of an envelope E(e,t) by the steady state solution p(e). Keeping in mind (7.19)-(7.21), one has

$$p(e,t)\sim E(e,t)\{N(e)\sum_{-\infty}^{\infty}\delta(e-k2\pi)*\tilde{J}(e)\}. \qquad (7.23)$$

Equation (7.23) helps the interpretation of the underlying phenomenology. It provides an heuristic consubstantiation of the error decomposition in equation (6.2). This is made rigorous for the case where a=0. Even in the case when a<0, (7.23) says that, as the probability mass diffuses outwarddly, the density function is of the multimodal type. In steady state, this shape is obtained by a weighted congruent repetition of J every 2π-cycles. The tappering N(e) is Gaussian. During the transient, the envelope E(e,t) spreads from an impulse located at the origin, to a constant. Further, it is interesting to observe that, due to the exponential character of p(e) and to the fact that the diffusion coefficient is constant, the envelope E(e,t) satisfies the Backward-Kolmogoroff equation

$$dE(e,t)=LE(e,t)dt,\ t>0, \qquad (7.24)$$

where L has been given in (7.11). Notice the change of sign in the left hand side, as well as the corresponding time reversal of equation (7.24) in relation to (7.13). The computational interest of (7.24) lies on the smoother behavior of the envelope, $E(e,t)$ when compared to the shape of the pdf $p(e,t)$ itself.

Equations (7.14) or (7.24) are parabolic partial differential equations (pde). To integrate them, grid algorithms, eg. implicit algorithms, may be used. This has been done in [26]. Here, an alternative method is proposed. The time parameter is discretized using a sufficiently small sample interval. The generator is decomposed as the sum

$$L=L_1+L_2 \tag{7.25}$$

where

$$L_1(.)=[ae-\sin e]\partial(.)/\partial e \tag{7.26}$$

and

$$L_2(.)=(\ /2)\partial^2(.)/\partial e^2. \tag{7.27}$$

One associates to each of these, the corresponding integral operators

$$T_t^1 f(e)=f(\alpha^{-1}(\alpha(e)+t)) \tag{7.28}$$

$$T_t^2 f(e)=f*Q \tag{7.29}$$

where Q is a Gaussian Kernel. In (7.28)

$$\alpha(e)=\int^e 1/[o(x)]dx \tag{7.30}$$

where

$$o(e)=ae-\sin e \tag{7.31}$$

For sufficiently small Δ

$$\tilde{\alpha}^{-1}(\alpha(e)+\Delta)=e+(ae-\sin e)\Delta, \tag{7.32}$$

and the time recursion that propagates the solution to (7.24) is

$$p_{n+1}^{1}(e)=p_n[e+(ae-\sin e)\Delta] \tag{7.33}$$

$$p_{n+1}(e)=p_{n+1}^{1}*Q \tag{7.34}$$

where the variance of Q is $\sigma\Delta$. Equation (7.33) corresponds to a rearrangement of the density on the space grid. It accounts for the drift in the PLL error equation (7.9). It is a crude approximation to (7.28). More sophisticated approximations may be tried, or (7.28) may be used directly. Equations (7.33) and (7.34) or (7.28) and (7.29) present the obvious advantage of preserving the positivity of the solution of the pde (7.24).

When a=0, besides the absolute PM problem studied before (see boundary conditions (7.16)), it is also of interest to consider the periodic or cyclic phase modulation problem.

By defining the cyclic density

$$\tilde{J}(e,t)=\sum_{k=-\infty}^{\infty} p(e+k2\pi,t) \tag{7.35}$$

where

$$\tilde{e}\in[-\pi,\pi], \tag{7.36}$$

it is easily shown that J satisfies equation (7.14), where L^* is given in (7.15) (with a=0), with boundary conditions

$$J(-\pi,t)=J(+\pi,t) \tag{7.37}$$

Due to the compact character of the solution domain (7.36), this cyclic problem is computationally less demanding.

8.OPTIMAL NONLINEAR DEMODULATORS

It was explained in sections 6 and 7, that the demodulators to be studied are designed via the methods of optimal nonlinear discrete filtering. Work has been reported on different design techniques for the PM cyclic problems. Reference [27] contains details of several expansion methods for density representation. It applies them to the cyclic phase demodulation problem.Only recently, have the absolute tracking and absolute acquisition PM problems been addressed by nonlinear methods. The present section gives indication of the approaches and results contained in [28]-[29] on absolute PM tracking and in [30]-[32] on absolute PM acquisition.

Absolute PM Tracking

Contrasting with the cyclic PM problem, the absolute tracking probability function has an unbounded state space domain. With positive probability, the error process exhibits a finite escape time from any bounded interval on the real line. This fact poses added difficulties to the nonlinear filter representation. In terms of computational effort, within certain constraints, the one dimensional absolute PM problem becomes equivalent to a two dimensional cyclic problem. The equivalence is in the sense that there are also two degrees of computational freedom: i) the number of points in the basic compact 2π-cell interval; and ii) the number of 2π-cell intervals.

The suitability of the demodulators with respect to absolute PM tracking is assessed in terms of the expected number of cycle slips. The counting of the slips is achieved via a type of hysteresis device [29]. To obtain accurate statistics requires that:

i) Each Monte Carlo run be long enough so that the demodulators experience a significant number of slips.

ii) A sufficiently large number of runs be carried out.

Larger number of slips requires larger probability functions' support. This augments the dimensionality of the nonlinear filter, as measured by the number of parameters representing the density functions (eg. the number of terms in the expansion representation or the number of points in the grid masses method). It is seen that long time runs and reliable functions' representations are competing requirements.

In practice, what is done is to consider the absolute phase tracking as a finite temporal horizon problem. This horizon is chosen as a compromise between the statistical environment (noise to signal ratio), which determines the average number of slips observed, and the dimensionality of the filter. The following sketches the design iteration:

i) Set a maximum value for the desired global density distortion.

ii) For a given statistical environment (noise to signal ratio) and for a selected expected number of slips, choose the horizon time of each Monte Carlo run.

iii) Experience on the filter dimensionality to meet constraint i).

iv) Adjust the time horizon (step ii) and the filter complexity (step iii) to the available computational infrastructure. This allows for a significant number of Monte Carlo runs to be carried out, giving statistical significance to the study.

To evaluate the value of density distortion, an estimate was computed. This corresponded to the total "loss" of density mass experienced at each Monte Carlo run by the finite realization of the optimal filter.

The filter implemented uses an adaptation of the point masses method. The state space is discretized. The probability functions are represented by masses located at the lattice points. The filter consists of the following basic steps:

i) Propagation of the line probability functions, through the recursion (5.5) - (5.7).

ii) Computation of estimates, slips, and statistics.

iii) Update of the grid corresponding to the support of the probability functions propagated by the filter.

These steps are within a Monte Carlo loop and a time loop. The filter floats on a fixed lattice of uniform mesh. The grid updating requires at each step the computation of its new center. This strategy takes full advantage of the periodicity of several of the functions in the filter structure. These functions, as well as the dynamics kernel, require trignometric and exponential evaluations. With the uniform mesh / floating center grid, they are computed once, at filter initialization.

The critical step of the filter is the convolution. The computational setup consisted of a minicomputer and an array processor. The convolution was implemented in the array processor. These processors are efficient when the implementation exploits the built in parallelism of arithmetic, memory fetches, and logic units, and the pipelining structure of the array operations (add, multiply, memory fetches). The array processor loops require an initial setup that puts in step the several stages of the different arithmetic and logic tasks. Also, synchronization may require dropping out of loops one step earlier, with added code to finish up outside the loop. To take full advantage of the array processor, one should take these facts into consideration and tune the algorithm and the program that implements the filter.

Best suited algorithms for the array processor exhibit:

i) Regularity on the algorithm flow, so that it can be implemented as a small number of loops. This cuts the overhead associated with loops' intros and early dropouts. This is a reason why, with array processors, the point masses is a viable alternative for filter's implementation.

ii) Operation on very large strings of data, to reduce the impact of the overhead weight associated to the loops.

To speed the implementation, one should be aware of further architecture constraints of the array processors, namely the existence of special registers with given length L. For the following convolution

$$P(j)=\sum_{i} F(i+j)Q(i), \quad j=1,\ldots,J \qquad (8.1)$$

assume that the support of Q is I, that of F is J, and that I<<J. This is usually true for nonlinear filtering problems. Break J in pieces of length L. For the case where L=J, figure 9 represents the algorithm structure for (8.1). It constructs each partial sum for every P(j), obtaining all P(j)'s at the last step of the i loop. The extra delay in obtaining P(1),..., is

immaterial in this application, since what is needed is the function P on its entire support. On the other hand, this structure minimizes the number of memory fetches (since each Q(i) is used once and for all), and exhibits in a regular fashion the longer loop. On actual implementation of the optimal nonlinear filter, the function F(i) is extended with zeros beyond its actual support (see figure 9, where F has been virtually extended with zeros up to F(L+I)). This avoids end problems with (8.1), preserving the regularity of the algorithm. Further details are in [19].

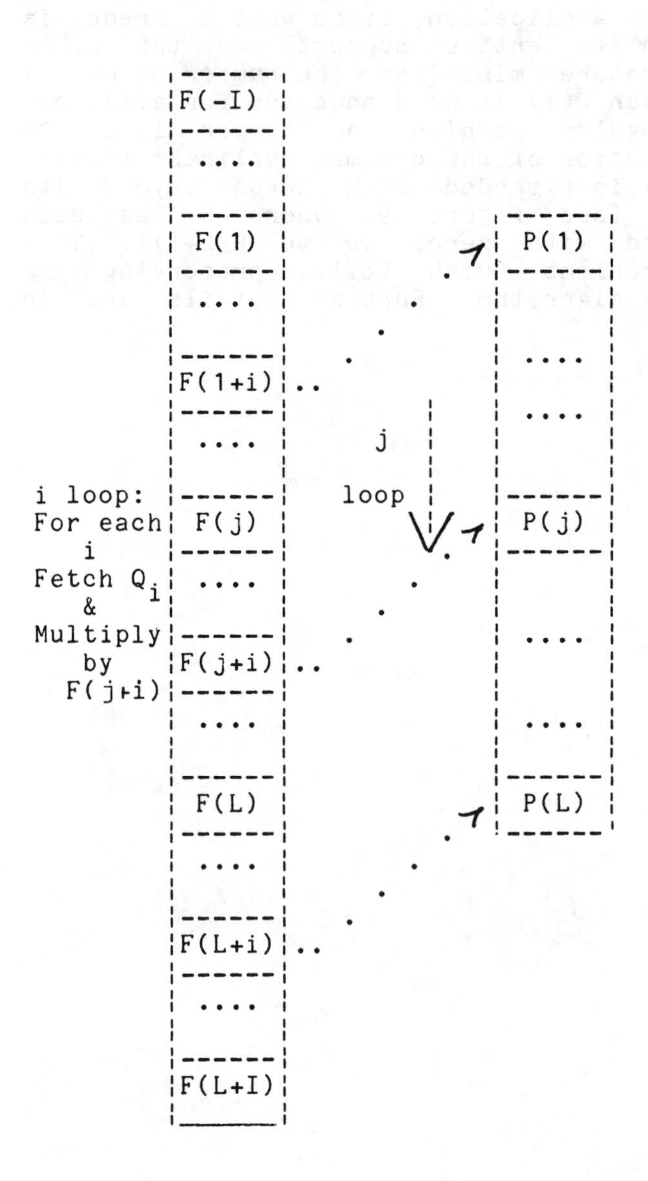

Figure 9: Interweaving of Loops in Convolution

Figures 10-12 present pictures of the line density evolution at different time steps. The pictures identify the corresponding iteration number (1357, 2108, and 7850 respectively). They are drawn for the same noise to signal ratio of - 3dB . The pictures

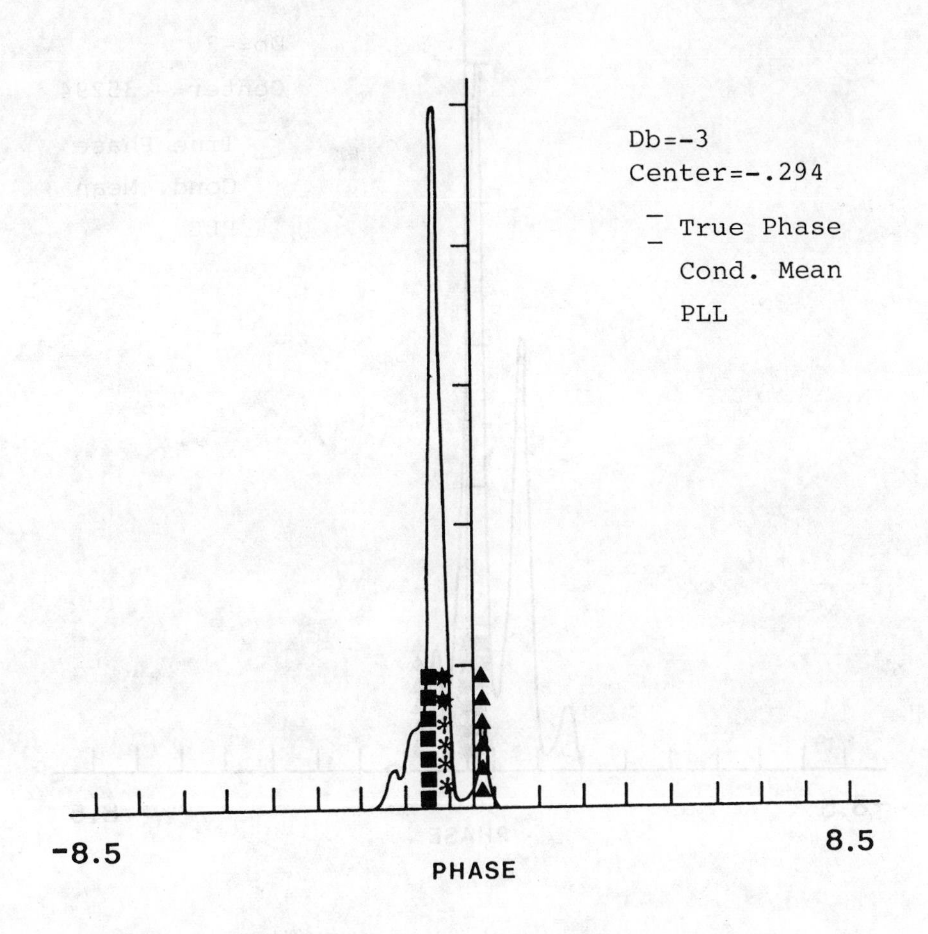

Figure 10: Line Density Plot no. 1357

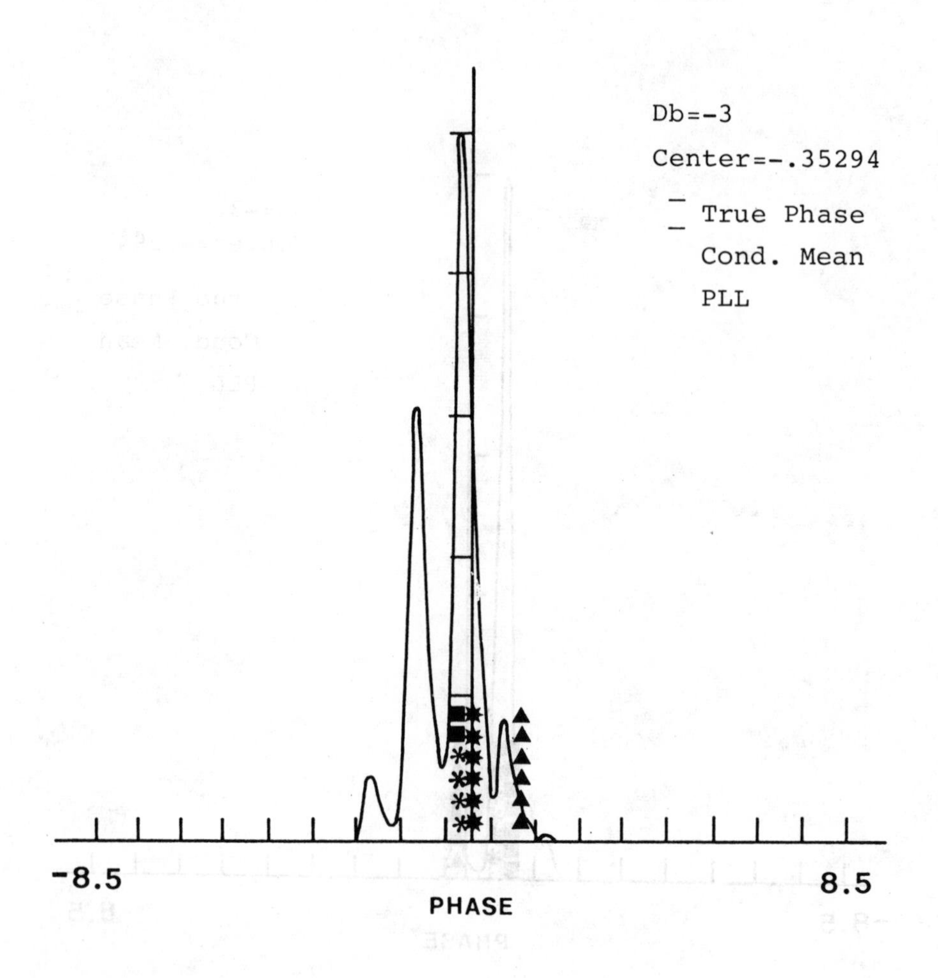

Figure 11: Line Density Plot no. 2108

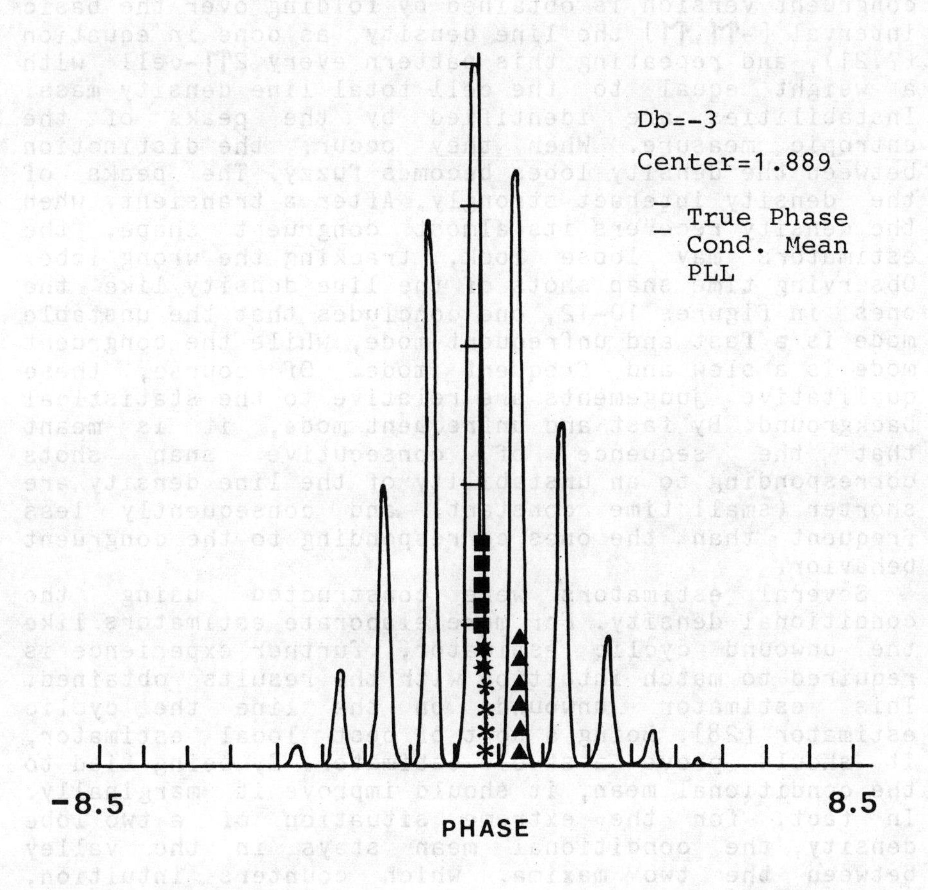

Figure 12: Line Density Plot no. 7850

exhibit the instantaneous location of the phase process x(t) and of several estimators (conditional mean of the line density and PLL).

With this type of line density plots, it was found [29], that the line density experienced two basic modes of behavior: a congruent mode and an unstable mode. An efficient detector of instabilities is provided by the entropic measure of instability [29], related to the difference between the line density entropy and the entropy of a corresponding congruent density. This congruent version is obtained by folding over the basic interval $[-\pi,\pi]$ the line density, as done in equation (7.21), and repeating this pattern every 2π-cell with a weight equal to the cell total line density mass. Instabilities are identified by the peaks of the entropic measure. When they occur, the distinction between the density lobes becomes fuzzy. The peaks of the density interact strongly. After a transient, when the density recovers its almost congruent shape, the estimators may loose lock, tracking the wrong lobe. Observing time snap shots of the line density like the ones in figures 10-12, one concludes that the unstable mode is a fast and unfrequent mode, while the congruent mode is a slow and frequent mode. Of course, these qualitative judgements are relative to the statistical background. By fast and unfrequent mode, it is meant that the sequence of consecutive snap shots corresponding to an unstability of the line density are shorter (small time constant) and consequently less frequent than the ones corresponding to the congruent behavior.

Several estimators were constructed using the conditional density. For more elaborate estimators like the unwound cyclic estimator, further experience is required to match intuition with the results obtained. This estimator unwounds on the line the cyclic estimator [28]. Being a sort of best local estimator, it should prove a stable estimator. By being tied to the conditional mean, it should improve it marginally. In fact, for the extreme situation of a two lobe density, the conditional mean stays in the valley between the two maxima, which counters intuition. However, problems have arisen on the slip counting with the unwound cyclic estimator. Whenever an unstable mode occurs, the unwound cyclic becomes hesitant between contiguous modes. Although, on the average, after the unstability, it slipped less times than the PLL, the slip counter implemented in the experiments counted a larger number. This discussion is presented with the purpose of illustrating some of the difficulties encountered when studying slip behavior. The question of constructing a "good" line estimator from the line density is still unresolved. Focussing attention on the conditional mean and on the PLL, it has been statistically found an improvement of the first over the second in the range of 25% to 50%, depending on the performance measure used, see [29].

Absolute PM Acquisition

Acquisition relates to the ambiguity associated with the location of the phase process at the origin of time t=0. In absolute PM problems, this ambiguity overlaps several 2π-cells. The periodic nature of the sensors makes this problem unsuitable for linearized structures like the EKB filter or the PLL. These filters follow the mode they have been initialized in, the initial global error persisting as an offset. In the long run, this picture changes due to the cycle slips. The model built in deterministic observability is lost in statistical terms because of the cycle slip phenomenon. Also, the cycle slip phenomenon does not represent a systematic recovery action of the filter. The acquisition is here considered as a short horizon problem.

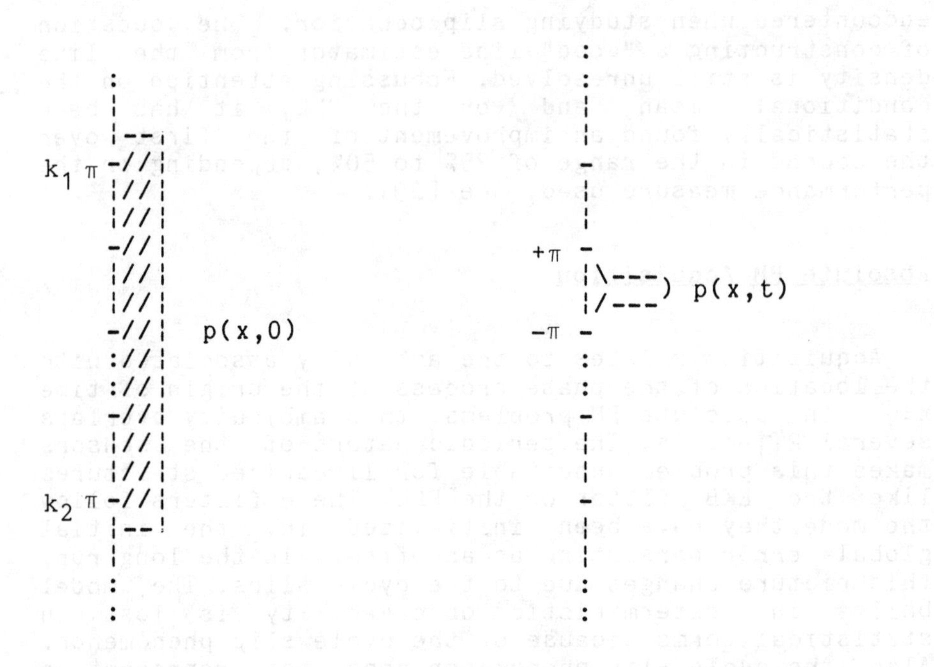

Figure 13: Acquisition: A Reduction of Density Spread by Filter Action

Figure 13 illustrates what the nonlinear filter achieves. Departing from a broad initial density, which models the ignorance on the initial value of the process, by processing the measurements, the filter narrows the spread of the line density.

The nonlinear filter was implemented using an approach which differs from the point masses previously considered. Reference [30] sketches the method. The basic idea behind it is to approximate the periodic exponential factor H of equation (5.8) or (5.9) by a suitable train $\tilde{H}$ of Gaussians. For PM problems, the (unnormalized) sensor factor H is circular-normal (Tikhonov density form). The filter [30] constructs H by substituting each period of H by an appropriate Gaussian function. The corresponding nonlinear filter recursion is equivalent to the following:

i) Bank of Discrete Kalman-Bucy predictors.

ii) Construction of $\tilde{H}$

iii) Bank of Discrete Kalman-Bucy filters operating on pseudo-measurements.

iv) Redimensioning of the nonlinear filter, followed by normalization.

The pseudo-measurements result from step ii), ie. the construction of $\tilde{H}$ approximating H. They correspond to the substitution of the nonlinear noisy measurements by the one-step cyclic estimate (compare with equation (6.6))

$$x_{cyc}=\mathrm{ATAN}(z_2/z_1). \tag{8.2}$$

It assumes that the measurement noise has a variance which is computed in a simple way from the H(n) function.

At each time step, and to keep the filter dimension low, in order not to tax the computational effort, the filter is redimensioned by two operations: lobe elimination and lobe agglutination. These two operations require the tuning of two parameters. Details on the construction of the filter, as well as a systematic evaluation of its performance for one and two dimensional PM cyclic and absolute PM problems are to be found in [32].

The filter has a built in scheme that decides when does acquisition occur. This scheme compares the density variance to a preset threshold. To assess the performance of the filter, Monte Carlo simulation runs were carried out. Besides several mean square error type statistics, the output of these studies consisted of histograms of correct and incorrect decisions. It was observed that, most of the time, incorrect decisions correspond to early decisions. Based on these histograms, several definitions for the acquisition time may be discussed. The dependence of the acquisition time, or the ratio between correct and incorrect decisions, on the statistical parameters and on the process bandwidth (pole a) is studied in [32]. The filter has been compared with a straight forward bank of EKB filters. This structure is based on the Gaussian sum concept of [33]-[34]. It linearizes the model nonlinearities about a local version of the conditional mean. The experimental evidence collected shows that the direct approximation of the density, preserving its qualitative shape as accomplished by the techniques of [30]-[32], besides reducing the computational effort, achieve a significant improvement of performance. The quantification of this improvement is in [32], and will be published elsewhere.

9. CONCLUSION

The paper has approached the phase modulation problems within the perspective of optimal nonlinear filtering. It has discussed several applications that fit the same mathematical framework. A state variable model has been considered. Concepts of continuous and discrete optimal nonlinear filtering theory have been recalled. The phase problems were grouped in cyclic (bounded support) or absolute (unbounded support), and in tracking (long term behavior) or acquisition (short term behavior). The Fokker-Planck theory of the PLL was briefly presented. An integration technique based on operator ideas was discussed. Finally, the paper considered the implementation and performance of the optimal nonlinear demodulators.

It is hard to present for optimal nonlinear filtering, general conclusions which are valid, irrespective of the nature of the problem. The implementation solutions are problem dependent. Each of the two problems analysed in section 8, provide however, two practical rules of sufficient generality to be worth to mention them explicitly:

i) For general purpose hardware with pipelining and parallelism, the algorithms implementing the filter should reflect a high degree of regularity in its structure.

ii) Instead of techniques using direct linearization of the model nonlinearities, it is better to try approximations directly on the space of the density functions. The filter of [30] is a very simple strategy that preserves qualitatively the shape of the sensor factor H . As mentioned, its behavior compares very favorably with respect to other types of implementation. For customed designed hardware, this is then a relevant issue and an important remark: do preferably, the approximations directly on the optimal filter, not on the starting model.

ACKNOWLEDGEMENTS

The work reported here on Absolute PM Tracking (see section 8), has been done with R. S. Bucy and A. J. Mallinckrodt (see references [28] and [29]), and on Absolute PM Acquisition with J. M. N. Leitao (see references [30] to [32]).

REFERENCES

[1]- Donald L. Snyder, "The State - Variable Approach to Continuous Estimation, with Applications to Analog Communication Theory" Research Monograph no.51, The MIT Press, 1969

[2]- Harry L. Van Trees, "Detection, Estimation and Modulation Theory: Part II Nonlinear Modulation Theory", John Wiley and Sons, 1971

[3]- José M. F. Moura, "Geometric Aspects in Array Processing", in Issues in Acoustic / Signal Image Processing and Recognition, ed C. H. Chen, Springer Verlag, 1983

[4]- José M. F. Moura, "Passive Systems Theory with Narrow Band and Linear Constraints: Part II-Temporal Diversity", IEEE Journal of Oceanic Eng., OE-4, no.1, pp 19-30, 1979; see also "Part I - Spatial Diversity" with A.B. Baggeroer, IEEE J. Oceanic Eng., OE-3, no.1, pp.5-13, 1978; and "Part III-Spatial / Temporal Diversity", IEEE J.Oceanic Eng., OE-4, no.3, pp.113-119, 1979

[5]- José M. F. Moura, "The Hybrid Algorithm: A solution to acquisition and tracking", J. Ac.Soc. of Am., vol.69, no.6, pp.1663-1672, June 1981

[6]- José M. F. Moura, "Modeling, Algorithmic and Performance Issues in Underwater Location Systems", in Underwater Acoustics and Signal Processing, ed. Leif Bjorno, D. Reidel, Holland, 1981

[7]- José M. F. Moura, "Recursive Techniques for Passive Source Location", IEEE ICASSP-82, Paris, France, May, 1982

[8]- G. Kallianpur, "Stochastic Filtering Theory", Springer-Verlag, 1980

[9]- B. L. Rozovskis, "On Stochastic Partial Differential Equations", Math. USSR-Sb., 25, 295-322, 1975

[10]- J. M. C. Clark, "The Design of Robust Approximations to the Stochastic Differential Equations of Nonlinear Filtering", in Communication Systems and Random Processes Theory, ed. J. K. Skwirzynski, Sijthoff and Noordhoff, Alphen aan den Rijn, 1978.

[11]- M. H. A. Davis, "A Pathwise Solution of the Equations of Nonlinear Filtering", Teoria Veroyatnostei i ee Prim., 1981

[12]- S. K. Mitter, "On the Analogy Between Mathematical Problems of Nonlinear Filtering and Quantum Physics", Richerche di Automatica, 1981; see also "Filtering Theory and Quantum Fields", Asterisque, 1980 (Bordeaux 1978)

[13]- E. Pardoux, "Stochastic Partial Differential Equations and Filtering of Diffusion Processes", Stochastics, vol. 3, pp. 127-167, 1979

[14]- M. Hazewinkel and J.C. Willems (eds.), "Stochastic Systems The Mathematics of Filtering and Identification and Applications", D. Reidel, 1981

[15]- S. K. Mitter, "Approximations for Nonlinear Filtering", this volume

[16]- E. Dieulesaint, "Elastic Waves in Solids: Applications to Signal Processing", John Wiley & Sons, 1980

[17]- R. S. Bucy, W. H. Steier, C. P. Christensen, A Sawchuk, D. Drake, "Feasibility Study of Device Synthesis of Nonlinear Filters", Technical Report, USC-EE no. 494, University of Southern Calif., 1978

[18]- R. S. Bucy and Kenneth D. Senne, "Nonlinear Filtering Algorithms for Vector Processing Machines", Comp. & Maths. with Appls., vol.6, pp. 317-338, Pergamon Press, 1980

[19]- R. S. Bucy, F. Ghovanlou, J. M. F. Moura, K. D. Senne, "Nonlinear Filetring and Array Computation ", submitted for publication.

[20]- B. Z. Bobrovsky and M. Zakai, "Asymptotic A Priori Estimates for the Error in the Nonlinear Filtering Problems", IEEE Trans. on Inf.TH., vol II-28, no.2, pp.371-376, March 1982

[21]- R. S. Bucy, "Distortion Rate Theory and Filtering", IEEE Trans. on Inf. Th., vol IT-28, no.2, pp.336-340, March 1982

[22]- Jorge I. Galdos, "A Rate Distortion Theory Lower Bound on Desired Function Filtering Error", IEEE Trans. on Inf. Th., vol IT-7, no. 3, pp.366-368, May 1981

[23]- R. S. Bucy, "Information and Filtering", Information Sciences, 18, pp. 179-187, 1979

[24]- Andrew J. Viterbi, "Principles of Coherent Communication", Mcgraw-Hill Comp., 1966

[25]- W.C.Lindsey, "Synchronization Systems", Prentice-Hall, Edgewood Cliffs, New Jersey, 1973

[26]- Carlos A. C. Belo, "Numerical Studies of the Line and Cyclic Fokker-Planck Equations", Proceedings Vigo Workshop on Signal Processing and Its Applications, July 1981

[27]- Richard S. Bucy, Calvin Hecht and Kenneth D. Senne, "An Engineer's Guide to Building Nonlinear Filters", vol 1 and vol 2, Frank J. Seiler Research Laboratory, May 1972

[28]- R. S. Bucy, J. F. Moura, A. J. Mallinckrodt, "Absolute Phase Demodulation", Mini and Micro Computers, 5, #2, 1980, pp. 116-119

[29]- R. S. Bucy, J. M. F. Moura, A. J. Mallinckrodt, "A Monte Carlo Study of Optimal Absolute Phase Demodulation", IEEE Trans. on Inf. Th., vol IT-29, no.4, July 1983

[30]- José M. N. Leitão and José M. F. Moura, "Nonlinear Filtering in Phase Acquisition", EUSIPCO-80, Signal Processing: Theories and Applications, eds. M.Kunt and F. De Coulon, North Holland, 1980

[31]- José M. F. Moura and José M. N. Leitão, "Experimantal Studies in Phase Acquisition", IEEE Inf. Th. Symp., Les Arcs, June, 1982.

[32]- José M. N. Leitão, forthcoming doctoral thesis, Jan., 1983.

[33]- D. L. Alspach, "A Bayesian Approximation Technique for Estimation and Control of Time Discrete Stochastic Systems", Ph. D. Dissertation, Univ. of California, San Diego, USA, 1970

[34]- D. L. Alspach and H. W. Sorenson, "Approximation of Density Functions by a Sum of Gaussians for Nonlinear Bayesian Estimation", Proc. Symp. on Nonlinear Estimation Theory and Its Applications, San Diego, Sept. 1970, 19-31

VOLTERRA SERIES AND FINITE DIMENSIONAL NONLINEAR FILTERING

Markku T. Nihtilä

Helsinki University of Technology
Department of Electrical Engineering
Otakaari 5 A, SF-02150 Espoo 15
Finland

Filtering in a class of nonlinear differential systems described by two subsystems connected with a polynomial link map is studied. The first subsystem is linear. The second one is given by a finite Volterra series driven by polynomials of the state of the preceeding system. The filtering problem is posed in a deterministic framework as a set of fixed interval optimization problems. All the systems studied are described by ordinary non-stochastic differential equations. A stochastic interpretation of the problem statement is discussed. It is shown that the given system structure admits finite dimensional representation for the optimal recursive filter. Explicit nonlinear filters are constructed in some examples. Comparisons with the corresponding stochastic systems and optimal finite dimensional conditional mean filters are performed. An extension to the observations described by an implicit state-linear equation is also discussed.

1. INTRODUCTION

Recently increased interest in the restudy of nonlinear filtering has its origin in the application of nonlinear stochastic filtering theory to problems of the real world. All the filters, approximate or exact, obtained via stochastic formulation are stochastic differential systems the input of which is the observation process, not one of its individual realizations. In practice, however, stochastic filters are always applied by substituting the observation process by its realization and by adding a correction term into the stochastic filter equation [1], which is then integrated as if it were an ordinary differential equation. Sussman [2] has shown that depending on the numerical approximations used in the integration different results can be obtained. In integrating ordinary

R. S. Bucy and J. M. F. Moura (eds.), Nonlinear Stochastic Problems, 387–396.

nonstochastic differential equations this kind of difficulty does not arise. The main reason for these application problems is the definition of the Brownian motion process which has no exact correspondence in the real world. Furthermore, it generates the state and observation process which have smooth, i.e. practical, realizations with probability zero [3].

A deeper analysis of the stochastic nonlinear filtering problem has been performed by Clark [4] and Davis [5] who obtained a so-called *pathwise solution* for the unnormalized conditional probability density of the estimate. A deterministic partial differential equation parametrized by the sample paths of the observation process defines the density. In essence, the solution has been obtained by extending the domain of the mapping which gives the filtering from the given observation into the whole of $C([0,T],\mathbf{R}^m)$ (continuous functions from the relevant time interval into m-vectors) in such a way that the continuity of the mapping is preserved. This extension technique in connection with stochastic differential equations dates back to the original papers of Krasnoselskii [6] and Sussmann [7]. Different type of stochastic calculus than that of Ito or Stratonovich have been developed by McShane [8] and Balakrishnan [9]. Their results are directly applicable also to smooth realizations of the observation process. McShane includes in the basic stochastic system equation a second order integral with respect to the input process. Balakrishnan defines elegantly a new finitely additive measure instead of the Wiener measure resulting in a directly instrumentable likelihood ratio.

By using a completely *deterministic problem formulation* stochastic concepts are not needed in the analysis. The deterministic approach, per se, does not reduce the finite dimensionality of the nonlinear optimal recursive filter. In this paper we show the existence of a finite dimensional optimal recursive filter for a class of nonlinear differential systems. The filtering problem is formulated as a set of fixed interval optimization problems (FIOP), the final values of the optimal trajectories of which give the filtering. The formulation is the same as originally was used by Detchmendy et al.[10] and Bellman et al.[11]. Via Pontryagin's maximum principle the FIOP is converted into a two-point boundary value problem (TPBVP) of ordinary differential equations. The solving of the final values of the optimal trajectories of the TPBVP in a recursive finite dimensional form for a class of nonlinear systems is the main result of this study.

The system class to be considered consists of two subsystems with a polynomial link map between them. The first subsystem is linear. The second one is given by a finite Volterra series, the input of which is a polynomial in the state of the linear subsystem. Observations are obtained only from the linear subsystem.

The finite dimensionality of the optimal least-squares filter is proved in the general case via successive differentiation in the Volterra series. For a special system class presented in the form of a differential system but admitting a finite Volterra series representation the optimal filter is obtained by applying a Riccati transformation technique. The equivalence of the finite Volterra series representation and the special differential system representation is due to the results given by Crouch [12]. Comparisons on the basis of some examples with the corresponding stochastic conditional mean filters developed by Marcus et al.[13] are given.

2. PROBLEM STATEMENT

The filtering problem is posed as a set of fixed interval optimization problems.

Filtering problem. For every final time $t \in [0,T]$ and for a given observation $y(\sigma)$ on the time interval $[0,t]$ find the final value of the optimal trajectory of the fixed interval optimization problem: Minimize the performance index

$$J(t,\xi,d) = \frac{1}{2}\|\xi(0)-\xi_0\|_J^2 + \frac{1}{2}\int_0^t [\|y(\sigma)-\tilde{H}(\sigma)\xi(\sigma)\|^2_{R(\sigma)} + \|d(\sigma)\|^2_{S(\sigma)}]d\sigma \tag{1}$$

with respect to the pair (ξ,d) subject to the constraint

$$\dot{\xi}(\sigma) = f(\sigma,\xi(\sigma)) + G(\sigma)d(\sigma). \tag{2}$$

J, $R(\sigma)$ and $S(\sigma)$ are positive definite weighting matrices of appropriate dimensions. $\|.\|_J$ denote a norm defined by $\|a\|_J^2 = a'Ja$. ξ, y and d has values in $(n+n_1)$-, m- and l-vectors, respectively. Denote the optimal trajectory (assumed it exists and is unique) by $\xi(.|t)$ for every final time t.

Definition. The filtering defined by $\eta(t) = \xi(t|t)$ is finite dimensionally computable if it is obtained from a finite dimensional differential system driven by the innovation $y(t)-\tilde{H}(t)\xi(t|t)$.

In the deterministic filtering we have two objectives: to disturb (by the pseudo-control d) the chosen state model as slightly as possible, and to keep the observation calculated via the chosen output model from the state of the disturbed system as close as possible to the true observation on the whole observation interval. The two goals are in disagreement if we have not a common measure for them. So, our selection is to minimize the weighted sum of the squared pseudo-control d and the observation error, or innovation $y-\tilde{H}\xi$ integrated over the observation interval. Almost whatever reasonable measure instead of the chosen quadratic measure could be used. However, the chosen measure leads into relatively simple algorithms; in the purely linear case into the finite dimensional formal Kalman - Bucy filter, and especially in the system class to be given into finite dimensional filters. It has to be emphasized that the selection of the quadratic measure corresponds in the stochastic formulation Gaussian assumptions of the process and observation noise. Furthermore, the weighting matrices in the cost functional can be interpreted as the inverses of the noise covariances of the stochastic formulation. From practical point of view the weighting matrices are tuning parameters in the filters.

3. A FINITE DIMENSIONAL CLASS

Theorem. Assume that the given filtering problem is decomposable as follows. $\xi = (x,z)$ where x satisfies a linear differential equation

$$\dot{x} = Ax + Bd\ , \quad x \in \mathbf{R}^n\ , \tag{3}$$

and z is given by a finite Volterra series of length ν

$$z(\sigma) = W_0(\sigma) + \sum_{i=1}^{\nu} \int_0^{\sigma}\!..\!\int_0^{\sigma} W_i(\sigma,\tau_1,..,\tau_i)(u(\tau_1),..,u(\tau_i))\,d\tau_1..d\tau_i \tag{4}$$

with symmetric separable kernels W_i (see Brockett [14] and Crouch [12]). The input $u(\tau) \in \mathbf{R}^N$ is a polynomial in x

$$u(\tau) = P^o(\tau) + \sum_{j=1}^{\nu_j} P^j(\tau)(x(\tau)^j)\ . \tag{5}$$

$\widetilde{H}$ picks only up the first component x

$$\widetilde{H} = [H \quad 0]\ , \tag{6}$$

and the pair $\{A,H\}$ is completely observable. Then the filtering (r,s) defined by

$$r(t) = x(t|t)\ , \quad s(t) = z(t|t) \tag{7}$$

is finite dimensionally computable.

The kernels W_i, as functions of u, in the Volterra series can be considered as multilinear mappings from $\mathbf{R}^N$ into $\mathbf{R}^{n1}$. Every term of order i in the vector-valued polynomial (5) in a vector-valued argument can also be interpreted as a multilinear mapping (of order i) from $\mathbf{R}^n$ into $\mathbf{R}^N$.

The proof of the theorem is given in [15]. It is shortly outlined in the following. It is trivial to see that r is obtained from the formal Kalman - Bucy filter. The finite dimensionality of the differential system giving s is shown by successively differentiating the Volterra series with respect to the final time t on the basis of the equation

$$\dot{s}(t) = z_{\sigma}(t|t) + z_{t}(t|t) \tag{8}$$

where the subindices σ and t denote partial derivatives with respect to the first and second argument position, respectively. In the differentiation the following result proved in [15] is needed

$$x_{t}(\sigma|t) = K(\sigma)\Psi(\sigma,t)H^{T}R[y(t) - Hx(t|t)] \tag{9}$$

where the n by n matrix valued Ψ satisfies

$$\Psi_{\sigma}(\sigma,t) = -[A^{T} - H^{T}RHK(\sigma)]\Psi(\sigma,t) \ , \tag{10}$$

$$\Psi(t,t) = I \ , \tag{11}$$

and K is the solution of the matrix Riccati differential equation of the linear subproblem. The series (4) and every differentiated term can be realized by a finite dimensional differential system on the basis of the results by Crouch [12] and of the equation (8). At last, the sequence of the successive differentiations is finite due to the polynomial structure of the series in $x(.|t)$.

The theorem can also be proved directly by using Crouch's realization results for finite Volterra series and a Riccati transformation technique in the differential system corresponding to the finite Volterra series. The Riccati transformation gives furthermore a constructive proof resulting in an explicit form of the optimal filter. This technique is used in the examples.

In our system class the linear observation equation corresponding to the second term in the performance index (1) can be substituted by an implicit observation equation of the following state-linear form

$$G_{0}(t,y) + G_{1}(t,y)x = e \ , \tag{12}$$

where G_0 and G_1 are appropriate smooth function of t and y with values in, say, l_1-vectors and l_1 by n matrices, respectively. The dimension l_1 of the generalized error need not be the same as m, the dimension of the observation. However, (12) must possess a feasible (not necessarily unique) solution for all relevant triples $(t,x(t),e(t))$. In this case the innovation term in the integrand of (1) has to be substituted by the generalized innovation term

$$\frac{1}{2}\,\|G_{0}(\sigma,y(\sigma)) + G_{1}(\sigma,y(\sigma))x(\sigma)\|^{2}_{R(\sigma)} \ . \tag{13}$$

Furthermore, the complete observability assumption in the theorem concerns now the pair $\{A, G_1(.,y(.))\}$ along the given observation. This state-linear observation problem (for $z(.)\equiv 0$) has been studied in [16] and [17].

4. EXAMPLES

The class of systems including the examples studied by Marcus et al.[13] in a stochastic framework is considered. Some mathematical notations used to keep the formulae compact are introduced in the appendix. An example concerning the on-line evaluation of the performance index in the state-linear filtering is also given.

The system class is given by

$$\dot{x} = Ax + Bd \,, \tag{14}$$

$$\dot{z} = Fz + \frac{1}{2}\sum_{i=1}^{n} x_i A_i x \,, \tag{15}$$

$$y = Hx + e \,, \tag{16}$$

where the coefficient matrix F is assumed stable. It is easily seen that (15) admits a finite Volterra series representation for z. By using the notations of the appendix the quadratic term in (15) can be given in a compact form

$$\frac{1}{2}(\mathsf{A}x)x \,, \qquad (\mathsf{A})_{ijk} = (A_j)_{ik} \,.$$

By applying the Riccati transformation

$$x(\sigma|t) = r(\sigma) - K(\sigma)p(\sigma|t) \,, \tag{17}$$

$$z(\sigma|t) = s(\sigma) - L(\sigma)p(\sigma|t) - \frac{1}{2}[M(\sigma)p(\sigma|t)]p(\sigma|t) \,, \tag{18}$$

in the TPBVP corresponding to (1), (14) and (15) where p is the co-state of the linear subproblem, K is the Riccati matrix appearing in (9), and L and M are mappings into $\mathbf{R}^{n1\times n}$ and $\mathbf{R}^{n1\times n\times n}$, respectively, the polynomial equation

$$\tilde{s}(\sigma) + \tilde{L}(\sigma)p(\sigma|t) + \frac{1}{2}[\tilde{M}(\sigma)p(\sigma|t)]p(\sigma|t) = 0 \tag{19}$$

is obtained. Eqn.(19) can be valid for all $\sigma \in [0,t]$ only if the zeroth order term and the coefficients $\tilde{L}$ and $\tilde{M}$ are all zero. Especially, for $\sigma = t$ we obtain in addition to the formal Kalman-Bucy filter for (r,K) the differential equations

$$\dot{s} = Fs + \frac{1}{2}(\mathsf{A}r)r + LH^T R(y - Hr) \,, \tag{20}$$

$$\dot{L} = FL + L(A^T - H^T RHK) + (\mathsf{A}r)K - MH^T R(y - Hr) \,, \tag{21}$$

$$\dot{M} = FM + M(A^T - H^T RHK) + M\cdot(A^T - H^T RHK) - (\mathsf{A}K)\cdot K \tag{22}$$

which actually give the optimal filtering. The initial conditions are obtained from the initial term of the performance index

$$s(0) = z_o, \quad L(0) = 0, \quad M(0) = 0 . \tag{23}$$

Marcus et al.[13] proved for the afore-mentioned system class the existence of a finite dimensional conditional mean filter. As an explicit example they considered the stochastic counterpart of the following system

$$\dot{x}_1 = -\alpha x_1 + d_1 ,$$

$$\dot{x}_2 = -\beta x_2 + d_2 , \quad \alpha, \beta, \gamma > 0 ,$$

$$\dot{z} = -\gamma z + x_1 x_2 ,$$

$$y = Hx + e ,$$

where H is the identity matrix, $J=0$, and R and S are diagonal (in (1)). By applying the general equations (20)-(22) we obtain the deterministic least-squares filter

$$\dot{s} = -\gamma s + r_1 r_2 + M_1 R(y_1 - r_1) + M_2 R(y_2 - r_2) ,$$

$$\dot{M}_1 = -(\alpha+\gamma+RK_{11})M_1 + K_{11} r_2 - M_3 R(y_2 - r_2) ,$$

$$\dot{M}_2 = -(\beta+\gamma+RK_{22})M_2 + K_{22} r_1 - M_3 R(y_1 - r_1) ,$$

$$\dot{M}_3 = -(\alpha+\beta+\gamma+RK_{11}+RK_{22})M_3 - K_{11}K_{22} ,$$

where K_{11} and K_{22} are the diagonal elements of the Riccati matrix of the linear subproblem. The nonlinear filter was obtained by using the Riccati transformation

$$z = s - M_1 p_1 - M_2 p_2 - M_3 p_1 p_2 .$$

A detailed inspection shows that stochastic and deterministic filters for this system are formally equivalent, although in our case we have one differential equation less than in the stochastic case. It has been proved by Fleming et al.[18] that the purely linear stochastic filtering problem has a formally equivalent deterministic counterpart which is, in fact, the same as our filtering problem of section 2.

It is seen in the next example that the formal equivalence breaks down due to the dependence of the components in the non-linear term. The example

$$\dot{x} = -\alpha x + d ,$$

$$\dot{z} = -\gamma z + x^2 ,$$

with linear observations from x gives for s the filtering equation

$$\dot{s} = -\gamma s + r^2 + M_1 R(y-r)$$

where M_1 is given by a two-dimensional system. The stochastic counterpart is now [13]

$$d\hat{z} = (-\gamma\hat{z} + \hat{x}^2 + K)dt + M_1 R(d\tilde{y} - \hat{x}dt)$$

where ^ stands for the conditional expectation, K is the covariance of $\hat{x}$ and $d\tilde{y}$ is given by $d\tilde{y} = xdt + v$ (v is a Wiener process with the incremental covariance $R^{-1}dt$). The obvious difference as compared with the first example is that instead of $\hat{x}^2$ in the stochastic filter there are $\hat{x}^2+K$, the conditional expectation of x^2. The stochastic filter of the first example can be considered to include the similar term $E\{x_1 x_2\}$ which is, however, equal to the product of the expectations of x_1 and x_2 resulting in the formal equivalence of the stochastic and the deterministic filter.

In this example we construct the differential system which gives on-line the values of the performance index (1) for every final time in the case of the state-linear observation equation. The TPBVP corresponding to the FIOP is given by

$$x_\sigma(\sigma|t) = Ax(\sigma|t) - BS^{-1}B^T p(\sigma|t) \quad , \tag{24}$$

$$p_\sigma(\sigma|t) = -A^T p(\sigma|t) - G_1^T(\sigma,y(\sigma))Rg(\sigma,y(\sigma),x(\sigma|t)) \quad , \tag{25}$$

$$z_\sigma(\sigma|t) = \tfrac{1}{2}\|g(\sigma,y(\sigma),x(\sigma|t))\|_R^2 + \tfrac{1}{2}\|S^{-1}B^T p(\sigma|t)\|_S^2 \tag{26}$$

$$x(0|t) = x_0 - J^{-1}p(0|t) \, , \tag{27}$$

$$p(t|t) = 0 \, , \tag{28}$$

$$z(0|t) = \tfrac{1}{2}\|x(0|t) - x_0\|_J^2 \, . \tag{29}$$

g denotes the observation equation, i.e. $g = G_0 + G_1 x$, z stands for the performance index. The quadratic Riccati transformation in addition to (17) is defined by

$$z = s + v^T p + \frac{1}{2} p^T M p \, , \tag{30}$$

where v is an n-vector, and M an n by n matrix. It turns out that $\dot{M}$ equals K, the Riccati matrix of the linear subproblem. Finally, for $\sigma = t$ we obtain

$$\dot{s} = \frac{1}{2}\|g(t,y,r)\|_R^2 \quad , \; s(0) = 0, \tag{31}$$

$$v(t) = 0. \tag{32}$$

On the basis of (31)-(33) we can evaluate the performance of the filtering based on the state-linear observation equation.

It is seen that by applying the technique introduced any performance index which is polynomial in the state and in the optimal pseudo-control can be finite dimensionally computed on-line.

5. CONCLUDING REMARKS

The formulation of the nonlinear filtering through deterministic concepts in the given system class resulted in the optimal filters which are finite dimensional. Furthermore, aside from finite dimensionality all the filters obtained via this formulation are ordinary nonstochastic differential systems the input of which is the observed signal which is only a function of time, not a stochastic process. The result is a counterpart of the one obtained by Marcus et al.[13] via stochastic formulation for the conditional mean of the state. Crouch's[12] realization results for finite Volterra series and the application of a polynomial type Riccati transformation give the optimal filters in an explicit differential system form.

APPENDIX

The multiplication operations used in the main text are shortly introduced here

$$(\mathrm{A}x)_{ij} = \sum_{k=1}^{n} \mathrm{A}_{ijk}x_k , \quad \mathrm{A} \in \mathbf{R}^{n_1 \times n \times n} , \quad x \in \mathbf{R}^n , \tag{A.1}$$

$$(\mathrm{A}K)_{ijk} = \sum_{l=1}^{n} \mathrm{A}_{ijl}K_{lk} , \quad K \in \mathbf{R}^{n \times n} , \tag{A.2}$$

$$(\mathrm{A}\cdot K)_{ijk} = \sum_{l=1}^{n} \mathrm{A}_{ilk}K_{lj} . \tag{A.3}$$

Some differentiation operations needed in the analysis are also given

$$[(\mathrm{A}x)x]_x = 2\mathrm{A}x \quad (\text{for } \mathrm{A}_{ijk} = \mathrm{A}_{ikj}), \tag{A.4}$$

$$[\mathrm{A}v(x)]_x = \mathrm{A}v_x(x) . \tag{A.5}$$

v is a mapping on n-vectors into itself. The subindex notation stands for differentiation.

REFERENCES

1. E.Wong, *Stochastic Processes in Information and Dynamical Systems*, McGraw-Hill, New York, 1971.
2. H.J.Sussmann, Ito, Stratonovich and other limits, *Proc. 19th IEEE Conf. on Decision & Control*, Albuquerque, U.S.A., Dec.1980, pp.57-61.
3. E.Wong and A.Zakai, On the convergence of ordinary integrals to stochastic integrals, *Ann.Math.Stat.* 36(1965), pp.1560-1564.
4. J.M.C.Clark, The design of robust approximations to the stochastic differential equations of nonlinear filtering. *Communication Systems and Random Process Theory*, J.K.Skwrzynski, ed., Sijthoff & Noordhoff, Alphen aan den Rijn, 1978, pp. 721-734.
5. M.H.A.Davis, Pathwise solutions and multiplicative functionals in nonlinear filtering, *Proc.18th IEEE Conf. on Decision & Control*, Fort Lauderdale, U.S.A., Dec.1979, pp.176-181.
6. M.A.Krasnoselskii and A.V.Pokrovskii, Natural solutions of stochastic differential equations, *Soviet Math.Dokl.* 19(1978), pp.578-582.
7. H.J.Sussmann, On the gap between deterministic and stochastic ordinary differential equations, *Ann.Probability* 6(1978), pp.19-41.
8. E.J.McShane, *Stochastic Calculus and Stochastic Models*, Academic Press, New York, 1974.
9. A.V.Balakrishnan, *Applied Functional Analysis*, Sringer-Verlag, New York, 1976.
10. D.Detchmendy and R.Sridhar, Sequential estimation of states and parameters in noisy nonlinear dynamical systems, *J.Basic.Eng.* 88(1966), pp.362-368.
11. R.Bellman, H.Kagiwada, R.Kalaba, and R.Sridhar, Invariant imbedding and nonlinear filtering theory, *J.Astronaut.Sci.* 13(1966), pp.110-115.
12. P.E.Crouch, Dynamical realization of finite Volterra series, *SIAM J.Control & Optimization* 19(1981), pp.177-202.
13. S.I.Marcus and A.S.Willsky, Algebraic structure and finite dimensional estimation, *SIAM J.Math.Anal.* 9(1978), pp.312-327.
14. R.W.Brockett, Volterra series and geometric control theory, *Automatica* 12(1976), pp.167-176.
15. M.T.Nihtilä, Finite dimensional nonlinear filters via Riccati and Volterra series, Submitted for publication, *SIAM J. Control & Optimization.*
16. M.T.Nihtilä, Optimal state-linear filtering through implicit output equation, *Preprints 8th IFAC Congress*, Kyoto, Japan, Aug.1981. Vol.V, pp.79-83.
17. M.T.Nihtilä, Optimal finite dimensional solution for a class of nonlinear observation problems, *J.Optimiz.Theory & Appl.* 38(1982) No.2 (to appear).
18. W.H.Fleming and R.W.Rishel, *Deterministic and Stochastic Optimal Control*, Springer-Verlag, New York, 1975.

CHAPTER VIII

STOCHASTIC PROCESSES

CAUSAL INVERTIBILITY: AN APPROACH TO THE INNOVATIONS PROBLEM

Dr. Deborah F. Allinger, Dept. of Mathematics,

M.I.T., Cambridge, MA 02139, USA

In this paper, the Innovations problem is studied by means of a method called causal invertibility. Both the problem and method are defined in Section I and examples are discussed in Section II.

I. The Innovations Problem. Let $(\Omega, \mathcal{F}, P)$ be a certain probability space ($T = 1$ for simplicity), $(\mathcal{F}_t, 0 \leq t \leq 1)$ be a non-decreasing family of the sub-σ-algebras of $\mathcal{F}$, and $W = (W_t, \mathcal{F}_t)$, $t \leq 1$ be a Wiener process. Denote by $(C_1, \mathcal{B}_1, \nu)$ the measurable space of continuous real-valued functions $x = (x_t, 0 \leq t \leq 1)$ on $[0,1]$ with the σ-algebra $\mathcal{B}_1 = (x : x_s, s \leq 1)$, and ν as Wiener measure. Also, set $\mathcal{B}_t = (x : x_s, s \leq t)$. Let $\alpha(t,x)$ and $b(t,x)$ be measurable, non-anticipative functionals (i.e., $\mathcal{B}_t$-measurable for each t).

Consider the stochastic differential equation (sde),

$$d\xi_t = \alpha(t,\xi)\,dt + b(t,\xi)\,dW_t, \xi_0 = 0 \tag{1.1}$$

If $b(t,x)$ is never zero, then one can write dW_t as an expression of ξ_t and conclude that the Wiener process is adapted to the family $(\mathcal{F}_t^{\xi})_{0 \leq t \leq 1}$ where $\mathcal{F}_t^{\xi}$ denotes $\sigma(\xi(w) : \xi_s(w), 0 \leq s \leq t)$. The innovations problem is to determine when the reverse inclusion holds; in this case, (ξ_t) is said to be a strong solution.

Methods for studying and solving the innovations problem are varied; my approach is called causal invertibility. This concept is the integration of ideas due to V.E. Beneš (3,4), and M.P. Ershov (6); it is not new. Nevertheless it is fundamental

R. S. Bucy and J. M. F. Moura (eds.), Nonlinear Stochastic Problems, 399–406.

to the study of strong solutions for sde and deserves a concrete formulation.

The proper framework for causal invertibility is the sde

$$d\xi_t = \alpha(t,\xi)dt + dW_t, \quad \xi_0 = 0 \tag{1.2}$$

Let us assume that $P(\int_0^1 |\alpha(s,\xi)|ds) = 1$. Corresponding to 1.2 I define a transformation, F, on C_1 by $F^*(x)(t) = s(t) - \int_0^t \alpha(s,x)ds$ and will say that F is causally invertible if there exists a transformation, f, on C_1 such that

$$F(f(y)) = y \quad \nu\text{-a.s.}$$

and

$$f^{-1}(\mathcal{B}_t) \subseteq \mathcal{B}_t \quad \nu\text{-a.s. for every } t. \tag{1.3}$$

Clearly, F is causally invertible if and only if 1.2 has a strong solution. F may be taken to be $\mathcal{B}_t/\mathcal{B}_t$-measurable for every t; in fact, F is causal in the sense that whenever $x(s) = y(s)$, $0 \leq s \leq t$, then $F(x)(s) = F(y)(s)$, $0 \leq s \leq t$ for x,y in the domain of F.

Proposition 1. F has a causal inverse if and only if there exists a Borel set, Ω for which

$$\nu(F(\Omega)) = 1 \tag{1.4}$$

and

whenever (x,y,t) satisfy $F(x)(s) = F(y)(s)$ for $0 \leq s \leq t$, then $x(s) = y(s)$ for $0 \leq s \leq t$ for x,y in . (1.5)

Proof: To prove necessity, first observe that 1.3 implies this condition:

There exists a set, D, with $\nu(D) = 1$ such that whenever (y,z,t) satisfy $y(v) = z(v)$, $0 \leq v \leq t$, then $f(y)(v) = f(z)(v)$, for $0 \leq v \leq t$ and y,z in D. (1.3a)

In other words, f is causal on D. The implication of 1.3a from 1.3 follows from Dudley's representation of Wiener functionals (5). This set, D, in 1.3a may be taken as a Borel set and in turn, $f(D)$ becomes an analytic set in C_1 within the domain of F. (See Sion (12).) Then a Borel set, Ω, satisfying 1.4,1.5 is extracted from $f(D)$.

For sufficiency, note that 1.4 and 1.5 force F to be causally injective on Ω. By a result of Sion (13), one can construct a Wiener-measurable transform, f, with domain $F(\Omega)$ and range, Ω, such that $F(f(y)) = y$; the rest of 1.3 follows from 1.5.

The following examples illustrate the strategy of Proposition 1.

Section II. Examples.

(a) A class of sde 1.1 with strong solution. Roughly speaking, a popular technique for obtaining strong solutions to 1.1 has been to use the strength of the diffusion and remove the drift (see (8),(14)). Here is a procedure for forming sde 1.1 with strong solution which uses the strength of both, thereby relaxing some restrictions on the diffusion. Take a functional $b(t,x)$ for which a system $(\Omega,F,P,F_t,\eta_t,W_t)$ is a strong solution of

$$d\eta_t = b(t,\eta)\,dW_t\,,\ \eta_0 = 0 \tag{1.6}$$

with

$$E(\int_0^1 (b(t,\eta))^2 dt) < \infty \text{ and } b^2(t,\eta) > 0 \quad \lambda\times P\text{-a.s.} \tag{1.6a}$$

Denote by μ_η, the measure induced on $(C_1,\mathcal{B}_1)$ by the process (η_t) and set

$$\hat{W}(x)(t) = \int_0^t 1/b(t,x)\,dx_t \quad \mu_\eta\text{-a.s.}$$

Then on $(C_1,\mathcal{B}_1,\mu_\eta)$, $(\hat{W}(X)_t,\mathcal{B}_t)$ is a Wiener process, and there is a non-anticipative functional, ψ, such that for $0\leq t\leq 1$,

$$x(t) = \psi(t,\hat{W}(x)) \quad \mu_\eta\text{-a.s.} \tag{1.7}$$

Define $\alpha(s,x)$ by $\alpha(s,x) = b(s,x)\,r(x,\hat{W}(x))$ where r determines both a causally invertible transform, F_r, and a solution of 1.2 with distribution equivalent to Wiener measure. For the latter, it is necessary and sufficient that

$$\nu(\int_0^1 (r(s,x))^2 ds < \infty \tag{1.7a}$$

and

$$E_\nu(\exp(\int_0^1 r(s,x)\,dx(s) - \tfrac{1}{2}\int_0^1 (r(s,x))^2 ds)) = 1$$

(see (9)). Also, α may be taken to be progressively measurable with respect to $(\mathcal{B}_t)$.

To see that the sde 1.1 with this choice of α and b has a strong solution, consider the transform, G given by

$$G(x)(t) = \hat{W}(x)(t) - \int_0^t r(s,\hat{W}(x))\,ds$$

which may be taken to be $\mathcal{B}_t/\mathcal{B}_t$-measurable. There exists a set, Ω, with $\mu_\eta(\Omega) = 1$ on which G satisfies 1.5 because of 1.7 and the choice of r..

First, the choice of r means that Proposition 1 is satisfied for a Borel set of Wiener measure 1. Thus, there is a set, $\Omega, \mu_\eta(\Omega) = 1$ 0 for which $G(x)(s) = G(y)(s)$, $0\leq s\leq t$ implies $\hat{W}(x)(s) = \hat{W}(y)(s)$, $0\leq s\leq t$ with x,y in Ω. Then from 1.7, the conclusion follows. The strength of both r and b combine to make G causally injective. Moreover, $(G(x)_t, \mathcal{B}_t)$ is a Wiener process on $(C_1, \mathcal{B}_1, \hat{\mu})$ where

$$\frac{d\hat{\mu}}{d\mu_\eta}(t,x) = \exp\left(\int_0^t r(s,\hat{W}(x))\,d\hat{W}(x)_s - \frac{1}{2}\int_0^t (r(s,\hat{W}))^2 ds\right)$$

and $\hat{\mu}$ is the equivalent to μ_η by 1.7a. Thus the aggregate $(x_t, G(x)_t, \mathcal{B}_t, C_1, \mathcal{B}_1, \hat{\mu})$ is a strong solution to

$$dx_t = \alpha(t,x)\,dt + b(t,x)\,dG(x)_t = b(t,x)\,[(r(t,\hat{W}(x))\,dt + dG(x)_t)]$$

It is path-wise unique among all solutions with distribution equivalent to μ_η. Finally, a strong solution of 1.6 is path-wise unique among all solutions which satisfy 1.6a and have equivalent distribution. This is due to the injective property of G. Some new examples of functionals, r, which satisfy the conditions above are given below.

(b). The algebra, H. With $(x_t, \mathcal{B}_t)$ as a Wiener process on $(C_1, \mathcal{B}_1, \nu)$, let H represent the algebra of real-valued functionals over $[0.1] \times C_1$ generated by bounded functions and stochastic integrals;

$$H = \sigma\left(h(s), \int_0^s g(u)\,dx(u),\ h \text{ in } Bdd[0,1],\ g \text{ in } L_2[0.1]\right)$$

Theorem 1. Every element, α, of H yields a transform, F, which satisfies 1.5 for a set of Wiener-measure 1.

Proof. The crux of this proof is to demonstrate Theorem 1 when α is a product of stochastic integrals and I begin with

$$\alpha(t,x) = \left(\int_0^t f(u)\,dx(u)\right)\left(\int_0^t g(u)\,dx(u)\right).$$

The hypothesis of 1.5 implies that for, $0 \leq s \leq t$

$$\begin{aligned}\frac{d}{dt}(x-y)(s) &= \alpha(s,x) - \alpha(s,y) \\ &= \int_0^s f(u)\,d(x-y)(u)\ \left(\int_0^s g(u)\,dx(u)\right) \qquad (1.8) \\ &\quad + \left(\int_0^s f(u)\,dy(u)\right)\left(\int_0^s g(u)\,d(x-y)(u)\right)\end{aligned}$$

By Hölder's inequality, $\int_0^t \left(\frac{d}{dt}(x-y)(s)\right)^2 ds$ is bounded above by

$$\psi(x,y)\ \int_0^t \left(\int_0^s \left(\frac{d}{dt}(x-y)(u)\right)^2 du\right) ds \quad \text{where} \quad \psi(x,y) \text{ is given}$$

$$\psi(x,y) = \|f\|_2^2 \sup_{0 \leq s \leq 1}\left|\int_0^s g(u)\,ds(u)\right| + \|g\|_2^2 \sup_{0 \leq s \leq 1}\left|\int_0^s f(u)\,dy(u)\right|$$

which is finite for each pair (x,y) in a set of full Wiener measure. Thus it follows that the left hand side of 1.8 is zero forcing $x(s) = y(s)$, $0 \leq s \leq t$. Notice that the analysis of 1.8 holds for any linear combination of products from H with the difference of stochastic products being linearized in $(x-y)$ according to the binomial expansion of $x^n - y^n$. Hence every element of H satisfies 1.5; Theorem 1 is proved.

Note: Multiple Wiener integrals are elements of H . (7). For $i=1,\ldots,n$, choose elements $f_i(u)$ in $L_2[0,1]$ and define

$$g(t_1,\ldots,t_n) = \sum_{\pi \in G} f_{\pi_1}(t_1)\ldots f_{\pi_n}(t_n)$$

where G is the symmetric group of permutations of n letters and $t_1 > t_2 > \ldots t_n$. The multiple Wiener integral, $I_n(g,t)$, $0 \leq t \leq 1$ is given by

$$I_n(g,t) = \int_0^t \int_0^{t_1} \ldots \int_0^{t_{n-1}} g(t_1,\ldots,t_n)\,dx(t_1)\ldots dx(t_n) .$$

(c). Roots. Unfortunately, most elements of H are too powerful to satisfy 1.7a; this is corrected by taking roots.

Theorem 2. Let $I_n(g,t)$ represent a multiple Wiener integral of degree n and set $\alpha(t,x) = |I_n(g,t)|^{1/n}$. Then 1.2 admits a solution with distribution equivalent to ν .

Proof. This follows from an estimate of McKean and a result of Novikov. From McKean (10), one can find a $\delta > 0$ such that

$$\sup_{0 \leq t \leq 1} E_\nu(\exp(\delta\alpha(t,x)^2)) < \infty \tag{1.9}$$

From Novikov's result (11), 1.9 is sufficient to insure that $\alpha(t,x)$ satisfies 1.7(a).

Corollary: For any finite stochastic product,

$$\alpha(t,x) = \prod_{i=1}^{n} \int_0^t f_i(u)\,dx(u)$$

$|\alpha(t,x)|^{1/n}$ satisfies 1.9 .

Proof: This is a consequence of Itô's theorem and an application of Hölder's inequality since any finite product can be written as a linear combination of multiple Wiener integrals of lower degree.

Combining Theorems 1 and 2 is Theorem 3.

Theorem 3. Let $\alpha(t,x)$ be a finite polynomial of H with stochastic degree, n . Denote by $r(s,x)$ the functional given by

$$r(s,x) = \begin{cases} |\alpha(s,x)| & \text{if } |\alpha(s,x)| \leq 1 \\ |\alpha(s,x)|^{1/n} & \text{if } |\alpha(s,x)| > 1 \end{cases} \tag{1.10}$$

Then r satisfies 1.7(a) and determines a causally invertible transform. Consequently, the sde 1.2 has a strong solution which is path-wise unique among all solutions with distribution equivalent to ν.

Proof: For any triple (x,y,u) such that $\frac{d}{dt}(x-y)(u) = r(u,x) - r(u,y)$, it follows that $|r(u,x) - r(u,y)| \leq |\alpha(u,x) - \alpha(u,y)|$; the case where both $|\alpha(u,x)|$ and $|\alpha(u,y)|$ are greater than one is evaluated by the Mean Value Theorem. Now apply the proof of Theorem 1.

Note that the functionals of Theorem 3 can be used in the procedure outlined in Example (a).

(e) Simple stochastic integrals. Any simple stochastic integral, α, yields a causally invertible transform. The proof is a straightforward application of Proposition 1. Nevertheless, this fact, along with the examples above, suggests the following question.

Question. On $(C_1, \mathcal{B}_1, \nu)$ let $(\alpha, \mathcal{B}_t)$ be a square integrable martingale with continuous paths. Denote by $d(s,x)$ the square root of the derivative of the α-increasing process, i.e.,

$$d(s,x) = \sqrt{\frac{d}{dt}\langle\alpha\rangle(s)}$$

Does α determine a transform satisfying 1.5 if and only if d does?

(f) "Harmonic Functionals." The following result suggests that functionals, α, which induce causally invertible transforms have some harmonic quality to them. I haven't been able to develop this notion rigorously so the reader should be prepared to use his own intuition.

Theorem 4. Let α determine a solution to 1.2 with distribution equivalent to ν. Then F is causally invertible if, ν-a.s., for every t, $0 \leq t \leq 1$,

$$\int_0^t (E_\nu(\alpha_s | \mathcal{F}_s^F)(x) - \alpha(s,x))\,ds = 0 \tag{1.11}$$

Corollary. If $\alpha(s,x)$ has a finite second moment in time for each x in C_1, then 1.11 is necessary and sufficient.

Proof. Both Theorem 4 and its Corollary are proved in Allinger (1), and they are not difficult.

The harmonic aspect of α comes from 1.11 since it follows that

$$\alpha(s,x) = \int_{C_1} \alpha(s,y)\, dP_{s,x}(y) \qquad \lambda\times\nu\text{-a.s.}$$

where $P_{s,x}$ is a regular conditional probability with respect to $\mathcal{F}_s^F = \sigma(F(x)(u),\ 0\leq u\leq s)$.

Going another direction, observe that 1.11 provides a Markov time, σ , for measuring causal invertibility. Let

$$\sigma(x) \;=\; \inf\,(t>0\colon\ |\int_0^t (E_\nu(\alpha_s|\mathcal{F}_s^F)(x) - (s,x))\,ds| > 0)$$

Then σ is a Markov time with respect to the ν-completed system, $(\mathcal{B}_t^\nu)$, and F is causally invertible if

$$\nu(x\colon \sigma(x) = 1) = 1 .$$

This Markov time is developed further in another paper, "A Spectral Model for transformations on C_1". (2) .

References

1. Allinger, D. A Note on Strong Non-anticipating Solutions for Stochastic Differential Equations: When is Path-wise uniqueness necessary? Proceedings of the Martingale Conference, Cleveland State University, July, 1981

2. _______. A Spectral Model for Transformations on C . (In preparation).

3. Beneš, V.E. Nonexistence of Strong Nonanticipating Solutions to Stochastic DEs: Implications for Functional DEs, Filtering, and Control. 1977, Stochastic Processes and Their Appl., 5, pp. 243-263.

4. ____. On Kailath's innovations conjecture. 1976. B.S.T.J. 55, pp. 981-1001.

5. Dudley, R.M. Wiener Functionals as Ito Integrals. 1977, Ann. of Prob., 5, pp. 140-141.

6. Ershov, M.P. Extension of Measures and Stochastic Equations. 1974, Theory of Prob. and Its Appl., 19, pp. 431-444.

7. Ito, K. Multiple Wiener Integral. 1951, Jour. Math. Soc. Japan, 3, pp. 157-169.

8. Krylov, N.V., and Veretennikov, A. Ju. On Explicit Formulas

for Solutions of Stochastic Equations. 1976, Math. USSR Sbornik, 29 pp. 239-256.

9. Lipster, R.S. and Shiryayev, A.N. Statistics of Random Processes, I. Springer-Verlag, New York, 1977.

10. McKean, H.P., Wiener's Theory of Nonlinear Noise, 1973, Stochastic Differential Equations, SIAM-AMS Proceedings, Amer. Math. Soc. pp. 191-209.

11. Novikov, A.A., On an identity for stochastic integrals, 1972, Teoria Verojatn. i Primenen, 27, pp. 761-765.

12. Sion, M. On Analytic Sets in Topological Spaces, 1960, Trans. Amer. Math. Soc., 96, pp. 341-354.

13. Sion, M. On Uniformization of Sets in Topological Spaces. 1960, Trans. Amer. Math. Soc., 96, pp. 237-245.

14. Zvonkin, A.K. A Transformation of the Phase Space of a Diffusion Process that Removes the Drift. 1974, Math. USSR Sbornik, 22 pp. 129-149.

SPECTRAL ANALYSIS OF NON LINEAR SEMI-MARKOV PROCESSES

Rodolfo De Dominicis

Istituto di Matematica - Facoltà di Economia and
Istituto Elettrotecnico - Facoltà di Ingegneria -
Università di Napoli - Italy

ABSTRACT

This note is primarily concerned with the spectral analysis of a time series associated to a homogeneous semi-markov process having a finite number of states . A statistical test for verifying the hypothesis on its semi-markovian nature will also be provided and the corresponding asymptotic power determined .

1. PRELIMINARIES

Our aim is concerned with the study of a two-states random system which evolves according to a homogeneous semi-markov process (SMP) . We start by introducing the essential notations.

Let (Z_t , $t \geq 0$) be a homogeneous semi-markov process having only two states . Let such a process defined by matrices $F = [F_{ij}(t)] = [F_j(t)]$, j= 1,2 (matrix of the distribution functions (d.f.) for waiting times at state j) and $P = [P_{kj}]$ (matrix of the transition probability for the inbedded Markov chain (MC)) .

According to notation by Cox-Lewis(1966) , we expressely note that matrix P has a form

$$P = \begin{bmatrix} \alpha_1 & 1-\alpha_1 \\ 1-\alpha_2 & \alpha_2 \end{bmatrix}$$

R. S. Bucy and J. M. F. Moura (eds.), Nonlinear Stochastic Problems, 407–415.

where α_1 and α_2 belongs to the interval $(0,1)$.

Denoting by $\pi = (\pi_1, \pi_2)$ the stationary probabilities vector of the inbedded MC, i.e. the unique solution of the system $\pi = \pi P$, and assuming $\beta = \alpha_1 + \alpha_2 - 1$, then $P_{kj} = \pi_j (1 - \beta) + \beta \delta_{kj}$, where $|\beta| < 1$, $0 < \pi_j < 1$ $(j = 1,2)$ and δ_{kj} is the Kronecker symbol.

Finally, let (X_r), $r = \pm 1, \pm 2, \ldots$, be the time series formed by a sequence of waits between consecutive events of PSM Z_t, and set $x_0, \ldots, x_{n-1}$ the observed values from the sequence (X_r).

2. SPECTRAL ANALYSIS OF TIME SERIES X_r

Let us start by determining the expression of spectral density $f(\lambda)$ of the process (X_r).

To this purpose, we assume the moments of any order of $F_i(t)$ $(i = 1,2)$ to exist and, under this assumption, we prove the following lemma:

Lemma 1 : The time series (X_r) is a strictly stationary process and the inequality

$$\sum_{-\infty}^{+\infty} | r_j \, S^{(k)}(0, r_2, \ldots, r_k) | < +\infty \qquad (1)$$

is valid for each $j = 1, \ldots, k-1$; $k = 2,3,\ldots$ and where

$S^{(k)}(r_1, \ldots, r_k) = S^{(k)}(\tau + r_1, \ldots, \tau + r_k)$, $\tau > 0$,

is the k-order cumulant of the process (X_r).

Proof . It is derived from results by Leonov-Shiyryaev (1959), which allow the cumulant of k-order to be expressed by means of the moments of process (X_r), to be, in turn, obtained by the characteristic function of vector $(X_{r_1}, \ldots, X_{r_k})$.

It can be easily proved that inequality (1) coincides with the summation of series $\beta + 2\beta^2 + \ldots + k\beta^k + \ldots$ which is convergent , as $|\beta| < 1$.

Inequality (1) on the basis of data found by Brillinger-Rosenblatt (1967) , demonstrates a spectral density of k - 1 order to exist as follows :

$$f^{(k-1)}(\lambda_1,\ldots,\lambda_{k-1}) = (2\pi)^{-k+1}\sum_{-\infty}^{+\infty} S^{(k)} \cdot$$
$$\cdot \exp\left(-i\sum_{j=1}^{k-1} r_j \lambda_j\right) \tag{2}$$

Furthermore , the spectral density $f^{(k-1)}$ is bounded and uniformly continuous within the domain defined by equation $\lambda_1 + \ldots\ldots + \lambda_k = 0$.

Let us now introduce the following additional notations :

$$\mu_i^{(k)} = \int_0^\infty x^k \, dF_i(x) \quad , \quad i = 1,2$$

$$\sigma_i^2 = \int_0^\infty (x-\mu)^2 \, dF_i(x) \; , \quad i = 1,2$$

$$\mu = \sum \pi_i \mu_i = E(X_r) \quad , \quad \gamma_o = \sum \pi_i \int_0^\infty x^2 \, dF_i(x) - \mu^2$$

$$p^2 = (\pi_2 - \pi_1)(\pi_1 \pi_2)^{-1} \quad , \quad \varepsilon^2 = \pi_1 \pi_2 (\mu_1 - \mu_2)^2 \; .$$

$$g = \pi_1 \pi_2 \left(\mu_1^{(2)} - \mu_2^{(2)} - 2\mu(\mu_1 - \mu_2)\right)$$

$$g_1 = g^2 p^{-2} \quad , \quad c = (\pi_1 \mu_1^{(2)} + \pi_2 \mu_2^{(2)} - 2\mu(\pi_2 \mu_1 + \pi_1 \mu_2) + \mu^2)$$

$$K_4 = \sum \pi_i \int_0^\infty (x - \mu)^4 \, dF_i(x) - 3\gamma_o^2$$

$$c_1 = \pi_1 \pi_2 (\mu_1 - \mu_2)\left(\mu_1^{(3)} - \mu_2^{(3)} - 3\mu(\mu_1^{(2)} - \mu_2^{(2)} + 3\mu^2(\mu_1 - \mu_2)\right) \; ; \quad \sigma^2 = \gamma_o - \varepsilon^2 \; .$$

It can be easily proved that function $f(\lambda)$ has the following expression :

$$f(\lambda) = (2\pi)^{-1} \sum_{-\infty}^{+\infty} \left| \varepsilon^2 \beta^{|s|} + \delta_{os} \sigma^2 \right| \tag{3}$$
$$= \sigma^2 (2\pi)^{-1} + \varepsilon^2 (1-\beta^2)(2\pi)^{-1} (1 - \beta\exp(i\lambda))^{-2}$$

We are now able to determine the spectral density of the 3rd order of process (X_r).

To this purpose, the following theorem holds :

<u>Theorem 1</u> : The spectral density of the 3rd order $f^{(3)}(\lambda_1, \lambda_2, \lambda_3)$ of the process (X_r) is given by :

$$f^{(3)}(\lambda_1,\lambda_2,\lambda_3) = 2(2\pi)^{-3}\ \mathrm{Re}\Big(\sum_i \sum_j \sum_k p^2 \varepsilon^4 \cdot$$
$$\cdot H(z_i,z_j)\,\Phi(\bar z_k)\ h(z_i z_j) + g\varepsilon^2\ h(z_i z_j)(H(z_i,z_j) +$$
$$+ \Phi(\bar z_k)) - 2\varepsilon^4\ k(z_i z_j)\ (H(z_i,z_j) + 1)(\Phi(\bar z_k) + 1)) +$$
$$+ 2\varepsilon^4\ k(z_1 z_2)\ h(z_1 z_2 z_3) - g\varepsilon^2\ h(z_1 z_2)\ h(z_1 z_2 z_3) +$$
$$+ (c - 2\varepsilon^2 - \gamma_o)\cdot\varepsilon^2\Big[\sum_i \sum_j h(z_i)h(\bar z_j) + h(z_3)\ h(z_1 z_2 z_3))\Big] +$$
$$+ (c_1 - 3\gamma_o)(\sum_i h(z_i) + h(z_1 z_2 z_3)) + g_1 \sum_i \sum_j h(z_i z_j) +$$
$$+ K_4\Big),$$

where $H(z_i,z_j) = h(z_i) + h(z_j)$; $z_j = \exp(-i\lambda_j)$

$h(x) = \beta x\,(1-\beta x)^{-1}$; $k(x) = h(\beta x)$

$\Phi(\bar z_j) = h(\bar z_j) + h(z_1 z_2 z_3)$.

<u>Proof</u>. It follows from (2) and from the expression of the 4th cumulant for the process (X_r), as hereafter reported :

$$S^{(4)}(s,t) = \begin{cases} p^2\gamma_s\gamma_t - \gamma_t^2 - \gamma_{t+s}\ \gamma_{t-s} & \text{if } s\neq t\neq 0 \\ r^2\gamma_{t+s} - 2\gamma_{t+s}^2 & \text{if } s=t=0,\ s\neq t \\ p^2\gamma_t^2 + \gamma_{2t}\,(\gamma_1^2 + \gamma_2^2 - 2\gamma_o) & \text{if } s=t\neq 0 \\ K_4 & \text{if } s=t=0 \end{cases}$$

where $r = \ell_o - 2\mu$, $\ell_s = \ell'_s(\mu_1 - \mu_2)^{-1}$

$\ell'_s = (\sigma_1^2 - \sigma_2^2)\delta_{os} + (\mu_1 - \mu_2)(\mu + \beta^s(\pi_2\mu_1 + \pi_1\mu_2))$

$\gamma_s = b_s - \mu^2$,

$$b_s = E(X_r X_{r+s}) = \mu^2 + \varepsilon^2 \beta^s + (\pi_1 \sigma_1^2 + \pi_2 \sigma_2^2)\delta_{os}$$

and s,t are positive integers . An estimator for γ_t is to be found in the Appendix .

Let us now consider , as an estimate of integral

$$\int_{-\pi}^{\pi} \psi(\lambda)\, f(\lambda)\, d\lambda$$

with $\psi(\lambda)$ being a weight function , the random variable (r.v.)

$$\int_{-\pi}^{\pi} \psi(\lambda)\, I_n(\lambda)\, d\lambda$$

where $I_n(\lambda) = (2n\pi)^{-1} \left| \sum_{\tau} (X_r - \bar{x} \exp(-i\lambda r)) \right|^2$ is the periodogram and $\bar{x} = (x_o + .. + x_{n-1})\, n^{-1}$.

On the basis of Lemma 1 it may be shown that :

Theorem 2 : Let $\psi_1(\lambda)$,.... , $\psi_m(\lambda)$ be given weight functions , with bounded variation , such that the following conditions

$$|\psi_j(\lambda)| < \infty \quad , \quad \text{Var}\, \psi_j(\lambda) < \infty , \quad j = 1,..,m$$

$$\int_{-\pi}^{\pi} |\psi_j(\lambda)|\, d\lambda < \infty$$

hold ; the m-dimensional random vector

$$\left(n^{\frac{1}{2}} \int_{-\pi}^{\pi} \psi_1(\lambda)(I_n(\lambda) - f(\lambda)\, d\lambda, ..., ... \right)$$

has , for $n \to \infty$, an m-dimensional normal distribution with zero mean and covariance matrix , whose j-th element is given by

$$4\pi \int_{-\pi}^{\pi} f^2(\lambda)\psi_j(\lambda)\psi_k(\lambda)\, d\lambda \; + \tag{5}$$
$$+ \; 2\pi \iint_{-\pi}^{\pi} f^{(3)}(\lambda_1, -\lambda_1, \lambda_2)\, \psi_j(\lambda_1)\, \psi_k(\lambda_2)\, d\lambda_1\, d\lambda_2$$

where j = 1,...,m and k = 1 ,.. , m and $f^{(3)}$ has (4) as an expression .

Proof. It easily follows from the main theorem by Bentkus (1972) .

We are now able to prove the following theorem :

Lemma 2 : The weight functions

$$\psi_j(\lambda) = \beta^3 |1 - \beta z|^2\; 2\, \text{Re}\,(z - \beta)\, z^{j+2} \tag{6}$$

where $j = 1,\ldots,m$ and $z = \exp(-i\lambda)$, cause the second term in (5) to vanish, and the covariance matrix of the random vector

$$Y = (Y_j) = \left(n^{\frac{1}{2}} \int_{-\pi}^{\pi} \psi_j(\lambda) \, (I_n(\lambda) - f(\lambda)) \, d\lambda \right),$$

where $j = 1,\ldots,m$, is the following

$$W = (W_{ij}) = 4\pi \int_{-\pi}^{\pi} f^2(\lambda) \psi_j(\lambda) \psi_k(\lambda) \, d\lambda \tag{7}$$

as $n \to \infty$.

Proof. The first part of the theorem can be proved by directly substituting (6) for (5). For the sake of brevity the second part is omitted.

We expressely note that matrix W is of the Toeplitz type, i.e. is such that its element W_{ij} is a function of $i - j$.

Matrix W^{-1} can be obtained by numerical methods reported by Akaike (1973) or Wold (1949).

Let us consider the quadratic form $Y'W^{-1}Y$, associated to matrix W.

The hypothesis H_o that process Z_t is a semi-Markov process can be proved by the customary method of using the statistics $Z^2 = Y'W^{-1}Y$. If the hypothesis H_o is valid for $n \to \infty$, statistics Z^2 will be distributed as an χ^2_m. This enables us to consider a critical region of the form $Z^2 > s_o$, corresponding to a given significance level α_o, s_o being defined by

$$\int_s^{\infty} \chi^2_m(x) \, dx = \alpha_o .$$

In order to evaluate the asymptotic power of our test, let us introduce the alternative hypothesis H_1, according to which the spectral density of process (X_r) is the following:

$$f_n(\lambda) = f(\lambda) + n^{-\frac{1}{2}} a(\lambda)$$

and the spectral density of the third order is

$$f^{(3)}(\lambda_1, -\lambda_1, \lambda_2) = g_n(\lambda_1, \lambda_2) - n^{-\frac{1}{2}} b(\lambda_1, \lambda_2)$$

where $a(\lambda)$ is a function integrable on $(-\pi, \pi)$ and $b(\lambda_1, \lambda_2)$

is a function integrable on the square $-\pi \leq \lambda_1, \lambda_2 \leq \pi$.

On the basis of Lemma 2 and Theorem 2 , the following theorem can be proved :

Theorem 3 : The m-dimensional random vector

$$Y_1 = (Y_{1j}) = (n^{\frac{1}{2}} \int_{-\pi}^{\pi} \psi_j(\lambda) (I_n(\lambda) - f_n(\lambda)) d\lambda)$$

where j = 1 ,... m , has a normal distribution whose mean vector is given by

$$L = (L_j) = (\int_{-\pi}^{\pi} \psi_j(\lambda) a(\lambda) d\lambda) , j = 1,..., m$$

and a covariance matrix W given by (7) , where the weight functions have a form like (6) .

Theorem 3 leads to the following corollary :

Corollary : Statistics $Z_1^2 = Y' W^{-1} Y_1$ has , for $n \to \infty$, a noncentral distribution χ^2 , with m degree of freedom and centrality parameter $L' W^{-1} L$.

The asymptotic power of the test to be used in choosing between the two hypotheses H_o and H_1 is :

$$\int_{s_o}^{\infty} \chi_m^2 (x / L' W^{-1} L) d x$$

where $\chi_m^2 (x / L' W^{-1} L)$ is the density of non central distribution χ^2 with m degrees of freedom and centrality parameter $L' W^{-1} L$.

3. APPENDIX

In the expression of the joint cumulant $S^{(4)}$ it is necessary to estimate the covariance function γ_t of the vector of observed values of the r.v. $(X_1, \ldots, X_n)$.

As an estimator for γ_t , let us consider a statistics of the form C_t , which is an observed value of the r.v.

$$C_t = n^{-1} \sum_{1}^{n-t} (X_s - \bar{\mu}) (X_{s+t} - \bar{\mu}) , t = 0 ,1,.., n-1$$

$$\bar{\mu} = (X_1 + \ldots + X_n) n^{-1} .$$

For the case in question , the following theorem is valid :

<u>Theorem A</u> : Under the above mentioned hypotheses , the m-dimensional random vector

$$(n^{\frac{1}{2}} (C_o - \gamma_o) , n^{\frac{1}{2}} (C_1 - \gamma_1), \ldots\ldots , n^{\frac{1}{2}}(C_{m-1} - \gamma_{m-1}))$$

has , for $n \to \infty$, a normal distribution with zero mean and covariance matrix $R = (r_{tu})$, $t,u = 0 , 1 , \ldots , m-1$

where $n \operatorname{Cov}(C_t , C_{t+u}) \cong r_{tu} =$

$$= (2 \gamma_o + \varepsilon^2 u - \varepsilon^2 (1 - 3 \varepsilon^2)(1 - \beta^2)^{-1}) \gamma_u +$$
$$+ p^2(u + (1 + \beta)(1 - \beta)^{-1}) \gamma_u \gamma_t + ((2 - 3\beta^2)(1 - \beta^2)^{-1} +$$
$$- 2t - u) \gamma_t \gamma_{t+u} + (\pi_2 - \pi_1) (\sigma_1^2 - \sigma_2^2) \gamma_{t+u} +$$
$$- (\sigma_1^2 + \sigma_2^2 - \gamma_o) \gamma_{2t + u} - (1 - \beta^2)^{-1} \gamma_{t+u} \gamma_{t+2u}$$

$t , u = 1 , \ldots. , m-1$, and

$$n D(C_o) \cong r_{oo} = E(X_\tau - \mu)^4 - \gamma_o^2 + 2 r^2 \varepsilon^2 \beta (1 - \beta)^{-1}$$
$$n D(C_t) \cong r_{to} = \gamma_o^2 + 2 \varepsilon^4 \beta^2 (1 - \beta^2)^{-1} +$$
$$+ ((1 - 3\beta^2)(1 - \beta^2)^{-1} - 2t) \gamma_t^2 +$$
$$+ (r^2 + 2p^2 \varepsilon^2 \beta (1 - \beta)^{-1} \gamma_t +$$
$$+ 2 (\sigma_1^2 + \sigma_2^2 - \gamma_o) \gamma_{2t} , \quad t = 1 , \ldots , m-1 ,$$
$$n \operatorname{Cov}(C_o ,C_u) \cong r_{ou} = 2 \theta_1 - \theta_2 ((\mu_1 - \mu_2) - 3 \mu(\ell_o +$$
$$- \mu)) \gamma_u - 2u \gamma_o \gamma_u + (u - 1) ((p^2 +$$
$$+ 2)\varepsilon^2 + \sigma_1^2 + \sigma_2^2) \gamma_u + 2\beta(1 - \beta)^{-1} (p^2 \varepsilon^2 +$$
$$+ (\pi_2 - \pi_1) (\sigma_1^2 - \sigma_2^2)) \gamma_u , \quad u = 1, \ldots ,m-1$$

$$\theta_j = \int_0^\infty x^3 f_j(x) \, dx , \quad j = 1 , 2 .$$

<u>Proof</u> . For proving theorem A , let us use the asymptotic Theorem 1.1 as stated by Bentkus (1972) , with respect to general processes (not necessary linear ones) to be defined stationary in a strict sense . The conditions of the theorem can be very easily shown to be valid . In particular , it should be noted that the random process (X_r) is not linear . Therefore , the formulas for the variance of estimator C_t , as advanced by Bartlett (1946)

may not be considered as valid in the present case . Nevertheless for t approaching infinity in expression (8) , it may be easily understood that (8) coincides with the formula for the variance of C_t, as provided by Cox-Lewis (1967) and computed employing the Bartlett's method .

4. CONCLUDING REMARK

In present paper the proofs of theorems and lemmas are confined to the essential for the sake of brevity ; complete proofs are available upon request from the author .

5. REFERENCES

1. Bentkus , R. : 1972 , Liet. Matem. Rink. 12 , pp. 319-328
2. Bloomfield , P. : Fourier analysis of time series , Wiley , New York , 1976
3. Brillinger , D.R. : 1969 , Biometrika 56 , pp. 375-390
4. Brillinger , D.R. , Rosenblatt,M. : Asymptotic theory of estimates of k-th order spectra , in Advenced Seminar on Spectral Analysis of Time Series , (B.Harris Editor) , Academic Press , New York , 1967 , pp. 153-188
5. Akaike , H : 1973 , SIAM J. Appl. Math. 24 , pp. 234-241
6. Cox , D., Lewis P. : Statistical analysis of series of events, Methuen , London , 1969
7. Leonov , V.P. , Shiryaev A.N. : 1959 , Theory Prob.Appl. 4 , pp. 319 - 328
8. Wold , H. : 1949 , J.Royal Stat.Soc. 11 , pp. 297-305

DIFFERENTIAL CALCULUS FOR GAUSSIAN RANDOM MEASURES

Ognian Enchev

Centre of Mathematics, P.O.Box 325,
7000 Rousse, Bulgaria

The goal of this exposition is to explain the link between the stochastic integrals and some notions from Quantum theory like second quantization and canonical free fields. Basing on this concept we come to a new comprehension of the random processes. A rule for differential calculus is also obtained.

1. PRELIMINARIES: CANONICAL FREE FIELDS, WICK POWERS, DECOMPOSITION OF (L^2)

Let $\mathcal{H}$ be a real Hilbert space and let H be its complexification. Everywhere $\circledast$ denotes the symmetrized tensor product. Consider the symmetric Fock space

$$\mathcal{F}(H) = \bigoplus_{n=0}^{\infty} H^n \tag{1}$$

where $H^n = H \circledast \dots \circledast H$ (n times), $n \geq 1$ and $H^0 = \mathbb{C}$. By F_0 we denote the family of all finitely-partial vectors in $\mathcal{F}(H)$, i.e. F_0 is the noncompleted direct product in (1). The vector $\Omega_0 = \langle 1,0,0,\dots \rangle \in F_0$ is called vacuum. For any $f \in H$ we define the birth and annihilation operators $(A\uparrow f)$ and $(A\downarrow f)$ as

$$(A\uparrow f) = \bigoplus_{n=0}^{\infty} (a_n \uparrow f) \qquad (A\downarrow f) = \bigoplus_{n=0}^{\infty} (a_n \downarrow f)$$

where $(a_n\uparrow f): H^n \to H^{n+1}$ and $(a_n\downarrow f): H^n \to H^{n-1}$ (=0 for n=1) are bounded operators such that for $\Psi \in H^n$, $\Psi = h_1 \circledast \dots \circledast h_n$, $h_i \in H$, $1 \leq i \leq n$

R. S. Bucy and J. M. F. Moura (eds.), Nonlinear Stochastic Problems, 417–424.

$$(a_n\uparrow f)(\Psi) = \sqrt{n+1}\,(f\otimes h_1\otimes \dots \otimes h_n) \in H^{n+1}$$

$$(a_n\downarrow f)(\Psi) = n^{-\frac{1}{2}}\sum_{i=1}^{n} (f,h_i)(h_1\otimes\dots\otimes \overset{\downarrow}{h_i} o \dots \otimes h_n) \in H^{n-1}$$

where the symbol $\downarrow$ means that the corresponding term is missing. For $f\in\mathcal{H}$ (not to H!) we put $\varphi(f) = (\sqrt{2})^{-1}[(A\uparrow f) + (A\downarrow f)]$. The map $f \to \varphi(f)$ is what is called canonical free field on $\mathcal{H}$. The most important properties of the canonical free fields are given in [2, Theorem X.43]. Let $\mathcal{U}$ be the family of all projectors of the operators $\{\varphi(f) \mid f\in\mathcal{H}\}$. Then $\mathcal{A} = \mathcal{U}''$ (the bicommutant of $\mathcal{U}$) is a commutative Von Neumann algebra. Let Δ be the family of all complex homomorphisms on $\mathcal{A}$ provided with the Gelfands topology under which Δ is a compact Hausdorff space. Then the linear map $\omega:\mathcal{A} \to \mathbb{C}$ given by $\omega(A) = (\Omega_o, A\Omega_o)$ determines according to [3, Theorem 11.32] a positive regular Borel (=Baire) measure P on Δ such that: $P(\Delta)=1$ and for any $A\in\mathcal{A}$, $\omega(A) = \int_\Delta \hat{A}(h)P(dh)$ where $\hat{}$ denotes the Gelfands transform. The Hilbert spaces $\mathcal{F}(H)$ and $\Gamma(\mathcal{H}) = L^2(\Delta,dP)$ are isomorphically equivalent under the unitary operator $\mathcal{T}: \mathcal{F}(H) \to \Gamma(\mathcal{H})$ such that $\mathcal{T}(A\Omega_o) = \hat{A} \in C(\Delta)\subset\Gamma(\mathcal{H})$ for any $A\in\mathcal{A}$. The map $\mathcal{H} \ni f \to \phi(f) = \mathcal{T}(\varphi(f)\Omega_o) \in \Gamma(\mathcal{H})$ is a Gaussian process on $\mathcal{H}$ (ϕ is unique to within isomorphism between the probability spaces). The mathematical expectation of the random variables in $\Gamma(\mathcal{H})$ we denote by $\langle\cdot\rangle$ and for $f_i\in\mathcal{H}$, $1 \le i \le n$, $:\phi(f_1)\dots\phi(f_n):$ denotes the Wick power of the random variables $\phi(f_1), \dots \phi(f_n)$.

<u>Proposition 1.1.</u> For any $f_i\in\mathcal{H}$, $0 \le i \le n$

(a) $:\phi(f_1)\dots\phi(f_n): = \frac{\sqrt{n!}}{(\sqrt{2})^n}\mathcal{T}(f_1\otimes \dots \otimes f_n)$;

(b) $\phi(f_o):\phi(f_1)\dots\phi(f_n): =$

$$=:\phi(f_o)\phi(f_1)\dots\phi(f_n): + \sum_{i=1}^{n} \langle\phi(f_o)\phi(f_i)\rangle :\phi(f_1)\dots\overset{\downarrow}{\phi}(f_i)\dots\phi(f_n):\ ;$$

(c) $\mathcal{T}^{-1}(\phi(f_o)\mathcal{T}(f_1\otimes\dots\otimes f_n)) = \varphi(f_o)[f_1\otimes\dots\otimes f_n]$. #

The statement (c) follows from (a) and (b) and it means that the multiplication in $\Gamma(\mathcal{H})$ with $\phi(f_o)$ is equivalent to the action of $\varphi(f_o)$ on $\mathcal{F}(H)$. By ${}^o\Gamma_n(\mathcal{H})$ usually is denoted the complex Hilbert subspace of $\Gamma(\mathcal{H})$ generated by the random variables $\{:\phi(f_1)\dots\phi(f_n): \mid f_i\in\mathcal{H}, 1\le i\le n\}$ ($\Gamma_o(\mathcal{H})=\mathbb{C}$ for n=0). Then we have that $\Gamma_n(\mathcal{H})$ and H^n are unitary equivalent and it follows from (1) that $\Gamma(\mathcal{H})$ is direct sum of the spaces $\Gamma_n(\mathcal{H})$.

2. BAXTER PROCESSES AND MULTIPLE INTEGRALS

Let us fix once and for all a locally compact Hausdorff space X which satisfies the second axiom of countability and a positive Borel measure q on X. X is the basic measure space on which we shall define the random measures. Therefore we treat X as a "time-parameter space" which role usually is played by the real line $\mathbb{R}$. However the special structure of $\mathbb{R}$ is not obligatory for the stochastic integration (see [6] and [7]). It follows that X is separable and metrizable and any Borel measure on X is regular (Borel measure on a locally compact space is called any signed measure ν on the Borel σ-field such that $|\nu|(K) < +\infty$ for any compact K). Here we denote by $L^2_0(X,dq)$ ($L^2_0(X,dq;\mathbb{R})$) the family of all complex (real) finite square integrable functions on X (finite function on X is those vanishing q-a.s. outside some compact set). Consider the diagonal $D = \{\langle x,x\rangle \in X^2 \mid x \in X\}$ and define for any Borel $B \subset X^2$ $q_*(B) = q(\pi(B \cap D))$ where π is the standard projection from D onto X. We define also the following Borel measure on X^2 : $q^2_* = q_* + q^2$ where $q^2 = q \otimes q$. Any signed Borel measure m on X^2 can be decomposed as $m = m_D + m_{\bar{D}}$ where $m_D(B) = m(B \cap D)$ and $m_{\bar{D}}(B) = m(B \cap \bar{D})$. Both q_* and m_D are concentrated on D and π transforms q_* into q and m_D into some Borel measure on X, which we denote by πm_D.

Definition 2.1. Baxter covariance measure on X^2 is called any signed Borel measure m on X^2 such that: 1) $|m_D| \ll q_*$ and $|m_{\bar{D}}| \ll q^2$; 2) for any compact $K \subset X$ there exist constants A_K, $B_K \geq 0$ such that $|d(\pi m_D)(x)/dq| \leq A_K$ for q-a.e. $x \in K$ and $|dm_{\bar{D}}(x,y)/dq^2| \leq B_K$ for q -a.e. $(x,y) \in K^2$; 3) for any $f \in C_0(X)$ (=the family of all continuous functions on X with compact support) we have

$$\int_{X^2} f(x)f(y)m(dx,dy) \geq 0 \quad . \qquad \#$$

Let m be a Baxter covariance measure on X^2. Then m determines a Gaussian random process as follows. Define for any $f,g \in L^2_0(X,dq;\mathbb{R})$ the sesquilinear form

$$c(f,g) = 2\int_{X^2} f(x)g(y)m(dx,dy) \quad .$$

Let $\mathcal{H}$ be the Hilbert space completion of $L^2_0(X,dq;\mathbb{R})/\{f \in L^2_0(X,dq;\mathbb{R}) \mid c(f,f)=0\}$ in the norm induced by the sesquilinear form c. The inner product in $\mathcal{H}$ is the unique natural extension of $c(.,.)$. As in the Section 1. we denote by H the complexification of $\mathcal{H}$. Let I be the canonical inclusion map of $L^2_0(X,dq)$ into H. Consider the Gaussian process $\phi : \mathcal{H} \to \Gamma(\mathcal{H})$ constructed in the previous section. It determines a Gaussian random measure $b(.)$ on X given by $b(A)=\phi(I(1_A))$ for any Borel set A such that $A\subset K\subset X$ for some compact K. The random measure $b(.)$ is called Baxter Gaussian Random Measure (BGRM). The normal random measure considered in [7] is a special case of BGRM. For $f\in L_0^2(X,dq)$ the following notations are equivalent to $\phi(I(f))\in\Gamma(\mathcal{H})$: $b(f)$, $\int f db$, $\int_X f(t)b(dt)$. The inclusion I naturally generates mappings $I^n:L_0^2(X^n,dq^n)\to H^n, n\geq 1$ $(I^1=I)$ which mappings can be characterized as multiple integrals with respect to the random measure $b(.)$. More precisely we define for any $f\in L_0^2(X^n,dq^n)$, $n\geq 1$ the integral

$$:b^n(f): = \int_{X^n} f(x_1,\dots,x_n):b(dx_1)\dots b(dx_n):$$
$$= \frac{\sqrt{n!}}{(\sqrt{2})^n}\mathcal{T}(I^n(f))$$

(for n=1 $:b^1:(.) = b(.)$). If the function f is such that $f(x_1,\dots,x_n)=f_1(x_1)\dots f_n(x_n)$, $f_i\in L_0^2(X,dq;\mathbb{R}), 1\leq i \leq n$, then we have $:b^n:(f) = :\phi(If_1)\dots\phi(If_n):$.

<u>Theorem 2.1.</u> Let K X be a compact and let $f,g\in L_0^2(K^n, dq^n)$. Then

$$|\langle :b^n:(f):b^n:(g)\rangle| \leq 2^n[A_K + B_K.q(K)]^n \|f\|\|g\| . \quad \#$$

Everywhere in the sequel the integration of a vector-valued function is considered in the sense [3,Definition 3.26]. Let Φ be a function of the variables $\langle x_1,\dots,x_n\rangle$. The meaning of the symbol $/x_1,\dots,x_p/\Phi$ $1\leq p\leq n$ is that the variables $\langle x_1,\dots,x_p\rangle$ are fixed and $/x_1,\dots,x_p/\Phi$ is viewed as a function of the variables $\langle x_{p+1},\dots,x_n\rangle$.

<u>Theorem 2.2.</u> (Fubini theorem) For any function $F(t,x_1,\dots,x_n)\in L_0^2(X^{n+1},dq^{n+1})$, $n\geq 1$ we have

$$:b^n:(\int_X F(t,x_1,\dots,x_n)q(dt))=\int_X :b^n:(/t/F)q(dt) \quad . \quad \#$$

3. STOCHASTIC INTEGRATION (QUANTIZATION)

In this section we consider the random functions as elements of $L^2(X,dq;\Gamma(\mathcal{H}))$, which is the Hilbert space of all measurable functions $f:X\to\Gamma(\mathcal{H})$ such that $\int_X\|f(t)\|^2 q(dt)<+\infty$ (about the measurability of a vector-valued function we reffer to [1,the addition to IV.5]). The inner product in $L^2(X,dq;\Gamma(\mathcal{H}))$ is given by $(f,g) = \int_X\langle\overline{f(t)}g(t)\rangle q(dt)$. Consider the following random function

(2) $X\ni t\to \mathfrak{Z}(t) = f(t):b(g_1)\dots b(g_n):\in\Gamma_n(\mathcal{H})$

where $f, g_i \in L^2_0(X, dq; \mathbb{R})$. The only reasonable definition of the stochastic integral $\int_X \xi(t) b(dt)$ is

$$\int_X \xi(t) b(dt) = :b(f) b(g_1) \ldots b(g_n): + \\ + \sum_{i=1}^n \langle b(f) b(g_i) \rangle :b(g_1) \ldots \overset{\downarrow}{b(g_i)} \ldots b(g_n): .$$

Under the transformation $\mathcal{T}^{-1}$ the mapping (2) is equivalent to the following mapping from X to H

$$X \ni t \to \tilde{\xi}(t) = f(t) \left[\frac{\sqrt{n!}}{(\sqrt{2})^n} Ig_1 \otimes \ldots \otimes Ig_n \right] \in H^n .$$

It is obvious that

$$\mathcal{T}^{-1} \left[\int_X \xi(t) b(dt) \right] = \varphi(f) \left[\frac{\sqrt{n!}}{(\sqrt{2})^n} Ig_1 \otimes \ldots \otimes Ig_n \right] .$$

The right hand side of the last equality we denote by $Q(\tilde{\xi})$. The expression $Q(\tilde{\xi})$ we call quantization of the map $\tilde{\xi}$. These considerations show that instead of developing the theory of the stochastic integrals $\int_X \xi(t) . b(dt)$ it can be equivalently developed the theory of quantizations $Q(\tilde{\xi})$. For any $f_0 \in \mathcal{H}$ and $\Psi_0 \in F_0$ we define $(\Psi_0 * df_0)$ to be the following bilinear form on $\mathcal{F}(H) \times H$

$$(\Psi_0 * df_0) \langle \Psi, f \rangle = (\Psi, \Psi_0)_{\mathcal{F}(H)} (f, f_0)_H ,$$

$\Psi \in \mathcal{F}(H)$, $f \in H$. Then we define $\mathcal{L}$ to be the family of all finite linear combinations of such forms. The quantization Q is defined as the linear mapping $Q: \mathcal{L} \to \mathcal{F}(H)$ given by $Q(\Psi_0 * df_0) = \varphi(f_0)(\Psi_0)$, $\Psi_0 \in F_0$, $f_0 \in \mathcal{H}$, which equality is extended by linearity to the whole $\mathcal{L}$. For $l_1, l_2 \in \mathcal{L}$ we define the sesquilinear form

$$c^*(l_1, l_2) = (Q(l_1), Q(l_2))_{\mathcal{F}(H)}$$

The Hilbert space completion of $\mathcal{L} / \{ f \in \mathcal{L} | c^*(l,l) = 0 \}$ in the norm induced by c* we denote by $\mathcal{F}(H)*dH$. The elements of $\mathcal{F}(H)*dH$ are referred to as quantified processes and there exists unique natural extension of the quantization Q to the whole $\mathcal{F}(H)*dH$. Our program in this section is to construct a linear topological space QP with topology τ on it such that: 1) there exists a linear map $J^*: QP \to \mathcal{F}(H)*dH$ which is injective and continuous (i.e. the quantization Q can be properly defined for the elements of QP); 2) there exists a many-to-one linear map $\theta: QP \to L^2(X, dq; \Gamma(\mathcal{H}))$ which is continuous (i.e. the elements of QP can reasonable be identified with random functions). For any $\Psi \in QP$ we identify the stochastic integral $\int \theta\Psi(t) b(dt)$ with $\mathcal{T}(Q(J^*(\Psi)))$. Note that the last quantity depends on Ψ , not on $\theta\Psi$.

Since θ is a many-to-one map one and the same random function might be identified under θ in general with many different elements of QP with different quantizations (=stochastic integrals) which elements we treat as versions of the random function. When a random function is given its stochastic integral would be well defined if there is a rule which of all possible versions to choose. Here we consider the product space X^{n+1} as being described by the variables $\langle t,x_1,\ldots,x_n\rangle$. Define the Borel measure Λ_{n+1} on X^{n+1}, $n\geq 0$ ($\Lambda_1=q$) by

$$\Lambda_{n+1}(dt,dx_1,\ \ldots\ ,dx_n) =$$
$$= \sum_{i=1}^{n} q(dx_1)\ldots q(dx_{i-1})q_*^2(dt,dx_i)q(dx_{i+1})\ldots q(dx_n).$$

For any $f\in L_0^2(X^{n+1},d\Lambda_{n+1})$, $n\geq 1$ we define τf to be the following function of the variables $\langle x_1,\ldots,x_{n-1}\rangle$

$$(\tau f)(x_1,\ \ldots\ ,x_{n-1}) =$$
$$= \sum_{i=1}^{n} \int_{X^2} f(t,x_1,\ldots,x_{i-1},s,x_i,\ldots,x_{n-1})m(dt,ds)\ .$$

Let the compact $K\subset X$ be fixed. On each of the spaces $L^2(K^{n+1},d\Lambda_{n+1})$, $n\geq 0$ we define the equivalent norm $\|\cdot\|_{n+1}^{\sim} = C(K,n)\|\cdot\|$, where $\|\cdot\|$ is the standard norm in $L^2(K^{n+1},d\Lambda_{n+1})$ and the constants $C(K,n)$ are apropriately chosen. The space $L^2(K^{n+1},d\Lambda_{n+1})$ provided with the norm $\|\cdot\|_{n+1}^{\sim}$ we denote by $\tilde{L}_{n+1}^2(K)$. For each $n\geq 0$ the bounded linear operator $J_n:\tilde{L}_{n+1}^2(K)\to \mathcal{F}(H)*dH$ is defined such that

$$J_n[f(t)g_1(x_1)\ldots g_n(x_n)] = [Ig_1\otimes\ldots\otimes Ig_n]*d(If)\ .$$

We construct the space $QP=QP(K)$ with the desired properties as $QP(K) = \oplus_{n\geq 0}\tilde{L}_{n+1}^2(K)$. The linear map J^* is defined as $J^*=\oplus_{n\geq 0}J_n$. For any $\Psi\in QP(K)$ such that $\Psi = \oplus_{n\geq 0}\Phi_n(t,x\ ,\ldots,x\ )$ the random function $\theta\Psi(t)$ is defined by

$$\theta\Psi(t) = \sum_{n=0}^{\infty} :b^n:(/t/\Phi_n)\quad .$$

For the quantity $\mathcal{T}(Q(J^*(\Psi)))$ which we identify with the stochastic integral $\int(\theta\Psi)db$ holds

$$\mathcal{T}(Q(J^*(\Psi))) = \tau\Phi_1 + \sum_{n=1}^{\infty} :b^n:(\Phi_{n-1}+\tau\Phi_{n+1})\ .$$

The relation

$$\int :b^n:(/t/\Phi_n)b(dt) = :b^{n+1}:(\Phi_n) + :b^{n-1}:(\tau\Phi_n)$$

is simply a more general form of (b) from Proposition 1.1.

4. DIFFERENTIAL CALCULUS

In this section we reduce X to the real line $\mathbb{R}$ and q to the Lebesgue measure on it. Suppose a Baxter covariance measure m on $X^2=\mathbb{R}^2$ is given. We denote by r and R respectively the Radon-Nykodime derivatives

$$r(x)=\frac{d(\pi m_D)}{dq}(x),\ x\in\mathbb{R};\quad R(x,y)=\frac{d m_{\bar{D}}}{dq^2}(x,y),\ (x,y)\in\mathbb{R}^2.$$

A very important case of BGRM on $\mathbb{R}$ is considered in [9]. In particular for the standard Wiener process on $\mathbb{R}$ we have: r=1, q-a.s. and R=0, q^2-a.s. . Let b(.) be the BGRM on $\mathbb{R}$ derived from the Baxter covariance measure m on $\mathbb{R}$. Consider the random function $B(t)=b([0,t])$, $0\le t\le T<+\infty$. One may suppose that B(t) is defined for all $t\in\mathbb{R}$, letting $B(t)=0$ for $t\bar{\in}[0,T]$. It can be proved that B(.) is L^2-continuous random function from $L^2(\mathbb{R},dq;\Gamma(\mathcal{H}))$. Let us denote by $\mathcal{C}$ the class of all complex functions f on $\mathbb{R}$ which satisfy the following three conditions: 1) $f\in L^2(\mathbb{R},[(2\pi)^{-\frac{1}{2}}\exp(-x^2/2)]dx)$; 2) f' exists and $f'\in L^2(\mathbb{R},[(2\pi)^{-\frac{1}{2}}.\exp(-x^2/2)]dx)$; 3) there exist a real constant $\alpha<\frac{1}{2}$ and a polynomial P(x) such that $|f'(x)|\le|P(x)|\exp(\alpha x^2)$, $x\in\mathbb{R}$.
Let us suppose now that $0<\langle B(t)^2\rangle\le 1$ for every $t\in[t_0,T]$, $0<t_0<T<+\infty$.

Theorem 4.1. Let f and g are two complex or real functions on $\mathbb{R}$ such that: $f,g\in\mathcal{C}$; g' exists; $g'\in\mathcal{C}$. Then there exist Ψ_f , $\Psi_g\in QP([t_0,T])$ which are versions of the random functions f'(B(.)) and g'(B(.)). The following stochastic integrals are well defined

$$(0)\int_{t_0}^{T} g'(B(t))dB(t)=\mathcal{T}(Q(J^*(\Psi_g)))$$

$$(\tfrac{1}{2})\int_{t_0}^{T} f'(B(t))dB(t)=\mathcal{T}(Q(J^*(\Psi_f))).$$

The versions Ψ_g and Ψ_f can be chosen such that

$$(0)\int_{t_0}^{T} g'(B(t))dB(t)=g(B(T))-g(B(t_0))-$$
$$-\tfrac{1}{2}\int_{t_0}^{T} g''(B(t))r(t)q(dt)\ ;$$

$$(\tfrac{1}{2})\int_{t_0}^{T} f'(B(t))dB(t)=f(B(t))-f(B(t_0))\quad .\qquad \#$$

5. CONCLUDING REMARKS

Remark 5.1. If we define the quantization Q as $Q(\Psi_0{*}df_0)=$

$(A\uparrow f_0)(\Psi_0)$ we come to the construction of the stochastic integral, given in [8]. Remind that here we have $Q(\Psi_0 * df_0) = (\sqrt{2})^{-1} \cdot [(A\uparrow f\ ) + (A\downarrow f\ )]$. #

Remark 2.2. The stochastic integral $(\frac{1}{2})\int_{[t_0,T]} B(t)dB(t)$ could be defined as a limit of Riemann sums of the type

$$\sum_{i=1}^{n} \tfrac{1}{2}(B(s_i)+B(s_{i-1}))(B(s_i)-B(s_{i-1})) ,$$

$t_0 = s_0 < s_1 < \ldots < s_n = T$. It is interesting to compare the last sum with the Riemann sum for the Stratonovich integral:

$$\sum_{i=1}^{n} B(\tfrac{1}{2}(s_i + s_{i-1}))(B(s_i) - B(s_{i-1})) . \quad \#$$

REFERENCES

1. M. Reed and B. Simon, Methods of Modern Mathematical Physics. vol.1: Functional Analysis, Academic Press, New York and London, 1972.

2. M. Reed and B. Simon, Methods of Modern Mathematical Physics. vol.2: Fourier Analysis, Self-adjointness, Academic Press, New York, San Francisco and London, 1975.

3. W. Rudin, Functional Analysis, McGraw - Hill, New York, St. Louis, ---, 1973.

4. B. Simon, The $P(\varphi)_2$ Euclidian (Quantum) Field Theory, Princeton Univ. Press, Princeton and New Jersey, 1974.

5. G. Ciucu and C. Tudor, Probabilităţi şi Procese Stocastice. vol.1, Editura Academiei Rep. Soc. Româna, Bucureşti, 1978. (Romanian)

6. R. Curtain and R. Keppler, A Survey of Stochastic Integration in Infinite Dimensions, Control Theory Centre Report No 34.

7. K. Ito, Multiple Wiener Integral, J. Math. Soc. Japan, 3 (1951), 157 - 169.

8. S. Huang and S. Cambanis, Stochastic and Multiple Wiener Integral for Gaussian Processes, The Annals of Probability, 6 (1978), 585 - 614.

9. J. Yeh, Stochastic Integral of L_2 - Functions with Respect to Gaussian Processes, Tohoku Math. J., 27 (1975), 175 - 186.

FINITE DIMENSIONAL CAUSAL FUNCTIONALS OF BROWNIAN MOTION*

Omar Hijab

Department of Mathematics and Statistics
Case Western Reserve University
Cleveland, Ohio 44106 U.S.A.

0. INTRODUCTION

This paper deals with a study of scalar nonanticipating functionals α of Brownian motion $t \to b(t)$ and in particular considers those that are finite dimensional.

A functional α is finite dimensional if there is a point x_0 in some Euclidean space $\mathbb{R}^N$, infinitely differentiable vector fields $f,g : \mathbb{R}^N \to \mathbb{R}^N$ and an infintely differentiable function $h : \mathbb{R}^N \to \mathbb{R}$ such that the solution $t \to x(t)$ of

$$x(t) = x_0 + \int_0^t f(x(s))ds + \int_0^t g(x(s))db(s) \tag{1}$$

exists for all $t \geq 0$ and satisfies

$$\alpha(t) = h(x(t)) \tag{2}$$

for all $t \geq 0$.

Of course, not every functional α is finite dimensional in the above sense. For example let $k(t)$ be an infinitely differentiable (sure) function of $t \geq 0$ and let

---

*Supported in part by DOE Contract EDE AC01-79-ET-29363.

R. S. Bucy and J. M. F. Moura (eds.), Nonlinear Stochastic Problems, 425–435.

$$\alpha(t) = \int_0^t k(t-s)db(s) \tag{3}$$

for $t \geq 0$. We shall see that if k is an analytic function of time then α is finite dimensional if and only if

$$k(t) = ce^{tA}b \ , \qquad t \geq 0 \quad , \tag{M}$$

for some $n \times 1$ vector b, $1 \times n$ vector c and $n \times n$ matrix A.

To each sufficiently "smooth" functional α we shall assign a nonnegative integer, its rank, in such a way that the following is true.

Theorem. If α satisfies (1) and (2) then the rank of α is well-defined and is less than or equal to N.

The ideas discussed here arose out of the study of the "nonlinear filtering" problem. Although we do not make any remarks about this problem here, the above theorem implies all rigorous results known on this problem as well as make rigorous the "homomorphism principle" of Brockett [1] who studied finite dimensional functionals in the context of filtering. In particular the results appearing in [2], [3], [4], [5] are implied by the general formulation given here.

There is a deterministic analog to the results outlined here due to M. Fliess [6]. To go from here to Fliess' results, simply "replace db by udt." The operator A that we define here has been used before [7]. Finally, another reason why this formalism "should work" is the beautiful "support theorem" of Stroock and Varadhan [10].

1. SMOOTH BROWNIAN FUNCTIONALS

Let $\Omega = C([0,\infty);\mathbb{R})$ and let $b(t) : \Omega \to \mathbb{R}$ be given by $b(t,\omega) = \omega(t)$ for $t \geq 0$. For each $t \geq 0$, let $\mathfrak{P}_t$ denote the σ-algebra on Ω generated by the maps $b(s)$, $0 \leq s \leq t$, and let $\mathfrak{P}$ denote the σ-algebra generated by the union of all of the $\mathfrak{P}_t$, $t \geq 0$. Let P denote Wiener measure on $(\Omega,\mathfrak{P})$. Then $t \to b(t)$ is a Brownian motion on $(\Omega,\mathfrak{P},P)$.

A causal functional of Brownian motion (a functional, for short) is a measurable map $\alpha : \Omega \times [0,\infty) \to \mathbb{R}$ that is progressively measurable, i.e. for each $t \geq 0$ $\alpha(t) : \Omega \to \mathbb{R}$ is $\mathfrak{P}_t$-measurable. We say that a functional α is continuous if α has continuous sample paths, almost surely. A continuous functional can then be thought of as a measurable map $\alpha : \Omega \to \Omega$. We say that a functional α is <u>differentiable</u> if there are functionals β and γ satisfying

$$P\left(\int_0^t |\beta(s)| + \gamma(s)^2 \, ds < \infty, \ t \geq 0\right) = 1$$

and

$$\alpha(t) = \alpha(0) + \int_0^t \beta(s)ds + \int_0^t \gamma(s)db(s), \quad t \geq 0 \tag{4}$$

almost surely. At no time in what follows will our functionals have differentiable sample paths so there should be no confusion about the above definition of "differentiable." If $\alpha^2 = \alpha$ then by computing the differential of α, we see that both β and γ are identically zero and thus β and γ are uniquely determined by α. We shall denote β by $A\alpha$, γ by $B\alpha$, and $\alpha(0)$ by $C\alpha$. Thus for any differentiable α, $A\alpha$ and $B\alpha$ are the unique functionals satisfying (4) and $C\alpha$ is the real number $\alpha(0)$.

A functional α is <u>continuously differentiable</u> if $A\alpha$ and $B\alpha$ are continuous. Inductively, we say α is (k+1)-times continuously differentiable if α is differentiable and $A\alpha$, $B\alpha$ are k-times continuously differentiable, for all $k \geq 1$. If α is k-times continuously differentiable for all $k \geq 0$, then α is a <u>smooth Brownian functional</u>.

For any smooth functional α one can thus form V_α, the smallest vector space of functionals containing α and closed under application of A and B. Some elements of V_α are, for example, α, $A\alpha$, $B\alpha$, $A^2\alpha = A(A\alpha)$, $AB\alpha = A(B\alpha)$, $BA\alpha = B(A\alpha)$, $B^2\alpha$, $A^3\alpha$, $A^4B^2A^3\alpha$, ..., and their linear combinations. If α is given by (3) with $k(t)$ an infinitely differentiable function of $t \geq 0$, then α is smooth and

$$A^i\alpha(t) = \int_0^t \partial^i k(t-s)/\partial t^i \, db(s)$$

$$BA^i\alpha(t) = k^{(i)}(0) \quad ,$$

with all other "derivatives" of α being equal to zero. Thus V_α is spanned by the constants and $A^i\alpha$ for $i \geq 0$, for this α.

2. FINITE DIMENSIONAL BROWNIAN FUNCTIONALS

One way to construct smooth functionals is via stochastic differential equations. If $f,g : \mathbb{R}^N \to \mathbb{R}^N$ are infinitely differentiable maps, let $t \to x(t)$ be the solution of

$$x(t) = x_0 + \int_0^t f(x(s)) + \frac{1}{2}\frac{\partial g}{\partial x}(x(s))g(x(s))ds$$

$$+ \int_0^t g(x(s))db(s) \quad , \qquad (1')$$

where x_0 is a fixed point in $\mathbb{R}^N$. It is well-known then that $t \to x(t)$ exists (and is unique) up to some explosion time $\tau \le \infty$. Assume that there is no finite escape time i.e. that $\tau = \infty$, almost surely. If $h : \mathbb{R}^N \to \mathbb{R}$ is an infinitely differentiable function, let α be given by (2). Then we have the

Lemma. If α satisfies (1') and (2) for all $t \ge 0$ then α is smooth.

Before we give the proof, we note that α satisfies (1) and (2) for some f,g,h if and only if α satisfies (1') and (2) for some f_1,g_1,h_1. Simply set

$$g_1 = g, \quad h_1 = h, \quad \text{and} \quad f_1 = f - \frac{1}{2}\frac{\partial g}{\partial x} g \quad .$$

Given a functional α, we say that α is finite dimensional if and only if α satisfies (1') and (2) for some choice of f, g, h, x_0, and N. By the above remark, this definition agrees with the one given in the introduction.

Let ϕ be an infinitely differentiable function on $\mathbb{R}^N$ and let g denote any vector field on $\mathbb{R}^N$. Let $g(\phi)(x)$ denote the directional derivative of ϕ in the direction of g at the point x:

$$g(\phi)(x) = \frac{\partial \phi}{\partial x_1}(x)g_1(x) + \ldots + \frac{\partial \phi}{\partial x_N}(x)g_N(x) \quad .$$

If g is infinitely differentiable, then so is $g(\phi)$ and we can form $g^2(\phi) = g(g(\phi))$ and

$$L(\phi) = (f + \frac{1}{2} g^2)(\phi) \quad .$$

L is then a second order differential operator and is the (backward) infinitesimal generator of the diffusion $t \to x(t)$ satisfying (1'). In what follows we shall denote f by g_0 and g by g_1. With the above notation, the proof of the lemma reduces to the Ito differential rule: If $\alpha(t) = \phi(x(t))$ and $t \to x(t)$ satisfies (1') then

$$\alpha(t) = \alpha(0) + \int_0^t L(\phi)(x(s))ds + \int_0^t g(\phi)(x(s))db(s), \quad t \geq 0 .$$

Thus for $\alpha(t) = \phi(x(t))$, we have

$$(A\alpha)(t) = L(\phi)(x(t)), \; (B\alpha)(t) = g(\phi)(x(t)), \; t \geq 0 \quad . \tag{5}$$

X_0 denote the operator given by

$$X_0\alpha = (A - \frac{1}{2} B^2)(\alpha) = A\alpha - \frac{1}{2} (B(B\alpha)) \quad ;$$

let X_1 denote B. By applying the result (5) repeatedly one sees that $\alpha(t) = h(x(t))$ is smooth and

$$X_{i_p} \ldots X_{i_2} X_{i_1} (\alpha)(t) = g_{i_p} \ldots g_{i_2} g_{i_1} (h)(x(t)) \qquad t \geq 0 \tag{6}$$

for all choices of $i_1, i_2, \ldots, i_p$ equal to zero or one, and all $p \geq 0$.

3. THE RANK OF A FUNCTIONAL

Let α be a smooth functional and let X_0, X_1 be the operators $V_\alpha \to V_\alpha$ described above. If X and Y are two linear operators taking V_α to V_α let

$$[X,Y](\alpha) = X(Y(\alpha)) - Y(X(\alpha)) = (XY - YX)(\alpha) \quad .$$

$[X,Y]$ is the <u>bracket</u> of X and Y. One can check that $[\ ,\ ]$ satisfies

$$\text{(i)}\quad [X,Y] = -[Y,X]$$

$$\text{(ii)}\quad [X+Y,Z] = [X,Z] + [Y,Z]$$

$$\text{(iii)}\quad [[X,Y],Z] + [[Y,Z],X] + [[Z,X],Y] = 0$$

for all X, Y, Z acting on V_α. A vector space upon which is imposed a map $[\ ,\ ]$ satisfying (i), (ii), (iii) above is called a <u>Lie algebra</u>, and thus the set of all linear operators on V_α is a Lie algebra.

Let $\mathfrak{F}$ denote the <u>free</u> Lie algebra generated by

$$X_0 = A - \frac{1}{2}B^2 \quad \text{and} \quad X_1 = B \quad .$$

This means $\mathfrak{F}$ is the set of all linear combinations of formal products of X_0, X_1 of the form

$$[X_{i_p},[\ldots,[X_{i_4},[X_{i_3},[X_{i_2},X_{i_1}]]]\ldots]] \quad .$$

Recall that for any functional β, $\tilde{C}\beta$ denotes the real number $\beta(0)$.

Now let α be a smooth Brownian functional and let V_α denote the vector space defined above. Define a pairing

$$\begin{aligned} &\mathfrak{F} \times V_\alpha \to \mathbb{R} \\ &(X,\beta) \to C(X\beta) \quad . \end{aligned} \tag{7}$$

<u>Definition</u>. The <u>rank of α</u> is the rank of the pairing (7).

We can now state the main result.

<u>Theorem</u>. If α is finite dimensional then $\text{rank}(\alpha) \leq N$.

The rank of a functional α can be computed in two ways. Since $\mathfrak{F}$ and V_α are linear spaces, let $\mathfrak{F}^*$ and V_α^* denote their duals (the spaces of all linear functions into $\mathbb{R}$ from $\mathfrak{F}$ and V_α). Let

$$\pi_1 : \mathfrak{F} \to V_\alpha^* \quad \text{and} \quad \pi_2 : V_\alpha \to \mathfrak{F}^*$$

be given by

$$\pi_1(X)(\beta) = C(X\beta) = \pi_2(\beta)(X) \quad .$$

Then π_1 and π_2 are linear and by linear algebra

$$\text{rank}(\alpha) = \dim(\pi_1(\mathfrak{F})) = \dim(\pi_2(V_\alpha)) \quad .$$

(Proof of the theorem.) Let α be a finite dimensional functional and let $\mathfrak{G}$ denote the Lie algebra of vector fields generated by $f = g_0$ and $g = g_1$. Equation (6) implies

$$C(X_{i_p} \ldots X_{i_1} \alpha) = g_{i_p} \ldots g_{i_1}(h)(x_0) \quad , \tag{8}$$

for any choice of $i_1, \ldots, i_p$ equal to zero or one. Let W_h be the smallest subspace of functions on $\mathbb{R}^N$ containing h and closed under differentiation by f and g. Let $\mathfrak{G}_0$ denote those elements a in $\mathfrak{G}$ satisfying

$$a(\phi)(x_0) = 0 \tag{9}$$

for all ϕ in W_h. Let $\mathfrak{F}_0$ denote those elements X in $\mathfrak{F}$ satisfying

$$C(X\beta) = 0 \tag{10}$$

for all β in V_α. Since $\mathfrak{G}$ is generated by 2 elements and $\mathfrak{F}$ is the free Lie algebra generated by 2 elements, there is a natural homomorphism $\mathfrak{F} \to \mathfrak{G}$ sending X_0 to g_0 and X_1 to g_1. We claim that under this homomorphism the inverse image of $\mathfrak{G}_0$ is exactly $\mathfrak{F}_0$. Indeed if a is in $\mathfrak{G}_0$ then (9) holds and then by (8) so does (10). Hence X is in $\mathfrak{F}_0$. Thus

$$\pi_1(\mathfrak{F}) \simeq \mathfrak{F}/\mathfrak{F}_0 \simeq \mathfrak{G}/\mathfrak{G}_0$$

and hence

$$\dim(\pi_1(\mathfrak{F})) = \dim(\mathfrak{G}/\mathfrak{G}_0) \leq N \quad ,$$

this last inequality following from the definition of $\mathfrak{G}_0$ (equation (9)). This completes the proof of the theorem.

Remark 1. It is possible that the rank of α be equal to 2 and yet the smallest N for which α satisfies (1) and (2) be arbitrarily large.

Because of this remark, one does not expect the converse to the theorem to hold unless some additional regularity assumptions are imposed.

Remark 2. Let W_α denote the smallest algebra containing α and closed under application of A and B. Then if we replace V_α by W_α in the definition of the rank, the rank is unchanged.

Remark 2 is due to the following product rules.

Lemma. If α and β are continuously differentiable then so is $\alpha\beta$ and

$$A(\alpha\beta) = A(\alpha)\beta + \alpha A(\beta) + (B\alpha)(B\beta)$$

$$B(\alpha\beta) = B(\alpha)\beta + \alpha B(\beta)$$

$$C(\alpha\beta) = C(\alpha)C(\beta) \quad .$$

Thus both $X_0 = A - \frac{1}{2}B^2$ and $X_1 = B$ satisfy

$$X(\alpha\beta) = X(\alpha)\beta + \alpha X(\beta) \quad .$$

Remark 3. Say that a functional α is locally finite dimensional if equations (1), (2) hold only up to some positive stopping time. Then the main result is still true. Thus (global) finite dimensionality ⇒ local finite dimensionality ⇒ infinitesimal finite dimensionality i.e. $\text{rank}(\alpha) < \infty$.

4. LINEAR FUNCTIONALS

Let α be given by (3) with k a real-analytic function of $t \geq 0$. Then

$$CBA^i\alpha = k^{(i)}(0)$$

and so the rank of α is equal to the rank of the (Hankel) matrix having (i,j)-th element given by

$$k^{(i+j)}(0) \quad .$$

Thus the rank of α is finite iff there is a N and $a_1,\ldots,a_N$ real numbers such that

$$k^{(j+N)}(0) + a_1 k^{(j+N-1)}(0) + \ldots + a_N k^{(j)}(0) = 0$$

for all $j \geq 0$. By analyticity this happens iff

$$k^{(N)}(t) + a_1 k^{(N-1)}(t) + \ldots + a_N k(t) = 0$$

for $t \geq 0$. Thus the rank of α is finite iff $k(t)$ is given by a matrix element (M). This shows that α is finite dimensional iff $k(t)$ is as advertised, provided $k(t)$ is real-analytic. It would be interesting to extend this result to any infinitely differentiable $k(t)$.

5. THE HOMOMORPHISM PRINCIPLE

Equation (8) can be used to prove more. Given a smooth functional α say that $\alpha \sim 0$ (is infinitesimally zero) if

$$CX_{i_p} \ldots X_{i_1} \alpha = 0$$

for all choices of $i_1,\ldots,i_p$ equal to zero or one. Fix a smooth functional α and let $\mathfrak{F}_\alpha$ be the set of all X in $\mathfrak{F}$ satisfying

$$X\beta \sim 0$$

for all β in V_α. Then $\mathfrak{F}_\alpha$ is an ideal in $\mathfrak{F}$ and, following Fliess,

$$\mathfrak{F}/\mathfrak{F}_\alpha$$

is the <u>syntactic Lie algebra</u> of α. The following is a general version of the "homomorphism principle" of R. W. Brockett.

<u>Theorem</u>. Let α be a finite dimensional functional with f, g and h all real-analytic. Let $\mathfrak{G}$ be the Lie algebra generated by f and g.

Assume that the system (1), (2) is <u>minimal</u> [8] i.e.

(i) $\dim \mathfrak{G}(x) = N$ for all x in $\mathbb{R}^N$ (controllability)

(ii) If $W = W_h$ is as defined in section 3, then

$$\dim \frac{\partial W}{\partial x}(x) = N \quad \text{for all } x \text{ in } \mathbb{R}^N .$$

(observability)

Then

$$\mathfrak{F}/\mathfrak{F}_\alpha \simeq \mathfrak{G} \quad .$$

The proof of this theorem rests on equation (8) and analyticity. Given a function ϕ in W, say that $\phi \sim 0$ iff

$$g_{i_p} \cdots g_{i_1}(\phi)(x_0) = 0$$

for all choices of $i_1,\ldots,i_p$ equal to zero or one. Let $\mathfrak{G}_\alpha$ be the set of all a in $\mathfrak{G}$ satisfying

$$a(\phi) \sim 0 \tag{11}$$

for all ϕ in W. Then $\mathfrak{G}_\alpha$ is an ideal in $\mathfrak{G}$ and by (8)

$$\mathfrak{F}/\mathfrak{F}_\alpha \simeq \mathfrak{G}/\mathfrak{G}_\alpha \quad .$$

But now analyticity and controllability imply that $\phi \sim 0 \Rightarrow \phi \equiv 0$ and thus a in $\mathfrak{G}_\alpha$ iff $a(\phi) \equiv 0$ for all ϕ in W. Now observability implies that $a \equiv 0$. Thus $\mathfrak{G}_\alpha \equiv \{0\}$ concluding the theorem.

REFERENCES

[1] R. W. Brockett, "Nonlinear Systems and Nonlinear Estimation Theory," in [9] below.

[2] O. Hijab, "Minimum Energy Estimation," Ph.D. Dissertation, University of California, Berkeley, December 1980.

[3] ________, "A Class of Infinite Dimensional Filters," Proceedings of the 19th IEEE Conference on Decision and Control, Albuquerque, New Mexico, December 1980.

[4] H. Sussmann, "Rigorous Results on the Cubic Sensor Problem," in [9] below.

[5] M. Hazewinkel and S. I. Marcus, "On the Role of Lie Algebras in Filtering," in [9] below.

[6] M. Fliess, "Realizations of Nonlinear Systems and Abstract Transitive Lie Algebras," Bull. Amer. Math. Soc., May 1980.

[7] T. Kurtz, "Semigroups of Conditional Shifts and Approximation of Markov Processes," Annals of Probability, 1975, v. 3, pp. 618-642.

[8] R. Hermann and A. J. Krener, "Nonlinear Controllability and Observability," IEEE Trans. Aut. Control, v. AC-22, 1977, pp. 728-740.

[9] M. Hazewinkel and J. C. Willems (editors), Stochastic Systems: The Mathematics of Filtering and Identification, D. Reidel (NATO-ASI series) 1981.

[10] D. W. Stroock and S. R. S. Varadhan, "On the Support of Diffusion Processes with Applications to the Strong Maximum Principle," Sixth Berkeley Symposium on Probability and Statistics, 1972.

TRANSIENCE, RECURRENCE AND INVARIANT MEASURES FOR DIFFUSIONS

W. Kliemann

Forschungsschwerpunkt "Dynamische Systeme"
University of Bremen, 2800 Bremen,
West Germany

Qualitative theory describes the long term behavior of stochastic systems without solving the equations. In this paper we concentrate on transience, recurrence and ergodic properties of Markov solutions of stochastic differential equations. For several approaches the relations between the regularity of the transition semigroup (strong Markov - Feller - strong Feller-nondegeneracy) and the adequat topology are discussed. Results are given for diffusions, i.e. Feller processes with continuous trajectories on some state space $X \subset \mathbb{R}^d$, endowed with the usual topology from $\mathbb{R}^d$, because this is the situation amenable for applications.

I. Introduction

For our purposes a stochastic system is just a classical dynamical system in $\mathbb{R}^d$ perturbed by some noise process

$$\dot{x}_t = f(x_t, \xi_t). \tag{1}$$

Here ξ_t may be e.g. white noise, colored noise, i.e. a diffusion process, or real noise, i.e. a stochastically continuous process. The first two cases lead to a systems description as a stochastic differential equation

$$dx_t = g(x_t)dt + G(x_t)dW_t \tag{2}$$

R. S. Bucy and J. M. F. Moura (eds.), Nonlinear Stochastic Problems, 437–454.

and

$$d\begin{pmatrix} x_t \\ \xi_t \end{pmatrix} = \begin{pmatrix} f(x_t,\xi_t) \\ h(\xi_t) \end{pmatrix} dt + \begin{pmatrix} 0 \\ H(\xi_t) \end{pmatrix} dW_t \qquad (3)$$

respectively, where W_t is a vector of m independent Wiener processes. Hence the state process x_t of (2) and the combined state-noise process (x_t,ξ_t) of (3) can be viewed as a Feller-Markov process with continuous trajectories (if for (3) the initial value x_o is nice, compare [1], chapter II.A for the development of the set up).

Qualitative theory for stochastic systems now means studying the nature of a solution (for $t \geq o$) without solving the equation. The concepts are just the stochastic analogues (in some sense, depending on the various topologies for random variables) of deterministic qualitative theory, cf. [7], [25], i.e. recurrence, transience, stationarity, invariant measures, ergodicity, stability... . The monographs of Kushner [22], Hasminskii [16] or Friedman [12] treat these questions via a Lyapunov function approach and for nondegenerate (or slightly degenerate) diffusions. A series of papers by Azéma et al. [3] - [6] uses potential theoretic methods for strong Markov processes.

In this paper we treat the problems of recurrence, stationarity and ergodicity for Feller processes with continuous trajectories, e.g. solutions of (2) or (3), without any further assumptions. This class of systems show up in a lot of applications to natural and engineering sciences, cf. e.g. [1], [2], [20].

We start with a brief review of Markov and diffusion processes in part II, and different approaches to qualitative theory in part III before we get to our results in part IV and V.

II. Markov and diffusion processes

The basic textbook on Markov processes still seems to be Dynkins two volumes [11] from 1965, though Williams's book [29] contains a nice and distinct introduction on the several ways of looking at a Markov process: the transition functions $P(t,x,A)$ on the state spate X, the family of probability measures P_x on the path space over X, the semigroup P_t on the bounded measurable functions $b(X)$, the resolvents R_λ and the generator G. In particular for Feller process (i.e. P_t maps continuous functions into continuous ones and is strongly continuous at $t = 0$) all these descriptions are equivalent (on a nice state space, such that the Riesz representation theorem holds, e.g. locally compact spaces with countable base). The associated process can be picked to be strongly Markovian and have cadlag (= right continuous with left hand limits) trajectories. Dynkin's characteristic operator for Feller processes, describing the infinitesimal space behavior, coincides with the generator of the corresponding semigroup.

For general Markov processes the behavior in state space gives rise to a new topology, the fine topology T_f: Let $(E,\mathcal{B}(E))$ be the state space of x_t. A set $A \subset E$ belongs to T_f, if for every $x \in A$ there exists a borel set B such that $x \in B \subset A$ and $P_x\{\tau_B > o\} = 1$, where τ is the first exit time from B. The given topology E on E is always contained in T_f. Since any invariant Borel set A is open in the fine topology, it contains even for continuous processes in general many open sets that have void interior in the given topology. Furthermore x_t is a Feller process in the state space $(E,\mathcal{B}(T_f))$. From this point of view the study of Markov processes reduces to that of Feller processes, but on a nasty topological space.

A diffusion on $\mathbb{R}^d$ is just a Feller process with continuous trajectories such that the domain of the generator G contains all C^∞ functions with compact support $C_k^\infty(\mathbb{R}^d)$. Dynkin's theorem tells us that the restriction of G to $C_k^\infty(\mathbb{R}^d)$ is a second order elliptic operator

$$Lf(x)=\frac{1}{2}\sum_i \sum_j a_{ij}(x)\frac{\partial^2}{\partial x_i \partial x_i}f(x)+\sum_i b_i(x)\frac{\partial}{\partial x_i}f(x)-c(x)f(x) \quad (4)$$

with continuous functions a_{ij}, b_i, c, the matrix (a_{ij}) is symmetric nonnegative definite and $c \geq 0$. If x_t is nonterminating, then $c = 0$. Or more precisely: if $\lim_{t\downarrow o} \frac{1}{t}(1-P(t,x,E))$ exists, it is equal to $c(x)$.
Consider the reversed problem: Given L on $C_k^\infty(\mathbb{R}^d)$ with continuous coefficients $(c=o)$, does there exist a unique Feller process whose generator extends L? The Stroock and Varadhan theory [28] tells us that this is the case for nondegenerate elliptic operators, i.e.

$a > o$, where uniqueness is ment for the laws P_x. A partial answer to the reversed problem is given via Ito's stochastic calculus [19]: if a has a Lipschitz square root σ and if b is Lipschitz too, then there exists a unique (pathwise) Feller process whose generator extends L.

Hypoellipticity now relates the Fokker-Planck or Kolmogorov's forward equation to the present set-up: if the adjoint parabolic operator $-\frac{\partial}{\partial t} + L^*$ is C^∞ and hypoelliptic, the fundamental solutions of the forward equation $-\frac{\partial p}{\partial t} + L^*p = o$ are C^∞ densities of the transition functions. Hörmanders theorem gives criteria for hypoellipticity in terms of the Lie algebra generated by the vector fields defining the stochastic differential equation. For this ideas and the generalization to manifolds Ikeda and Watanabe [18] is a book worthwile to read.

Let us have a brief look at the regularity properties of the semigroups defining a diffusion. By definition these are Feller, i.e. $P_t: C_b(\mathbb{R}^d) \to C_b(\mathbb{R}^d)$. This implies that for compact sets A $P(t,x,A)$ is upper semicontinuous in x, cf. [11], I p. 138. The strong Feller property, i.e. $P_t: b\mathcal{B} \to C_b$, is equivalent to continuous transition probabilities. In the C^∞ context Ichihara and Kunita [17] gave a nice characterization of strong Feller diffusions: Let L be given by

$$L = \frac{1}{2} \sum_{i=1}^{m} A_i^2 + A_o, \tag{5}$$

where A_j, $j = o - m$ are C^∞ vector fields on $\mathbb{R}^d$. Denote by $LA(A_o,A_1,..A_m)$ the Lie algebra generated by the vector fields A_j and by LA_o the ideal in $LA(A_o,A_1..A_m)$ generated by $A_1,..,A_m$. If LA_o is locally finitely generated (e.g. if all vector fields are analytic), then the diffusion x_t corresponding to L is strong Feller if and only if $-\frac{\partial}{\partial t} + L^*$ is hypoelliptic. Hence in particular any nondegenerate diffusion, i.e. the operator L is nondegenerate elliptic, is strong Feller.

The so called support theorems of Stroock and Varadhan [26], [27] are a bridge between diffusions and nonlinear (deterministic) control systems:
Consider the operator L in (4) with $c = o$ and the

related control system

$$\dot{\varphi}(t) = a(\varphi)u + \bar{b}(\varphi) \tag{6}$$

where $\bar{b} = b - \frac{1}{2}a'$ and u is a piecewise constant function $(u_1(t),\dots,u_m(t))$. Assume that $a \in C_b^2(\mathbb{R}^d)$, i.e. a has two continuous bounded derivatives and b is bounded and uniformly Lipschitz, then

$$\text{supp } P_x = \{\varphi(\cdot,x,u),\ u \text{ piecewise constant}\}^{cl}$$

where the closure ^{cl} has to be taken in the uniform compact topology in $C(\mathbb{R}^d)$. C^2 for a is necessary to get sufficiently smooth roots of a.
If the operator is given in Hörmander form (5), it suffices to consider the control system

$$\dot{\varphi}(t) = A(\varphi)u + A_o(\varphi) \tag{7}$$

where A is the dxm matrix $(A_1 .. A_m)$.
For nondegenerate diffusions we have $\text{supp } P_x = C_x(\mathbb{R}^d)$, the continuous functions in $\mathbb{R}^d$ starting at x .

III. Approaches to qualitative theory

Let us begin making the concepts of qualitative theory more precise:

Definition III.1. For a Markov process x_t on a topological state space $(X,\mathcal{B}(X))$ the point $x \in X$ is called transient, if there exists an open neighbourhood $U \in X$ of x such that $P_x\{T_U\} = 1$, $T_U = \{$there exists $t_o > o$ such that $x_t \notin U$ for all $t \geq t_o\}$.
The point $x \in X$ is recurrent, if for all open neighbourhoods U of x $P_x\{R_U\} = 1$, $R_U = \{$there exists $t_n \uparrow \infty$ such that $x_{t_n} \in U\}$.
The point $x \in X$ is positive w.r.t. $U \in X$ if $E_x \sigma_U < \infty$, where $\sigma_U = \inf\{t>o, x_t \in U\}$ is the first entry time of U.
A diffusion is transient, recurrent, positive, if all points $x \in X$ are transient, recurrent, positive with respect to all $U \in X$.

Since we restrict ourselves to Markov processes, stationarity means the existence of an invariant probability.

Definition III 2. A measure μ on X is invariant for x_t, if

$$\mu(A) = \int P(t,x,A)\mu(dx) \text{ for all } t > 0, \text{ all } A \in \mathcal{B}(X).$$

Except for the Lyapunov techniques, the proof of recurrence is accomplished via construction of an imbedded chain:
let $x \in X$ and U a neighbourhood of x. Recurrence means that we have to construct a sequence $t_n \uparrow \infty$ such that $x_{t_n} \in U$. This can be set on as follows: Let V be in X such that $U \subset V$ and $\partial U \cap \partial V = \emptyset$. Define a sequence of stopping times

$$\eta_1 = \inf\{t>0, x_t \notin V\}$$
$$\eta_2 = \inf\{t>\eta_1, x_t \in U\}$$
$$\vdots$$
$$\eta_{2n} = \inf\{t>\eta_{2n-1}, x_t \in U\}$$
$$\eta_{2n+1} = \inf\{t>\eta_{2n}, x_t \notin V\}$$
$$\vdots$$

One has to prove $\eta_{2n} \uparrow \infty$ and $P_x\{\eta_{2n}<\infty\} = 1$ for all $n \in \mathbb{N}$. This obviously requires some continuity properties for the function $\Phi_B(x) = P_x\{\sigma_B<\infty\}$. Observe that $P_t\, \Phi_B(x) = P_x\{\sigma_B \cdot \Theta_t < \infty\} = P_x\{\lambda_B > t\}$, λ_B the last exit from B, which gives the connection to transience. Now for strong Markovian, cadlag processes x_t the function Φ_B is excessive (hence superharmonic, even harmonic for continuous x_t). Hence the first approach is

<u>(a) strong Markov processes in the fine topology.</u>

For the fine topology $\mathcal{T}_f$ is exactly the coarsest one, making all α-excessive functions continuous. This approach is worked out e.g. by Azema et al. [3] - [6] or Getoor [14]. The main results are:

- <u>dichotomy for points:</u> $x \in X$ is finely recurrent for x_t iff $R(x,U) = \int_0^\infty P(t,x,U)\,dt = \infty$ for all neighbourhoods $U \in \mathcal{T}_f$ of x. x is finely transient iff there exists $V \in \mathcal{T}_f$, a neighbourhood of x, such that $R(x,V) < \infty$.
- <u>continuation of recurrence:</u> let $x,y \in X$ and $P_x\{\sigma_U<\infty\} > 0$ for all fine neighbourhoods U of y, then if x is finely recurrent, so is y and $P_y\{\sigma_V<\infty\} > 0$ for all fine neighbourhoods V of x.
- <u>invariant measures:</u> with a finely recurrent point x we call $C(x)$, the set of points y, such that $P\,\{\sigma_U<\infty\} > 0$ for all fine nbhd's U of y, a finely recurrent class for x_t. If any two recurrent classes

are contained in disjoint finely closed stochastically invariant sets D_α, then there exists a unique invariant σ-finite measure μ_α concentrated on D_α. This of course entails ergodic properties of the process x_t, cf. [4], [6], [10].

To give an idea of the specific properties of the fine topology let us show an example where the continuation of recurrence does not hold in the given topology:

Example III 1: Consider the Ornstein-Uhlenbeck process on $\mathbb{R}^1$

$$dx_t = -a\,x\,dt + b\,dW_t,\quad a,b > 0 \text{ constant},$$

which has an invariant probability μ on $\mathbb{R}^1$. Transform $x_t \longrightarrow \arctan\, x_t = \hat{x}_t$, which yields for $\hat{x}_t$ the differential equation in $(-\frac{\pi}{2}, \frac{\pi}{2})$

$$d\hat{x}_t = \frac{-(a+b^2)\tan\hat{x}_t - a\tan^3\hat{x}_t}{(1+\tan^2\hat{x}_t)^2}dt + \frac{b}{1+\tan^2\hat{x}_t}dW_t$$

and $\hat{x}_t$ has an invariant probability in $(-\frac{\pi}{2}, \frac{\pi}{2})$, hence any point $x \in (-\frac{\pi}{2}, \frac{\pi}{2})$ is recurrent for $\hat{x}_t$. Now continue the coefficients to $\mathbb{R}$ such that a solution y_t exists on $\mathbb{R}$ and is unique. The points $-\frac{\pi}{2}$ and $\frac{\pi}{2}$ are absorbing for y_t and so $\{-\frac{\pi}{2}\}$ and $\{\frac{\pi}{2}\}$ are open in the fine topology for y_t. On the other hand considering the given topology on $\mathbb{R}$, $P_x\{\sigma_U<\infty\} > 0$ for all open neighbourhoods U of $-\frac{\pi}{2}$ (or $\frac{\pi}{2}$ respectively), but $P_{\pm\frac{\pi}{2}}\{\sigma_V<\infty\} = 0$ for all open sets $V \subset (-\frac{\pi}{2}, \frac{\pi}{2})$.

What is really needed in the proofs of the above results is the lower semicontinuity of α-excessive functions, so we have the second approach

(b) strong Feller processes in the given topology.

For strong Feller processes any α-excessive function is lower semicontinuous in the given topology X. For processes with continuous trajectories the harmonic functions, in particular the Φ_B's, are even continuous. Hence we get the same results as in (a), now for the topology X. The assumption for the existence of an invariant measure is fufilled under Meyer's hypothesis (L), which is valid for strong Feller processes, so there is no restriction in this case.

To get an invariant measure for recurrent strong Feller processes we may set on a chain construction: consider the imbedded chain obtained by stopping the process on the boundary B of some bounded open set. The chain on the compact state space B has an invariant probability, ν say. Continuity of the transition probabilities gives a σ-finite measure on X

$\mu(A) = \int_B E_x \rho_A \nu(dx)$, $\rho_A = \int_0^\tau \chi_A(x_t)dt$, the time spent in A

during one cycle of the process from B to B. Of course the invariant measure is unique on recurrent classes.

Still there may be several recurrent classes and recurrent as well as transient states for a strong Feller process:

Example III 2: Consider a strong Feller process given on $[0,3] \subset \mathbb{R}^1$ by the operator $L = \frac{1}{2}(\cos x \frac{\partial}{\partial x})^2 + (\sin x)\frac{\partial}{\partial x}$. Use Hörmander's theorem to see hypoellipticity of $-\frac{\partial}{\partial t} + L^*$ and hence the strong Feller property. Notice by Stroock and Varadhan's support theorem that $[0,1]$ and $[2,3]$ are compact S-invariant sets, which carry invariant probabilities μ_1 and μ_2 such that supp $\mu_1 = [0,1]$ and supp $\mu_2 = [2,3]$. Hence all points in $[0,1] \cup [2,3]$ are recurrent, but all points in $(1,2)$ are transient, compare [1], theorem 3.5.

To get a characterization of the diffusion process as being either recurrent or transient, Stroock and Varadhan's (nondegenerate) support theorem may be used:

(c) nondegenerate diffusions

Nondegenerate diffusions are strong Feller with continuous trajectories, hence any harmonic function is continuous in the given topology. The continuation of recurrence now is assured by the maximum principle. The nondegenerate support theorem shows that $P_x\{x_t \in U(y)\} > 0$ for all $y \in \mathbb{R}^d$ and all open neighbourhoods U of y (even for all $t > 0$). So from (a) we have:

- the diffusion x_t is either recurrent or transient,
- if there exists an invariant measure μ for x_t, then $\operatorname{supp} \mu = \mathbb{R}^d$.

Criteria for nonexplosion, recurrence, existence of an invariant probability one can get via the observation, that a nondegenerate diffusion is nonexplosive,... if and only if the equations $Lu-u=0$, $Lu=0$, $Lu=-1$ have suitable solutions. Now this means the study of harmonic functions for the operator L. This can be done explicitly for the 1-dimensional case. So, following Hasminskii [15], Bhattacharya [8] gave radial test functions $(F(|x-z|), x,z \in \mathbb{R}^d$ (notice that a nondegenerate diffusion spreads into all directions, probably with different speed) for recurrence, transience... (compare also Friedman [12]). Via Dynkin's formula, which is just the martingale formulation of a diffusion and optional stopping, these test functions are related to the probabilities of hitting times.

(d) Lyapunov techniques

Dynkin's formula also is the key to sufficient conditions for recurrence, transience, existence of invariant measures, which rely on some kind of boundedness of the process. This boundedness can be given by some Lyapunov function, indicating the inward tendency of the process on some level surface. Dynkin's formula again relates the hitting probabilities to the behavior of the Lyapunov function. For results see Hasminskii [16], Wonham [30] or Miyahara [23], [24]. (In [16] Lemma III.8.2 of course deals with transience and not with recurrence.)

Remaining questions:

Given a diffusion, i.e. a Feller process with continuous trajectories, what is its recurrence and transience behaviour (in the given topology on $\mathbb{R}^d$) and how is it related to invariant measures, i.e. stationarity and ergodicity. We will answer these questions for diffusions in $\mathbb{R}^d$.

IV. Transience and recurrence for diffusions

Let us consider a diffusion x_t in its state space $X \subset \mathbb{R}^d$, for all $x \in X$ $P_x\{x_t \in X \text{ f.a. } t \geq 0\} = 1$, i.e. X is S-invariant for x_t. The first result states the dichotomy for points in X w.r.t. x_t:

Theorem IV 1. Let x_t be a Feller process with continuous trajectories in its state space $X \subset \mathbb{R}^d$. Then any point $x \in X$ is either recurrent or transient for x_t. Furthermore

(i) x is recurrent iff for all

$$\text{nbhds } U(x)\ R(x,U) = \int_0^\infty P(t,x,U)\,dt = \infty,$$

(ii) x is transient iff there exists U such that

$$R(x,u) < \infty.$$

We give the proof to show how a chain construction and the continuity properties of the transition functions of Feller processes give results. The first lemma is taken from [3], p. 196.

Lemma IV 1. If for $x,y \in X$ $P_y\{R_U\} > 0$ for all nbhds U of x, then $R(x,U) = \infty$ for all U.

Lemma IV 2. If there exists a nbhd V of x such that $P_x\{R_V\} < 1$, then there exists a nbhd W of x with $R(x,W) < \infty$.

Proof: By assumption x is no absorbing point, hence there exists a nbhd U of x with compact closure such that $\sup_{y \in X} E_y \tau_U \leq c < \infty$ by [11], lemma 5.5. Denote $G = U \cap V$, then $P_x\{T_G\} > \alpha > 0$ and the function $P_t\Phi_G(x): t \longrightarrow P_{x_t}\{\sigma_G<\infty\}$ tends to 0 on T_G P_x-a.s. by [3] p. 189. Choose a nbhd G' of x such that $\overline{G'} \subset G$. Then there exists $t_o > 0$ and $E_x(P_{x_{t_o}}\{\sigma_{G'}<\infty\}) = P_x\{\sigma_{G'} \cdot \Theta_{t_o} < \infty\} < 1 - \frac{\alpha}{2}$. By [11] lemma 5.4 the set $H = \{z, P_z\{\sigma_{G'} \cdot \Theta_{t_o} < \infty\} < 1 - \frac{\alpha}{4}\}$ is an open nbhd of x, so there exists a nbhd W of x with compact closure $\overline{W} \subset H$.

Now let us consider the following chain construction:

$$\begin{aligned}
\eta_1 &= t_o + \sigma_W \cdot \Theta_{t_o} \\
\eta_2 &= \eta_1 + (\tau_U \vee t_o) \cdot \Theta_{\eta_1} \\
&\vdots \\
\eta_{2n} &= \eta_{2n-1} + (\tau_U \vee t_o) \cdot \Theta_{\eta_{2n-1}} \\
\eta_{2n+1} &= \eta_{2n} + \sigma_W \cdot \Theta_{\eta_{2n}} \\
&\vdots
\end{aligned}$$

where $\tau_U \vee t_o = \sup\{\tau_U, t_o\}$, and estimate

$$R(x,W) \le t_o + \sum_{n=1}^{\infty} E_x(1_{\{\eta_{2n-1}<\infty\}}(\eta_{2n}-\eta_{2n-1}))$$

$$\le t_o + \sum_{n=1}^{\infty} E_x(1_{\{\eta_{2n-1}<\infty\}} E_{x_{\eta_{2n-1}}}(\tau_U \vee t_o))$$

by the strong Markov property [10], I. (3.34),

$$\le t_o + (t_o+c) \sum_{n=1}^{\infty} E_x(1_{\{\eta_{2n-1}<\infty\}}).$$

We compute

$$\begin{aligned}
P_x\{\eta_{2n+1}<\infty\} &= P_x\{\eta_{2n-1}<\infty,\ \sigma_W \circ \Theta_{\tau_U \vee t_o} \cdot \Theta_{\eta_{2n-1}} < \infty\} \\
&= E_x(1_{\{\eta_{2n-1}<\infty\}} P_{x_{\eta_{2n-1}}}\{\sigma_W \cdot \Theta_{\tau_U \vee t_o} < \infty\}).
\end{aligned}$$

Since $\overline{W}$ is compact and $x_{\eta_{2n-1}} \in \overline{W}$

$$P_{x_{\eta_{2n-1}}}\{\sigma_W \cdot \Theta_{\tau_U \vee t_o} < \infty\} \le P_{x_{\eta_{2n-1}}}\{\sigma_W \cdot \Theta_{t_o} < \infty\} < 1-\frac{\alpha}{4}$$

and by induction

$$P_x\{\eta_{2n+1}<\infty\} < (1-\tfrac{\alpha}{4}) P_x\{\eta_{2n-1}<\infty\} < (1-\tfrac{\alpha}{4})^n .$$

Putting this together yields

$$R(x,W) \leq t_o + (t_o+c) \sum_{n=1}^{\infty} (1-\frac{\alpha}{4})^n \leq t_o + (t_o+c)\frac{4}{\alpha} < \infty$$

□

To get a dichotomy not only for points, but for the diffusion itself, let us introduce the following notation:

Definition IV 1. A point $x \in X$ is called central, if there exists $y \neq x$ such that for all $\varepsilon > 0$ $P_x\{\sigma_{B(y,\varepsilon)} < \infty\} > 0$ and $P_y\{\sigma_{B(x,\varepsilon)} < \infty\} > 0$, where $B(z_o,\varepsilon) = \{z, |z-z_o| < \varepsilon\}$.

Here we choose $x \neq y$ to exclude the trivial case. Let us define central sets as the maximal (with respect to set inclusion) subsets of X containing only central points, and those one point sets that consist of absorbing states. Notice that any recurrent point belongs to some central set. For the diffusions we have

Theorem IV.2. The following dichotomy holds for diffusions on central sets c:

(i) if some $x_o \in C$ is recurrent, then all $x \in C$ are recurrent,

(ii) if some $x_o \in C$ is transient, then all $x \in C$ are transient.

For a proof of this theorem see [21]. Using the support theorem of Stroock and Varadhan we have an immediate corollary:

Corollary IV.1. Let the diffusion x_t be given by the operator L with a C_b^2, b bounded and uniformly Lipschitz. If the associated deterministic control system $\dot{\varphi}(t) = a(\varphi)u + \bar{b}(\varphi)$ is completely (approximately) controllable in X, then x_t is either recurrent or transient in X.

For the special case of nondegenerate diffusions this is theorem 3.2 in [8]. In general a diffusion may have several central sets, some of which may be recurrent, some transient, see example III.2. But for the recurrent parts we may take any central set C as a new state space:

Proposition IV.1. If C is a central recurrent set, then C is S-invariant.

Recurrent diffusions are classified according to the expectation of the hitting time of open sets:

Definition IV.2. A recurrent process x_t on X is called positive, if $E_x \sigma_U < \infty$ for all $x \in X$, $U \subset X$ open. Otherwise x_t is call null recurrent.

The following proposition simplifies the study of positive recurrence (and generalizes lemma IV.3.1 in [16] to diffusions).

Proposition IV.2. If for a recurrent process there exists some open set U_o such that $E_x \sigma_{U_o} < \infty$ for all $x \in X$, then $E_x \sigma_U < \infty$ for all open sets $U \subset X$.

The assumption of this proposition can be checked e.g. using Lyapunov techniques, we refer to [16], [23].

V. Ergodic properties of diffusions

Invariant probabilities for a diffusion can only live on central recurrent sets and these are S-invariant by proposition IV.1. For strong Feller diffusions these sets are even closed, cf. [21]. This fact is not true in general for Feller diffusions, as example III.1. shows: the intervall $(-\frac{\Pi}{2}, \frac{\Pi}{2})$ is a central recurrent set there, carrying an invariant probability. We see that in this case the Lie algebra generated by the vector fields has rank 0 in $\frac{-\Pi}{2}, \frac{\Pi}{2}$ and the diffusion is not strong Feller by the Ichihara-Kunita classification.

Let us assume without loss of generality that x_t is recurrent in X . Then we have the following picture:

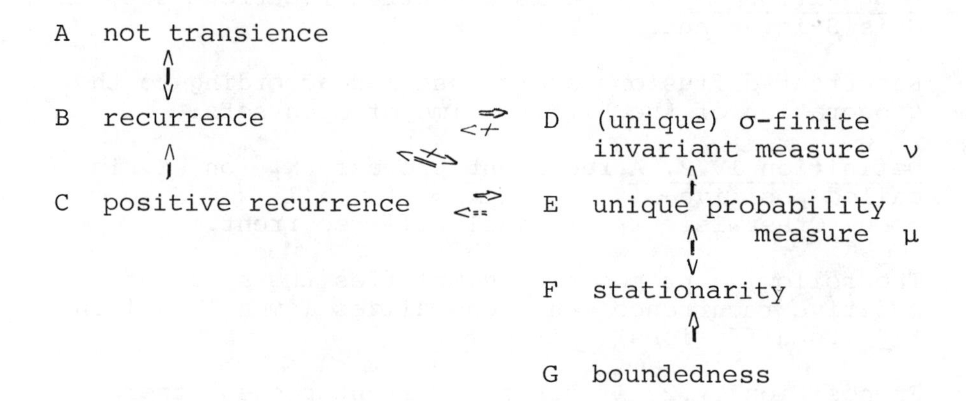

In D unique means up to a multiplicative constant. In a little bit more detail this scheme means:

A $\Leftrightarrow$ B This is theorem IV.1.

B $\Rightarrow$ D See [21].

D $\not\Rightarrow$ B The Wiener process in $\mathbb{R}^d$ has the Lebesgue measure as an invariant measure, but for $d \geq 3$ the Wiener process is not recurrent.

C $\Rightarrow$ E See [21].

E $\Rightarrow$ C If x_t is Feller, C holds for μ-almost $x \in X$, if x_t is strong Feller, C holds for all $x \in X$, see [21].

E $\Rightarrow$ B See [1].

B $\not\Rightarrow$ E The Wiener process in $\mathbb{R}^d$, $d \leq 2$ is recurrent, but has no invariant probability.

E $\Leftrightarrow$ F See [16], p. 71.

G $\Rightarrow$ F (Ultimate) boundedness of the process (i.e. dissipativity, see [16], p. 15) or of its moments imply the existence of stationary solutions via Lyapunov techniques, see [16], p. 57 ff or [24]. In particular in compact central sets the process is stationary.

The existence of a unique invariant probability μ implies ergodic properties of the process x_t. We give two examples:

<u>Law of large numbers:</u> For any μ-integrable function g

$$P_x\{\lim_{T\to\infty} \frac{1}{T}\int_o^T g(x_t)dt = \int_X g(x)\mu(dx)\} = 1$$

for μ-almost all $x \in X$

(see [1]) and even for all $x \in X$ if x_t is strong Feller, see [15].

<u>Convergence of transition functions:</u> For Feller processes the Cesaro limits converge: For any

measurable set $A \subset X$ $\lim_{T\to\infty} \frac{1}{T}\int_o^T P(t,x,A)dt = \mu(A)$,

f.a. x X, if x_t is strong Feller we even have the convergence of the transition functions themselves: $\lim_{t\to\infty} P(t,x,A) = \mu(A)$

f.a. $x \in X$ (compare [1], [15] and [17]).

Remember for a stochastic system in the C^∞-context the characterization of the strong Feller diffusions as the hypoelliptic ones. Then one has C^∞ densities for the transition probabilities and also for an invariant probability if it exists. For nondegenerate diffusions the density is strictly positive, see [17].

References

[1] Arnold, L.; Kliemann, W. Qualitative theory of stochastic systems. to appear in: Probabilistic analysis and related topics, Vol. 3, Bharucha-Reid, A. T. (ed.), New York: Academic Press, 1982

[2] Arnold, L.; Lefever, R. (eds.) Stochastic nonlinear systems. Berlin: Springer Verlag, 1981

[3] Azéma, J.; Kaplan-Duflo, M.; Revuz, D. Récurrence fine des processus de Markov. Ann. Inst. Henri Poincaré, 2 (1966) p. 185-220

[4] -, -, -. Mesure invariante sur les classes recurrentes des processus de Markov. Z. Wtheorie verw. Gebiete 8(1967) p. 157-181

[5] -, -, -. Note sur la mesure invariante des processus de Markov recurrents. Ann. Inst. Henri Poincaré, 3(1967) p. 397-402

[6] -, -, -. Propriétés relatives des processus de Markov recurrents. Z. Wtheorie verw. Gebiete 13(1969) p. 286-314

[7] Bhatia, N. P.; Szegö, G. P. Stability theory of dynamical systems. Berlin: Springer Verlag, 1970

[8] Bhattacharya, R. N. Criteria for recurrence and existence of invariant measures for multidimensional diffusions. Ann. Prob. 6 (1978) p. 541-553 and 8 (1980) p. 1194-1195

[9] Blumenthal, R. M.; Getoor, R. K. Markov processes and potential theory. New York: Academic Press, 1968

[10] Duflo, M.; Revuz, D. Propriétés asymptotiques des probabilités de transition des processus de Markov recurrents, Ann. Inst. Henri Poincaré, 5(1969) p. 233-244

[11] Dynkin, E. B. Markov processes, I and II, Berlin: Springer Verlag, 1965

[12] Friedman, A. Wandering out to infinity of diffusion processes. Trans. AMS 184(1973) p. 185-203

[13] - . Stochastic differential equations and applications, I and II. New York: Academic Press, 1975 and 1976

[14] Getoor, R. K. Transience and recurrence of Markov processes. Lecture Notes in Mathematics 784, Séminaire de Probabilité XIV (1980) p. 397-409

[15] Hasminskii (Khasminkii), R. Z. Ergodic properties of recurrent diffusion processes and stabilization of the solution to the Cauchy problem for parabolic equations. Theor. Prob. Appl. 5 (1960) p. 179-195

[16] -. Stochastic stability of differential equations. Alphen aan den Rijn: Sijthoff and Noordhoff, 1980, (russian: 1969)

[17] Ichihara, K.; Kunita, H. A classification of the second order degenerate elliptic operators and its probabilistic characterization. Z. Wtheorie verw. Gebiete 30(1974) p. 235-254 and 39(1977) p. 81-84

[18] Ikeda, N.; Watanabe, S. Stochastic differential equations and diffusion processes. Amsterdam: North Holland. 1981

[19] Ito, K. On stochastic differential equations. Mem. Amer. Math. Soc. 4; 1951

[20] Kliemann, W. Qualitative theory of stochastic dynamical systems applications to life sciences, to appear in: Bull. Math. Bio. (1982)

[21] -. Diffusions on manifolds. Report "Forschungsschwerpunkt Dynamische Systeme" University of Bremen, 1982

[22] Kushner, H. Stochastic stability and control. New York: Academic Press, 1967

[23] Miyahara, Y. Ultimate boundedness of the systems governed by stochastic differential equations. Nagoya Math. J. 47(1972) p. 111-144.

[24] -. Invariant measures of ultimately bounded stochastic processes. Nagoya Math. J. 49(1973) p. 149-153

[25] Nemytskii, V. V.; Stepanov, V. V. Qualitative theory of differential equations. Princeton, 1960

[26] Stroock, D. W.; Varadhan, S. R. S. On the support of diffusion processes with applications to the strong maximum principle. Proc. Sixth Berkeley Symp. Vol. III (1972) p. 333-359

[27] -, -. On degenerate elliptic parabolic operators of second order and their associated diffusions. Comm. Pure Appl. Math. 25 (1972) p. 651-713

[28] -, -. Multidimensional diffusion processes. Berlin: Springer Verlag, 1979

[29] Williams, D. Diffusions, Markov processes and martingales. New York: Wiley, 1979

[30] Wonham, W. M. Lyapunov criteria for weak stochastic stability. J. Diff. Equ. 2 (1966) p. 365-377

EIGENFUNCTION EXPANSION FOR THE NONLINEAR TIME DEPENDENT BROWNIAN MOTION

Claudio Maccone

Istituto Matematico - Politecnico di Torino
Corso Duca degli Abruzzi 24 - 10129 Torino
Italy

ABSTRACT. The nonlinear time dependent Brownian motion, $B(g(t))$, is the object of the present paper. The time-rescaling function, $g(t)$, may be any function continuous over a finite positive time interval ranging from 0 to a fixed T, and such that $g(0)=0$. We firstly prove $B(g(t))$ to be equivalent to a white-noise integral having zero mean and a time-dependent variance. Secondly we find the eigenfunction (Karhunen-Loève) expansion of $B(g(t))$. The eigenfunctions are Bessel functions of the first kind, with a suitable time function for argument, and they are multiplied by another time function. The eigenvalues virtually are determined by the zeros of the Bessel functions.

1. WHITE-NOISE INTEGRALS.

Let $B(t)$ be the Brownian motion ($t \geq 0$), having zero mean, variance t, and initial condition $B(0)=0$. Its derivative $B'(t)$, is the (stationary) white noise, with autocorrelation

$$\langle B'(t_1)\, B'(t_2)\rangle = \delta(t_1 - t_2). \tag{1.1}$$

We call WHITE-NOISE INTEGRAL any integral of the form

$$X(t) = \int_0^t f(s)\, dB(s) = \int_0^t f(s)\, B'(s)\, ds \tag{1.2}$$

where $f(t)$ may be any function, continuous, with $f'(t)$, over a finite positive time interval ranging from zero to a fixed T.

Of course $X(t)$ is a stochastic process, and has zero mean. To find its autocorrelation, we interchange the average brackets with the integral sign, and apply (1.1); the UNIT STEP FUNCTION $U(t)$ (equal to zero for $t<0$ and to unity for $t>0$) yields

R. S. Bucy and J. M. F. Moura (eds.), Nonlinear Stochastic Problems, 455–466.

$$\left\langle X(t_1)\, X(t_2) \right\rangle = \int_0^{t_1} ds\, f(s) \int_0^{t_2} dt\, f(t)\, \delta(t-s) =$$

$$= \int_0^{t_1} ds\, f(s) \left[f(s)\, U(t_2 - s) \right] = \int_0^{t_1 \wedge t_2} f^2(s)\, ds. \qquad (1.3)$$

In the last step the MINIMUM (= the smaller) of t_1 and t_2, denoted $t_1 \wedge t_2$, was introduced: it equals t_1 for $t_1 < t_2$ and t_2 for $t_2 < t_1$. Eq.(1.3) is the AUTOCORRELATION of the white--noise integral (1.2); upon setting $t_1 = t_2 = t$ its VARIANCE reads

$$\sigma^2_{X(t)} = \left\langle X^2(t) \right\rangle = \int_0^t f^2(s)\, ds \equiv g(t). \qquad (1.4)$$

2. NONLINEAR TIME DEPENDENT BROWNIAN MOTION.

In this section we prove that the white-noise integral $X(t)$ of eq.(1.2), and the NONLINEAR TIME DEPENDENT BROWNIAN MOTION

$$B\big(g(t)\big) = B\left(\int_0^t f^2(s)\, ds \right) \qquad (2.1)$$

are just the same thing, namely a Gaussian stochastic process with zero mean and variance given by eq.(1.4).

For the proof, let us firstly notice that the two processes (1.2) and (2.1) evidently fulfill the same initial condition :

$$X(0) = 0 \qquad \text{and} \qquad B(0) = 0 \quad ; \qquad (2.2)$$

secondly, they have the same mean value (zero) :

$$\left\langle X(t) \right\rangle = 0 \qquad \text{and} \qquad \left\langle B\big(g(t)\big) \right\rangle = 0 \quad ; \qquad (2.3)$$

thirdly, we claim they also have the same autocorrelation :

$$\left\langle X(t_1)\, X(t_2) \right\rangle = \left\langle B\big(g(t_1)\big)\, B\big(g(t_2)\big) \right\rangle. \qquad (2.4)$$

In fact from the autocorrelation (1.3) one gets

$$\left\langle X(t_1)\, X(t_2) \right\rangle = \int_0^{t_1 \wedge t_2} f^2(s)\, ds = \left[\int_0^{t_1} f^2(s)\, ds \right] \wedge \left[\int_0^{t_2} f^2(s)\, ds \right] =$$

by virtue of the autocorrelation of the Brownian motion, $\left\langle B(a)\, B(b) \right\rangle = a \wedge b$, we have

$$= \left\langle B\left(\int_0^{t_1} f^2(s)\, ds \right) B\left(\int_0^{t_2} f^2(s)\, ds \right) \right\rangle = \left\langle B\big(g(t_1)\big)\, B\big(g(t_2)\big) \right\rangle$$

and eq.(2.4) is thusly proved.

To sum up, we have proven that the processes (1.2) and (2.1) have same initial condition, same mean value, and same autocorrelation. Since the (time-rescaled) Brownian motion (2.1) is a GAUSSIAN process, all its higher-moments' properties are fully determined by its first two moments' properties (for this see p.299 of ref.(1)). But, then, the same fact must hold good for the process (1.2) as well, and so THE TWO PROCESSES COINCIDE COMPLETELY. It follows that the probability density of $X(t)$ is just that of $B(g(t))$, namely the TIME-RESCALED GAUSSIAN

$$f_{X(t)}(x) = \frac{1}{\sqrt{2\pi}\sqrt{\int_0^t f^2(s)\,ds}} \cdot \exp\left[-\frac{x^2}{2\int_0^t f^2(s)\,ds}\right]. \tag{2.5}$$

Of course, in the case $f(t) = 1$, eq.(2.5) reduces to the well-known Gaussian probability density of the Brownian motion

$$f_{B(t)}(x) = \frac{1}{\sqrt{2\pi}\sqrt{t}} \cdot \exp\left[-\frac{x^2}{2t}\right]. \tag{2.6}$$

Next we are going to find the DIFFUSION PARTIAL DIFFERENTIAL EQUATION fulfilled by the time-rescaled Gaussian density (2.5). To this end, let us recall the result (proved in Problem 15.5, p.552, of ref.(1)) stating that the solution of the diffusion partial differential equation with time-dependent coefficients

$$\frac{\partial p(x,t)}{\partial t} = \frac{\beta(t)}{2}\frac{\partial^2 p(x,t)}{\partial x^2} - \alpha(t)\frac{\partial p(x,t)}{\partial x} \tag{2.7}$$

fulfilling the INITIAL CONDITION $p(x,0) = \delta(x)$, is given by

$$p(x,t) = \frac{1}{\sqrt{2\pi}\sqrt{\int_0^t \beta(s)\,ds}} \cdot \exp\left[-\frac{\left(x - \int_0^t \alpha(s)\,ds\right)^2}{2\int_0^t \beta(s)\,ds}\right]. \tag{2.8}$$

Then, we may evidently let (2.8) and (2.5) coincide by setting

$$\alpha(t) = 0 \qquad \text{and} \qquad \beta(t) = f^2(t) \tag{2.9}$$

and so the diffusion partial differential equation is given by

$$\frac{\partial f_{X(t)}(x)}{\partial t} = \frac{f^2(t)}{2}\frac{\partial^2 f_{X(t)}(x)}{\partial x^2}. \tag{2.10}$$

Several more results along these lines could be derived, but we have little space for them here. We merely confine ourselves to stating without proof that the FIRST- PASSAGE TIME, namely the amount of time necessary to reach the value $x = a$ (starting from the origin at $t = 0$) has probability density

$$f_{T_a}(t) = \frac{|a|}{\sqrt{2\pi}} \cdot \frac{f^2(t)}{\left[\int_0^t f^2(s)\,ds\right]^{3/2}} \cdot \exp\left[-\frac{a^2}{2\cdot\int_0^t f^2(s)\,ds}\right]. \tag{2.11}$$

After these preliminaries, we may now turn to this paper's bulk.

3. EIGENFUNCTION (KARHUNEN-LOÈVE) EXPANSION OF A PROCESS.

Let the time t range from zero to a fixed positive T : $0 \leq t \leq T$. Let $X(t)$ be a second-order stochastic process with continuous autocorrelation over the above time interval, and with IDENTICALLY ZERO MEAN. Then the expansion holds good

$$X(t) = \sum_{n=1}^{\infty} Z_n \varphi_n(t) \quad . \tag{3.1}$$

This is guaranteed by the KARHUNEN-LOÈVE THEOREM, that is proved, for instance, at the pages 262-279 of ref.(2), and will not be proved here. The functions $\varphi_n(t)$ (with $n=1,2,...$) are the EIGEN-FUNCTIONS OF THE AUTOCORRELATION OF $X(t)$, namely the solutions of the integral equation (having a symmetric kernel)

$$\int_0^T \left\langle X(t_1)\, X(t_2) \right\rangle \varphi_n(t_2)\, dt_2 = \lambda_n\, \varphi_n(t_1) \tag{3.2}$$

where the constants λ_n (with $n=1,2,...$) are the EIGENVALUES. The eigenfunctions $\varphi_n(t)$ are ORTHONORMAL, namely one has

$$\int_0^T \varphi_n(t)\, \varphi_m(t)\, dt = \delta_{mn} \tag{3.3}$$

and this may be restated by saying that the $\varphi_n(t)$ form an orthonormal base spanning the Hilbert space $L^2(0,T)$.

The Z_n in (3.1) are RANDOM VARIABLES, formally given by

$$Z_n = \int_0^T X(t)\, \varphi_n(t)\, dt \tag{3.4}$$

as one can derive from (3.1) and (3.3). Then, from (3.2), (3.3) and (3.4), a few steps yield the STOCHASTIC ORTHOGONALITY of Z_n

$$\left\langle Z_m Z_n \right\rangle = \lambda_n\, \delta_{mn} \tag{3.5}$$

meaning that the random variables Z_n are INDEPENDENT OF EACH OTHER. Moreover from (3.1) one can infer that the mean value of Z_n is zero, and, from (3.5) in the case $m=n$, that the variance of each Z_n is just the corresponding eigenvalue λ_n. Hence THE EIGENVALUES ARE NONNEGATIVE.

A further consequence is the EXPANSION OF THE AUTOCORRELATION

$$\left\langle X(t_1)\, X(t_2) \right\rangle = \sum_{n=1}^{\infty} \lambda_n\, \varphi_n(t_1)\, \varphi_n(t_2) \tag{3.6}$$

following from (3.1) and (3.5). Setting $t_1=t_2=t$, (3.6) yields

$$\sigma^2_{X(t)} = \sum_{n=1}^{\infty} \lambda_n\, \varphi^2_n(t) \tag{3.7}$$

which is the EXPANSION OF THE VARIANCE. An important consequence is obtained by integrating both sides of (3.7) with respect to t from zero to T : using (3.3) with $m=n$, and (1.4), one finds

$$\sum_{n=1}^{\infty} \lambda_n = \int_0^T \sigma^2_{X(t)}\, dt = \int_0^T dt \int_0^t f^2(s)\, ds \quad . \tag{3.8}$$

This is the SUM OF THE SERIES OF THE EIGENVALUES, showing that, if $\sigma_{X(t)}$ is square-integrable over $[0,T]$, the series (3.8) converges. In turn, this means that the sequence of the λ_n must decrease for increasing n, namely $\lim_{n\to\infty} \lambda_n = 0$. And finally we reach the conclusion : since the mean value of each Z_n is zero, and its variance is the eigenvalue λ_n, the limit holds good

$$\lim_{n\to\infty} Z_n = 0$$

showing that the eigenfunction (Karhunen-Loève) expansion (3.1) CONVERGES (it might be added "UNIFORMLY IN t ", see ref.(2),p.278).

We complete the present section by pointing out that the probability distribution of the random variables Z_n is GAUSSIAN if that of the stochastic process $X(t)$ is so. Evidently, this is the case if $X(t)$ is the white-noise integral (1.2), namely the time-rescaled Brownian motion (2.1). Thus, THE Z_n ARE GAUSSIAN RANDOM VARIABLES HAVING ZERO MEAN AND VARIANCE EQUAL TO THE EIGENVALUE λ_n.

4. FROM AN INTEGRAL EQUATION TO A DIFFERENTIAL EQUATION.

This paper's aim is the expansion (3.1) of the white-noise integral (1.2), i.e. the time-rescaled Brownian motion (2.1). The main difficulty lies in solving the integral equation for the eigenfunctions $\varphi_n(t)$, that, inserting (1.3) into (3.2), reads

$$\int_0^T dt_2\, \varphi_n(t_2) \int_0^{t_1\wedge t_2} f^2(s)\, ds = \lambda_n\, \varphi_n(t_1) \quad . \tag{4.1}$$

In order to solve this integral equation, we shall firstly convert it into a differential equation (and this will be done in the present section), and secondly solve the differential equation thusly obtained (what will be done in the sections to come).

Before starting any manipulation of the integral equation (4.1), let us remark that, if we set $t_1 = 0$, the minimum $0\wedge t_2 = 0$ makes the inner integral vanish, yielding the INITIAL CONDITION

$$\varphi_n(0) = 0 \tag{4.2}$$

which must be fulfilled by the eigenfunctions $\varphi_n(t)$.

Next we turn to a useful lemma about the MINIMUM, stating that

$$\frac{d(t_1\wedge t_2)}{dt_1} = U(t_2 - t_1) \quad . \tag{4.3}$$

Its proof is based on the following self-evident properties :

$$\frac{d(t_1\wedge t_2)}{dt_1} = \frac{d}{dt_1}\Big[t_1\, U(t_2-t_1) + t_2\, U(t_1-t_2)\Big] =$$
$$= U(t_2-t_1) - t_1\,\delta(t_2-t_1) + t_2\,\delta(t_1-t_2) =$$
$$= U(t_2-t_1) + \underbrace{(t_2-t_1)\delta(t_2-t_1)}_{0} = U(t_2-t_1)\,.$$

Now we are ready to differentiate both sides of (4.1) with respect to t_1 ; by virtue of Leibniz's theorem for differentiation of an integral (see p.11 of ref.[3], for instance), we get :

$$\lambda_n \varphi_n'(t_1) = \int_0^T dt_2\, \varphi_n(t_2) \frac{\partial}{\partial t_1} \int_0^{t_1 \wedge t_2} f^2(s)\, ds = \int_0^T dt_2\, \varphi_n(t_2) \cdot$$

$$\cdot \left[f^2(t_1) \frac{d(t_1 \wedge t_2)}{dt_1} \right] = f^2(t_1) \int_0^T dt_2\, \varphi_n(t_2)\, U(t_2 - t_1).$$

That is :

$$\lambda_n \varphi_n'(t_1) = f^2(t_1) \int_{t_1}^T \varphi_n(t_2)\, dt_2 . \qquad (4.4)$$

An immediate consequence of (4.4) is found by setting $t_1 = T$:

$$\varphi_n'(T) = 0 \qquad (4.5)$$

This is the FINAL CONDITION to be fulfilled by the eigenfunctions, and (4.2) and (4.5) together form the two BOUNDARY CONDITIONS.

Now we merely have to differentiate (4.4) with respect to t_1 (replacing it by t) to get the DIFFERENTIAL EQUATION FOR $\varphi_n(t)$

$$\frac{d}{dt}\left[\frac{1}{f^2(t)} \varphi_n'(t)\right] + \frac{1}{\lambda_n} \varphi_n(t) = 0 . \qquad (4.6)$$

This differential equation (with its solution) is the most important result of the present paper.

5. OSCILLATORY BEHAVIOUR OF THE EIGENFUNCTIONS.

It is remarkable that the qualitative behaviour of the eigenfunctions, $\varphi_n(t)$, solutions of eq.(4.6), can be understood even WITHOUT actually SOLVING the equation. In fact, on the one hand (4.6) is a linear differential equation of the second order put in its self-adjoint form already ; and, on the other hand, to such a differential equation the well-known SONINE-POLYA THEOREM, that we shall later state, may be applied. Let us firstly set :

$$\xi = \int_0^t f^2(s)\, ds \quad \text{whence} \quad \frac{1}{dt} = f^2(t) \frac{1}{d\xi} . \qquad (5.1)$$

Then a few steps show that, in terms of the new independent variable ξ , the differential equation (4.6) becomes

$$\frac{d^2 \varphi_n(\xi)}{d\xi^2} = - \frac{1}{\lambda_n f^2(\xi)} \varphi_n(\xi) . \qquad (5.2)$$

Since the eigenvalues λ_n are positive, and so is $f^2(\xi)$, we infer that the second order derivative of $\varphi_n(t)$ always has the OPPOSITE SIGN to that of $\varphi_n(t)$, meaning that THE EIGENFUNCTIONS OSCILLATE ABOVE AND BELOW THE POSITIVE TIME AXIS. Eq.(5.2) shows that the inflexion points of the eigenfunctions coincide with their zeros, and both lie just on the time axis, also. Moreover the initial condition (4.2) says that the origin is a zero of the eigenfunctions, while the final condition (4.5) reveals that the

final instant $t = T$ is a maximum or a minimum of $\varphi_n(t)$.
The Sonine-Polya theorem will now be applied to (5.2) : it states that the successive maxima and minima of $\varphi_n(t)$ form an increasing (respectively decreasing) sequence if one has (respectively)

$$\frac{d}{dt}\left[\frac{1}{\lambda_n f^2(t)}\right] \lesseqgtr 0 \text{ namely if } -2\frac{1}{f^3(t)}\cdot\frac{df(t)}{dt} \lesseqgtr 0$$

and, finally, since $f^2(t)$ is positive, the last formula becomes

$$\frac{d \ln f(t)}{dt} \gtreqless 0 . \tag{5.3}$$

In conclusion : IF THE LOGARITHMIC DERIVATIVE OF $f(t)$ IS POSITIVE (NEGATIVE) OVER A CERTAIN TIME INTERVAL, THERE THE MAXIMA AND MINIMA OF THE EIGENFUNCTIONS $\varphi_n(t)$ FORM AN INCREASING (DECREASING) SEQUENCE, as shown on figure 1 (figure 2) hereafter :

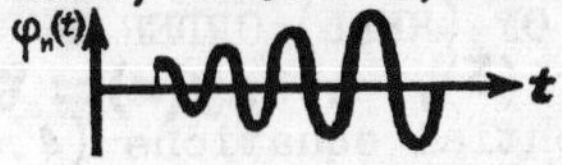

Figure 1, 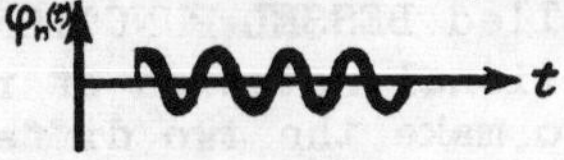 Figure 3, 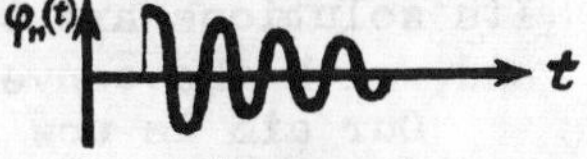 Figure 2.

The case "in the middle", shown on Figure 3, evidently corresponds to the logarithmic derivative of $f(t)$ equalling zero. Thus $f(t) = a$ const., and the white-noise integral (1.2), with an integration, shows that $X(t) = C B(t)$, where C is any constant with respect to time. So, this is virtually the BROWNIAN CASE, with NO TIME RESCALING. Now we claim that the eigenfunctions in figure 3 are SINES. In fact (4.6) becomes the HARMONIC OSCILLATOR equation, whose general integral is notoriously given by (A and B = constants)

$$\varphi_n(t) = A \sin\left(\frac{C}{\sqrt{\lambda_n}}\, t\right) + B \cos\left(\frac{C}{\sqrt{\lambda_n}}\, t\right) . \tag{5.4}$$

Invoking then the two boundary conditions (4.2) and (4.5), we get $B = 0$, and, by differentiating (5.4), the EIGENVALUES are found

$$\lambda_n = \left[\frac{2CT}{\pi(2n-1)}\right]^2 \qquad (n = 1, 2, \ldots) . \tag{5.5}$$

Inserting (5.5) into (5.4), the EIGENFUNCTIONS are in the end

$$\varphi_n(t) = A \sin\left[\frac{\pi(2n-1)}{2T}\, t\right] \qquad (n = 1, 2, \ldots) . \tag{5.6}$$

where the normalization condition (3.3) with $n = m$ yields $A = \sqrt{\frac{2}{T}}$.

6. EXPLICIT EXPRESSION OF THE EIGENFUNCTIONS.

In the present section we are going to SOLVE THE DIFFERENTIAL EQUATION (4.6) BY REDUCING IT TO THE WELL-KNOWN BESSEL EQUATION. Unfortunately the relevant calculations are rather lenghty.

The first step is made by CHANGING THE UNKNOWN FUNCTION in (4.6) into the product of a function $\chi(t)$ INDEPENDENT OF THE INDEX n , by a new n -depending function $y_n(t)$, namely :

$$\varphi_n(t) = \chi(t) \cdot y_n(t). \tag{6.1}$$

Differentiating (6.1), and inserting into (4.6), one gets

$$\frac{d}{dt}\left[\frac{\chi'(t)}{f^2(t)} y_n(t) + \frac{\chi(t)}{f^2(t)} y_n'(t)\right] + \frac{\chi(t)}{\lambda_n} y_n(t) = 0. \tag{6.2}$$

Differentiating (6.2) once again, one arrives at a linear second order differential equation that, in SELF-ADJOINT FORM, reads

$$\frac{d}{dt}\left[\frac{\chi^2(t)}{f^2(t)} y_n'(t)\right] + \chi(t)\left\{\frac{d}{dt}\left[\frac{\chi'(t)}{f^2(t)}\right] + \frac{\chi(t)}{\lambda_n}\right\} y_n(t) = 0. \tag{6.3}$$

Now we leave (6.3) and turn to the BESSEL DIFFERENTIAL EQUATION, that, in SELF-ADJOINT FORM (see p.4, Vol.2 of ref.(4)), is

$$\frac{d}{dx}\left[x \frac{dy(x)}{dx}\right] + \left[x - \frac{\nu^2}{x}\right] y(x) = 0. \tag{6.4}$$

Its solutions are called BESSEL FUNCTIONS OF (REAL) ORDER ν, and, with the conventional notation of ref.(5), p.98, $y(x) = \mathscr{C}_\nu(x)$.

Our aim is now to make the two differential equations (4.6) and (6.4) coincide. As we already performed on (4.6) a change of function (given by (6.1)), now we must perform on (6.4) the change

$$x = \psi(t) \tag{6.5}$$

of independent variable. We call $\psi(t)$ the TIME-RESCALING FUNCTION, and the new form of (6.4) TIME-RESCALED BESSEL EQUATION

$$\frac{d}{dt}\left[\frac{\psi(t)}{\psi'(t)} y'(t)\right] + \left[\psi^2(t) - \nu^2\right] \frac{\psi'(t)}{\psi(t)} y(t) = 0. \tag{6.6}$$

In order to let (6.6) coincide with (6.3) we have to put the n not just to $y(t)$ but to $\psi(t)$ also, because $y_n(t) = \mathscr{C}_\nu(\psi_n(t))$:

$$\begin{cases} \dfrac{\chi^2(t)}{f^2(t)} = \dfrac{\psi_n(t)}{\psi_n'(t)} & (6.7) \\[2ex] \chi(t)\left\{\dfrac{d}{dt}\left[\dfrac{\chi'(t)}{f^2(t)}\right] + \dfrac{\chi(t)}{\lambda_n}\right\} = \left[\psi_n^2(t) - \nu^2\right] \dfrac{\psi_n'(t)}{\psi_n(t)} & (6.8) \end{cases}$$

this is the system found by checking (6.6) against (6.3). Since the left of (6.7) is n-independent, so must be its right, whence

$$\psi_n(t) = \ell_n \vartheta(t) \tag{6.9}$$

where ℓ_n is a constant and $\vartheta(t)$ a new function. These yield

$$\begin{cases} \dfrac{\chi^2(t)}{f^2(t)} = \dfrac{\vartheta(t)}{\vartheta'(t)} & (6.10) \\[2ex] \chi(t) \dfrac{d}{dt}\left[\dfrac{\chi'(t)}{f^2(t)}\right] + \dfrac{\chi^2(t)}{\lambda_n} = \ell_n^2 \vartheta(t) \vartheta'(t) - \nu^2 \dfrac{\vartheta'(t)}{\vartheta(t)}. & (6.11) \end{cases}$$

Now we observe that there are two n-depending terms in (6.11) and two n-independent ones. Since (6.11) holds for any $n=1,2,\ldots$ it is to be regarded as the sum of two equations : one n-independent and one n-dependent, and the latter in turn splits down :

$$
\begin{cases}
\dfrac{\chi^2(t)}{f^2(t)} = \dfrac{\vartheta(t)}{\vartheta'(t)} & (6.12) \\[2ex]
\chi(t)\,\dfrac{d}{dt}\left[\dfrac{\chi'(t)}{f^2(t)}\right] = -\nu^2\,\dfrac{\vartheta'(t)}{\vartheta(t)} & (6.13) \\[2ex]
\chi^2(t) = \vartheta(t)\,\vartheta'(t) & (6.14) \\[1ex]
\dfrac{1}{\lambda_n} = \ell_n^2\,. & (6.15)
\end{cases}
$$

Eliminating $\chi^2(t)$ between (6.12) and (6.14), and integrating :

$$\vartheta(t) = \int_0^t f(s)\,ds\,. \tag{6.16}$$

From (6.15) and (6.16) the TIME- RESCALING FUNCTION of (6.9) reads

$$\psi_n(t) = \frac{1}{\sqrt{\lambda_n}}\cdot\int_0^t f(s)\,ds \tag{6.17}$$

where the positive sign was taken because the (rescaled) time is non-negative. Similarly, from (6.14) and (6.16) we get for $\chi(t)$

$$\chi(t) = \sqrt{f(t)\cdot\int_0^t f(s)\,ds}\,. \tag{6.18}$$

As for the ORDER, (6.13) and (6.12) show it to be a TIME FUNCTION

$$\nu^2 = -\frac{\chi^3(t)}{f^2(t)}\cdot\frac{d}{dt}\left[\frac{\chi'(t)}{f^2(t)}\right] \tag{6.19}$$

where $\chi(t)$ is given, in terms of the known $f(t)$, by (6.18). Finally we have to decide which KIND of Bessel function fits to our needs. Evidently we require an OSCILLATORY KIND. And if we set $f(t) = 1$ (BROWNIAN CASE) whence $\chi(t) = \sqrt{t}$ and, from (6.19), $\nu = \pm\frac{1}{2}$, the only Bessel function yielding the SINE of (5.6) is

$$J_{\frac{1}{2}}(t) = \sqrt{\frac{2}{\pi t}}\,\sin t\,. \tag{6.20}$$

We infer that the Bessel functions of the FIRST KIND are correct.

In conclusion, (6.1) (with (6.18) inserted) and (6.5) (with (6.17) inserted) yield the EIGENFUNCTIONS, SOLUTION OF (4.6) :

$$\varphi_n(t) = N_n\sqrt{f(t)\int_0^t f(s)\,ds}\cdot J_{\nu(t)}\left(\frac{1}{\sqrt{\lambda_n}}\int_0^t f(s)\,ds\right). \tag{6.21}$$

7. EIGENVALUES AND NORMALIZATION CONSTANTS.

In (6.21) the eigenvalues λ_n and the normalisation constants N_n still have to be determined. To this end, let us write the orthogonality condition (3.3) with $m \neq n$ for the $\varphi_n(t)$ of (6.21):

$$0 = \int_0^T \varphi_m(t)\,\varphi_n(t)\,dt = N_n^2\int_0^T f(t)\cdot\int_0^t f(s)\,ds\cdot J_{\nu(t)}\left(\frac{1}{\sqrt{\lambda_m}}\int_0^t f(s)\,ds\right)\cdot$$

$$\cdot J_{\nu(t)}\left(\frac{1}{\sqrt{\lambda_n}}\int_0^t f(s)\,ds\right) dt. \tag{7.1}$$

Upon setting $x\cdot\left[\int_0^T f(s)\,ds\right] = \int_0^t f(s)\,ds$ and $\gamma_n = \dfrac{\int_0^T f(s)\,ds}{\sqrt{\lambda_n}}$ (7.2)

one gets :

$$\int_0^1 x\, J_{\nu(x)}(\gamma_m x)\, J_{\nu(x)}(\gamma_n x)\,dx = 0 \tag{7.3}$$

as the new ORTHOGONALITY CONDITION for the Bessel functions. The particular case of (7.3) where ν is a CONSTANT, namely is independent of x, is well known, and is formula (48), Vol.2, on p.70 of ref.(4). Moreover the EIGENVALUES are found from (7.2)

$$\lambda_n = \left[\int_0^T f(s)\,ds\right]^2 \cdot \frac{1}{(\gamma_n)^2} \tag{7.4}$$

as expressed in terms of the new quantities γ_n. But what is the meaning of the γ_n ? To work it out we must resort to the two BOUNDARY CONDITIONS (4.2) and (4.5), so far never used. On the one hand (4.2) is IDENTICALLY fulfilled by the $\varphi_n(0)$ of (6.21), and yields no valuable information. But (4.5) does help : in fact, by differentiating (6.21) once with respect to t, and then replacing t by T, and equalling to zero as requested by (4.5), inserting the γ_n defined by (7.2), one gets :

$$\chi'(T)\, J_{\nu(T)}(\gamma_n) + \chi(T)\left[\frac{f(T)\cdot\gamma_n}{\int_0^T f(s)\,ds}\, J'_{\nu(T)}(\gamma_n) + \frac{\partial J_{\nu(T)}(\gamma_n)}{\partial \nu}\,\nu'(T)\right] = 0$$

that we call eq.(7.5). Thus, THE γ_n ARE THE ROOTS OF (7.5), of course to be found numerically on a computer, and we conclude that the eigenvalues λ_n of (7.4), or the related quantities γ_n, are virtually the ZEROS of suitable combinations of Bessels.

Finally we want the NORMALIZATION CONSTANTS N_n given by

$$1 = \int_0^T \left[\varphi_n(t)\right]^2 dt = N_n^2 \left[\int_0^T f(s)\,ds\right]^2 \cdot \int_0^1 x\left[J_{\nu(x)}(\gamma_n x)\right]^2 dx. \tag{7.6}$$

Once again the last integral has to be found numerically on a computer, and the normalization constants are thus determined.

8. THE FRACTIONAL BROWNIAN MOTIONS.

As an application of all the previous results, in this section we consider the FRACTIONAL BROWNIAN MOTIONS, denoted $B_H(t)$, and defined as the Riemann-Liouville fractional integral/derivative of order $H-\frac{1}{2}$ of the Brownian motion (see refs.(6),(7)) :

$$B_H(t) = \int_0^t \frac{(t-s)^{H-\frac{1}{2}}}{\Gamma(H+\frac{1}{2})}\, B'(s)\,ds = \int_0^t \frac{s^{H-\frac{1}{2}}}{\Gamma(H+\frac{1}{2})}\, B'(s)\,ds \tag{8.1}$$

(in the last step we inverted the convolution's order and used the property $B'(t-s) = B'(s)$ of the STATIONARY white-noise). Checking (8.1) against (1.2), they coincide by setting

$$f(t) = \frac{t^{H-\frac{1}{2}}}{\Gamma(H+\frac{1}{2})} \quad (0 \leq t \leq T). \tag{8.2}$$

The variable H is real, and (8.1) is a fractional derivative of the Brownian motion for $H < 1/2$, while it is a fractional integral for $H > 1/2$. Moreover the VARIANCE (1.4) of (8.2) is

$$\sigma^2_{B_H(t)} \equiv g(t) = \frac{t^{2H}}{2H[\Gamma(H+\frac{1}{2})]^2} \tag{8.3}$$

showing that H must be positive (actually : $0 < H < 1$)

Replacing (8.2) into (6.18) and (6.19), respectively, one has

$$(8.4) \quad \chi(t) = \frac{t^H}{\sqrt{H+\frac{1}{2}}\;\Gamma(H+\frac{1}{2})} \quad \text{and} \quad \nu = \frac{2H}{2H+1} \quad (8.5)$$

with the basic result that THE ORDER ν OF THE BESSEL FUNCTIONS IS INDEPENDENT OF THE TIME, depending on the constant H only. Therefore $\nu'(T) = 0$ in (7.5), and the γ_n are determined by

$$\nu\, J_\nu(\gamma_n) + \gamma_n J'_\nu(\gamma_n) = 0 \quad \text{namely} \quad \gamma_n J_{\nu-1}(\gamma_n) = 0 \tag{8.6}$$

so that THE γ_n ARE THE (POSITIVE) ZEROS OF $J_{\nu-1}(x)$ (and these are known to be infinite in number). The eigenvalues are given by

$$(8.7) \quad \lambda_n = \frac{T^{2H+1}}{[\Gamma(H+\frac{3}{2})]^2} \cdot \frac{1}{(\gamma_n)^2}$$

as from (7.4) and (8.2), while the condition (5.3) shows that the maxima and minima of the $\varphi_n(t)$ form an increasing (decreasing) sequence if $H \gtrless 1/2$ (respectively). Also, the normalization constants (7.6) (ref.(4), Vol.2, p.71) may be computed, and appear on p.248 of ref.(6). In conclusion, (3.1) and (6.21) yield the EIGENFUNCTION EXPANSION OF $B_H(t)$ of (8.1)

$$B_H(t) = t^H \sum_{n=1}^{\infty} Z_n N_n J_\nu\left(\gamma_n \frac{t^{H+\frac{1}{2}}}{T^{H+\frac{1}{2}}}\right) \tag{8.7}$$

which converges for $0 \leq t \leq T$, and is the main result of ref.(6).

9. CONCLUSION : EIGENFUNCTION EXPANSION FOR THE NONLINEAR TIME DEPENDENT BROWNIAN MOTION.

In this final section we get the explicit formula for the eigenfunction (or Karhunen-Loève) expansion of the time-rescaled (or nonlinear time dependent) Brownian motion $B(g(t))$.

The function $f(t)$ is assigned for $0 \leq t \leq T$ and is continuous there with $f'(t)$. The final instant T is assigned also. Alternatively, the function $g(t)$ may be assigned, but one must have $g(0) = 0$ because of (1.4), the definition of $g(t)$:

$$g(t) = \int_0^t f^2(s)\, ds \quad \text{whence} \quad g'(t) = f^2(t). \tag{9.1}$$

Then $B(g(t))$ can be expanded in a (convergent) series of the form (3.1). The eigenfunctions $\varphi_n(t)$ are expressed by (6.21), where the generic eigenvalue λ_n is to be replaced by (7.4). Thus, in terms of the function $g(t)$ of (9.1), the EIGENFUNCTION EXPANSION OF $B(g(t))$ READS :

$$B(g(t)) = \sqrt{\sqrt{g'(t)}\int_0^t \sqrt{g'(s)}\,ds} \cdot \sum_{n=1}^{\infty} Z_n N_n J_{\nu(t)}\left(\gamma_n \frac{\int_0^t \sqrt{g'(s)}\,ds}{\int_0^T \sqrt{g'(s)}\,ds}\right)$$

that we call eq.(9.2). Here :

1) The order $\nu(t)$ of the Bessel function is expressed, in terms of the known function $f(t)$, by (6.19) with $\chi(t)$ of (6.18).
2) The constants γ_n $(n=1,2,...)$ are the real positive zeros of (7.5), arranged in ascending order of magnitude, and must be found numerically by aid of a computer.
3) The normalization constants N_n are expressed by (7.6), where the integral may, in general, be computed numerically only.
4) The Z_n are a set of orthogonal (namely stochastically independent) Gaussian (normal) random variables. Each of them has mean zero and variance equal to the eigenvalue λ_n of (7.4) :

Also, eq.(9.2) may be regarded as the output of any linear system having the (stationary) white noise for input and just the function $f(t)$ as IMPULSE RESPONSE. In fact, one merely has to insert the white-noise property $B'(t-s) = B'(s)$ into (1.2) :

$$B(g(t)) = X(t) = \int_0^t f(s)\,B'(s)\,ds = \int_0^t f(s)\,B'(t-s)\,ds \qquad (9.3)$$

showing that the time-rescaled Brownian motion is the output, the stationary white-noise the input, and the function $f(t) = h(t)$ the impulse response, of the linear system.

References

(1) Papoulis, A.: "Probability, Random Variables, and Stochastic Processes", McGraw-Hill, New York, 1965.
(2) Ash, R.: "Information Theory", Interscience, New York, 1965.
(3) Abramowitz, M., and Stegun, I. (Editors): "Handbook of Mathematical Functions", Dover, New York, 1965.
(4) Erdélyi, A., Magnus, W., Oberhettinger, F., and Tricomi, F.G.: "Higher Transcendental Functions", McGraw-Hill, New York, 1953.
(5) Watson, G.N.: "Theory of Bessel Functions", Cambridge, 1966.
(6) Maccone, C.: Il Nuovo Cimento, Vol. 61 B (1981) pp. 229 - 248.
(7) Maccone, C.: Il Nuovo Cimento, Vol. 65 B (1981) pp. 259 - 271.

POINT PROCESS DIFFERENTIALS WITH EVOLVING INTENSITIES

Philip Protter*

Mathematics and Statistics Departments,
Purdue University

*These results were obtained jointly with Jean Jacod while the author was visiting the Université de Rennes.

1. Introduction and Statement of the Problem.

The theory of stochastic differential equations customarily assumes one is a priori given known coefficients and known driving terms (e.g., white noise, Poisson white noise, etc.). Motivated by simple models in microeconomics ([3], [5]), in this article we investigate stochastic differential equations where one of the driving terms is not a priori given but evolves in a fashion depending on the paths of the solution. For example, a driving term may be a point process $N = N[\lambda(X)]$ with an instantaneous stochastic intensity $\lambda_s = \lambda(X)_s$, where X is a solution of the equation:

$$X_t = K_t + \int_0^t F(X)_s dY_s + \int_0^t G(X)_s dN[\lambda(X)]_s \; . \tag{1.1}$$

Here $F(X), G(X), \lambda(X)$ are predictable processes depending on the solution X, K is a given exogeneous term, and Y is a given semimartingale.

A more general framework would be to consider the equation

$$X_t = K_t + \int_0^t F(X)_s dY_s + \int_0^t G(X)_s dZ[C(X)]_s \tag{1.2}$$

R. S. Bucy and J. M. F. Moura (eds.), Nonlinear Stochastic Problems, 467–472.

where $Z[C(X)]$ denotes a semimartingale having local characteristics depending on X. Due to space limitations we will not discuss (1.2) here; the reader may consult [3], which in addition contains all the results of this article.

In this article we show that establishing the existence of a solution of (1.1) is equivalent to showing the existence of a solution of a "classical" equation, namely (3.2). Unfortunately the coefficients of (3.2) are not Lipschitzian and hence do not fall into the existing theory. In Theorem (3.6) we show that nevertheless (3.2) has unique solutions.

Solutions of (1.1) are of necessity "weak" solutions, since if $K_t = X$, $dY_t = dt$, for example, one has $N[\lambda(X)]$ alone providing the random input. This also explains our need to assume that the underlying probability space supports a nontrivial stochastic process (e.g., a Poisson random measure). Furthermore, any meaningful unicity results will be statements concerning the law of the solution process X. In this paper we establish only the existence of a solution; the uniqueness statement in Theorem (3.6) should be viewed, thus, within the framework of the particular method we are using for establishing existence.

2. Preliminaries and Notation.

Undefined terms and notations can be found in any of the recent books [1], [2], or [4]. Stochastic integrals here are semimartingale integrals of predictable processes as defined in these books. (That is, they are extensions of pathwise Stielties integrals and/or Itô-type martingale integrals, etc.).

We are always working on a fixed filtered probability space $(\Omega,\underline{\underline{F}},\underline{\underline{F}}_t,P)$. We denote by $\underline{\underline{D}}$ the space of adapted processes whose paths have left limits and are right continuous. We let $\underline{\underline{P}}$ denote the space of predictable processes.

Poisson random measures are defined and discussed in [2, pp. 80-83]. They are "basically" random measures with "jumps" of size one, deterministic compensators, and a random measure analogue of independent increments.

(2.1) Definition. An operator H from $\underline{\underline{D}}$ to stochastic processes will be called a predictable operator if H: $\underline{\underline{D}} \to \underline{\underline{P}}$ and if X, Y are in $\underline{\underline{D}}$ and T is a stopping time then $X^{T-} = Y^{T-}$ implies that $H(X)^T = H(Y)^T$.

Recall $X_t^T = X_{t\wedge T}$ and $X_t^{T-} = X_t 1_{(t<T)} + X_{T-}1_{(t\leq T)}$.

3. Theorems and Proofs.

We assume we are given predictable operators F, G, and λ from $\underline{\underline{D}}$ into $\underline{\underline{P}}$ such that $\lambda(X) \geqq 0$ and $\int_0^t \lambda(X)_s ds < \infty$ for all $X \in \underline{\underline{D}}$ and all $t \geq 0$. The next theorem transforms equation (1.1) into a "classical" stochastic differential equation:

(3.1) Theorem. *Suppose the space* $(\Omega, \underline{\underline{F}}, \underline{\underline{F}}_t, P)$ *supports a Poisson random measure* m *on* $\mathbb{R}_+ \times \mathbb{R}$ *with compensator* $\bar{m}(dt\,dx) = dt \otimes dx$. *Then every solution process of the following equation is a solution process of* (1.1):

$$X_t = K_t + \int_0^t F(X)_s dY_s + \int_0^t \int_{\mathbb{R}} G(X)_s 1_{\{0 \leqq x \leqq \lambda(X)_s\}} m(ds\,dx) \quad (3.2)$$

and for $N[\lambda(X)]$ *one can take the process*

$$N_t = \int_0^t \int_{\mathbb{R}} 1_{\{0 \leqq x \leqq \lambda(X)_s\}} m(ds\,dx). \quad (3.3)$$

Proof. Formula (3.3) defines a counting process for each $X \in \underline{\underline{D}}$, with an eventual explosion time $T = \inf\{t: N_t = \infty\}$. Define $T_n = \inf\{t: \int_0^t \lambda(X)_s ds \geqq n\}$. Then

$$E(N_{T_n}) = E\{\int_0^{T_n} \int_{\mathbb{R}} 1_{\{0 \leqq x \leqq \lambda(X)_s\}} \bar{m}(ds\,dx)\}$$

$$= E\{\int_0^{T_n} \lambda(X_s) ds\} \leqq n . \quad (3.4)$$

Thus $N_{T_n} < \infty$ a.s. and $T_n \leqq T$ a.s. Since T_n increases to ∞ a.s. by the hypothesis that $\int_0^t \lambda(X)_s ds < \infty$ for all $X \in \underline{\underline{D}}$, we have $T = \infty$ a.s. Thus N is a point process without an explosion and has $\int_0^t \lambda(X)_s ds$ for its compensator. That (3.2) gives a solution to (1.1) is clear. □

We now give hypotheses on F, G and λ that ensure the existence and uniqueness of a solution of (3.2). Note that the coefficient $G(X)1_{\{0\leq X\leq\lambda(X)\}}$ is not Lipschitz in X, in general. We will call the coefficient <u>Y-acceptable</u> if the equation

$$X_t = K_t + \int_0^t F(X)_s dY_s \tag{3.5}$$

has a unique solution for any initial condition $K \in \underline{\underline{D}}$. Very general conditions are known that imply F is Y-acceptable; c.f., e.g., [2] or [4].

<u>(3.6) Theorem.</u> *Let* F,G, *and* λ *be predictable operators from* $\underline{\underline{D}}$ *into* $\underline{\underline{P}}$. *Assume that* F *is* Y*-acceptable and that* $\lambda(X) \geq 0$ *and* $\int_0^t \lambda(X)_s ds < \infty$, *for all* $X \in D$, $t \in \mathbb{R}_+$. *Then equation* (3.2) *has a unique solution up to a strictly positive predictable explosion time* T.

By a solution up to a predictable explosion time T we mean the following: for any sequence of stopping times (T_n) announcing T we have $X^{T_n} \in \underline{\underline{D}}$ and X^{T_n} is a solution of (3.2) on the stochastic interval $[\![0,T_n]\!] = \{(t,\omega): 0\leq t\leq T_n(\omega)\}$. Moreover, X_{T-} does not exist or is infinite on $\{T<\infty\}$.

<u>Proof.</u> For each $X \in \underline{\underline{D}}$ let N^X denote the point process given in (3.3). We saw in Theorem (3.1) that N^X is finite-valued. Inductively define processes $(X(n))_{n\geq 0}$ in $\underline{\underline{D}}$ and an increasing sequence of stopping times $(T_n)_{n\geq 0}$ as follows:

$X(0) = 0$ and $T_0 = 0$;

$$\begin{cases} X(n+1)_t = K_t + \int_0^t F[X(n+1)]_s dY_s + \sum_{i=1}^n G[X(n)]_{T_i} 1_{\{t\geq T_i\}} \\ T_{n+1} = \inf\{t: N_t^{X(n+1)} > N_{T_n}^{X(n+1)}\} . \end{cases} \tag{3.7}$$

Given X(n), the Y-acceptability of F implies the existence of a unique process X(n+1). Note also that $T_{n+1} > T_n$ on $\{T_n < \infty\}$. Using induction and the unicity of solutions, one easily verifies:

$$X(n)_t = K_t + \int_0^t F[X(n)]_s dY_s + \sum_{i=1}^{n-1} G[X(n)]_{T_i} 1_{\{t \geq T_i\}} \tag{3.8}$$

$$X(n) = X(n-1) \text{ on } [\![0,T_{n-1} [\![. \tag{3.9}$$

Relation (3.9) implies $\lambda[X(n)] = \lambda[X(n-1)]$ on $[\![0,T_{n-1}]\!]$, since λ is a predictable operator. Thus $N^{X(n)} = N^{X(n-1)}$ on $[\![0,T_{n-1}]\!]$, and hence $N^{X(n)} = N^{X(p)}$ on $[\![0,T_p]\!]$, all $p \leq n$. Hence the first n jump times of $N^{X(n)}$ are precisely $T_1,\ldots,T_n$; thus (3.8) can as well be written:

$$X(n)_t = K_t + \int_0^t F[X(n)]_s dY_s + \int_0^t \int_{\mathbb{R}} G[X(n)]_s 1_{\{0 \leq x \leq \lambda[X(n)]_s\}} m(ds\,dx). \tag{3.10}$$

Define $T = \lim T_n = \sup T_n$, and clearly T is predictable and strictly positive. Relations (3.8) and (3.9) allow us to define X on $[\![0,T_n [\![$, and then (3.10) shows that X is a solution of (3.2) on $[\![0,T [\![$.

The reader can easily check that T is indeed an explosion time for X (cf. [3] for details). As for unicity: if Z is another solution of (3.2) with explosion time R, let R_p announce R; then by the unicity already established in (3.8) through (3.10), one easily sees by induction on n that for all n, $p \geq 1$ we have $X(n)^{R_p} = Z^{T_n \wedge R_p}$. It then follows that T = R and that X = Z. □

<u>(3.11) Corollary.</u> *With the hypotheses of Theorem* (3.6) *suppose in addition that there exists a finite-valued increasing process* A *such that* $\int_0^t \lambda(X)_s ds \leq A_t$ *for all* $X \in \underline{\underline{D}}$ *and* $t \in \mathbb{R}_+$. *Then the equation* (3.2) *has a unique solution in* $\underline{\underline{D}}$ *(i.e., the explosion time* T *is* a.s. *infinite).*

<u>Proof.</u> Let $S_p = \inf\{t: A_t \geq p\}$. Then $A_{S_p-} \leq p$ and hence $\int_0^{S_p} \lambda(X)_s ds \leq p$ for all $X \in \underline{\underline{D}}$. Using the notation established in the proof of Theorem (3.6), this implies, as in (3.4), that

$N^{X(n)}_{T_n \wedge S_p} = n$ on the set $\{T_n < S_p\}$; hence $nP\{T_n < S_p\} \leq p$. Letting n tend to ∞ yields $P(T < S_p) = 0$; and since S_p increases to ∞, this gives $P(T < \infty) = 0$. □

(3.12) Conclusion. *Suppose the space* $(\Omega, \underline{F}, \underline{F}_t, P)$ *supports a Poisson random measure* m *with compensator* $\bar{m}(dtdx) = dt \otimes dx$. *If* F, G, *and* λ *are predictable operators from* $\underline{D}$ *into* $\underline{P}$ *and* F *is* Y-*acceptable, then there exists a solution of* (1.1) *on this space, at least up to a strictly positive explosion time.*

References.

1. Dellacherie, C., Meyer, P.A.: 1980, Probabilités et Potentiel (Chapitres V á VIII). Paris: Hermann.
2. Jacod, J.: 1979, Calcul Stochastique et Problemes de Martingales. Springer Lecture Notes in Math., 714.
3. Jacod, J., Protter, P.: Quelques Remarques sur un Nouveau Type d'Equations Differentielles Stochastiques. To appear in Seminaire de Probabilites XVI, Springer Lect. Notes in Math.
4. Metivier, M., Pellaumail, J.: 1980, Stochastic Integration. New York: Academic Press.
5. Wernerfelt, B.: Private Communication.

This research was supported in part by NSF Grant #0464-50-13955.

TRANSFORMATION PROPERTIES OF STOCHASTIC DIFFERENTIAL EQUATIONS

K.Seitz

Technical University of Budapest

ABSTRACT

Consider the following vector-differential equation

(1) $\dot{\underline{x}}(t) = \underline{f}(\underline{x}(t), \underline{\xi}(t), t)$,

where $t \in [0, T] = \mathfrak{I}$,

(2) $\underline{x}(t) = \begin{bmatrix} x_1(t) \\ x_2(t) \\ \vdots \\ x_n(t) \end{bmatrix}$, $\underline{\xi}(t) = \begin{bmatrix} \xi_1(t) \\ \xi_2(t) \\ \vdots \\ \xi_n(t) \end{bmatrix}$, $\underline{f}(\underline{x}(t), \underline{\xi}(t), t) = \begin{bmatrix} f_1(\underline{x}, \underline{\xi}, t) \\ f_2(\underline{x}, \underline{\xi}, t) \\ \vdots \\ f_n(\underline{x}, \underline{\xi}, t) \end{bmatrix}$

We assume that in $\mathfrak{I}$ $\underline{\xi}(t)$ is a differentiable stochastic vector process and $\underline{f}(\underline{x}(t), \underline{\xi}(t), t)$ is a differentiable function of $\underline{x}$, $\underline{\xi}$ and t.

In this lecture we shall investigate the structure of transformations

(3) $\underline{x} = \underline{\varphi}(\hat{\underline{x}}(t), \hat{\underline{\xi}}(t), t)$,

$t \in \mathfrak{I}$, which map (1) to

R. S. Bucy and J. M. F. Moura (eds.), Nonlinear Stochastic Problems, 473–480.

$$(4) \qquad \dot{\underline{\hat{x}}}(t) = \underline{f}(\underline{\hat{x}}(t), \underline{\eta}(t), t)$$

where $\underline{\eta}(t)$ does not depend from $\underline{\hat{x}}(t)$, and

$$(5) \quad \underline{\hat{x}}(t) = \begin{bmatrix} \hat{x}_1(t) \\ \hat{x}_2(t) \\ \vdots \\ \hat{x}_n(t) \end{bmatrix}, \quad \underline{\eta}(t) = \begin{bmatrix} \eta_1(t) \\ \eta_2(t) \\ \vdots \\ \eta_n(t) \end{bmatrix}, \quad \underline{\hat{\xi}}(t) = \begin{bmatrix} \hat{\xi}_1(t) \\ \hat{\xi}_2(t) \\ \vdots \\ \hat{\xi}_n(t) \end{bmatrix},$$

further in $\underline{\varphi}(\underline{\hat{x}}, \underline{\hat{\xi}}, t)$ is a differentiable function of $\underline{\hat{x}}, \underline{\hat{\xi}}, t$, and $\underline{\hat{\xi}}(t)$ is a differentiable stochastic vector process.

The algebraic structure of transformations (3) is investigated in any important special cases on a s i m p l e w a y .

The material of this paper is strongly related to the works [1], [2], [3].

1. General remarks and basic lemmas

Now we suppose that in $\mathfrak{J}$, $\left(\frac{\partial \underline{\varphi}}{\partial \underline{\hat{x}}}\right)^{-1}$ exists.

From (3) we have

$$(6) \qquad \dot{\underline{x}} = \frac{\partial \underline{\varphi}}{\partial \underline{\hat{x}}} \dot{\underline{\hat{x}}}(t) + \frac{\partial \underline{\varphi}}{\partial \underline{\hat{\xi}}} \dot{\underline{\hat{\xi}}} + \frac{\partial \underline{\varphi}}{\partial t},$$

$$t \in \mathfrak{J}.$$

Since (3) and (4) hence it follows that

$$(7) \quad \frac{\partial \underline{\varphi}}{\partial \underline{\hat{x}}} \underline{f}(\underline{\hat{x}}, \underline{\eta}, t) + \frac{\partial \underline{\varphi}}{\partial \underline{\hat{\xi}}} \dot{\underline{\hat{\xi}}} + \frac{\partial \underline{\varphi}}{\partial t} = \underline{f}\left[\underline{\varphi}(\underline{\hat{x}}, \underline{\hat{\xi}}, t), \underline{\xi}, t\right].$$

Therefore if transformations (3) map (1) to (4), then $\underline{\varphi}(\underline{\hat{x}}(t), \underline{\hat{\xi}}(t), t)$ satisfies (7).

If $\underline{\varphi}(\underline{\hat{x}}(t), \underline{\hat{\xi}}(t), t)$ satisies (7) and

$$(8) \qquad \dot{\hat{\underline{x}}}(t) = \underline{f}(\hat{\underline{x}}(t), \underline{\eta}(t), t)$$

then

$$(9) \qquad \frac{\partial \underline{\varphi}}{\partial \hat{\underline{x}}} \dot{\hat{\underline{x}}}(t) + \frac{\partial \underline{\varphi}}{\partial \hat{\underline{\xi}}} \dot{\hat{\underline{\xi}}}(t) + \frac{\partial \underline{\varphi}}{\partial t} = \underline{f}[\underline{\varphi}(\hat{\underline{x}}, \hat{\underline{\xi}}, t), \underline{\xi}, t]$$

from which

$$(10) \qquad \dot{\underline{x}}(t) = \underline{f}(\underline{x}(t), \underline{\xi}, t),$$

follows.

Lemma 1. If

$$(11) \qquad \underline{f}(\hat{\underline{x}}, \underline{\eta}, t) = \underline{\underline{A}}^{-1} \underline{f}(\underline{\underline{A}} \hat{\underline{x}}, \underline{\xi}, t)$$

where $\underline{\underline{A}}$ is an invertible constant $(n \times n)$ matrix, then $\underline{x} = \underline{\underline{A}} \hat{\underline{x}}$ maps (1) to (4).

Proof. If $\underline{x} = \underline{\underline{A}} \underline{x}$, then $\frac{\partial \underline{\varphi}}{\partial \hat{\underline{x}}} = \underline{\underline{A}}$ and in this case (7) has the following form

$$(12) \qquad \underline{\underline{A}} \underline{f}(\hat{\underline{x}}, \underline{\eta}, t) = \underline{f}(\underline{\underline{A}} \hat{\underline{x}}, \underline{\xi}, t),$$

which is equivalent with (11).

Denote by G the set of all invertible $(n \times n)$ matrices for which (11) holds.

Lemma 2. If $\underline{\eta} = \underline{\xi}$ and for all $\hat{\underline{x}}$, $(\dim \hat{\underline{x}} = n)$ (11) holds, then $(G, \cdot)$ is a group.

Proof. If $\underline{\underline{A}}, \underline{\underline{B}} \in G$ then since (11)

$$(13) \qquad \underline{\underline{A}} \underline{f}(\hat{\underline{x}}, \underline{\xi}, t) = \underline{f}(\underline{\underline{A}} \hat{\underline{x}}, \underline{\xi}, t)$$

and

$$(14) \qquad \underline{\underline{B}} \underline{f}(\hat{\underline{x}}, \underline{\xi}, t) = \underline{f}(\underline{\underline{B}} \hat{\underline{x}}, \underline{\xi}, t).$$

From (12), (14) we have

$$(15) \qquad \underline{\underline{A}} \cdot \underline{\underline{B}} \underline{f}(\hat{\underline{x}}, \underline{\xi}, t) = \underline{\underline{A}} \underline{f}(\underline{\underline{B}} \hat{\underline{x}}, \underline{\xi}, t) =$$

$$= \underline{f}\,(\underline{\underline{A}}\cdot\underline{\underline{B}}\;\hat{\underline{x}},\underline{\xi},t)\,.$$

Therefore $\underline{\underline{A}}\cdot\underline{\underline{B}}\in G$.

Since

$$(16)\qquad \underline{f}(\hat{\underline{x}},\underline{\xi},t)=\underline{\underline{E}}^{-1}\underline{f}(\underline{\underline{E}}\hat{\underline{x}},\underline{\xi},t)\,,$$

$\underline{\underline{E}}$ belongs to G .

If $\underline{\underline{A}}\in G$, then $\underline{\underline{A}}^{-1}\in G$, ,because

$$(17)\qquad \underline{\underline{A}}\,\underline{f}(\underline{\underline{A}}^{-1}\hat{\underline{x}},\underline{\xi},t)=\underline{f}(\hat{\underline{x}},\underline{\xi},t)$$

from which

$$(18)\qquad (\underline{\underline{A}}^{-1})^{-1}\underline{f}(\underline{\underline{A}}^{-1}\hat{\underline{x}},\underline{\xi},t)=\underline{f}(\hat{\underline{x}},\underline{\xi},t)\,,$$

follows.

2. Transformations of linear equations

Consider the following differential equation

$$(19)\qquad \dot{\underline{x}}(t)=\underline{\underline{S}}(t)\,\underline{x}(t)+\underline{\xi}(t)\,,$$

and the

$$(20)\qquad \underline{x}(t)=\underline{\underline{A}}(t)\,\hat{\underline{x}}(t)+\hat{\underline{\xi}}(t)$$

transformation, where $\underline{\underline{A}}(t)$ is a differentiable $(n\times n)$ matrix $(t\in\mathfrak{I})$.

We want to transform (19) to

$$(21)\qquad \dot{\hat{\underline{x}}}(t)=\underline{\underline{S}}(t)\,\hat{\underline{x}}(t)+\underline{\eta}(t)$$

by (20).

Theorem 1. If $\underline{\underline{A}}^{-1}(t)$ exists in $\mathfrak{J}$ further

$$(22) \qquad \dot{\underline{\underline{A}}}(t) = \underline{\underline{S}}(t)\underline{\underline{A}}(t) - \underline{\underline{A}}(t)\underline{\underline{S}}(t),$$

and

$$(23) \qquad \dot{\hat{\underline{\xi}}}(t) = \underline{\underline{S}}(t)\hat{\underline{\xi}}(t) + \underline{\xi}(t) - \underline{\underline{A}}\,\underline{\eta}(t),$$

then

$$(24) \qquad \underline{x}(t) = \underline{\underline{A}}(t)\hat{\underline{x}}(t) + \hat{\underline{\xi}}(t)$$

maps (19) to (21).

Proof. From (19) and (24) we have

$$\dot{\hat{\underline{x}}}(t) = [\underline{\underline{A}}^{-1}(t)\underline{\underline{S}}(t)\underline{\underline{A}}(t) - \underline{\underline{A}}^{-1}(t)\dot{\underline{\underline{A}}}(t)]\hat{\underline{x}}(t) +$$

$$(25) \qquad + \underline{\underline{A}}^{-1}(t)\underline{\underline{S}}(t)\hat{\underline{\xi}}(t) + \underline{\underline{A}}^{-1}(t)\underline{\xi}(t) -$$

$$- \underline{\underline{A}}^{-1}(t)\dot{\hat{\underline{\xi}}}(t).$$

If (22) holds, then

$$(26) \qquad \underline{\underline{A}}^{-1}\dot{\underline{\underline{A}}} = \underline{\underline{A}}^{-1}\underline{\underline{S}}(t)\underline{\underline{A}} - \underline{\underline{S}}(t)$$

from which

$$(27) \qquad \underline{\underline{S}}(t) = \underline{\underline{A}}^{-1}(t)\underline{\underline{S}}(t)\underline{\underline{A}}(t) - \underline{\underline{A}}^{-1}(t)\dot{\underline{\underline{A}}}(t).$$

If (23) holds, then

$$(28) \qquad \underline{\eta}(t) = -\underline{\underline{A}}^{-1}(t)\dot{\hat{\underline{\xi}}}(t) + \underline{\underline{A}}^{-1}(t)\underline{\underline{S}}(t)\hat{\underline{\xi}}(t) + \underline{\underline{A}}^{-1}(t)\underline{\xi}(t).$$

Denote by $\mathfrak{F}$ the set of all matrices which satisfie (22) in $\mathfrak{J}$.

It is easy to see that $(\mathfrak{F}, \cdot)$ is a semigroup.

If $\underline{\underline{A}}(t)$, $B(t) \in \mathfrak{F}$, then

$$(29) \qquad \dot{\underline{\underline{A}}}(t)\underline{\underline{B}}(t) = \underline{\underline{S}}(t)\underline{\underline{A}}(t)\underline{\underline{B}}(t) - \underline{\underline{A}}(t)\underline{\underline{S}}(t)\underline{\underline{B}}(t),$$

and

$$(30)\qquad \underline{\underline{A}}(t)\,\dot{\underline{\underline{B}}}(t) = \underline{\underline{A}}(t)\,\underline{\underline{S}}(t)\,\underline{\underline{B}}(t) - \underline{\underline{A}}(t)\,\underline{\underline{B}}(t)\,\underline{\underline{S}}(t).$$

From (29) and (30) we have

$$(31)\qquad [\underline{\underline{A}}(t)\underline{\underline{B}}(t)]^{\bullet} = \underline{\underline{S}}(t)(\underline{\underline{A}}(t)\underline{\underline{B}}(t)) - (\underline{\underline{A}}(t)\underline{\underline{B}}(t))\underline{\underline{S}}(t).$$

Since $\dot{\underline{\underline{E}}} = \underline{\underline{S}}(t)\,\underline{\underline{E}} - \underline{\underline{E}}\,\underline{\underline{S}}(t) = \underline{\underline{0}}$, $\underline{\underline{E}} \in \mathfrak{F}$.
If $\underline{\underline{A}}^{-1}(t)$ exists in $\mathfrak{J}$, then

$$-\underline{\underline{A}}^{-1}(t)\dot{\underline{\underline{A}}}(t)\underline{\underline{A}}^{-1}(t) = -\underline{\underline{A}}^{-1}(t)\underline{\underline{S}}(t) + \underline{\underline{S}}(t)\underline{\underline{A}}^{-1}(t),$$

from which

$$(32)\qquad (\underline{\underline{A}}^{-1}(t))^{\bullet} = \underline{\underline{S}}(t)\underline{\underline{A}}^{-1}(t) - \underline{\underline{A}}^{-1}(t)\underline{\underline{S}}(t)$$

follows.

If for $\underline{\underline{A}}(t)$ from $\mathfrak{F}$, $\underline{\underline{A}}^{-1}(t)$ exists then $(\mathfrak{F}_1 \cdot)$ is a group.

3. Transformations of quadratic equations

Consider the following differential equation

$$(33)\qquad \dot{\underline{x}}(t) = [\underline{x}^*(t)\,\underline{\underline{P}}(t)\underline{x}(t)]\underline{\alpha}(t) + \underline{\underline{S}}(t)\underline{x}(t) + \underline{\xi}(t),$$

where $\underline{\underline{P}}(t)$ is an $(n \times n)$ matrix, $(t \in \mathfrak{J})$ and $\dim \underline{\alpha}(t) = n$.

We want to transform (33) to

$$(34)\qquad \dot{\hat{\underline{x}}}(t) = [\hat{\underline{x}}^*(t)\underline{\underline{P}}(t)\hat{\underline{x}}(t)]\underline{\alpha}(t) + \underline{\underline{S}}(t)\hat{\underline{x}}(t) + \underline{\eta}(t)$$

by (20).

<u>Theorem 2</u>. If $\underline{\underline{A}}^{-1}(t)$ exists in $\mathfrak{J}$, further

$$\underline{\underline{A}}^{-1}(t)\,\underline{\alpha}(t) = \mu(t)\,\underline{\alpha}(t), \tag{35}$$

$$\mu(t)\,\underline{\underline{A}}^*(t)\,\underline{\underline{P}}(t)\,\underline{\underline{A}}(t) = \underline{\underline{P}}(t), \tag{36}$$

$$\begin{aligned}\dot{\underline{\underline{A}}}(t) &= \underline{\underline{S}}(t)\underline{\underline{A}}(t) - \underline{\underline{A}}(t)\underline{\underline{S}}(t) + \\ &+ \underline{\alpha}(t)\,\hat{\underline{\xi}}^*(t)\,(\underline{\underline{P}}(t) + \underline{\underline{P}}^*(t))\,\underline{\underline{A}}(t),\end{aligned} \tag{37}$$

and

$$\begin{aligned}\dot{\hat{\underline{\xi}}}(t) &= -\underline{\underline{A}}(t)\,\underline{\eta}(t) + [\hat{\underline{\xi}}^*(t)\,\underline{\underline{P}}(t)\,\hat{\underline{\xi}}(t)]\,\underline{\alpha}(t) + \\ &+ \underline{\underline{S}}(t)\,\hat{\underline{\xi}}(t) + \underline{\xi}(t)\end{aligned} \tag{38}$$

then

$$\underline{x}(t) = \underline{\underline{A}}(t)\,\hat{\underline{x}}(t) + \hat{\underline{\xi}}(t) \tag{39}$$

maps (33) to (34).

Proof. From (33) and (39) we have

$$\begin{aligned}\dot{\hat{\underline{x}}}(t) &= [\hat{\underline{x}}^*(t)(\underline{\underline{A}}^*(t)\underline{\underline{P}}(t)\underline{\underline{A}}(t))\hat{\underline{x}}(t)]\,\underline{\underline{A}}^{-1}(t)\underline{\alpha}(t) + \\ &+ \underline{\underline{A}}^{-1}(t)\underline{\alpha}(t)\cdot[\hat{\underline{x}}^*(t)\underline{\underline{A}}^*(t)\underline{\underline{P}}(t)\hat{\underline{\xi}}(t)] + \underline{\underline{A}}^{-1}(t)\underline{\alpha}(t)\,\cdot \\ &\cdot [\hat{\underline{\xi}}^*(t)\underline{\underline{P}}(t)\underline{\underline{A}}(t)\hat{\underline{x}}(t)] - \underline{\underline{A}}^{-1}(t)\dot{\underline{\underline{A}}}(t)\hat{\underline{x}}(t) - \underline{\underline{A}}^{-1}(t)\dot{\hat{\underline{\xi}}}(t) + \\ &+ \underline{\underline{A}}^{-1}(t)(\hat{\underline{\xi}}^*(t)\underline{\underline{P}}(t)\hat{\underline{\xi}}(t))\underline{\alpha}(t) + \underline{\underline{A}}^{-1}(t)\underline{\underline{S}}(t)\underline{\underline{A}}(t)\hat{\underline{x}}(t) + \\ &+ \underline{\underline{A}}^{-1}(t)\,\underline{\underline{S}}(t)\,\hat{\underline{\xi}}(t) + \underline{\underline{A}}^{-1}(t)\,\underline{\xi}(t).\end{aligned}$$

It is easy to see, that

$$\underline{\hat{x}}^*(t) \cdot [\underline{\underline{A}}^*(t)\,\underline{\underline{P}}(t) \cdot \underline{\hat{\xi}}(t)] =$$

$$(41) \qquad = [\underline{\underline{A}}^*(t)\,\underline{\underline{P}}(t)\,\underline{\hat{\xi}}(t)]^*\,\underline{\hat{x}}(t) =$$

$$= \underline{\hat{\xi}}^*(t)\,\underline{\underline{P}}^*(t)\,\underline{\underline{A}}(t) \cdot \underline{\hat{x}}(t).$$

Since (41),(40) has the following form

$$(42) \qquad \begin{aligned} \dot{\underline{\hat{x}}}(t) &= [\underline{\hat{x}}^*(t)(\underline{\underline{A}}^*(t)\,\underline{\underline{P}}(t)\,\underline{\underline{A}}(t))\,\underline{\hat{x}}(t)]\,\underline{\underline{A}}^{-1}(t)\,\underline{\alpha}(t) + \\ &+ [\underline{\underline{A}}^{-1}(t)(\underline{\underline{S}}(t)\underline{\underline{A}}(t) - \dot{\underline{\underline{A}}}(t)) + \underline{\underline{A}}^{-1}(t)\,\underline{\alpha}(t)\,\underline{\hat{\xi}}^*(t) \cdot \\ &(\underline{\underline{P}}(t) + \underline{\underline{P}}^*(t))\,\underline{\underline{A}}(t)] \cdot \underline{\hat{x}}(t) + \underline{\underline{A}}^{-1}(t)\,\underline{\hat{\xi}}^*(t)\,\underline{\underline{P}}(t)\,\underline{\hat{\xi}}(t)\,\underline{\alpha}(t) + \\ &+ \underline{\underline{A}}^{-1}(t)[\underline{\underline{S}}(t)\,\underline{\hat{\xi}}(t) + \underline{\xi}(t) - \dot{\underline{\hat{\xi}}}(t)]. \end{aligned}$$

If (35), (36), (37) and (38) satisfie then from (42) we have

$$(43) \qquad \dot{\underline{\hat{x}}}(t) = [\underline{\hat{x}}^*(t)\,\underline{\underline{P}}(t)\,\underline{\hat{x}}(t)]\,\underline{\alpha}(t) + \underline{\underline{S}}(t)\,\underline{\hat{x}}(t) + \underline{\eta}(t).$$

REFERENCES

[1] R.W. Brockett: Finite Dimensional Linear Systems, John Wiley and Sons, New York, 1970.

[2] R.E.Kalman, P.L.Falb, M.A.Arbib: Topics in mathematical system theory, Mc Graw Hill New York, 1969.

[3] A.Tannenbaum: Invariance and System Theory, Springer-Verlag Berlin-Heidelberg-New York 1981.

SOME APPLICATIONS OF STOCHASTIC CALCULUS ON THE NUCLEAR SPACES TO THE NONLINEAR PROBLEMS

A. S. USTUNEL

CNET/PAA/ATR/SST, 38-40, rue de General Leclerc
92131 Issy Les Moulineaux, FRANCE

Abstract:This work consists of the applications of the infinite dimensional stochastic calculus to the following problems:Construction of the C^∞-flows of the finite dimensional diffusions with regular coefficients, a simple and short proof of the generalized Ito-Stratonovitch formula,probabilistic solutions of the heat and Schrodinger's equations in the space of the distributions,diffusions depending on a parameter and the reduction of the unnormalized density equation of the filtering of non-linear diffusions into a deterministic Cauchy problem depending on a parameter.

R. S. Bucy and J. M. F. Moura (eds.), Nonlinear Stochastic Problems, 481–508.

INTRODUCTION

This paper is devoted to some applications of the stochastic calculus on the nuclear spaces to the finite dimensional stochastic flows and infinite dimensional stochastic evolution equations.

In order to prove Feynman-Kac formula one constructs first the semi-group associated to the heat equation and then applies Trotter product formula. If the potential is positive then there is also a more probabilistic method consisting of killing a Wiener process by the multiplicative functional associated to the potential function. Having recalled the basic tools in the first section, we give another method in Section 2, which is inspired from these two methods but the result is extended to the space of the distributions. Inspite of the fact that the Feynman-Kac formula is meaningless on $\mathcal{D}'(\mathbb{R}^d)$, for any distribution in $\mathcal{D}'(\mathbb{R}^d)$, there exists a semimartingale starting from this distribution which is the unique solution of a stochastic partial differential equation and if it is integrable, then its mathematical expectation solves the heat equation with the given potential function. We show that this semimartingale is a Markov process and calculate its infinitesimal generator. The important difference from the finite dimensional case is that there is no leak to infinity. The same method is applied to Schrödinger's equation in the third section.

In the fourth section we study the generalized Ito's formulae for the flows of the diffusion processes with regular coefficients. We show that the formula given in (2) is a special case of the integration by parts formula for the infinite dimensional semimartingales and give another one for the exponential semimartingales. In the following section we show the existence and the uniqueness in the space of the distributions, of the solutions of the stochastic evolution equations associated to the flows of the diffusion processes which are perturbed by the exponential semimartingales. In the sixth section the integration by parts formula is applied to the reduction of the unnormalized density equation of the nonlinear filtering of the diffusion processes into a deterministic Cauchy problem.

In the last section we prove a general result which is a sufficient condition for the smoothness of the semimartingale valued random fields and then give an extension of the generalized Itô's formula to the diffusion processes depending on a parameter.

1. PRELIMINARIES AND NOTATIONS

In the following we shall study exclusively on the space of the distributions, however, the techniques that we use are valuable for a large class of nuclear spaces, hence the definitions are given in an abstract frame.

Φ denotes a complete nuclear space whose continuous dual Φ' is nuclear under its strong topology $\beta(\Phi', \Phi)$, denoted by Φ'_β. If U is an absolutely convex neighbourhood (of zero) in Φ, $\Phi(U)$, denotes the quotient set $\Phi/p_U^{-1}(0)$ completed with respect to the gauge function p_U of U and k(U) represents the canonical mapping from Φ onto $\Phi(U)$. If $V \subset U$ is another such neighbourhood, $k(U,V) : \Phi(V) \longrightarrow \Phi(U)$ is defined by $k(U) = k(U,V) \circ k(V)$. Let us recall that Φ is called nuclear if there exists a neighbourhood base $\mathcal{U}$ in Φ such that, for any $U \in \mathcal{U}$, there exists $V \in \mathcal{U}, V \subset U$, for which k(U,V) is a nuclear mapping. If B is a bounded, absolutely convex subset of Φ, we note by $\Phi[B]$ the completion of the subspace (of Φ) spanned by B with respect to the norm P_B (i.e. the gauge function of B). It is well known that (c.f. [7]) in each nuclear space Φ, there exists a neighbourhood base $\mathcal{U}_h(\Phi)$ such that, for any $U \in \mathcal{U}_h(\Phi)$, $\Phi(U)$ is a separable Hilbert space whose dual can be identified by $\Phi'[U°]$, U° being the polar of U and Φ is the projective limit of

$$\{(\Phi(U), k(U,V)) ; V, U \in \mathcal{U}_h(\Phi), V \subset U\}$$

We denote by $\mathcal{K}_h(\Phi)$ the set

$$\{U° : U \in \mathcal{U}_h(\Phi'_\beta)\}$$

and $\mathcal{K}_h(\Phi'_\beta)$ is defined by interchanging Φ and Φ'_β.

$(\Omega, \mathcal{F}, P)$ denotes a complete probability space with a right continuous increasing filtration $\{\mathcal{F}_t ; t \geq 0\}$ of the sub-σ-algebras of $\mathcal{F}$. $\mathcal{F}_0$ is supposed to contain all the P-negligeable subsets of $\mathcal{F}$. S° represents the space of the equivalence classes (modulo evanescent processes) of real-valued semimartingales. We recall that one can define a distance on S° under which it is a (non-locally convex) Fréchet space (c.f. [3]).

If $\mathcal{X}$ is a separable Hilbert space and Z is a semimartingale in $\mathcal{X}$, we denote by $[\![Z,Z]\!]_t$ the following process :

$$[\![Z,Z]\!]_t = \langle\!\langle Z^c, Z^c \rangle\!\rangle_t + \sum_{s \leq t} \|\Delta Z_s\|^2$$

where $\Delta Z_s = Z_s - Z_{s^-}$, Z^c is the continuous local martingale part of Z and $\langle\!\langle Z^c, Z^c \rangle\!\rangle$ is the unique previsible process of finite variation such that $\|Z^c\|^2 - \langle\!\langle Z^c, Z^c \rangle\!\rangle$

is a local martingale (c.f. [6]). The set of $\mathcal{X}$ - valued semimartingales is denoted by $S^\circ(\mathcal{X})$. If $Z \in S^\circ(\mathcal{X})$ is a special semimartingale with its canonical decomposition

$$Z = M + A, \; A_0 = 0,$$

(defined as in the finite dimensional case) we define

$$\|Z\|_p = \Big\| \, [\![M,M]\!]_\infty^{1/2} + \int_0^\infty \|dA_s\| \, \Big\|_{L^p(\Omega,\mathcal{F},P)}, \quad p \geq 1.$$

As in the finite dimensional case the elements of $S^\circ(\mathcal{X})$ for which $\|\cdot\|_p$ is finite is a Banach space under this norm and their set is denoted by $S^p(\mathcal{X})$ (simply S^p if $\mathcal{X} = \mathbb{R}$).

The concept of stochastic process will be extended to Φ' in the following manner :

Definition 1.1.

Let X be the set : $\{X^U : U \in \mathcal{U}_h(\Phi'_\beta)\}$

where X^U is a stochastic process with values in $\Phi'(U)$. X is called a projective system of stochastic processes (on Φ') if, for any $U, V \in \mathcal{U}_h(\Phi'_\beta)$, $V \subset U$, $k(U,V) \circ X^V$ and X^U are undistinguishable.

Definition 1.2.

Let X be a projective system of stochastic processes on Φ'. We say that X has a limit in Φ' if there exists a mapping X' on $\mathbb{R}_+ \times \Omega$ with values in Φ' such that, for any $t \geq 0$, X'_t is a random variable from $(\Omega, \mathcal{F})$, into $(\Phi', \mathcal{C}(\Phi'))$ where $\mathcal{C}(\Phi')$ is the cylindrical σ-algebra of Φ, and for any $U \in \mathcal{U}_h(\Phi'_\beta)$,

$$X_t^U = k(U) \circ X'_t \quad \text{a.e.}$$

where X^U is the element of X corresponding to U.

In [9] the following result is proved :

Proposition 1.1.

Suppose that Φ is a nuclear Fréchet space or strict inductive limit of a sequence of such spaces. Then, any projective system of stochastic processes on Φ' has a limit in Φ'.

Definition 1.3.

Let X be a projective system of stochastic processes on Φ' with limit X' in Φ'. The pair (X,X') is called a g-process on Φ'. We say that a g-process (X,X') or a projective system X possesses the property π if any element X^U of X possesses the property π in $\Phi'(U)$. In particular, (X,X') is called a semimartingale if X^U belongs to $S^\circ(\Phi'(U))$ for any $U \in \mathcal{U}_h(\Phi'_\beta)$.

for any U. Similarly (X,X') is called an S^p-semimartingale, $p \geqslant 1$, if X^U belongs to $S^p(\Phi'(U))$ for any $U \in \mathcal{U}_h(\Phi'_\beta)$.

The proof of the following result can be found in (9) and [12] :

Theorem 1.1.

Suppose that X' is a stochastic process with values in $(\Phi', \mathcal{C}(\Phi'))$ such that, for any $\varphi \in \Phi$, the process

$$(t,\omega) \longmapsto \langle X'_t(\omega), \varphi \rangle$$

has a modification wich is a real-valued semimartingale. Then there exists a unique projective system of semimartingales

$$X = \{X^U \,;\, U \in \mathcal{U}_h(\Phi'_\beta)\}$$

such that (X,X') is a semimartingale on Φ'.

The following statement is an easy consequence of the closed graph theorem (c.f. [9] and [12] :

Theorem 1.2.

Suppose that (X,X') belongs to $S^1(\Phi')$. Then there exists a stochastic process $\hat{X}$ with values in Φ' having right continuous trajectories with left limits such that, for any $U \in \mathcal{U}_h(\Phi'_\beta)$, $k(U) \circ \hat{X}$ and X^U are undistinguishable. Furthermore there exists an element K of $\mathcal{K}_h(\Phi'_\beta)$ and a semimartingale Y in $S^1(\Phi'[K])$ whose image under the injection $i_K : \Phi'[K] \hookrightarrow \Phi'$ is undistinguishable from $\hat{X}$ in the sense of Definition 1.3.

If Φ'_β is metrizable we have also :

Theorem 1.3.

Suppose that Φ'_β is a nuclear Fréchet space and let X' be a Φ'-valued stochastic process such that, for any $\varphi \in \Phi$, the mapping

$$(t,\omega) \longmapsto \langle X'_t(\omega), \varphi \rangle$$

has a modification which is a semimartingale. Then for any finite time interval, there exists a probability measure equivalent to P for which the conclusion of Theorem 1.2. holds on this time interval.

The following result is very important for the applications :

Theorem 1.4. (Integration by parts formula)

Suppose that Z is in $S^1(\Phi')$ and X is in $S^\circ(\Phi)$ (instead of (Z,Z') and (X,X') we write respectively Z and X if there is no confusion). Then the mapping :

$$(t,\omega) \mapsto \langle Z_t(\omega), X_t(\omega)\rangle$$

has a modification which is a semimartingale and if $K \in \mathcal{K}_h(\Phi'_\beta)$ absorbs Z as a $\Phi'[K]$-valued semimartingale (c.f. Theorem 1.2.), then

$$(1.1.) \quad \langle Z_t, X_t\rangle = \int_0^t (dZ_s | X^{K^\circ}_{s-}) + \int_0^t (Z_{s-} | dX^{K^\circ}_s) + [Z, X^{K^\circ}]_t \text{ a.e.}$$

Where X^{K° is the element of $\{X^U : U \in \mathcal{U}_h(\Phi)\}$ for $U = K^\circ$. Moreover the right hand side of (1.1.) is independent of any particular choice of K.

Remark :

If Φ'_β is metrizable then the theorem holds for any $Z \in S^\circ(\Phi')$.

2. STOCHASTIC FEYNMAN-KAC FORMULA

Let φ be an element of $\mathcal{D}(\mathbb{R}^d)$ and V be a C^∞-function on $\mathbb{R}^d$. If $W = (W_t)$ denotes the standard Wiener process with values in $\mathbb{R}^d$, outside a fixed negligeable set, the mapping

$$x \mapsto \varphi(x+W_t(\omega)) \exp\left(-\int_0^t V(x+W_s(\omega))\,ds\right)$$

is an element of $\mathcal{D}(\mathbb{R}^d)$. Hence for any $g \in \mathcal{D}'(\mathbb{R}^d)$

$$\langle X'_t(\omega), \varphi\rangle \equiv \langle g, \varphi(\cdot+W_t(\omega)) \exp\left(-\int_0^t V(\cdot+W_s(\omega))ds\right)\rangle = \langle g, Z_t\,\varphi(\cdot+W_t)\rangle$$

is well-defined. Moreover, from Ito's formula, we have for fixed $x \in \mathbb{R}^d$:

$$\varphi(x+W_t)Z_t(x) = \varphi(x) + \int_0^t D_i\varphi(x+W_s)Z_s(x)\,dW^i_s +$$

$$+ \frac{1}{2}\int_0^t \Delta\varphi(x+W_s)Z_s(x)\,ds -$$

$$- \int_0^t V(x+W_s)\,\varphi(x+W_s)\,Z_s(x)\,ds\,.$$

Since the mapping $y \mapsto \varphi(\cdot+y)$ is continuous from $\mathbb{R}^d$ into $\mathcal{D}(\mathbb{R}^d)$, $(t,\omega) \mapsto \varphi(\cdot+W_t(\omega))$ is a stochastic process with almost surely continuous trajectories in $\mathcal{D}(\mathbb{R}^d)$, moreover, stopping (W_t) on the increasing compact subsets of $\mathbb{R}^d$, we see that it is locally bounded. By theorem 1.3. and by the hypocontinuity of the multiplication operation on $\mathcal{E}(\mathbb{R}^d)\times\mathcal{D}(\mathbb{R}^d)$ with values in $\mathcal{D}(\mathbb{R}^d)$, all the integrands of the above formula are again locally bounded, previsible processes with values in $\mathcal{D}(\mathbb{R}^d)$ (c.f [9] and [11]). Hence the integrals can be modified as $\mathcal{D}(\mathbb{R}^d)[K]$-valued stochastic integrals for some $K \in \mathcal{K}_h(\mathcal{D}(\mathbb{R}^d))$. If g is any element of $\mathcal{D}'(\mathbb{R}^d)$, we have then

$$\langle g, \varphi(\cdot+W_t)Z_t\rangle = \langle g,\varphi\rangle + \int_0^t \langle g, D_i\varphi(\cdot+W_s)Z_s\rangle\, dW_s^i +$$

$$+ \frac{1}{2}\int_0^t \langle g, \Delta\varphi(\cdot+W_s)Z_s\rangle\, ds -$$

$$- \int_0^t \langle g, \varphi(\cdot+W_s)V(\cdot+W_s)Z_s\rangle\, ds \quad a.e.$$

Hence, for any $\varphi\in\mathcal{D}(\mathbb{R}^d)$, $(t,\omega) \longmapsto \langle X_t'(\omega),\varphi\rangle$ has a modification which is a semimartingale i.e. it induces a linear mapping $\tilde{X}$ from $\mathcal{D}(\mathbb{R}^d)$ into S°, such that

$$\tilde{X}(\varphi)_t = \langle g,\varphi\rangle + \int_0^t \tilde{X}(D_i\varphi)_s\, dW_s^i + \frac{1}{2}\int_0^t \tilde{X}(\Delta\varphi)_s\, ds - \int_0^t \tilde{X}(V\varphi)_s\, ds \qquad \text{a.e}$$

An application of the closed graph theorem shows that X is sequentially continuous. Consequently there exists a stochastic process $Y_t(\omega)$ with values in $\mathcal{D}'(\mathbb{R}^d)$ such that, for any $t \geq 0$, we have

$$\langle Y_t(\omega),\varphi\rangle = \tilde{X}(\varphi)_t(\omega) \qquad \text{a.e}$$

Theorem 1.1. implies the existence of a projective system of (continuous) semimartingales. In fact we have more :

Theorem 2.1.

There exists an ordinary sense stochastic process X with values in $\mathcal{D}'(\mathbb{R}^d)$, having almost surely continuous trajectories (in the usual sense) which satisfies the following stochastic partial differential equation :

$$(2.1.)\quad dX_t(\varphi) = -D_i X_t(\varphi)\, dW_t^i + \tfrac{1}{2}\Delta X_t(\varphi) - VX_t\, dt$$
$$X_0(\varphi) = \langle g, \varphi\rangle\,,\quad \varphi \in \mathcal{D}(\mathbb{R}^d).$$

Proof : Let (K_n) be an increasing sequence of compact subsets of $\mathbb{R}^d$ such that $\bigcup_n K_n = \mathbb{R}^d$. Stopping (W_t) on K_n by the stopping time T_n, define $\tilde{X}^n(\varphi)$ as the equivalence class defined by

$$\langle g, \varphi(\cdot + W_{t\wedge T_n}) \exp(-\int_0^{t\wedge T_n} V(\cdot + W_s)\, ds)\rangle\,,\quad \varphi \in \mathcal{D}(\mathbb{R}^d).$$

Then $\tilde{X}^n$ is a linear mapping from $\mathcal{D}(\mathbb{R}^d)$ into S^1, obviously it is sequentially continuous, hence it is continuous. From Theorem 1.2., there exists an ordinary sense stochastic process X^n with continuous trajectories in $\mathcal{D}'(\mathbb{R}^d)$. Since T_n increases to infinity and if $m < n$

$$X^n_{t\wedge T_m}(\varphi) = \tilde{X}^n(\varphi)_{t\wedge T_m} = \tilde{X}(\varphi)_{t\wedge T_m} = \tilde{X}^m(\varphi)_t = X^m_t(\varphi) \text{ a.e.},$$

the proof is completed. || Q.E.D.

Theorem 2.2.

$X = (X_t)$ contructed above is the unique solution of the stochastic partial differential equation (2.1.).

Proof : Let $X = (X_t)$ be any solution of (2.1). Using the integration by parts formula, we shall calculate first

$$\langle X_t, \varphi(\cdot - W_t)\rangle\,,\quad \varphi \in \mathcal{D}(\mathbb{R}^d).$$

Stopping (W_t) as above we see that $\{\varphi(.-W_t)\,;\,t\geq 0\}$ belongs locally to $S^1(\mathcal{D}(\mathbb{R}^d))$, hence the integration by parts formula is applicable. We have

$$\langle X_{t\wedge T_n},\varphi(.-W_{t\wedge T_n})\rangle=\langle g,\varphi\rangle+\int_0^{t\wedge T_n}\langle dX_s,\varphi(.-W_s)\rangle+$$

$$+\int_0^{t\wedge T_n}\langle X_s,d\varphi(.-W_s)\rangle+[\![X,\varphi(.-W)]\!]_t$$

$$=\langle g,\varphi\rangle+\frac{1}{2}\int_0^{t\wedge T_n}\langle\Delta X_s,\varphi(.-W_s)\rangle ds-$$

$$-\int_0^{t\wedge T_n}\langle D_i X_s,\varphi(.-W_s)\rangle dW_s^i-$$

$$-\int_0^{t\wedge T_n}\langle VX_s,\varphi(.-W_s)\rangle ds-$$

$$-\int_0^{t\wedge T_n}\langle X_s,D_i\varphi(.-W_s)\rangle dW_s^i+$$

$$+\frac{1}{2}\int_0^{t\wedge T_n}\langle X_s,\Delta\varphi(.-W_s)\rangle ds+$$

$$+\int_0^{t\wedge T_n}\langle D_i X_s,D^i\varphi(.-W_s)\rangle ds$$

$$=\langle g,\varphi\rangle-\int_0^{t\wedge T_n}\langle VX_s,\varphi(.-W_s)\rangle ds\quad\text{a.e.}$$

Since T_n increases to infinity we have

$$\langle X_t,\varphi(.-W_t)\rangle=\langle g,\varphi\rangle-\int_0^t\langle VX_s,\varphi(.-W_s)\rangle ds\quad\text{a.e.}$$

Define $Y_t(\varphi)$ as $X_t(\varphi(.-W_t))$. It is not difficult to see that (Y_t) has almost surely continuous trajectories in $\mathcal{D}'(\mathbb{R}^d)$ and it satisfies the following equation :

(2.2.)
$$dY_t(\varphi) = -Y_t(V(\cdot+W_t)\varphi)\,dt$$
$$Y_0(\varphi) = g(\varphi), \quad \varphi \in \mathcal{D}(\mathbb{R}^d).$$

We pretend that (2.2.) has one and only one solution. To see this we use again the integration by parts formula :

$$\langle Y_t, Z_t^{-1}\varphi\rangle = g(\varphi) - \int_0^t \langle V(\cdot+W_s)Y_s, \varphi Z_s^{-1}\rangle ds + \int_0^t \langle Y_s, Z_s^{-1}\varphi V(\cdot+W_s)\rangle ds$$
$$= g(\varphi) \text{ a.e.}$$

i.e. (Z_t^{-1}) is a first integral of the equation (2.2.) and this shows the uniqueness of (X_t). || Q.E.D.

Remark :

Suppose that
$$E\left\{\left(\int_0^t |X_s(\varphi)|^2 ds\right)^{1/2}\right\} < +\infty$$
for any $t \geq 0$ and $\varphi \in \mathcal{D}(\mathbb{R}^d)$ and define H_t as $H_t(\varphi) = E[X_t(\varphi)]$.

H_t is then a solution of
$$\frac{\partial H_t}{\partial t} = \frac{1}{2}\Delta H_t - VH_t, \quad H_0 = g,$$
in $\mathcal{D}'(\mathbb{R}^d)$. Because of this result equation (2.1.) is called stochastic Feynman-Kac formula.

Let us denote by $\{X_t^s(\varpi,g); s \leq t, \omega \in \Omega\}$ the solution of (2.1.) starting from g at time s. Suppose that $(\Omega, \mathcal{F}, (\mathcal{F}_t), P)$ is the completed canonical Wiener space. Since

$$X_t(\omega,g) = (g Z_t(\omega,\cdot)) * \delta_{W_t(\omega)} \quad \text{a.e.}$$

we see that $(\varpi,g) \longmapsto X_t^s(\omega,g)$ is $\mathcal{F}_t^s \times \mathcal{B}(\mathcal{D}'(\mathbb{R}^d))$ measurable where $\mathcal{B}(\mathcal{D}'(\mathbb{R}^d))$ denotes Borel σ-algebra of $\mathcal{D}'(\mathbb{R}^d)$ and $\mathcal{F}_t^s$ is the completed σ-algebra generated by $\{W_r - W_s : s \leq r \leq t\}$. From the uniqueness of the solution of (2.1.) we have
$$X_{t+h}(\omega,g) = X_{t+h}^t(\omega, X_t(\omega,g)) \quad \text{a.e.}$$

Using the monotone class theorem (c.f. [2]) and the fact that $\mathcal{F}_t^s$ is independent of $\mathcal{F}_{s}$, it is easy to that

$$E[f(\langle X_{t+h}(g),\varphi\rangle)\mid \mathcal{F}_t] = E[f(\langle X_{t+h}(g),\varphi\rangle)\mid X_t(g)] \text{ a.e.}$$

hence $(X_t(g))$ is a Markov process with values in $\mathcal{D}'(\mathbb{R}^d)$.

Theorem 2.3.

$\{X_t(g)\ ;\ t \geq 0,\ g\in\mathcal{D}'(\mathbb{R}^d)\}$ is a strong Markov process in $\mathcal{D}'(\mathbb{R}^d)$. If V has bounded derivatives of all orders than $\{X_t(g);\ t\geq 0,\ g\in\mathcal{S}'(\mathbb{R}^d)\}$, where $\mathcal{S}'(\mathbb{R}^d)$ denotes the space of the tempered distributions, is also a Markov process and its infinitesimal generator on the class of $\mathcal{C}^2$-functions on $\mathcal{S}'(\mathbb{R}^d)$ with bounded Fréchet derivatives is of the following form :

$$AF(g) = d/dt\,(E(F(X_t(g))))|_{t=0} =$$

$$= 1/2\langle DF(g), \Delta g\rangle + 1/2\ \langle D^2F(g), D^i g\otimes D_i g\rangle -$$

$$-\langle DF(g), Vg\rangle .$$

Proof :

The strong Markov property follows from the fact that

$$(t, \varpi, g)\longmapsto X_t(\varpi, g)$$

is $\mathcal{B}(\mathbb{R}_+)\times\mathcal{F}\times\mathcal{B}(\mathcal{D}')$-measurable and that

$$X_{T+t}(g) = X^T_{t+T}(\varpi, X_T(\varpi, g)) \quad \text{a.e.},$$

for any bounded stopping time T.

To calculate A we shall use Ito's formula for the Hilbert space-valued semimartingales (c.f. [6]). In fact, since X(g) is locally in $S^1(\mathcal{D}'(\mathbb{R}^d))$, there exists a sequence of stopping times (T_n) increasing to infinity such that $\{X^n_t(g)\ ;\ t\geq 0\}$, where $X^n_t(g) = X_{t\wedge T_n}(g)$, is undistinguishable from the image, under the natural injection i_n, of a $\mathcal{D}'(\mathbb{R}^d)[K_n]$-valued semimartingale Y^n for some $K_n\in\mathcal{K}_h(\mathcal{D}'(\mathbb{R}^d))$. Hence

$$F(X^n_t) = (F\circ i_n)(Y^n_t) = f_n(Y^n_t)$$

and

$$f_n(Y^n_t) = F(g) + \int_0^t (Df_n(Y^n_s), dY^n_s) + 1/2 \int_0^t D^2 f_n(Y^n_s).d\langle Y^n, Y^n\rangle_s \cdot$$

As we have shown in (11) and (12), the above integrals are independent of any particular choice of K_n, hence we have

$$F(X^n_t) = F(g) + 1/2 \int_0^t \langle DF(X^n_s), \Delta X^n_s\rangle ds - 1/2 \int_0^t \langle DF(X^n_s), VX^n_n\rangle ds -$$

$$- \int_0^t \langle DF(X^n_s), D_i X^n_s\rangle \; dW^{i,n}_s +$$

$$+1/2 \int_0^t \langle D^2F(X^n_s), D_i X^n_s \otimes D^i X^n_s\rangle ds \qquad \text{a.e.}$$

Whenever $g \in \mathcal{S}'(\mathbb{R}^d)$ we have

$$E \int_0^t |X_s(\varphi)|^2 ds < \infty \quad , \varphi \in \mathcal{D}(\mathbb{R}^d),$$

for any $t \geq 0$, therefore all the components of $\{F(X^n_t); n \in \mathbb{N}\}$ are uniformly integrable and the theorem follows from Fatou's Lemma.

‖Q.E.D.

3. SCHRÖDINGER'S EQUATION

Let $g \in \mathcal{E}'(\mathbb{R}^d)$ (i.e. a distribution of compact support), then, for any $\varphi \in \mathcal{D}(\mathbb{R}^d)$

$$(t,\omega) \longmapsto \langle g, (\exp i^{3/2} W_t(\omega).x)\varphi\rangle \; , \; i = (-1)^{1/2}$$

has a modification which a semimartingale with continuous trajectories with values in $\mathcal{E}'(\mathbb{R}^d)$ which is denoted by $(X_t(\varphi))$. Moreover (X_t) satisfies the following equation

$$\text{(3.1.)} \quad dX_t(\varphi) = i^{3/2} X_t(x_j \varphi)\, dW^j_t - \frac{i}{2} X_t(|x|^2 \varphi)\, dt$$
$$X_0(\varphi) = g(\varphi)$$

If we take the Fourier transformation of (X_t) we find that

$$\text{(3.2.)} \quad d\hat{X}_t(\varphi) = i^{1/2} \hat{X}_t(D_j \varphi)\, dW^j_t + \frac{i}{2} \hat{X}_t(\Delta\varphi)\, dt \, , \hat{X}_0(\varphi) = \hat{g}(\varphi)$$

i.e. $(\hat{X}_t)$ satisfies an evolution such that $\psi_t(\varphi)$ defined by

$$\psi_t(\varphi) = E[\hat{X}_t(\varphi)]$$

is a weak solution of Schrödinger's equation with the initial condition $\hat{g}$. These results are rather limited since the initial condition $\hat{g}$ is restrained to be an analytic function (c.f. [8]). The difficulty comes from the fact that we must make a coordinate transformation by the vector $\sqrt{i}\, W_t$ to obtain (3.2.) or equivalently we should calculate $(g \exp i^{3/2} W_t \cdot x)^\wedge$, but even if g is a tempred distribution, the product is not necessarily a tempered distribution hence its Fourier transformation is not defined. Instead of working on $\mathcal{D}(\mathbb{R}^d)$ we shall work on $\mathcal{Z}(\mathbb{R}^d)$ which is the image of $\mathcal{D}(\mathbb{R}^d)$ under the Fourier transformation (which will be denoted by $\mathcal{F}$ or "$\hat{\ }$"). On $\mathcal{Z}(\mathbb{R}^d)$ we shall put the finest topology for which $\mathcal{F}: \mathcal{D}(\mathbb{R}^d) \to \mathcal{Z}(\mathbb{R}^d)$ is continuous.

Proposition 3.1.

Let $g \in \mathcal{S}'(\mathbb{R}^d)$ then $(t,\omega) \mapsto \langle g, \mathcal{F}(e^{\sqrt{-i}\, W_t \cdot x}\varphi)\rangle$ defines a semimartingale with almost surely continuous trajectories in $\mathcal{Z}'(\mathbb{R}^d)$, denoted by $X_t(\hat{\varphi})$. Moreover (X_t) satisfies the following evolution equation :

$$(3.3.) \qquad \begin{aligned} dX_t(\hat{\varphi}) &= \sqrt{i}\, X_t(D_j\hat{\varphi})\, dW_t^j + \frac{i}{2} X_t(\Delta\hat{\varphi})\, dt \\ X_0(\hat{\varphi}) &= g(\hat{\varphi}). \end{aligned}$$

Moreover, for any $g \in \mathcal{S}'(\mathbb{R}^d)$, (3.3.) has a unique solution.

<u>Proof</u> : Evidently $\mathcal{Z}(\mathbb{R}^d)$ and $\mathcal{Z}'_\beta(\mathbb{R}^d)$ are both nuclear spaces hence the results of the first section can be applied. To prove the first part of the proposition one proceeds as in the second section. The uniqueness can be proved by applying the integration by parts formula to prove

$$\langle X_t, \hat{\varphi}(\cdot - \sqrt{i}\, W_t)\rangle = \langle g, \hat{\varphi}\rangle \qquad \text{a.e.}$$

|| QED

If we note $X_t \circ \mathcal{F}$ by α_t, (α_t) becomes a $\mathcal{D}'$-valued semimartingale such that

$$(3.4.) \qquad \begin{aligned} d\alpha_t(\varphi) &= (-i)^{1/2} \alpha_t(x_j\varphi)\, dW_t^j - \frac{i}{2} \alpha_t(|x|^2\varphi)\, dt \\ \alpha_0(\varphi) &= \hat{g}(\varphi), \quad \varphi \in \mathcal{D}(\mathbb{R}^d). \end{aligned}$$

Of course (3.4.) has also a unique solution. If $\varphi \mapsto E[\alpha_t(\varphi)]$ is a tempered distribution and if

$$E\left\{\left(\int_0^t |\alpha_s(\varphi)|^2 ds\right)^{1/2}\right\} < +\infty, \text{ for } t \geq 0, \varphi \in \mathcal{D}(\mathbb{R}^d),$$

then $H_t(\varphi) = E[\alpha_t(\mathcal{F}^{-1}\varphi)]$ is a solution of Schrödinger's equation

$$\frac{\partial H_t}{\partial t} = \frac{i}{2}\Delta H_t, \quad H_0 = g \in \mathcal{S}'(\mathbb{R}^d).$$

Proposition 3.2.

Suppose that $g \in \mathcal{E}'(\mathbb{R}^d)$ and V is a C^∞-function of polynomial growth having an analytic extension. Then there exists a semimartingale (X_t) with continuous trajectories in $\mathcal{Z}'(\mathbb{R}^d)$ such that, for any $\hat{\varphi} \in \mathcal{Z}(\mathbb{R}^d)$, one has

$$X_t(\hat{\varphi}) = \langle g \exp -iS_t, \mathcal{F}(\varphi.\exp(\sqrt{-i}W_t.x))\rangle \text{ a.e.}, t \geq 0,$$

where $S_t(x) = \int_0^t V(x+\sqrt{i}W_s)ds$. Moreover (X_t) satisfies the following evolution equation

$$(3.5) \quad \begin{aligned} dX_t(\hat{\varphi}) &= [\tfrac{i}{2}X_t(\Delta\hat{\varphi}) - iX_t(V\hat{\varphi})]dt + \sqrt{i}\, X_t(D_j\hat{\varphi})\, dW_t^j \\ X_0(\hat{\varphi}) &= g(\hat{\varphi}) \end{aligned}$$

Moreover equation (3.5.) has a unique solution for any $g \in \mathcal{E}'(\mathbb{R}^d)$.

Proof : The proof of the first part is similar to the proof of Feynman-Kac formula. To show the uniqueness it is sufficient to apply the integration by parts formula twice as in the proof of Theorem 2.2. ‖ QED.

Remark : Suppose that $\hat{\varphi} \mapsto E[X_t(\hat{\varphi})]$ is continuous under the topology induced by $\mathcal{S}(\mathbb{R}^d)$ and the local martingale part of (X_t) is a martingale. Then H_t defined by $H_t(\varphi) = E[X_t(\mathcal{F}^{-1}\hat{\varphi})]$ is a solution of Schrodinger's equation

$$D_t H = \frac{i}{2}\Delta H_t - i\, VH_t$$

$$H_0 = g \in \mathcal{E}'(\mathbb{R}^d).$$

4. GENERALIZED FORMULAE OF ITO

In this section we give two applications of the integration by parts formula for the nuclear space-valued semimartingales. The first one is given for a flow of a diffusion process with regular coefficients and the second is for the exponential semimartingale associated to the same diffusion process. Let us note that one can multiply this kind of formulae using Theorem 7.1. of Section 7, the integration by parts formula and the following proposition :

Proposition 4.1.

Suppose that k is a semimartingale with values in $\mathbb{R}^d$, $d \geq 1$, and R is a distribution on $\mathbb{R}^d$. Let X be defined on the space of the infinitely differentiable functions with compact support in $\mathbb{R}^d$ in the following way

$$\tilde{X}(\varphi)_t = R(\varphi(. + k_t)), \quad \varphi \in \mathcal{D}(\mathbb{R}^d)$$

i.e., we make a random, linear coordinate transformation of $\mathbb{R}^d$ defined by the random vector k.

Then $\tilde{X}$ defines a linear, sequantially continuous mapping from $\mathcal{D}(\mathbb{R}^d)$ into S^0. Moreover there exists a right continuous stochastic process $\hat{X}$ with left limits which is a semimartingale such that, for any $\varphi \in \mathcal{D}(\mathbb{R}^d)$, one has :

$$\hat{X}_t(\varphi) = \hat{X}_0(\varphi) - \int_0^t D_i \hat{X}_{s-}(\varphi)\, dk_s^i + \frac{1}{2}\int_0^t D_{ij}\hat{X}_{s-}(\varphi)\, d\langle k^{i,c}, k^{j,c}\rangle_s +$$

$$+ \sum_{0<s\leq t} \Big[\hat{X}_s(\varphi) - \hat{X}_s(\varphi) + \sum_{i=1}^{d} D_i \hat{X}_{s-}(\varphi)\, \Delta k_s^i \Big]$$

up to an evancescent process and $\hat{X}.(\varphi)$ belongs to the equivalence class $\tilde{X}(\varphi)$, for any $\varphi \in \mathcal{D}(\mathbb{R}^d)$. If R is a distribution of compact support one can read the proposition replacing $\mathcal{D}'$ by $\mathcal{E}'$ (i.e. the distributions of compact support) and $\mathcal{D}(\mathbb{R}^d)$ by $\mathcal{E}(\mathbb{R}^d)$ (i.e. the space of C^∞-functions on $\mathbb{R}^d$).

Let us denote by $\{x_t(\omega, x) : t \geq 0, x \in \mathbb{R}^d\}$ the solution of the following Ito's equation :

$$dx_t(x) = b(x_t(x))\, dt + \sigma(x_t(x))\, dW_t$$

$$x_0(x) = x$$

where b and σ are C^∞-vector fields with values respectively in $\mathbb{R}^d$ and $\mathbb{R}^d \otimes \mathbb{R}^d$, having bounded derivatives of all orders and (W_t) is the d-dimensional Wiener process starting from zero.

Using the theory of vector measures, we have shown in [9] the following.

Proposition 4.2.

There exists a semimartingale π with values in $\mathcal{E}(\mathbb{R}^d, \mathbb{R}^d)$ (i.e. C^∞-functions with values in $\mathbb{R}^d$), having continuous trajectories such that, for any $x \in \mathbb{R}^d$, $\pi(x)$, and $\{x_t(x) ; t \geq 0\}$ are undistinguishable.

The following result is an application of the integration by parts formula (cf. Theorem I.4.) and Proposition I.1. :

Proposition 4.3.

Suppose that k is in $S^o(\mathbb{R}^d)$. Then $\{\pi_t(k_t) ; t \geq 0\}$ is a semimartingale with values in $\mathbb{R}^d$ satisfying the following relation :

$$\begin{aligned}\pi_t^\ell(k_t) = k_0^\ell &+ \int_0^t b_\ell(\pi_s(k_s))ds + \int_0^t \sigma_{\ell i}(\pi_s(k_s))dW_s^i + \int_0^t (D_i \pi_s^\ell)(k_{s-})dk_s^i + \\ &+ \frac{1}{2}\int_0^t (D_{ij}\pi_s^\ell)(k_{s-})\, d\langle k^{i,c}, k^{j,c}\rangle_s + \\ &+ \int_0^t (D_m \sigma_{\ell i})(\pi_s(k_{s-}))(D_j \pi_s^m)(k_{s-})\, d\langle W^i, k^{j,c}\rangle_s + \\ &+ \sum_{0<s\leq t} [\pi_s^\ell(k_s) - \pi_s^\ell(k_{s-}) - (D_i \pi_s^\ell)(k_{s-})\Delta k_s^i]\end{aligned}$$

for $\ell = 1, 2, \ldots, d$, up to an evanescent process.

Proposition 4.4.

i) Suppose that σ^{-1} exists and it has bounded derivatives of all orders. Then the martingale :

$$Z_t(x) = \exp\left\{\int_0^t (\sigma^{-1}b)(x_s(x)).dW_s - \frac{1}{2}\int_0^t (a^{-1}b, b)(x_s(x))ds\right\}, \; a = \sigma\sigma^t$$

can be modified as an $\mathcal{E}(\mathbb{R}^d)$ -valued martingale with continuous trajectories (it will be denoted by the same letter). Moreover for any $F \in \mathcal{E}'(\mathbb{R}^d)$ one has :

$$\langle F, Z_t\rangle = \langle F, 1\rangle + \int_0^t \langle F, Z_s((\sigma^{-1}b)(\pi_s(.)), dW_s\rangle$$

up to an evanescent process.

ii) Suppose that k is a semimartingale with values in $\mathbb{R}^d$. Then $(t,\omega)\mapsto Z_t(\omega, k_t(\omega))$ is a semimartingale such that :

$$\begin{aligned}
Z_t(k_t) = 1 &+ \int_0^t D_i Z_s(k_{s-})\, dk_s^i + \frac{1}{2}\int_0^t D_{ij} Z_s(k_{s-})\, d\langle k^{i,c}, k^{j,c}\rangle_s + \\
&+ \sum_{0<s\leq t} [Z_s(k_s) - Z_s(k_{s-}) - D_i Z_s(k_{s-})\Delta k_s^i] + \\
&+ \int_0^t Z_s(k_{s-})\, ((\sigma^{-1} b)(\pi_s(k_{s-})), dW_s) + \\
&+ \int_0^t [Z_s(k_{s-})\, D_\ell \sigma^{-1}_{im}(\pi_s(k_s))\, D_j \pi_s^\ell(k_{s-})\, b^m(\pi_s(k_{s-})) + \\
&\quad + Z_s(k_{s-})\, (\sigma^{-1}_{im} D_\ell b^m)(\pi_s(k_{s-}))\, D_j \pi_s^\ell(k_{s-}) + \\
&\quad + D_j Z_s(k_{s-})\, (\sigma^{-1} b)_i(\pi_s(k_{s-}))]\, d\langle W^i, k^{j,c}\rangle_s\,,
\end{aligned}$$

up to an evanescent process, where σ^{-1}_{ij} denotes $(\sigma^{-1})_{ij}$.

Proof : The first part of the proposition follows from Proposition I.2. and the equation (I.2.) is a consequence of the continuity of the trajectories of (Z_t) and Theorem I.3. For the second part it is sufficient to apply the integration by parts formula to $\langle \delta_{k_t}, Z_t(\cdot)\rangle$ by the help of Proposition I.1. and the equation (I.2.). Note that, by Theorem I.3., for any $T>0$, there exists a probability measure equivalent to P such that $\{Z_t\,;\, t\leq T\}$ belongs to $S_T^1(\mathcal{E}(\mathbb{R}^d))$ for this measure and this justifies the applicability of the integration by parts formula.

|| QED.

5. CAMERON-MARTIN TRANSFORMATION AND STOCHASTIC EVOLUTION EQUATIONS

Let T be any element of $\mathcal{D}'(\mathbb{R}^d)$, b a C^∞-vector field on $\mathbb{R}^d$ with bounded deirvatives. Define $(X_t\,;t\geq 0)$ as :

$$\tilde{X}(\varphi)_t = \langle T, \exp\left(\int_0^t (b(\cdot+W_s), dW_s) - \frac{1}{2}\int_0^t (b,b)(\cdot+W_s)ds\right)\varphi(\cdot+W_t)\rangle$$

for $\varphi\in\mathcal{D}(\mathbb{R}^d)$. If we denote the exponential martingale by $(Z_t(x);t\geq 0)$, by Ito's formula and integration by parts formula we have :

$$\begin{aligned}\varphi(x+W_t)Z_t(x) = \varphi(x) &+ \int_0^t Z_s(x)\, D_i\varphi(x+W_s)dW_s^i + \\ &+ \frac{1}{2}\int_0^t \Delta\varphi(x+W_s)Z_s(x)ds + \\ &+ \int_0^t \varphi(x+W_s)Z_s(x)\,(b(x+W_s), dW_s) + \\ &+ \int_0^t D_i\varphi(x+W_s)\, b^i(x+W_s)Z_s(x)ds\,.\end{aligned}$$

Each of these integrals can be modified as a $\mathcal{D}(\mathbb{R}^d)[K]$-valued integral for some $K\in\mathcal{K}_h(\mathcal{D}(\mathbb{R}^d))$ (cf. [9]), hence we have :

$$\tilde{X}(\varphi)_t = T(\varphi) + \int_0^t \tilde{X}(D_i\varphi)_s dW_s^i + \frac{1}{2}\int_0^t \tilde{X}(\Delta\varphi)_s ds + \int_0^t \tilde{X}(b_i\varphi)_s dW_s^i + \int_0^t \tilde{X}(b_i D^i\varphi)_s ds$$

Hence $\tilde{\tilde{X}}$ is a linear, sequantially continuous mapping from $\mathcal{D}(\mathbb{R}^d)$ into S^o and Theorem I.1. implies the existence of a g-process which is a semimartingale on $\mathcal{D}'(\mathbb{R}^d)$. Moreover all the discontinuities are due to the discontinuities of (W_t) and (Z_t), hence there exists a stochastic process (X_t) with values in $\mathcal{D}'(\mathbb{R}^d)$ having almost surely continuous trajectories such that, for any $U\in\mathcal{U}_h(\mathcal{D}'_\beta)$, $k(U)\circ X$ is a $\mathcal{D}'(\mathbb{R}^d)(U)$ -valued semimartingale. Hence, for any $\varphi\in\mathcal{D}(\mathbb{R}^d)$, we have :

$$\text{(5.1.)}\quad \begin{aligned} dX_s(\varphi) &= [-D_iX_s(\varphi)+(b_iX_s)(\varphi)]dW_s^i + [\tfrac{1}{2}\Delta X_s(\varphi) - D_i(b^iX_s)(\varphi)]ds \\ X_0(\varphi) &= T(\varphi),\end{aligned}$$

up to an evanescent process.

<u>Remark</u> : if b = 0, then (X_t) satisfies

$$dX_t(\varphi) = \frac{1}{2}\Delta X_t(\varphi)dt - D_iX_t(\varphi)dW_t^i\,, \quad X_0(\varphi) = T(\varphi)\,.$$

Hence, Cameron-Martin transformation changes the drift and the diffusion coefficients of (X_t) .

Theorem 5.1.

For any $T \in \mathcal{D}'(\mathbb{R}^d)$, the equation (5.1.) has a unique solution.

Proof : First, note that, by Theorem I.3. $\{\bar{Z}_t^{-1}(x) : t \geq 0, x \in \mathbb{R}^d\}$ is an element of $S^\circ(\mathcal{E}(\mathbb{R}^d))$. Let us suppose that (X_t) is any solution of (II.1.). Define Y_t as :

$$Y_t(\varphi) = X_t(\varphi(.-W_t)).$$

Using the integration by parts formula we have :

$$\begin{aligned} Y_t(\varphi) = T(\varphi) &+ \int_0^t (\langle -D_i Y_s, \varphi\rangle + \langle (b_i X_s) * \delta_{-W_s}, \varphi\rangle) dW_s^i + \\ &+ \int_0^t \Delta Y_s(\varphi) ds - \int_0^t \langle D_i(b^i X_s) * \delta_{-W_s}, \varphi\rangle ds + \\ &+ \int_0^t \langle D_i Y_s, \varphi\rangle dW_s^i - \int_0^t \Delta Y_s(\varphi) ds + \\ &+ \int_0^t \langle D_i(b^i X_s) * \delta_{-W_s}, \varphi\rangle ds . \end{aligned}$$

Hence :

$$Y_t(\varphi) = T(\varphi) + \int_0^t \langle b_i X_s, \varphi(.-W_s)\rangle dW_s^i .$$

Moreover we have :

$$\begin{aligned} d\bar{Z}_t^{-1}(x) &= \tfrac{1}{2} (b,b)(x+W_t) dt - b_i(x+W_t) dW_t^i \\ \bar{Z}_0^{-1}(x) &= 1 . \end{aligned}$$

Applying the integration by parts formula we have :

$$\begin{aligned} \langle Y_t, \varphi \bar{Z}_t^{-1}\rangle = T(\varphi) &+ \int_0^t \langle (b_i X_s) * \delta_{-W_s}, \varphi \bar{Z}_s^{-1}\rangle dW_s^i + \\ &+ \int_0^t \langle X_s * \delta_{-W_s}, \varphi \bar{Z}_s^{-1} (b,b)(.+W_s)\rangle ds - \\ &- \int_0^t \langle X_s * \delta_{-W_s}, \bar{Z}_s^{-1} \varphi\, b_i(.+W_s)\rangle dW_s^i - \\ &- \int_0^t \langle (b_i X_s) * \delta_{-W_s}, \varphi \bar{Z}_s^{-1} b_i(.+W_s)\rangle ds \end{aligned}$$

Hence :

$$Y_t \bar{Z}_t^{-1} = (X_t * \delta_{-W_t}) \bar{Z}_t^{-1} = T \quad \text{a.e.},$$

so

$$X_t = (T Z_t) * \delta_{W_t} \quad \text{a.e.},$$

since both sides are continuous (in $\mathcal{D}'(\mathbb{R}^d)$), they are undistinguishable.

// QED.

We shall apply the same method to the non-linear flows : Let $\{x_t(x) ; t \geq 0\}$ be the solution of following equation :

$$dx_t(x) = \sigma(x_t(x))dW_t$$
$$x_0(x) = x \in \mathbb{R}^d,$$

where σ and σ^{-1} are C^∞ and they have bounded derivatives of all orders. It is known that (cf. [2]) a.s. for any $t \geq 0$, $x \mapsto \pi_t(\omega, x)$ is a diffeomorphism (cf. Proposition 4.2.). We denote by $\{Z_t(x) ; t \geq 0\}$ the $\mathcal{E}(\mathbb{R}^d)$-valued martingale obtained from :

$$\exp\left\{ \int_0^t ((\sigma^{-1}b)(x_s(x)), dW_s) - \frac{1}{2}\int_0^t (a^{-1}b, b)(x_s(x))\, ds \right\}.$$

If $T \in \mathcal{D}'(\mathbb{R}^d)$, define $\{X_t ; t \geq 0\}$ as

$$X_t(\varphi) = \langle T, Z_t(\cdot)\varphi \circ \pi_t(\cdot) \rangle .$$

It is not difficult to show that $\{X_t ; t \geq 0\}$ is a semimartingale with values in $\mathcal{D}'(\mathbb{R}^d)$, having almost surely continuous trajectories, satisfying the following equation :

$$(5.2.)\quad dX_t(\varphi) = \left[\frac{1}{2} D_{ij}(a^{ij}X_t)(\varphi) - D_i(b^i X_t)(\varphi)\right] dt -$$

$$- \left[D^j(\sigma_{ij} X_t)(\varphi) + (b^j \sigma_{ij}^{-1} X_t)(\varphi)\right] dW_t^j ,$$

$$X_0(\varphi) = T(\varphi),$$

where σ_{kl}^{-1} denotes $(\sigma^{-1})_{kl}$. We shall show that the equation (5.2.) has one and only one solution for any $T \in \mathcal{D}'(\mathbb{R}^d)$. First we recall that the inverse of the diffeomorphism $x \mapsto \pi_t(\omega, x)$, denoted by $y_t(\omega, x)$ can be modified as an $\mathcal{E}(\mathbb{R}^d, \mathbb{R}^d)$-valued semimartingale (cf. [9]) and for any $x \in \mathbb{R}^d$,

$$y_t(x) = x + M_t(x) + A_t(x)$$

where :

$$M_t(x) = -\int_0^t (J\pi_s)^{-1}(y_s(x))\sigma(x)\, dW_s ,$$

$$A_t^i(x) = -\frac{1}{2}\sum_{kmjl}\int_0^t (J\pi_s)^{-1}(y_s(x))_{ik}\,(D^2\pi_s)_{kml}(y_s(x))\,[(J\pi_s)^{-1}(y_s(x))\sigma(x)]_{ml}\cdot [(J\pi_s)^{-1}(y_s(x))\,\sigma(x)]_{jl} +$$

$$+\sum_{kml}\int_0^t (J\pi_s)^{-1}_{ik}(y_s(x))\,\sigma_{km}(x)\,D_l\sigma_{ki}(x)\,ds\ ,\ i=1,\dots,d,$$

where $J\pi_t$ is the Jacobian marix of π_t .

Theorem 5.2.

For any $T\in\mathcal{D}'(\mathbb{R}^d)$, the equation (5.2.) has a unique solution.

Proof : For the notational simplicity we shall work with d = 1. Using the integration by parts formula we shall calculate $\langle X_t, \varphi\circ y_t(\cdot)\rangle$ where (X_t) is any solution of (5.2.). Using the finite dimensional Ito's formula, Theorem I.3. and integration by parts formula we have (cf. (14) for further details) :

$$X_t(\varphi\circ y_t) = T(\varphi) + \int_0^t X_s(\sigma^{-1}b.\varphi\circ y_s)\,dW_s\ ,$$

(c.f. Appendix).

Define H_t as

$$H_t(\varphi) = X_t(\varphi\circ y_t)$$

For any $\varphi\in\mathcal{D}(\mathbb{R}^d)$, $(H_t(\varphi)\ ;\ t\geq 0)$ is a semimartingale. Hence from theorem I.1, there exists a projective system of continuous semimartingales $\{H^U;\ U\in\mathcal{U}_h(\mathcal{D}'(\mathbb{R}))\}$ admitting H as a limit. Let $T>0$ and choose $K\in\mathcal{K}_h(\mathcal{D}(\mathbb{R}))$ such that $\varphi\in\mathcal{D}(\mathbb{R})[K]$ and $\{\varphi.(\sigma^{-1}b)(\pi_t)\ ;t\leq T\}\in S_T^\circ(\mathcal{D}[K])$. Then

$$H_t(\varphi) = (H_t^{K^\circ}\,|\,\varphi) = T(\varphi) + \int_0^t (H_s^{K^\circ}|\varphi(\sigma^{-1}b)\circ\pi_s)\,dW_s \quad \text{a.e.}$$

for $t\leq T$. Choosing K sufficiently large we can also suppose that $\{Z_t^{-1}\varphi\ ;t\leq T\}$, $\{\varphi.\int_0^t(\sigma^{-1}b)(\pi_s)ds\ ;\ t\leq T\}$ are also in $S_T^\circ(\mathcal{D}(\mathbb{R})[K])$. Then an application of integration by parts formula gives that

$$\langle H_t, Z_t^{-1}\varphi \rangle = \langle T, \varphi \rangle \text{ a.e.}, \qquad \text{a.e.},$$

hence

$$\langle Z_t T, \varphi \circ \pi_t \rangle = \langle X_t, \varphi \rangle \qquad \text{a.e.}$$

Since (X_t) has almost surely continuous trajectories the proof is completed.

‖ Q.E.D.

6. ON THE FILTERING OF DIFFUSION PROCESSES

In the filtering of diffusion processes, the unnormalied conditional probability density function (p_t) satisfies the following stochastic evolution equation :

(6.1.) $$dp_t = Lp_t \; dt + hp_t \cdot dW_t$$

where L is an elliptic operator, h is a bounded function and (W_t) is a finite dimensional Wiener process. In the following we shall assume that the coefficients of (6.1) are smooth.

Lemma 6.1.

Let $M_t(\varpi, x)$ be defined as

$$M_t(\varpi, x) = \exp(-h(x) \cdot W_t(\varpi) + |h|^2(x)\, t/_2).$$

Then M is a continuous semimartingale with values in $\mathcal{E}(\mathbb{R}^d)$.

The proof is omitted since it is exactly the same as the proof of Proposition 4.4.

Theroem 6.1.

Let K be the operator defined by

$$K_t f = M_t L(f/M_t) \;,\; f \in \mathcal{D}(\mathbb{R}^d).$$

If (p_t) is any solution of (6.1) then $M_t p_t = y_t$ satisfies

(6.2.) $$dy_t = Ky_t \; dt - |h|^2 y_t \; dt$$

$$y_0 = p_0$$

Proof :

Let f be any element of $\mathcal{D}(\mathbb{R}^d)$. By the integration by parts formula we have

$$\langle p_t, fM_t\rangle = p_o(f)+\int_o^t \langle dp_s, fM_s\rangle + \int_o^t \langle p_s, fdM_s\rangle + [\![p, fM]\!]_t$$

$$= p_o(f)+\int_o^t \langle Lp_s, fM_s\rangle ds+\int_o^t \langle h_i p_s, fM_s\rangle\ dW_s^i -$$

$$- \int_o^t \langle p_s, h_i fM_s\rangle\ dW_s^i - \int_o^t \langle |h|^2\ p_s, fM_s\rangle\ ds\ ,$$

hence we have

$$\langle p_t, fM_t\rangle = p_o(f) + \int_o^t \langle M_s L p_s, f\rangle\, ds - \int_o^t \langle |h|^2 M_s p_s, f\rangle\, ds.$$

|| Q.E.D.

Remark

If L is of the form (for d=1)

$$\frac{a}{2}\ D^2 + b\ D$$

then the principle part of K can be written as

$$K_\theta\ f = \frac{1}{2}\ a\ D^2 f + (b+Dh\ W_\theta\ (\varpi)\ + h\ Dh)\ \ Df$$

hence the equation (6.1) is reduced to a Cauchy problem for the operator K depending measurably on ϖ .

7. THE REGULARIZATION OF SEMIMARTINGALE-VALUED RANDOM FIELDS

The main result of this section is the following theorem :

Theorem 7.1.

Let X be a random field on $[0,1] \times \mathbb{R}^d$ such that, for any $x \in \mathbb{R}^d$, X(x) is a semimartingale in S^1. Suppose that the following conditions are verified :

a) For any bounded, previsible process h, the mapping

$$x \longmapsto E\int_o^1 h_s dX_s(x)$$

is infinitely differentiable ;

b) For any $H \in L^{\infty}(\Omega, \mathcal{F}, P)$, the mapping

$$x \longmapsto E\,(HX_t(x))$$

is infinitely differentiable.

Then X has a modification which is a semimartingale with values in $\mathcal{E}(\mathbb{R}^d) \tilde{\otimes} \mathbb{R}^d$.

Proof :

Let us denote by $\mathcal{X}$ the Banach space of the bounded, previsible processes under the topology of uniform convergence. Then for any $x \in \mathbb{R}^d$, $C \longmapsto E \int_0^1 1_C dX_s(x)$ defines a measure on the previsible σ-algebra of $[0,1] \times \Omega$. Hence, if h^n converges to zero in $\mathcal{X}$, then $E \int_0^1 h_s^n dX_s(x)$ converges to zero in $\mathcal{E}(\mathbb{R}^d) \tilde{\otimes}_\pi \mathbb{R}^d$ as one can see using the closed graph theorem. Suppose that $|h^n| \leq 1$ and converges to zero pointwise. If F is a distribution of compact support, we pretend that

$$\langle F, E \int_0^1 h_s^n \, dX_s(.)\rangle \xrightarrow[n\to\infty]{} 0.$$

Since $(h^n ; n \in \mathbb{N})$ is bounded in $\mathcal{X}$, $(E \int_0^1 h_s^n \, dX_s(.) ; n \in \mathbb{N})$ is bounded in $\mathcal{E}(\mathbb{R}^d) \tilde{\otimes} \mathbb{R}^d$, hence it is relatively compact. Let q be any element of its closure. Let $(E \int_0^1 h_s^{n_k} dX_s(.) ; k \in \mathbb{N})$

be a subsequence converging to q. However, it is a trivial fact that the sequence $(E \int_0^1 h_s^n \, dX_s(.) ; n \in \mathbb{N})$ converges to zero in $\mathcal{D}'(\mathbb{R}^d) \tilde{\otimes} \mathbb{R}^d$ hence $q = 0$, i.e., the sequence has only one adherence point $q = 0$, therefore it converges to zero. Consequently $h \longmapsto E \int_0^1 h_s dX_s(x)$ induces a vector measure on the previsible subsets of $[0,1] \times \Omega$ with values in $\mathcal{E}(\mathbb{R}^d) \tilde{\otimes} \mathbb{R}^d$, which does not charge the evanescent sets. Hence there exists a previsible process B of integrable variation with values in $\mathcal{E}(\mathbb{R}^d) \tilde{\otimes} \mathbb{R}^d$ having right continuous trajectories with left limits (c.f. (11)) such that, for any distribution of compact support F, one has

$$\langle F, E \int_0^1 h_s \, dX_s(.)\rangle = E \int_0^1 h_s d \langle F, B_s\rangle$$

for any $h \in \mathcal{X}$. SImilarly, the condition (b) implies that the mapping

$$A \longmapsto E(1_A \, X_t(x)) = m_t(x,A)$$

is a vector measure on $(\Omega,\mathcal{F})$ with values in $\mathcal{E}(\mathbb{R}^d)\tilde{\otimes}\mathbb{R}^d$ which is absolutely continuous with respect to P, hence it has a Radon-Nikodyme density Z_t (ω,x) with values in $\mathcal{E}(\mathbb{R}^d)\tilde{\otimes}\mathbb{R}^d$. Let L_t be defined as $Z_t - B_t$. If $s < t, A\in\mathcal{F}_s$ and $F = D^k g$, where g is any continuous function of compact support, we have

$$E(1_A \langle F,L_t\rangle) = (-1)^k E\int 1_A g(x)\, D^k L_t\,(x)\; dx$$

$$=(-1)^k \int g(x) E(1_A D^k(Z_t(x)-B_t(x)))\, dx$$

$$=(-1)^k \int g(x)\, D^k(E(1_A(Z_t(x)-B_t(x)))\, dx$$

$$=(-1)^k \int g(x)\, D^k(E(1_A(X_t(x)-B_t(x)))\, dx$$

$$= E(1_A \;\langle F,L_s\rangle),$$

since $(X_t(x)-B_t(x)\ ;\ t\in[0,1])$ is a martingale. Hence for any distribution of compact support F, $(\langle F,Z_t\rangle\ ,\ t\in[0,1])$ has a modification which is a semimartingale. Then the theorem follows from Theorem 1.3. Q.E.D.

As an application of Theorem 7.1. and the integration by parts formula, we have

Theorem 7.2.

Let x_t^α be the solution of the following equation :

$$dx_t^\alpha = b(t,\alpha\ ,\ x_t^\alpha)\quad dt + \sigma(t,\alpha\ ,\ x_t^\alpha)\quad dW_t$$

where W is a one dimensional Wiener process, b and σ are defined on $\mathbb{R}_+ \times \mathbb{R} \times \mathbb{R}$ with real values such that they have bounded derivatives of all orders with respect to x and α. Then there exists a semimartingale π with values in $\mathcal{E}(\mathbb{R})$, such that, for any $\alpha\in\mathbb{R}$, $\pi(\alpha)$ is undistinguishable from x^α and if z is any semimartingale, then one has

$$\Pi_t(z_t) = \Pi_0(z_0) + \int_0^t D_\alpha \Pi_s(z_{s-})\, dz_s + \frac{1}{2}\int_0^t D_\alpha^2 \Pi_s(z_{s-})\, d\langle z^c, z^c\rangle_s +$$

$$+ \int_0^t b(s, z_{s-}, \Pi_s(z_{s-}))\, ds + \int_0^t \sigma(s, z_{s-}, \Pi_s(z_{s-}))\, dW_s +$$

$$+ \int_0^t \left[D_\alpha \sigma(s, z_{s-}, \Pi_s(z_{s-})) + D_x \sigma(s, z_{s-}, \Pi_s(z_{s-}))\, D_\alpha \Pi_s(z_{s-})\right] \cdot d\langle W, z^c\rangle_s +$$

$$+ \sum_{0<s\leq t} \left[\Pi_s(z_s) - \Pi_s(z_{s-}) - D_\alpha \Pi_s(z_{s-})\, \Delta z_s\right] .$$

Proof :
The exsitence of Π follows from Theorem 7.1. and the formula is then a consequence of Proposition 4.1. and the integration by parts formula.

|| Q.E.D.

APPENDIX TO THEOREM 5.2

$$
\begin{aligned}
\langle X_t, \varphi \circ y_t \rangle = \langle T, \varphi \rangle & - \int_0^t X_s \big(D\varphi(y_s)(D\pi_s^{-1})(y_s)\sigma \big) \, dW_s - \\
& - \frac{1}{2} \int_0^t X_s \big(D\varphi(y_s)(D\pi_s^{-3})(y_s) D^2\pi_s(y_s)\, a \big) \, ds + \\
& + \int_0^t X_s \big(D\varphi(y_s)(D\pi_s^{-1})(y_s)\, \sigma D\sigma \big) \, ds + \\
& + \frac{1}{2} \int_0^t X_s \big(D^2\varphi(y_s)\, a\, (D\pi_s^{-2})(y_s) \big) \, ds + \\
& + \int_0^t X_s \big(\sigma D\varphi(y_s)(D\pi_s^{-1})(y_s) \big) \, dW_s + \\
& + \frac{1}{2} \int_0^t X_s \big(a\, D^2\varphi(y_s)(D\pi_s^{-2})(y_s) \big) \, ds - \\
& - \frac{1}{2} \int_0^t X_s \big(a\, D\varphi(y_s)(D\pi_s^{-3})(y_s)(D^2\pi_s)(y_s) \big) \, ds + \\
& + \int_0^t X_s \big(\sigma^{-1} b \varphi(y_s) \big) \, dW_s + \int_0^t X_s \big(b D\varphi(y_s)(D\pi_s^{-1})(y_s) \big) \, ds \\
& - \int_0^t X_s \big(D^2\varphi(y_s)(D\pi_s^{-2})(y_s)\, a \big) \, ds + \\
& + \int_0^t X_s \big(a\, D\varphi(y_s)(D\pi_s^{-3})(y_s)(D^2\pi_s)(y_s) \big) \, ds - \\
& - \int_0^t X_s \big(D\varphi(y_s)(D\pi_s^{-1})(y_s)\, \sigma D\sigma \big) \, ds - \\
& - \int_0^t X_s \big(b\, D\varphi(y_s)(D\pi_s^{-1})(y_s) \big) \, ds \\
= T(\varphi) + \int_0^t X_s \big(\sigma^{-1} b\, \varphi \circ y_s \big) \, dW_s \quad \text{a.e.}
\end{aligned}
$$

REFERENCES

(1) A. Badrikian : "Séminaire sur les Fonctions Aléatoires Linéaires et les Mesures Cylindriques". Lect. Notes in Math. Vol. 139. Springer-Verlag. Berlin-Heidelberg-New York, 1970.

(2) J-M. Bismut : "A generalized formula of Ito and some other properties of stochastic flows". Zeitschr-für Wahrsch. verw. Geb. 55, 331-350 (1981).

(3) C. Dellacherie and P.A. Meyer : Probalilités et Potentiel, Chapitres I-IV. Hermann, Paris, 1975.

(4) C. Dellacherie and P.A. Meyer : Probabilités et Potentiel, Chapitres V-VIII. Hermann, Paris, 1980.

(5) A. Grothendieck : "Produits tensoriels topologiques et espaces nucléaires". Mem. Amer. Math. Soc. No. 6 (1955)

(6) M. Métivier and J. Pellaumail : Stochastic Integration. Academic Press, 1980.

(7) H.H. Schaefer : Topological Vector Spaces. Graduate Texts in Math. Springer-Verlag. Berlin-Heidelberg-New York, 1970.

(8) L. Schwartz : Théorie des Distributions. Hermann, Paris, 1973.

(9) A.S. Üstünel : "Calcul stochastique sur les espaces nucléaires et ses applications". Thèse de Doctorat d'Etat, Université de Paris VI, 1981.

(10) A.S. Üstünel : "Formule de Feynman-Kac stochastique". C.R. Acad. Sc. Paris, t. 292, p. 595-597 (1981).

(11) A.S. Üstünel : "Stochastic integration on nuclear spaces and its applications". Preprint, to appear in A.I.H.P.

(12) A.S. Üstünel : "A characterization of semimartingales on the nuclear spaces". Preprint, to appear in Zeit. für Wahrs. und verw Geb.

(13) A.S. Üstünel : "Some applications of stochastic integration in infinite dimension I". Preprint. to appear in Stochastics.

(14) A.S. Üstünel : "Some applications of stochastic integration in infinite dimension II". Preprint. To appear in Stochastics.

CHAPTER IX

APPLICATIONS

Y. BARAM, M. MARGALIT
Reconstruction and compression of two dimensional fields from sampled data by pseudo-potential functions

A. BARKANA
Optimum perturbation signal that makes nonlinear systems approach to linear systems

K. G. BRAMMER
Stochastic filtering problems in multiradar tracking

C. A. BRAUMANN
Population extinction probabilities and methods of estimation for population stochastic differential equation models

P. T. K. FUNG, Y. L. CHEN, M. J. GRIMBLE
Dynamic ship positioning control systems design including nonlinear thrusters and dynamics

F. K. GRAEF
Joint optimization of transmitter and receiver for cyclostationary random signal processes

S. C. NARDONE, A. G. LINDGREN, K. F. GONG
Fundamental properties and performance of nonlinear estimators for bearings-only target tracking

RECONSTRUCTION AND COMPRESSION OF TWO DIMENSIONAL FIELDS FROM SAMPLED DATA BY PSEUDO-POTENTIAL FUNCTIONS

Yoram Baram and Moshe Margalit

Department of Electronic Systems, School of Engineering, Tel-Aviv University, Ramat-Aviv, Tel-Aviv, Israel.

ABSTRACT

A method for representing two dimensional data for purposes of smoothing, interpolation and compression is presented. The underlying model is composed of a finite sum of pseudo-potential elements which play a role similar to that of sources and sinks in potential field theory. The local behaviour of the data is exploited along with the structure of the pseudo-potential function in converting the non-linear parameter estimation problem into linear subproblems, which are solved by standard techniques. A criterion for selecting the number of field elements to be used in the representation is proposed. The resolution of the proposed technique as a function of the number of elements used and the data size is illustrated by solving a problem of topographical mapping from radar altimetry data.

1. INTRODUCTION

This paper is concerned with the problem of reconstructing two dimensional fields from sampled data for such purposes as reducing the uncertainty in the data, interpolating (or extrapolating) the data and compressing it for storage and transmission purposes. Two dimensional fields are encountered in such areas as topography, meteorology, oceanography, electromagnetic propagation, aerodynamics, elasticity, thermal propagation and radiation. Several methods for interpolating two dimensional data have been previously suggested in the literature. A distinction can be made between local interpolation techniques, which determine the field value at a point from the values at neighboring points (e.g. [1]) and global representation techniques, which are

R. S. Bucy and J. M. F. Moura (eds.), Nonlinear Stochastic Problems, 511–524.

based on fitting a single function to the entire data set. The latter approach seems generally more suitable for data compression. Several functions, motivated by physical (e.g. [2], [3]) or structural (e.g. [4], [5]) arguments, have been suggested in the literature. Typically, such functions consist of a uniform finite set of "base functions" - simple polynomial, exponential or logarithmic elements, centered about different points. The placement problem itself has been solved either by non-linear optimization using some error criterion (e.g., least - squares) or by arbitrarily placing the base functions at grid nodes or at the data points themselves. While the effectiveness of non-linear optimization techniques is often limited to cases involving relatively few parameters (few base functions), arbitrary placement may introduce significant inaccuracies, as shown in this work.

The underlying field model proposed in this paper is based on the viewpoint that many physical fields can be described as being induced by a combination of "sources" and "sinks" and the data obtained from such fields consists, in essence, of measurements of their potential function or its derivatives. This viewpoint is motivated by potential field theory, where field phenomena are reproduced by superposition of basic elements; sources, sinks, dipoles, vorteces etc. However, while the potential function of a "pure" (mathematical) source is singular at the source location, we find it necessary to "trim" the function so as to make it more suitable for describing real physical fields, which do not display singularities. We call the resulting function a "pseudo potential" and the corresponding field element (which replaces the source or the sink) - a "zero".

By exploiting the particular structure of the pseudo potential function along with the local behaviour of the data, the zero placement problem is constructed as a set of linear problems. These and the remaining linear problem of estimating the zero powers are solved by standard least - squares procedures. We discuss the resolution of the proposed technique as a function of the number of zeros used and propose a criterion for selecting this number. The effectiveness of the proposed method is demonstrated by constructing a topographical map from radar altimetry data. We examine the mapping error behaviour as a function of the number of parameters used and of the amount of data and show that the proposed zero placement technique gives significantly better results than placement at grid nodes.

2. THE UNDERLYING FIELD FUNCTION

The field potential induced by a source (or a sink) of discharge (or intake) p_k, located at a point z_k in the plane is given by

$\Phi_k(z)=p_k \ln r_k(z)$, where $r_k(z)=|z-z_k|$, and the sign p_k determines whether $\Phi_k(z)$ represents a source (positive p_k) or a sink (negative p_k). Clearly, the potential function $\Phi_k(z)$ is singular at z_k. However, in real physical fields, "sources" and "sinks" have finite geometrical sizes and no points with infinite potential. In order to describe such fields, we "trim" the source (or sink) potential function as

$$\Phi_k(z) = q_k(z)\, p_k \ln r_k(z) = q_k(z) \ln r_k(z)^{p_k} \tag{2.1}$$

with

$$q_k(z) = 1 - e^{-r_k(z) d_k}$$

where d_k is a scale factor. While $\Phi_k(z)$ is not strictly a potential function (i.e., it does not satisfy the Laplace equation), it approaches the potential function $\Phi_k(z)$ as $r_k(z)$ increases, at a rate which can be controlled by d_k. As $r_k(z)$ decreases, $\Phi_k(z)$ approaches zero. For these reasons we call the function $\Phi_k(z)$ a pseudo-potential and the "element" which induces it (replacing the source or the sink) a "zero". The term p_k will be called the zero's "power" for both physical and mathematical reasons. It is particularly important to note that since $\phi_k(z)$ is a function of $r_k(z)$ alone, at any point z the gradient of $\phi_k(z)$ (which, for instance, represents the velocity vector in the case of a flow field) is directed at z_k, i.e., for $z = x+iy$ and $\theta_k(z) = \tan^{-1} [(y - y_k)/(x - x_k)]$ we have

$$v_k^{\theta}(z) = \frac{\partial \phi_k(z)}{\partial \theta_k(z)} = 0 \tag{2.2}$$

while

$$v_k^{r}(z) = \frac{\partial \phi_k(z)}{\partial r_k(z)} = d_k e^{-r_k(z) d_k} \ln r_k(z) + \frac{p_k q_k(z)}{r_k(z)} \tag{2.3}$$

A uniform field (e.g., a constant flow) is characterized by the potential function

$$\phi^{u}(z) = ax + by + c$$

where a and b are the potential gradients (e.g., the flow velocities) in the x and y directions and c is some constant.

Our underlying proposition is that in many applications the observed field may be adequately represented by superposition of a uniform field and a finite set of zeros. The observations themselves are, normally, samples of the (pseudo) potential function of the field(e.g. altitude in topographical mapping) or of its derivatives (e.g., wind velocities in meteorological mapping). The zeros represent local "highs" and "lows" of the pseudo potential function, whose global level is represented by the uniform field plane. For K zeros the underlying field function is then

$$\phi(z) = \sum_{k=1}^{K} \phi_k(z) + \phi^u(z)$$

$$= \sum_{k=1}^{K} q_k(z) \ln r_k(z)^{p_k} + ax + by + c \tag{2.4}$$

Let $\tilde{\phi}(z)$ denote an observation of a field potential, which is approximately represented by the pseudo potential function $\phi(z)$. The difference between the values of $\tilde{\phi}(z)$ and $\phi(z)$ may be caused by inaccuracies both in the model $\phi(z)$ and in the observation $\tilde{\phi}(z)$. Denoting this difference by $\rho(z)$, we have

$$\tilde{\phi}(z) = \phi(z) + \rho(z) \tag{2.5}$$

3. ESTIMATING THE FIELD PARAMETERS.

3.1. Problem Description

The main problem at hand is to determine the parameters of the field function $\phi(z)$. These are the number of zeros, K, their locations, (x_k, y_k), and powers p_k, k=1,...K, and the constants a,b and c. It can be seen that the function $\phi(z)$ is linear in p_k, a,b and c, but not in K and (x_k, y_k). Defining an approximation criterion (e.g., least - squares or maximum likelihood), solution of the parameter determination problem would require a non-linear optimization procedure. However, for the number of parameters involved in describing even relatively simple fields, such procedures have been found to be ineffective. An alternative approach is to reconstruct the problem as two separate subproblems, namely placing the field zeros and determining their powers along with the plane constants. Once the zeros are placed, the powers and the plane constants can be readily estimated from the data by solving a linear set of equations. One possible solution for the zero placement problem is selecting arbitrary, equally spaced points, such as grid- nodes. This approach suffers, however, from the fact that in order to reduce the inaccuracies caused by the arbitrary zero placement, the number of zeros used in the model must be increased, thus increasing the dimension of the linear equation that must be solved for the powers. Instead of placing the zeros at arbitrary points, the proposed approach exploits the particular nature of the function $\phi(z)$ as well as the local behaviour of the data in placing the zeros. A final adjustment of $\phi(z)$ to the data is then obtained by finding the set of powers and the plane constants that minimize the mean square deviation between the observations and $\phi(z)$ over the entire data set.

3.2. Placing The Field Zeros: The Equivalent Zero Method.

Suppose that the entire available data is given within a domain B, which we divide into subdomains B_k, k=1...,K. For each B_k we would like to find the location of an equivalent zero that best represents the field in B_k, assuming that it alone induces this field. By (2.2), the gradient of $\phi_k(z)$ at each point z is directed at z_k. Denoting

$$v_k^x(z) = \frac{\partial}{\partial x}\phi_k(z) \qquad v_k^y(z) = \frac{\partial}{\partial y}\phi_k(z)$$

we have

$$\frac{v_k^y(z)}{v_k^x(z)} = \frac{y - y_k}{x - x_k} \tag{3.1}$$

Let us first consider the case where gradient data is directly available in the form

$$\tilde{v}^x(z) = v_k^x(z) + \xi_k^x(z)$$

and (3.2)

$$\tilde{v}^y(z) = v_k^y(z) + \xi_k^y(z)$$

where $\tilde{v}^x(z)$ and $\tilde{v}^y(z)$ are the given (measured) gradient values at z, $v_k^x(z)$ and $v_k^y(z)$ are the values corresponding to the model $\phi_k(z)$ and $\xi_k^x(z)$ and $\xi_k^y(z)$, the differences between $\tilde{v}^x(z)$ and $v_k^x(z)$ and between $\tilde{v}^y(z)$ and $v_k^y(z)$, respectively, represent the inaccuracies in the measurements and in the model. Substituting (3.2) into (3.1), we get

$$\frac{\tilde{v}^y(z) - \xi_k^y(z)}{\tilde{v}^x(z) - \xi_k^x(z)} = \frac{y - y_k}{x - x_k}$$

or

$$[\,-\tilde{v}^y(z) \quad \tilde{v}^x(z)\,] \begin{bmatrix} x_k \\ y_k \end{bmatrix} = g_k(z) + w_k(z) \tag{3.3}$$

where

$$g_k(z) = \tilde{v}^x(z)y - \tilde{v}^y(z)x$$

and

$$w_k(z) = \xi_k^x(x - x_k) - \xi_k^y(y - y_k)$$

Since $w_k(z)$ depends on the difference $z-z_k$, its covariance appears to be unbounded. However, the zero location error becomes less significant as the difference $z-z_k$ grows, since the potential gradient decreases, as can be seen from (2.3). It makes sense, then, to bound the domain within which zeros are to be placed and discard zeros which are found to be outside this domain, thereby also bounding the uncertainties and simplifying the function $\phi(z)$.

Suppose that gradient measurements are available at N_k points in the domain B_k. Then we have a set of linear equations

$$A_k \alpha_k = g_k + w_k \tag{3.4}$$

where

$$\alpha_k = \begin{bmatrix} x_k \\ \\ y_k \end{bmatrix} \qquad A_k = \begin{bmatrix} -\tilde{v}^y(z_1) & \tilde{v}^x(z_1) \\ \cdot & \\ \cdot & \\ -\tilde{v}^y(z_{N_k}) & \tilde{v}^x(z_{N_k}) \end{bmatrix} \tag{3.5a}$$

$$g_k = \begin{bmatrix} g_k(z_1) \\ \cdot \\ \cdot \\ g_k(z_{N_k}) \end{bmatrix} \qquad w_k = \begin{bmatrix} w_k(z_1) \\ \cdot \\ \cdot \\ w_k(z_{N_k}) \end{bmatrix} \tag{3.5b}$$

Suppose that the uncertainty variables $\rho_k(z)$, $\xi_k^x(z)$ and $\xi_k^y(z)$ are zero mean, then so is $w_k(z)$. The least - squares criterion based on (3.4) suggests that the best estimate of α_k is the value that gives

$$\min_{\alpha_k} \{(A_k\alpha_k - g_k)^T (A_k\alpha_k - g_k)\} \tag{3.6}$$

The solution of (3.6) is known to be

$$\hat{\alpha}_k = (A_k^T A_k)^{-1} A_k^T g_k \tag{3.7}$$

This estimate can be written more explicitly by noting that

$$A_k^T A_k = \begin{bmatrix} A1_k & A2_k \\ A2_k & A3_k \end{bmatrix}$$

where

$$A1_k = \sum_{j=1}^{N_k} \tilde{v}^y(z_j)^2$$

$$A2_k = \sum_{j=1}^{N_k} \tilde{v}^x(z_j)\, \tilde{v}^y(z_j)$$

$$A3_k = \sum_{j=1}^{N_k} \tilde{v}^x(z_j)^2$$

Also note that

$$A_k^T g_k = \begin{bmatrix} E1_k \\ E2_k \end{bmatrix}$$

where

$$E1_k = - \sum_{j=1}^{N_k} \tilde{v}^y(z_j)\, g_{k,j}$$

$$E2_k = \sum_{j=1}^{N_k} \tilde{v}^x(z_j)\, g_{k,j}$$

Then (3.7) can be rewritten as

$$\hat{\alpha}_k = \frac{1}{A1_k A3_k - (A2_k)^2} \begin{bmatrix} E1_k\ A3_k - E2_k\ A2_k \\ -E1_k\ A2_k + E2_k\ A1_k \end{bmatrix}$$

For later reference we call the above zero placement procedure the "equivalent zero" method. Performing the procedure for all the subdomains B_i, $i=1,\ldots,K$ we get the locations of the K zeros that compose the function $\phi(z)$. The number and the size of the subdomains depend on the desired resolution. The division may be such that the subdomains cover equal areas (division by a grid) or contain equal numbers of data points. When prior knowledge on the observed field is available (e.g. in compressing a given map) the division may be variable, in accordance with the variability of the observed field. Since the next step is finding the zero powers from a set of linear equations, K may be taken, in principle, to be the maximal rank of the linear equation that can be solved.

While the least squares criterion seems most suitable for the problem at hand, other criteria may also be used. For instance, under the assumption that $\rho_k(z)$, $\xi_k^x(z)$ and $\xi_k^y(z)$ are normally distributed, the probability density function for g_k given α_k is readily obtained and the maximum likelihood criterion may be used. However, since the covariances of $w_k(z)$ and $\bar{w}_k(z)$ depend on α_k, the maximum likelihood criterion leads to a non-linear optimization problem.

3.3. Gradient Calculation

When potential gradient measurements are not directly available, the gradient components in the x and y directions must be calculated from potential measurements. The gradient at $z=x+iy$ is normal to the local tangent plane, which is approximated by the plane passing through $(x,y,\tilde{\phi}(z))$ and two close neighbouring points $(x_1,y_1,\tilde{\phi}(z_1))$ and $(x_2,y_2,\tilde{\phi}(z_2))$, provided that the three points are not on the same straight line. The gradient is then obtained as the cross product of the vectors $(x_1,y_1,\tilde{\phi}(z_1)) - (x,y,\tilde{\phi}(z))$ and $(x_2,y_2,\tilde{\phi}(z_2)) - (x,y,\tilde{\phi}(z))$. The calculated gradient components in the x and y directions are then

$$\tilde{v}^x(z) = (y_1 - y)\,[\tilde{\phi}(z_2) - \tilde{\phi}(z)] - (y_2 - y)[\tilde{\phi}(z_1) - \tilde{\phi}(z)]$$

and (3.8)

$$\tilde{v}^y(z) = (x_2 - x)\,[\tilde{\phi}(z_1) - \tilde{\phi}(z)] - (x_1 - x)[\tilde{\phi}(z_2) - \tilde{\phi}(z)]$$

when the measurements are either continuous or sufficiently dense so that $\tilde{\phi}(z)$ may be adequately differentiated by small changes in the x and y directions, the gradient components may be obtained as

$$\tilde{v}^x(z) = \frac{\tilde{\phi}(x+\Delta x,y) - \tilde{\phi}(x,y)}{\Delta x}$$

and (3.9)

$$\tilde{v}^y(z) = \frac{\tilde{\phi}(x,y+\Delta y) - \tilde{\phi}(y)}{\Delta y}$$

In principle, equations of the type (3.8) or (3.9) may be written for a number of points neighbouring to z and solved simultaneously (by, say, a least-squares procedure) to provide improved gradient estimates.

3.4. Estimating the Zero Powers and the Plane Parameters

Given the locations of the field zeros, it remains to find their powers p_k, k=1...K, and the plane parameters a,b,c. The measurement equations (of the form (2.5)) are linear in the p_k's, and are written in vector form for N measurements as

$$\tilde{\phi} = AP + R \tag{3.10}$$

where

$$\tilde{\phi} = \begin{bmatrix} \tilde{\phi}_1 \\ \cdot \\ \cdot \\ \cdot \\ \tilde{\phi}_N \end{bmatrix} \quad P = \begin{bmatrix} p_1 \\ \cdot \\ \cdot \\ p_K \\ a \\ b \\ c \end{bmatrix} \quad R = \begin{bmatrix} \rho(z_1) \\ \cdot \\ \cdot \\ \rho(z_N) \end{bmatrix}$$

and

$$A = \begin{bmatrix} q_1(z_1)\ \ln r_1(z_1) & \cdot\ \cdot & q_K(z_1)\ \ln r_K(z_1) & x_1 & y_1 & 1 \\ \cdot & & & & & \\ \cdot & & & & & \\ q_1(z_N)\ \ln r_1(z_N) & \cdot\ \cdot & q_K(z_N)\ \ln r_K(z_N) & x_N & y_N & 1 \end{bmatrix}$$

with

$$r_k(z_j) = |z_j - z_k|$$

where z_j is the location of the j'th measurement and z_k is the location of the k'th zero. The least squares criterion for estimating P from (3.10) is

$$\min_P \{ (\tilde{\phi} - AP)^T (\tilde{\phi} - AP) \}$$

yielding the least - squares estimate

$$\hat{P} = (A^T A)^{-1} A^T \tilde{\phi}$$

3.5. The Resolution: How Many Zeros to Use?

The resolution (or accuracy) of the proposed technique may be defined by some measure of the difference between the constructed surface and the one corresponding to the actual field (e.g., the field potential). Let us define the deviation between the two surfaces at the measurement points by

$$\sigma(K) = \left(\frac{1}{N} \sum_{j=1}^{N} [F(z_j) - \phi(z_j,K)]^2 \right)^{\frac{1}{2}}$$

where $F(z_j)$ is the true field value and $\phi(z_j,K)$ is the value of the surface constructed from K zeros, at z_j. The value of $F(z_j)$, which is generally unknown, is represented by the observation $\tilde{\phi}(z_j)$. The deviation between the constructed surface and the measurements is given by

$$\tilde{\sigma}(k) = \left(\frac{1}{N} \sum_{j=1}^{N} [\tilde{\phi}(z_j) - \phi(z_j,K)] \right)^{\frac{1}{2}}$$

The value of $\tilde{\sigma}(K)$ can be expected to decrease to zero as K is increased. However, in the presence of measurement noise the resolution is not necessarily improved by increasing K indefinitely. In fact, while $\tilde{\sigma}(K)$ will still approach zero as K is increased, $\sigma(K)$ will approach the measurement uncertainty r. Beside losing the error reduction (smoothing) effect at the measurement points themselves, perfectly matching the surface to the measurements may produce an irregular surface with even larger errors at intermediate points. To avoid this difficulty, we suggest that the field representation procedure be started with a small number of zeros, which will then be increased until the deviation of the constructed surface from the measurements equals the expected deviation of the true surface from the measurements. This ensures that the smallest number of zeros (i.e. the simplest function $\phi(z)$) that gives the prescribed resolution r (which is the only available information about the difference between the actual surface and the data) will be used. Also noting that the number of zeros used cannot exceed the maximal rank M of a linear equation that can be solved on the available computer, we obtain the selected number of zeros to be used as

$$K^o = \inf \{K : \tilde{\sigma}(K) = r, K = M\}$$

In fact, since $\tilde{\sigma}(K)$ will normally be monotonically decreasing with K, there will be a single value of K that gives $\tilde{\sigma}(K) = r$. The above considerations are well illustrated by the following example.

4. EXAMPLE

The proposed method has been used to obtain topographical maps of the Antarctic ice shell, using radar altimetry data obtained in the TWERLE program [6]. The specified standard deviation of the measurements is 60 meters. For illustration purposes we consider the section between latitudes 90° and 75° and longtitudes 0° and 90° and a data base consisting of 1027 altitude measurements. Various numbers of zeros have been used in reconstructing the observed field. The deviations between the maps obtained and the measurements are plotted against the number of field elements in Figure 1. Graph a corresponds to the equivalent zero placement technique. It can be seen that the measurements deviation reduces significantly with the number of zeros used, particularly when this number is relatively small. The measurement inaccuracy crossover point (i.e., the point where e = 60m) is at K = 85, so that, for reasons discussed in the previous section, the maximal number of zeros that should be used is 85. Graph b of Figure 2 corresponds to zero placement at grid points. It can be seen that the measurement deviation is significantly larger than that obtained by the equivalent zero technique, particularly for a small number of zeros. For instance, for 25 zeros the measurement deviation produced by the equivalent zero technique is about 120m, while that produced by placement at grid nodes is about 200m. In addition, the measurement accuracy crossover point for placement at grid nodes is at K = 115, implying that 30 more zeros must be used in order to obtain the same critical accuracy as with the equivalent zero technique. This means that 30 more equations must be solved simultaneously for the powers. Of course, the placement process itself, which is trivial in the grid node approach requires additional computation (i.e., solving K linear equations of rank 2) in the equivalent zero technique. However, more crucial issues are the rank of the linear equation that need be solved for the powers to provide a specified accuracy, or the accuracy that can be obtained for a given number of zeros. In both respects the equivalent zero technique is clearly superior.

Next, the effect of the data size on the map reconstruction accuracy is examined. Using the entire data set of 1027 measurements, 27 zeros have been placed using the proposed technique. The powers (and the plane parameters) are then estimated using smaller subsets of the data. The resulting error (average deviation calculated for the entire data set) is plotted in Figure 2 against the number of data points used in the estimation. It can be seen that the reduction in accuracy caused by reducing the data set is very mild. This means that once the zeros are placed, the powers can be estimated using significantly smaller data sets. When the entire estimation procedure including zero placements is performed with reduced data sets, the reduction in accuracy is more significant, as can be seen from graph b of Figure 2. This means that the accuracy of the representation procedure is highly dependent on the zero placement accuracy which, in turn, can be improved by increasing the amount of data. These results again emphasize the crucial importance of the zero placement stage in the representation process.

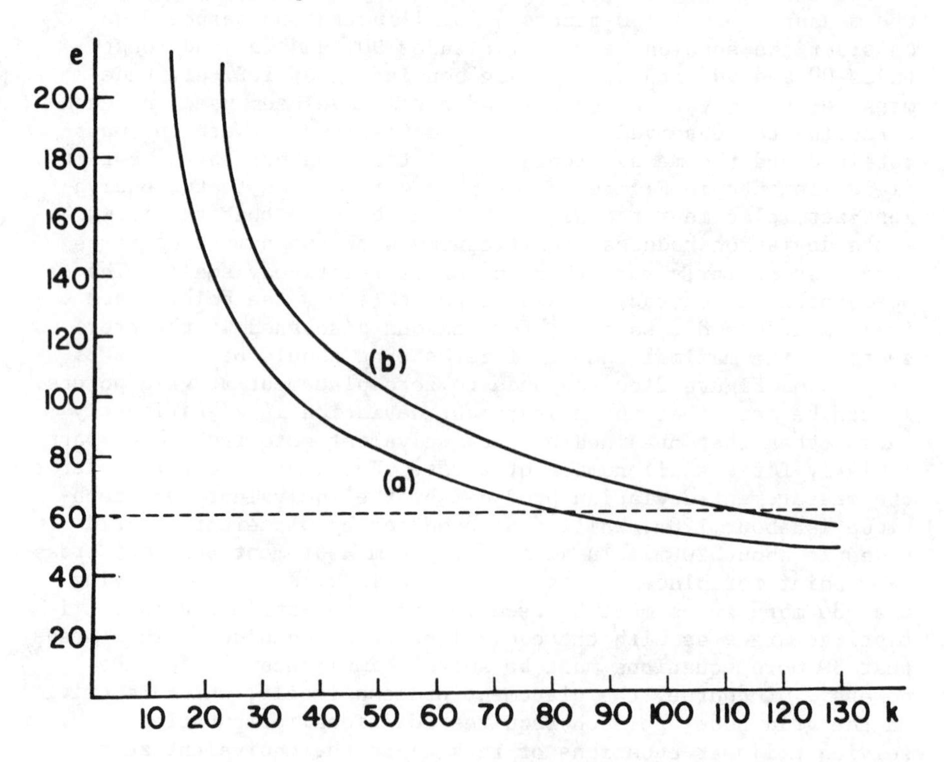

Figure 1. Measurement deviation from reconstructed surface plotted against the number of field elements used for (a) equivalent zero placement and (b) grid node placement.

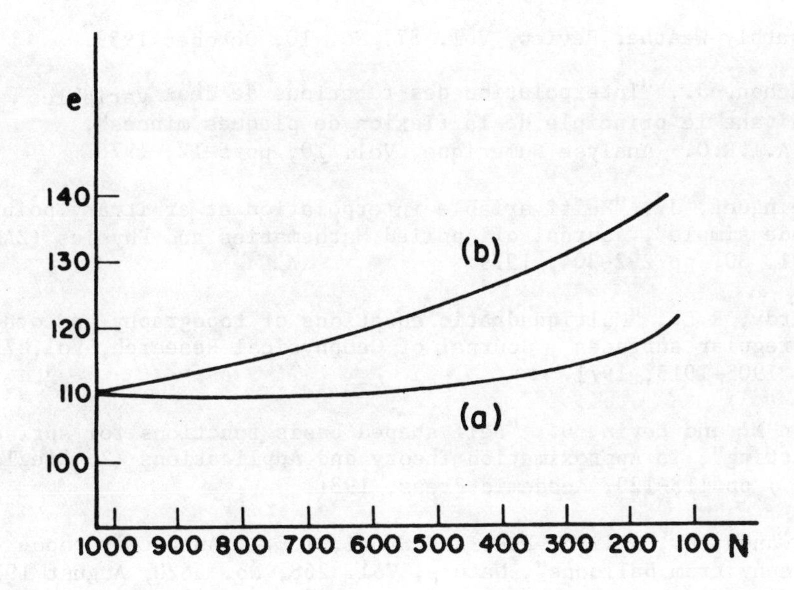

Figure 2. Mapping deviation as a function of data set reduction for (a) zero powers estimation and (b) entire representation, including zero placement.

5. CONCLUSION

A new method for efficient representation and reconstruction of two dimensional fields from sampled data has been presented. The main advantage of the proposed technique is that by exploiting the structure of the pseudo-potential function and the local behaviour of the data, the non-linear parameter estimation problem is converted into a set of linear sub-problems, which are solved by standard technique, thus avoiding non-linear optimization, on the one hand and arbitrary placement of the base functions, on the other. Over-parameterization may be avoided by using the proposed criterion for selecting the number of field zeros. The proposed zero placement technique is particularly useful for data compression purposes, since a relatively small number of parameters (compared to, say, grid-node placement) may be used for representing and reconstructing the field. Finally, we note that while in this work we have chosen to motivate and use the pseudo-potential base functions, the proposed technique will also apply, in principle, if other radial functions are used.

REFERENCES

(1) Cressman, G.P., "An operational objective analysis system",

Monthly Weather Review, Vol. 87, No. 10, October 1959.

(2) Duchon, J., "Interpolation des functions de deux variables suivant le principle de la flexion de plaques minces", R.A.I.R.O. Analyse Numerique, Vol. 10, pp 5-12, 1976.

(3) Meinguet, J., "Multivariable interpolation at arbitrary points made simple", Journal of Applied Mathematics and Physics (ZAMP) Vol. 30, pp 292-304, 1979.

(4) Hardy, R.L., "Multiquadratic equations of topography and other irregular surfaces", Journal of Geophysical Research, Vol. 76, pp 1905-1915, 1971.

(5) Dyn N. and Levin, D., "Bell shaped basis functions for surface fitting", in Approximation Theory and Applications (Z. Ziegler ed.) pp 113-129, Academic Press, 1981.

(6) Levanon, N., Julian, P.R. and Suomi, Y.E., "Antarctic topography from balloons", Nature, Vol. 268, No. 5620, August 1977.

OPTIMUM PERTURBATION SIGNAL THAT MAKES NONLINEAR SYSTEMS APPROACH TO LINEAR SYSTEMS

Atalay Barkana

D.M.M.A., Eskişehir, Turkey

A b s t r a c t :

It is usually desired to linearize the characteristics of a nonlinear system. One of methods applied for linearizing the nonlinear system characteristics is to inject a perturbation signal on to the input signal. Physical systems generally have built-in low pass filters and hence they filter the injected signals and yield averaged outputs. The relation between the input signal and the average output signal depends on the shape and amplitude of the perturbation signal. The wave shape and amplitude of the perturbation signal which makes the nonlinear characteristics approach to linear characteristics best have been searched. Using a suitable polynomial to approximate a nonlinear characteristics and using the first few terms of the Fourier Series of the perturbation signal yielded some analytical results. This method is superior to the methods that use total numerical simulation.

1. INTRODUCTION

There are two main reasons for wanting the linearization of the characteristics of a nonlinear system. One is to obtain similar principles for system analysis and design as in linear systems, the other is to obtain a better system response than what the nonlinear system has for an input signal. One of the methods for linearization of the nonlinear characteristics is the use of the perturbation signal. The perturbation signal is injected on to the input signal of the nonlinear system as shown in Figure 1 (1), (2), (3). The output m, is a function of the input e in the

R. S. Bucy and J. M. F. Moura (eds.), Nonlinear Stochastic Problems, 525–532.

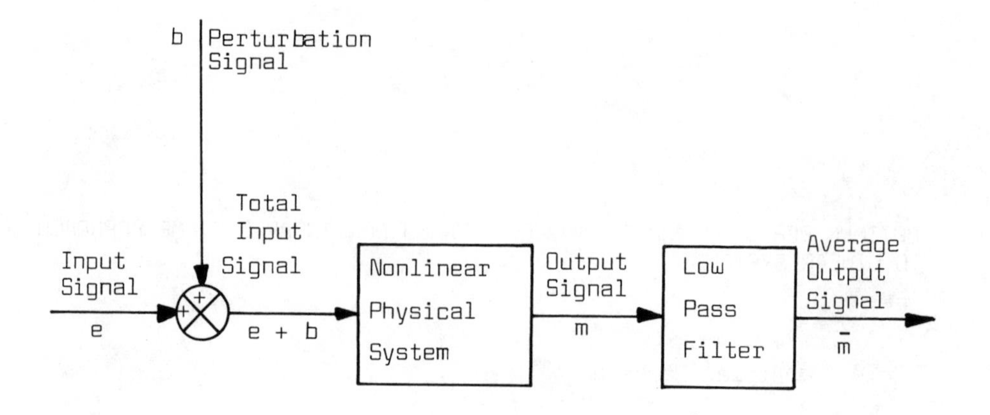

Figure 1. The block diagram of the nonlinear system with the perturbation signal and low-pass filter.

nonlinear system characteristic of interest,

$$m = f(e) \tag{1}$$

A high frequency perturbation signal b(t) is added to the input e of the nonlinear system. Since most of the pyhsical systems have built-in low-pass filters, the new total input signal e + b whose frequency is a lot higher than the cut - of frequency of the low - pass filter is averaged at the output of the overall system. This average output, $\bar{m}$, can be expressed as :

$$\bar{m} = \frac{1}{T}\int_0^T f[e + b(t)]\,dt \tag{2}$$

where T is the period of the perturbation signal. The original input e is considered to be a constant in the above expression since it changes a lot slower than the perturbation signal b(t). As it is obvious from the above equality, the average output $\bar{m}$ depends on the amplitude and the shape of the perturbation signal b(t). For this reason, researchers looked for the best amplitude of the perturbation signal for position detecting systems (4), (5), (6) and for increasing the resolution of A/D converters (7), (8). They tried to find the best wave shape of the perturbation signal either by intuition or by plotting a prespecified cost function with respect to the amplitude of a predetermined signal.

2. DEFINITION OF THE COST FUNCTION

In order to determine the best or optimum perturbation signal, first a cost function has to be defined. For this purpose, the system with input e and output $\bar{m}$ in Figure 1 must be considered. The relation between the average output $\bar{m}$ and the input e is given by equation (2). The characteristic of this new system is different than of the original system's and depends on the choice the perturbation signal b(t). Since the characteristic of the new system is desired to be linear, the desired output may be defined as :

$$\bar{m}_d = Ke \tag{3}$$

where K is a constant. After defining the desired characteristic by the above expression, the cost function can be written as :

$$J = \int_{e_{min}}^{e_{max}} (\bar{m} - \bar{m}_d)^2 \, de \tag{4}$$

where e_{min} and e_{max} are the physical lower and upper bounds on the input respectively. Minimization of the cost function means the minimization of the absolute difference bet ween the real output $\bar{m}$ and the desired output $\bar{m}_d$.

By assuming $e_{max} = E$ and $e_{min} = -E$, where E is a positive number, and by subtituting Equations (2) and (3) in Equation (4), the cost function becomes

$$J = \int_{-E}^{E} \left[\frac{1}{T}\int_{0}^{T} [f(e + b)]\, dt - Ke\right]^2 de \tag{5}$$

If the linear characteristic and the constant K are given, then the cost function will become a function of the perturbation signal b(t) only, that is,

$$J = J\,[b(t)] \tag{6}$$

One may try to obtain the minimum of the cost function by taking the first derivative of the cost function with respect to the perturbation signal and by equating it to zero :

$$\frac{dJ}{db} = \frac{2}{T}\int_{-E}^{E} \frac{d\int_{0}^{T} f(e + b)\, dt}{db} \left[\frac{1}{T}\int_{0}^{T} f(e + b)\, dt - Ke\right] de = 0 \tag{7}$$

But generally this does not yield any solution for the perturbation signal. Some assumptions must be made in order to be able to solve Equation (7) in terms of the perturbation signal $b(t)$.

3. ASSUMPTIONS ON THE NONLINEAR SYSTEM AND THE PERTURBATION SIGNAL

It is assumed that the characteristic of the nonlinear system passes through origin and is symmetric with respect to the origin and can be approximated by a fifth degree polynominal, that is, from Equation (1),

$$m = f(e) = c_1\, e + c_3\, e^3 + c_5\, e^5 \tag{8}$$

Even terms disappear due to symmetry.

The perturbation signal $b(t)$ is assumed to be periodic, single valued, continuous over one period and symmetric with respect to the origin and can be approximated closely with the first three of its Fourier Sine Series,

$$b(t) = b_1 \sin \omega t + b_2 \sin 2\omega t + b_3 \sin 3\omega t \tag{9}$$

where b_1, b_2, b_3 are the Fourier coefficients, ω is the angular frequency. In fact the coefficients b_1, b_2, b_3 are the same coefficients that are in the infinite Fourier Series since this is the best way to approximate the infinite Fourier Series with a finite sum (9).

By inserting Equation (8) and (9) into (2), the difference $(\bar{m} - \bar{m}_d)$ can be written as :

$$\bar{m} - \bar{m}_d = \bar{m} - Ke = a_1 e + a_3 e^3 + a_5 e^5 \tag{10}$$

where a_1, a_3, a_5 are given in terms of the b and c coefficients as below :

$$a_1 = c_1 + \frac{3}{2}(b_1^2 + b_2^2 + b_3^2)c_3 + \frac{15}{8}(b_1^4 + b_2^4 + b_3^4 + 4b_1^2 b_2^2 + 4b_1^2 b_3^2 - \frac{4}{3} b_1^3 b_3$$
$$+ 4b_2^2 b_3^2 + 4b_1 b_2^2 b_3)\, c_5 - K$$

$$a_3 = c_3 + 5(b_1^2 + b_2^2 + b_3^2)\, c_5 \tag{11}$$

$$a_5 = c_5$$

From equation (4), in terms of a_1, a_3 and a_5 the cost function becomes.

$$J = \frac{E^3}{2}\left(\frac{a_1^2}{3} + \frac{2a_1a_3}{5}E^2 + \frac{a_3^2 + 2a_1a_5}{7}E^4 + \frac{2a_3a_5}{9}E^6 + \frac{a_5^2}{11}E^8\right) \tag{12}$$

Since the coefficients c_1, c_3, c_5 of the polynomial and the constant K are assumed to be given, the cost function will be a function of the Fourier coefficients b_1, b_2, b_3 as can be seen easily from Equations (10) and (11). Then the necessary condition for finding the optimum Fourier coefficients can be obtained by taking the first partial derivatives of Equation (5), that is,

$$\frac{\partial J}{\partial b_n} = 2\int_{-E}^{E} (\bar{m} - Ke)\frac{\partial \bar{m}}{\partial b_n}\, de = 0 \qquad \text{for } n = 1,2 \text{ and } 3. \tag{13}$$

The partial derivatives of the average output $\bar{m}$ with respect to the Fourier coefficients can be found from Equations (10) and (11), and can generally be expressed as :

$$\frac{\partial \bar{m}}{\partial b_n} = d_{1n}\, e + d_{3n}\, e^3 \qquad \text{for } n = 1,2 \text{ and } 3. \tag{14}$$

Then by combining Equation (14) with (13), the following set of equations can be written.

$$\frac{a_1d_{1n}}{3} + \frac{a_1d_{3n} + d_{1n}a_3}{5}E^2 + \frac{a_3d_{3n} + d_{1n}a_5}{7}E^4 + \frac{d_{3n}a_5}{9}E^6 = 0 \tag{15}$$

$$n = 1,2 \text{ and } 3.$$

This is a complicated result.

Gradient method can be applied to find the optimum perturbation signal numerically by using Equations (13), (14) and (15).

One example is solved by using this method shown in Figure 2. The program is written in BASIC language, and run at Interdata 7/16 computer. The program can be obtained from the author.

Example : Let the nonlinear characteristic be given by the following equation :

$$m = -\,0.283\, e + 0.853\, e^3 - 0.181\, e^5$$

characteristic defined by the following equations in the region $-2 < e < +2$. In fact, this is the best curve fit for the nonlinear

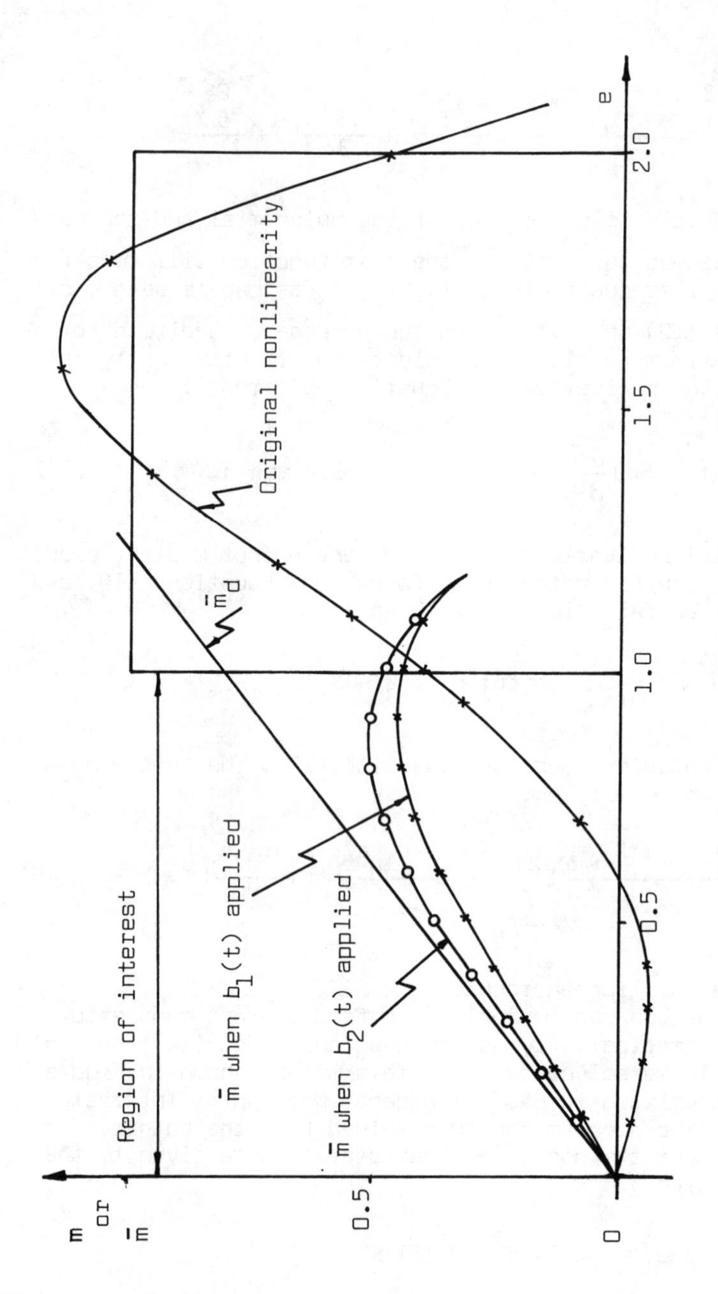

Figure 2. The nonlinear characteristic given by the polynomial and the average outputs obtained when the optimum perturbation signals are applied.

characteristic defined by

$$m = -1 \quad \text{for} \quad -2 < e < -1$$
$$m = 0 \quad \text{for} \quad -1 < e < +1$$
$$m = +1 \quad \text{for} \quad +1 < e < +2$$

These characteristics are shown in Figure 2. Let the desired relation between the input and the average output be :

$$\bar{m}_d = 0.8\, e$$

The region of interest for the characteristic shown in Figure 2 is taken as E=1. If only one term is taken from the Fourier series, the optimum value of the perturbation signal is found as :

$$b_1(t) = 0.969 \sin\omega t$$

and if three terms are taken, the optimum perturbation signal becomes :

$$b_2(t) = 0.988 \sin\omega t - 0.004 \sin 2\omega t + 0.272 \sin 3\omega t$$

The average outputs obtained with these perturbation signals are shown in Figure 2.

4. CONCLUSION

The equations (13), (14) and (15) do not yield easy solutions for the Fourier coefficients. A numerical method is essential to solve these coefficients. The gradient method is chosen and used for this purpose.

The optimum perturbation signals are found for one nonlinear characteristic by using the gradient method. It is also shown in this example how close the application of the optimum perturbation signals makes the nonlinear characteristics approach to linear characteristics.

The number of the Fourier coefficients that should be chosen to find the optimum wave shape are limited in the real case by the critical angular frequency of the low - pass filter, ω_F, and by the critical angular frequency of the nonlinear system, ω_E, since any stable system is bound to have a critical frequaency. If ω is the fundamental angular frequency, then the following inequality should hold for n :

$$\omega_F \ll n\omega \ll \omega_E$$

Considering this inequality the number of terms taken from

Fourier Series can be increased if desired. In that case, equation (11) has to be calculated again and the gradient method has to be adopted to the new case. The degree of the polynomial in Equation (8) can be increased for a better approximation of the real nonlinear characteristic. However, it should be kept in mind that the optimum perturbation signal found by using the approximate polynomial can only be suboptimal for the real nonlinear characteristic. Therefore care should be taken in finding the approximate polynomial.

The cost function and its partial derivatives with respect to the Fourier coefficients can be numerically calculated from equations (5) and (7) without deriving the equations (13) and (15). But since the cost function has double integrals, this method causes considerable numerical errors and/or it takes too much computer time to find the optimum value of the perturbation signal.

REFERENCES

(1) E.S. McVey, P.F. Chen (Improvement of position and velocity detecting accuracy by signal perturbation), IEEE Trans. on IECI, Vol. 16, No.1, 1969.
(2) P.F.Chen, E.S. McVey (Improvement of area, position and velocity sensing accuracy by a threshold setting method),Project THEMIS, University of Virginia, 1969
(3) E.A. Parrish, Jr., J.H. Aylor (Comment on "Improvement of position and velocity detecting accuracy by signal perturbation") IEEE Trans. on IECI, Vol. 18, No.y, 1971.
(4) J.H. Aylor, E.A. Parrish, Jr., and G. Cook (Optimum desing of a position detection system with a sinusoidal perturbation signal), IEEE Trans. on IECI, Vol. 19, No.4, 1972.
(5) E.A. Parrish, Jr. (On the improvement of a position detection accuracy signal perturbation theory), IEEE Trans. on IECI, Vol. 17, No.1, 1970.
(6) E.A. Parrish, Jr., and J.W. Stougton (Achieving perfect position detecting using a modified triangular perturbation signal), Project THEMIS, University of Virginia, 1969.
(7) G.Cook (Some practical limitations on the use of perturbation signals for resolution enhancement), IEEE Trans. on IECI, Vol. 20, No.3, 1973.
(8) G. Demas, A. Barkana, G. Cook (Experimental verification of the improvement of resolution when applying perturbation theory to a quantizer), IEEE Trans. on IECI, Vol.20, No. 4, 1973.
(9) G.B. Thomas, Jr. (Calculus and analytic geometry) Addison - Wesley Publishing Company, Inc., PP. 821-825, 1960

Acknowledgement : The author wishes to express his gratitute to Dr. Gerald Cook from University of Virginia for his start and help through this research paper. In fact he should have been the co-author of this paper.

STOCHASTIC FILTERING PROBLEMS IN MULTIRADAR TRACKING

Karl G. Brammer

ESG Elektronik-System-Gesellschaft
Vogelweideplatz 9, 8000 München 80,
W.-Germany

ABSTRACT: The paper deals with the linear and non-linear aspects of the filter algorithms used for tracking of aircraft trajectories on the basis of measurements made by surveillance-type radars (track-while-scan). The mathematical models for aircraft motion and radar measurements are reviewed and the possible simplifications are discussed. Effective suboptimal filter algorithms are needed if the number of tracked aircraft is large and computing capacity is limited. This holds in particular for multiradar networks with overlapping coverage, where the plot rate per track is proportional to the number of sensors seeing the same aircraft.

1. THE MULTIRADAR TRACKING TASK

Conventional surveillance radars with vertical fan beams scan the surrounding airspace by rotating their antenna in azimuth at a constant rate. While the beam passes over an aircraft it is hit by a sequence of radar pulses, the echoes of which are returned to the receiver of the radar antenna. From this echo sequence, a digital target extractor determines range and azimuth values of the aircraft while suppressing unwanted echoes from ground, clouds, birds etc. as good as possible. Modern phased array surveillance radars feature narrow pencil beams, scanned electronically in elevation and mechanically in azimuth. Electronic azimuth scan requires at least three antennas mutually displaced back to back at 120°.

R. S. Bucy and J. M. F. Moura (eds.), Nonlinear Stochastic Problems, 533–552.

The passing of the antenna beam across an aircraft thus results in a digital target report containing range, azimuth, possibly elevation and other data such as correlation with secondary surveillance radar (SSR) or identification friend or foe (IFF) responses. Such a report issued by the radar data extractor is also called a plot, which may be drawn directly on a plan position graph or display.

Monoradar tracking with surveillance radars consists of joining together (associating) the plots of a particular aircraft from a single radar, antenna rotation after antenna rotation, smoothing the position and estimating the current velocity vector. This information essentially constitutes the radar track, which mostly contains additional data such as track number, identification, correlation with flight plan data, track quality etc. Association is mainly done by looking for the next plot in the vicinity of the extrapolated position. This process can present major problems for maneuvering aircraft, crossing trajectories, formation flights and clutter (see Fig. 1).

Multiradar systems with overlapping coverage areas have become increasingly important in civil and military applications, although obviously the overall coverage area per given number of radars is decreased. The main advantage is reduction of the areas in radar shadow and increased detection probability in the overlaps. For instance, experiments performed in the Washington - Baltimore area and in the Los Angeles Basin (1) show that in the common coverage area of 2 primary radars only about 30% of the aircraft are detected by both sensors simultaneously, while ca. 2 x 25% of them are seen by a single one of the two sensors. (Some 20% were detected by neither of them but only by SSR. The unknown number that escaped detection by the two primary radars and the two SSR, is disregarded.) Addition of a second radar sensor thus increased the percentage of detected aircraft from 55% to 80%. The increase will be the more marked the higher the terrain profile and the lower the flight altitude will be.

Further advantages of multiradar systems are:
- Fail soft property of the surveillance system
- Altitude estimation for the 2D-radars
- Better track quality
- Smooth track transition from one radar to the next.

Multiradar tracking is the process of forming one single system track per aircraft from the corresponding plots originating at the extractors of the different radars in the multiradar system that see the aircraft. Four methods are feasible.

For the mosaic method the total surveillance area is divided into mutually exclusive subareas. For each of these subareas, the radars covering it are allotted a priority order. Tracking in each subarea is normally performed exclusively with the plots from the highest priority radar. Only if data from this radar are missing, plots from a lower priority radar are used (2). This method could be labelled piecewise monoradar tracking. The advantage is computational austerity, but drawbacks lie in the association process and track roughness at switching from one radar to another.

A variant of the mosaic method uses continuous monoradar tracking for each radar. From these local tracks the system track is selected according to the mosaic priority. This avoids the association difficulties but shows track discontinuities at the switching points.

A more sophisticated approach - being studied by EUROCONTROL - uses all available plots from every radar seeing the aircraft in question to construct the system track directly (3). This method is conceptually the most appealing but suffers from association and smoothing problems due to the systematic errors of the individual radars, e.g.:
- Radar range bias and scale factor
- Slant range projection
- Radar azimuth bias
- Radar station coordinate bias.

The latter two errors arise from imperfect survey or - in case of mobile stations - from navigation and azimuth alignment errors or even from erroneous data insertion. Since it is feasible to estimate all the systematic errors off-line or even adaptively on-line, this multiradar tracking method, although apparently not yet applied, seems to have promise. Note that plot time synchronisation and coordinate grid harmonization are important requirements here, too.

The fourth method is based on continuous monoradar tracking for each radar, as in the second mosaic method. Now, however, all the available local tracks

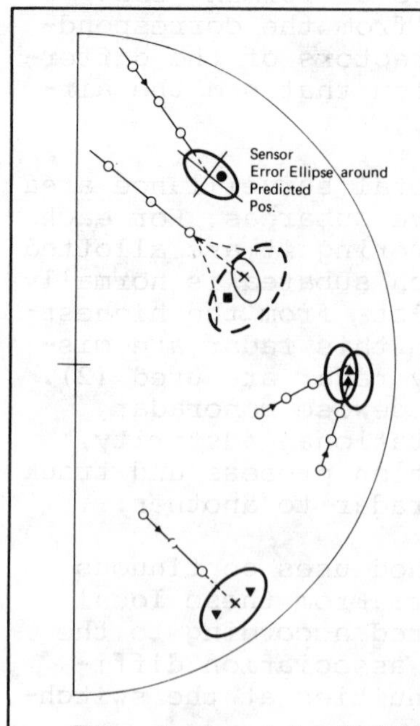

- Unique Association:

  Most Common for Local Plot and isolated Well-Established Straight Local Track

- Missed Association:
  - Target started Maneuver (Switch Filter)
  - Tentative New Track (Initialize)

- Ambiguous Association:
  - Targets Close Together
  - Large Sensor Errors
  - Targets Highly Maneuvering
  - Low Track Quality (Missed Plot History)
  - If only 1 Plot: Sensor Resolution

- Multiple Association:
  - Bifurcation (Group Splitting)
  - False Alarm (Clutter)
  - New Incoming Target

Figure 1: Association Problems

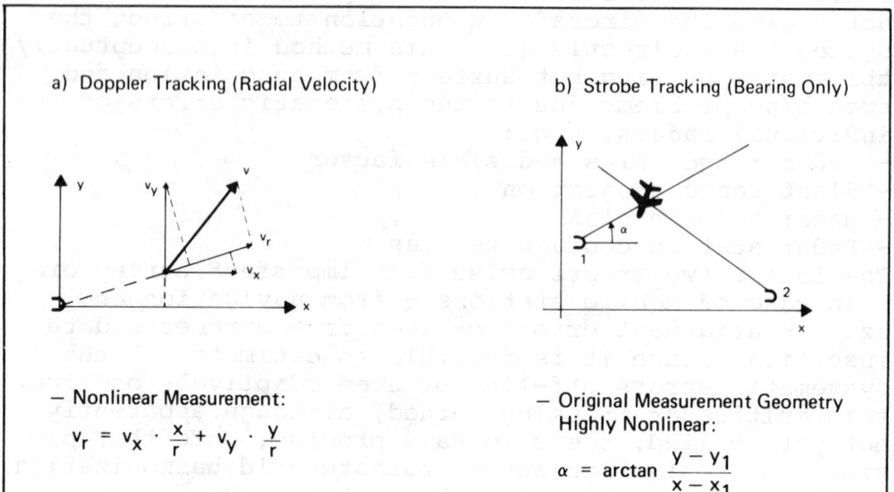

a) Doppler Tracking (Radial Velocity)

- Nonlinear Measurement:

$$v_r = v_x \cdot \frac{x}{r} + v_y \cdot \frac{y}{r}$$

b) Strobe Tracking (Bearing Only)

- Original Measurement Geometry Highly Nonlinear:

$$\alpha = \arctan \frac{y - y_1}{x - x_1}$$

Figure 2: Unconventional Multiradar Tracking

of a given aircraft are associated on a track-to-track basis and combined into the system track. Among the true multiradar tracking schemes this is the one most widely accepted since it yields an accurate track of high quality, while largely eliminating the radar-to-radar association and smoothing problems. For the local monoradar track the systematic radar errors cancel and a smooth velocity estimate is possible. This provides an additional criterion for association of the data from different radars. This method is being introduced e.g. by EUROCONTROL (3), by ESG (4), (5) and by the new French ATC system CAUTRA IV (6).

2. UNCONVENTIONAL MULTIRADAR TRACKING

The primary radar sensor outputs mentioned so far were position measurements. Some radar sensors can provide radial velocity (Doppler) measurements in addition. Expressed in terms of carthesian coordinates, the radial velocity measurement is highly nonlinear, see Figure 2a. The nonlinearity can be reduced by using rv_r as the measurement or it can be circumvented by treating x/r and y/r as known parameters. In any case the x- and y- axes are coupled by the measurement. On the other hand, if two radars take Doppler measurements of the target from different directions, then the two equations can be preprocessed to solve for v_x and v_y, not only avoiding nonlinearity as before but also eliminating the coupling between axes (7). This is a significant advantage as will be shown in Chapters 3 and 4.

Another example of measurements that are awkward with monoradar operation and comparatively easy in a multiradar system is shown in Figure 2b. In this case no range measurement is available but only bearing (passive radar mode, direction finding etc.). The measurement made by a single radar is again highly nonlinear in terms of the carthesian position coordinates of the aircraft. Furthermore, the latter can only be determined by observing a substantial portion of the trajectory. With two radars, position can be tracked by triangulation (8). The hardest problem is the ambiguity in the presence of multiple strobe targets, but this can be resolved by using additional measurements, e.g. from a third radar.

3. MATHEMATICAL MODEL OF THE OBSERVED PROCESS

The construction of a monoradar (local) track from associated plots is currently implemented by the use of alpha-beta-filters, Wiener filters or Kalman filters, although a nonlinear filtering situation is actually present. Similarly, the synthesis of a multiradar (system) track from associated local tracks is usually performed by a second filter using the smoothed position and estimated velocity values from the local tracks as "measurement" inputs.

This chapter reviews the mathematical models used to describe the motion of aircraft and the measurements taken by surveillance radars. Since the complexity of these models largely determines the bulk of the tracking filter, a great deal of effort is spent on carrying the simplifications as far as possible without sacrificing essential performance characteristics. Simple yet effective filter algorithms are especially needed if the number of tracked aircraft is large with respect to the available computing capacity. Neither in a civil air traffic control nor in a tactical air defense environment is it tolerable to let a part of the aircraft go untracked during a possible period of computer overload.

A basic issue is the type of coordinates to be chosen for the state variables. Figures 3 and 4 show that for the dynamics part of the model carthesian coordinates are excellent, but that for the measurement part polar coordinates would be optimal. Problems arising vice versa are nonlinearities, coupling between the coordinate axes and correlation between axes. Figure 5 indicates the feasible remedies to avoid nonlinearities in the carthesian and polar axes, together with the remaining difficulties. We conclude that for most of the applications carthesian coordinates are preferred. This choice features a purely linear differential equation with zero forcing input and exact 0's and 1's in the dynamics matrix for the linear motion, constant velocity case, yielding a high degree of robustness of this model (see Fig. 3, lower left hand corner).

A further question is the modelling of flight path disturbances and deliberate aircraft maneuvers. They give rise to input forcing terms to the dynamics part of the model. In Figure 6 two approaches are out-

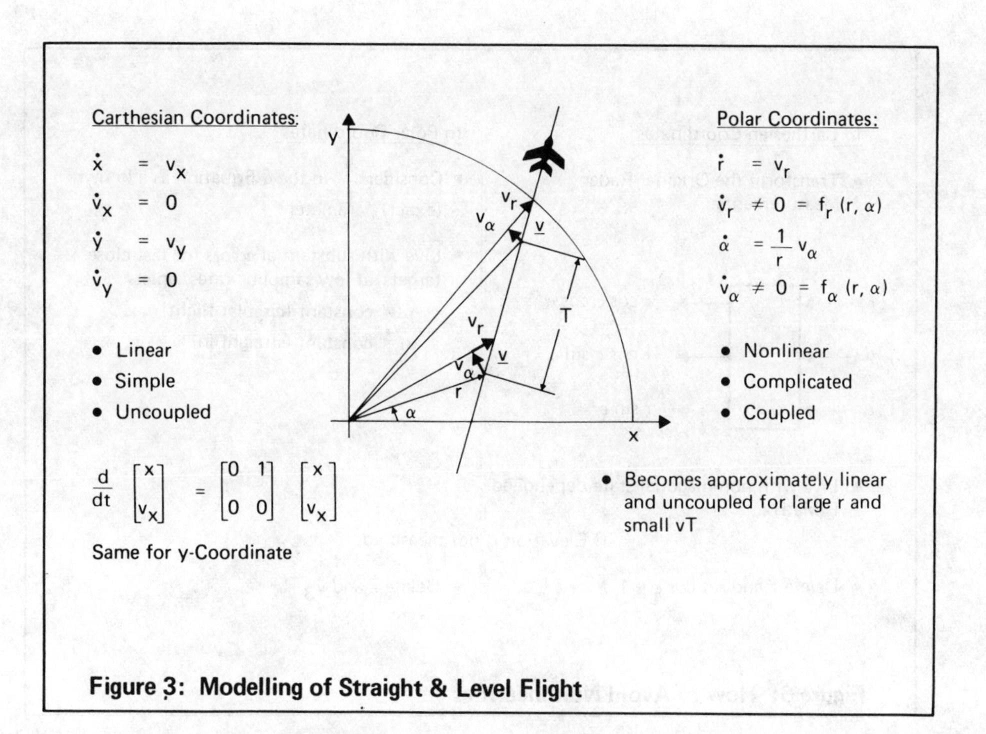

Carthesian Coordinates:
$\dot{x} = v_x$
$\dot{v}_x = 0$
$\dot{y} = v_y$
$\dot{v}_y = 0$
• Linear
• Simple
• Uncoupled
$\frac{d}{dt}\begin{bmatrix} x \\ v_x \end{bmatrix} = \begin{bmatrix} 0 & 1 \\ 0 & 0 \end{bmatrix}\begin{bmatrix} x \\ v_x \end{bmatrix}$
Same for y-Coordinate
Polar Coordinates:
$\dot{r} = v_r$
$\dot{v}_r \neq 0 = f_r(r, \alpha)$
$\dot{\alpha} = \frac{1}{r} v_\alpha$
$\dot{v}_\alpha \neq 0 = f_\alpha(r, \alpha)$
• Nonlinear
• Complicated
• Coupled
• Becomes approximately linear and uncoupled for large r and small vT
y
x
v_r
v_α
v
T
r
α

Figure 3: Modelling of Straight & Level Flight

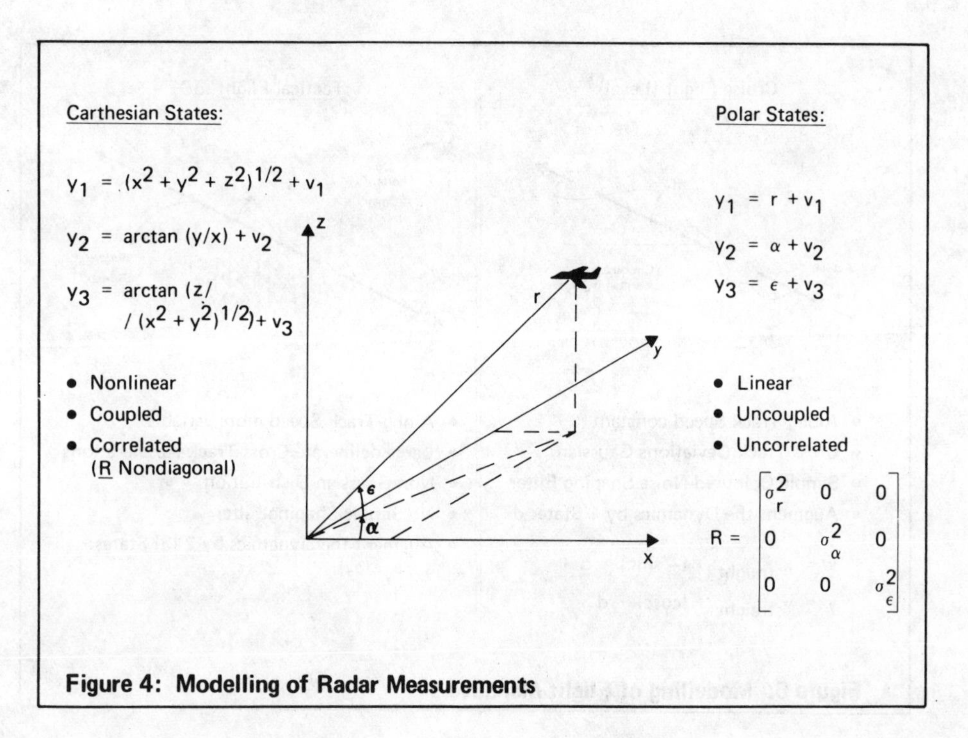

Carthesian States:
$y_1 = (x^2 + y^2 + z^2)^{1/2} + v_1$
$y_2 = \arctan(y/x) + v_2$
$y_3 = \arctan(z/(x^2 + y^2)^{1/2}) + v_3$
• Nonlinear
• Coupled
• Correlated ($\underline{R}$ Nondiagonal)
Polar States:
$y_1 = r + v_1$
$y_2 = \alpha + v_2$
$y_3 = \epsilon + v_3$
• Linear
• Uncoupled
• Uncorrelated
$R = \begin{bmatrix} \sigma_r^2 & 0 & 0 \\ 0 & \sigma_\alpha^2 & 0 \\ 0 & 0 & \sigma_\epsilon^2 \end{bmatrix}$
z
y
x
r
ϵ
α

Figure 4: Modelling of Radar Measurements

In Carthesian Coordinates:

- Transform the Original Radar Measurements:

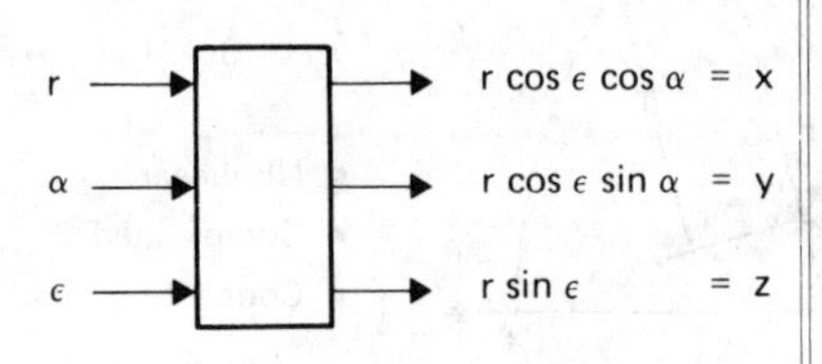

- Live with non-diagonal state-dependend R-Matrix

In Polar Coordinates:

- Consider $\frac{1}{r}$ in the $\dot{\alpha}$-Equation as a known (exact) parameter
- Live with substantial errors for fast close targets at low sampling rates, unless
  r = constant (circular flight)
  α = constant (straight in)

If Elevation is not measured:

- Delete z and set $\cos \epsilon = 1$
- Delete ϵ and y_3

Figure 5: How to Avoid Nonlinearities

Cruise Flight (Level):

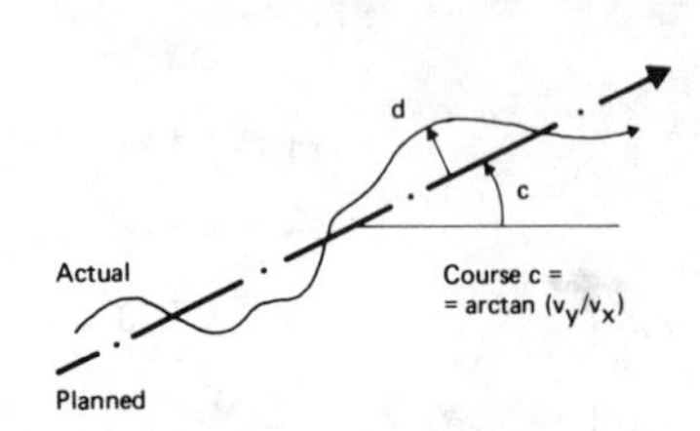

- Along-Track Speed constant
- Cross-Track Deviations Gaussian
- Simple Coloured Noise Shaping Filter
- Augment the Dynamics by 1 State: d

$$x = x_{straight} + (-\sin c) \cdot d$$

$$y = y_{straight} + (\cos c) \cdot d$$

Tactical Flight (3D):

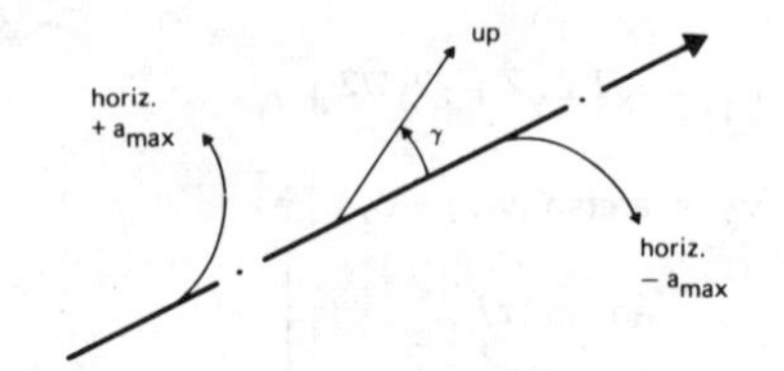

- Along-Track Speed more variable
- Large deliberate Cross-Track Accelerations
- Non-Gaussian Distribution
- No Simple Shaping Filter
- Augment the Dynamics by 2 (3) States: $a_x, a_y, (a_z)$

Figure 6: Modelling of Flight Maneuvers

lined. The first one requires merely one additional state variable for both geographic axes, namely the horizontal cross track deviation (9). But this model holds only for small disturbances by turbulence and piloting during straight cruise flight and introduces coupling between x- and y-coordinates. The second model is more general and uses the carthesian accelerations as augmenting state variables. This model is generally applied for heavily maneuvering aircraft in civil and tactical environments. The accelerations are treated as being decoupled in the coordinate axes and each of them is modelled as a first order autoregressive process in continuous time (10).

In order to arrive at the numerical parameters necessary to specify this model, we construct the distribution to obtain the mean and variance, and inspect the autocorrelation function to derive the time constant. This is done here in a heuristic fashion and the steps are illustrated in Figures 7 and 8. Note that along-track accelerations are an order of magnitude smaller than cross-track accelerations for heavily maneuvering aircraft and can be neglected in the horizontal plane. For a continuously maneuvering target a uniform distribution may be adequate for each of the carthesian accelerations, with zero mean and with a standard deviation depending on the maximum acceleration. - Ideally the model and with it the filter would have to be made adaptive with respect to the aircraft type and the intensity of its maneuvers. To this end, maneuver detectors and estimators are employed that switch the tracker structure and/or parameters from benign flights to maneuvering flights (11), (12), (13), (14). This implies stepping up the filter gain for the measurement inputs, thereby putting more weight on recent observations and increasing the filter bandwidth. -

The continuous-time model (shaping filter) for the accelerations is given in Figure 8b, with numerical values for its parameters (10) specified in Figures 7e and 8c. The augmented overall dynamics model for the geographic x-axis is shown in Figure 10 (top part), featuring the 3 state variables position, velocity and acceleration in the x-direction.

The next step in modelling is time discretization, which is performed in the center part of Figure 10. Although the continuous-time covariance matrix of the

a) Single Turn:

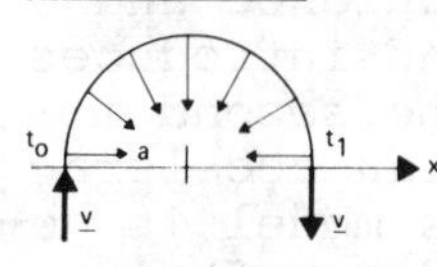

$$v_x(t_o) = v_x(t_1) = 0$$

$$v_y(t_o) = -v_y(t_1)$$

$$a_x(t) = a\cos\pi\frac{t - t_o}{t_1 - t_o}$$

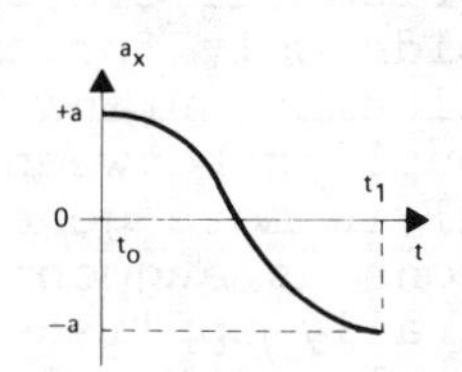

b) Maneuver Ensemble of Single Aircraft:

Tactical Maneuvers consist of circle segments with:

- Different Duration
- Different Orientation
- Different Radii

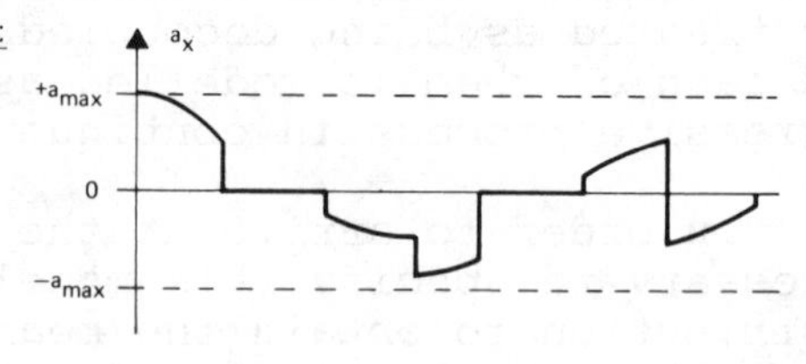

c) Probability Density for Single Type of A/C:

- Nonzero Probability for "No Maneuver"
- Rest is Uniformly Distributed (approxim.)
- Some authors add spikes at $\pm a_{max}$

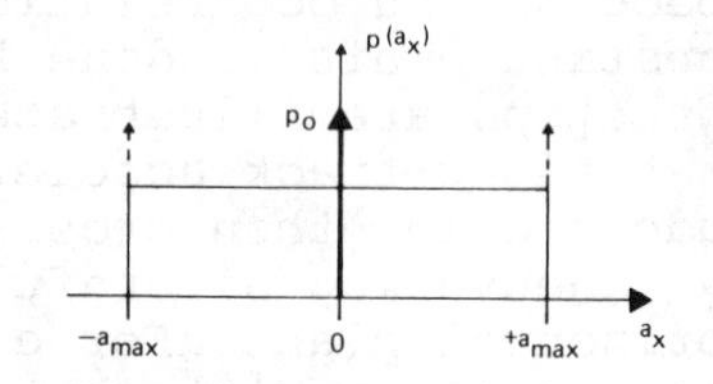

d) Probability Density for Aircraft Mix:

- For single type a/c the probability of "no maneuver" can vary.
- For mix of a/c types the maneuver capability a_{max} can vary.
- For a mixed air situation the Density approaches the Gaussian Distribution.

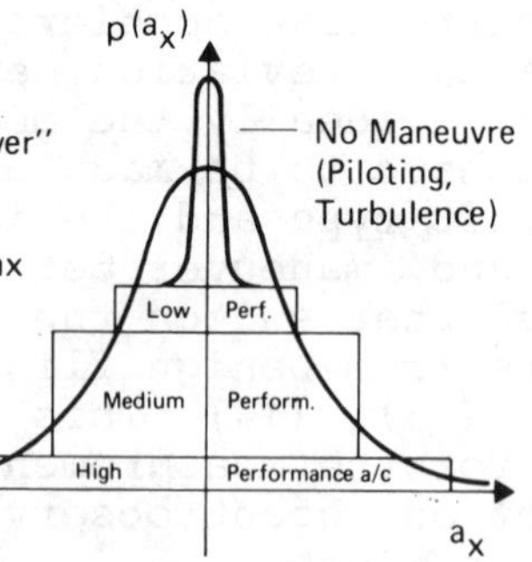

e) Parameters:

- Mean Acceleration Component: $E\{a_x\} = 0$
- Maximum Acceleration of Manned Aircraft: 7 ... 8 g
- Standard Deviations:
  - for Fully Maneuvering High Perf. A/C: $\sigma_{max} = 8/\sqrt{3}\,g$
  - for Tactical Air Situation: $\sigma_{a_x} \approx 1 \ldots 3\,g$
  - for Benign Air Situation: $\sigma_{a_x} \approx 0.1 \ldots 0.2\,g$

Figure 7: Distribution of Tactical Maneuvers (in Carthesians)

white input noise has only a single nonzero element, the time-discretization process results in an input covariance matrix which is completely filled. The largest term is given in Figure 10, the other terms can be found in (10).

The measurement equation carries directly over from the continuous to the discrete time case (see Fig. 10, bottom part). Usually only position measurements y_1 are present. Measurements y_2 of the carthesian velocity component appear e.g. in case of multiradar Doppler (see Chapter 2) or when the filter considered is combining local tracks into a system track. Note that in the latter case the "measurement" covariances are the state variable estimation error covariances of the monoradar tracking filters.

The mathematical model of the observed process, which was set up in this chapter, results for the horizontal situation (geographic x- and y-axes) in an overall system with

2 x 3 state variables and
2 x 1 position measurement equations.

The equations themselves are uncoupled between the x- and y- coordinates, but a coupling is introduced by the transformation of the measurement equation from the original polar form to the desired carthesian form (see Fig. 5). This coupling gives rise to a combined 2-axes filter of 6th order.

4. FURTHER SIMPLIFICATIONS IN RADAR TRACKING

Figure 9 summarizes by the broken lines the modelling results obtained so far. In this chapter the ultimate simplifications are discussed, which are feasible without intolerable degradations in tracking performance. To this end two contrivances are introduced:

a) The acceleration components are assumed to be white noise.
b) The cross coupling terms $r_{12} = r_{21}$ in the measurement covariance matrix are neglected.

Assumption (a) reduces the order of the observed system model from 6 to 4, and assumption (b) splits

a) Autocorrelation Function of Maneuver Ensemble

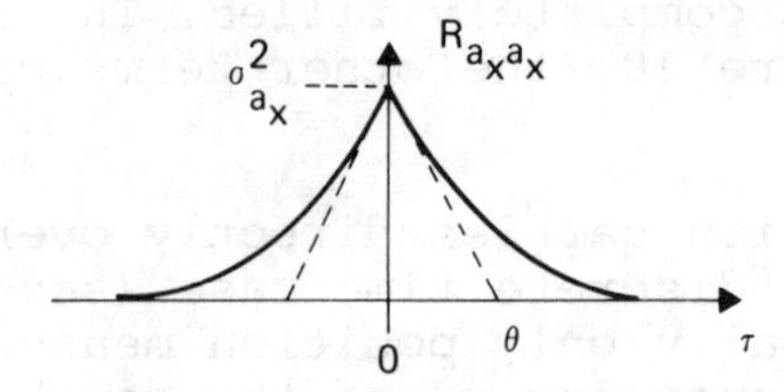

$$R_{a_x a_x}(\tau) = E\left\{ a_x(t)\, a_x(t+\tau) \right\} =$$

$$= \sigma^2_{a_x} \exp\left\{ -\frac{|\tau|}{\theta} \right\}$$

θ: Acceleration Time Constant

b) Shaping Filter

$$da_x(t) = -\frac{1}{\theta}\, a_x(t)\, dt + d\beta(t)$$

$$E\left\{ d\beta^2(t) \right\} = 2\,\frac{1}{\theta}\,\sigma^2_{a_x}\, dt$$

$\beta(t)$: Brownian Motion (Gaussian)

c) Time Constant (Example Values)

θ = 10 ... 20 sec for Tactical Maneuver, 60 ... 100 sec for Cruise Piloting, 1 sec for Turbulence

Figure 8: Dynamics of Acceleration Component (in Carthesians)

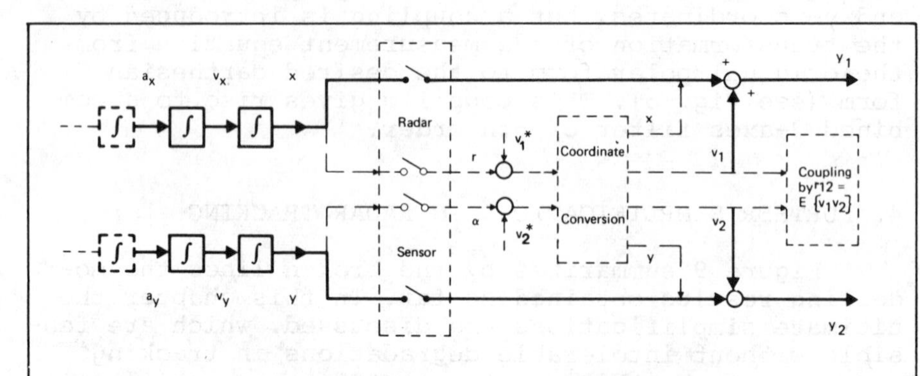

- Remove 2 of 6 state Variables by assuming a_x, a_y to be (discrete-time) white noise
- Remove Coupling between x- and y-Coordinates by neglecting r_{12}

- Result: 2 Filters of 2nd Order instead of 1 Filter of 6th Order

Figure 9: Simplifications in Radar Tracking

the fourth order model into two second order models. The resulting one-axis model is shown in the upper part of Figure 14. It is almost surprising, that despite these crude measures the filter performance remains acceptable. A valid justification is that these simple filters have been extensively used in practice.

A comparative performance analysis with simulations has been performed by Schumacher. Two of his conclusions are illustrated in Figures 11 and 12. Concerning the acceleration model he determined the value of the input noise covariance to be used for best suboptimal performance, if actually coloured noise is replaced by white noise (11, Chapter 5.3), see the upper part of Figure 11. With regard to the measurement model, neglection of the cross-coupling terms has the maximum effect in the intercardinal directions (Fig. 11, lower part). For an incoming trajectory of this kind, Figure 12 renders simulation runs (11, Chapter 8) with a radar sampling period of 1 sec., acceleration levels of 1g (std. dev.) and radar measurement errors of 100m in range and 1 degree in azimuth (both values are std. dev.). The simulation was performed with a 4-state model, i.e. incorporating position and velocity in the x- and y- axes. (The filter estimates were finally retransformed into polar coordinates simply for better comparison with the measurements.) The bottom graph in Figure 12 shows, that angular accuracy is barely affected by the artificial decoupling and that range accuracy is affected only for the initial period, where distance is large and, consequently, coupling is significant due to a very flat error ellipse of the radar measurements.

The discrete-time version of the observed system model, as simplified by the above assumptions, is summarized in Figure 13. The same model, but with different values of $\underline{Q}$ is obtained if the acceleration is modelled as discrete-time white noise with constant values within the sampling intervals (15). In both cases the corresponding Simple Kalman Filter (SKF) turns out to have the same form as the widely used conventional alpha-beta filter. The difference lies in the way the filter gains $k_1 = \alpha$ and $k_2 = \beta/T$ are determined (see Fig. 13).

In the early days of radar tracking the values of alpha and beta were found by trial and error. Later on they were computed by straight forward (naive) application of the basic method of least squares. Since

- Equations of Motion (Carthesian, in one Dimension)

$$d \begin{bmatrix} x \\ v_x \\ a_x \end{bmatrix} = \begin{bmatrix} 0 & 1 & 0 \\ 0 & 0 & 1 \\ 0 & 0 & -\frac{1}{\Theta} \end{bmatrix} \begin{bmatrix} x \\ v_x \\ a_x \end{bmatrix} dt + \begin{bmatrix} 0 \\ 0 \\ 1 \end{bmatrix} d\beta(t)$$

- Time Discretisation:

The Surveillance Radar takes samples at rate T $\left(\frac{T}{\Theta} \ll 1\right)$

$$\begin{bmatrix} x(k+1) \\ v_x(k+1) \\ a_x(k+1) \end{bmatrix} = \begin{bmatrix} 1 & T & T^2/2 + T^2 \cdot o(T/\theta) \\ 0 & 1 & T \quad + T \cdot o \quad (T/\theta) \\ 0 & 0 & \exp(-T/\theta) \end{bmatrix} \begin{bmatrix} x(k) \\ v_x(k) \\ a_x(k) \end{bmatrix} + \underline{w}(k)$$

where $\underline{w}(k)$ is discrete-time white noise and $\underline{Q}(k) = E\left\{\underline{w}(k)\,\underline{w}'(k)\right\}$ is obtained by integrating the inhomogeneous solution integral.
The most significant term is

$$E\left\{w_3^2(k)\right\} = \int_0^T e^{-2(T-\tau)/\theta}\, E\left\{d\beta^2(\tau)\right\} = \sigma_{a_x}^2 \left(1 - e^{-2T/\theta}\right)$$

- Measurement Equation

$$\begin{bmatrix} y_1(k) \\ y_2(k) \end{bmatrix} = \begin{bmatrix} 1 & 0 & 0 \\ 0 & 1 & 0 \end{bmatrix} \begin{bmatrix} x(k) \\ v_x(k) \\ a_x(k) \end{bmatrix} + \underline{v}(k)$$

Figure 10: Observed System Model

- Acceleration Model:

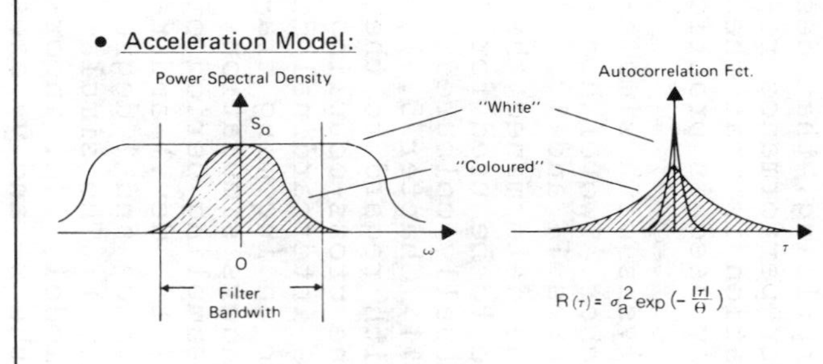

- For suboptimal 2nd order filter the best choice for $q\delta(o) = E\{\overline{w}_1^2(t)\}$ is:
  - for wide band "white" noise: $q = S_o = 2\sigma_a^2\Theta$ (from theory)
  - for narrow band "coloured" noise: $q = 2 \ldots 3 \cdot \sigma_a^2$ shown in (11)

- Measurement Model:
  - Cross-Coupling Term: $r_{12} = \frac{1}{2}\left(\sigma_r^2 - r^2\sigma_\alpha^2\right) \cdot \sin 2\alpha$
  - Neglecting r_{12} has no effect for:
    - $\alpha = n\pi/2$ (Cardinal Directions)
    - $r = \sigma_r/\sigma_\alpha$ (Circular Error)
  - Effect is maximal in the intercardinal directions

Figure 11: Sensitivity of Simplifications

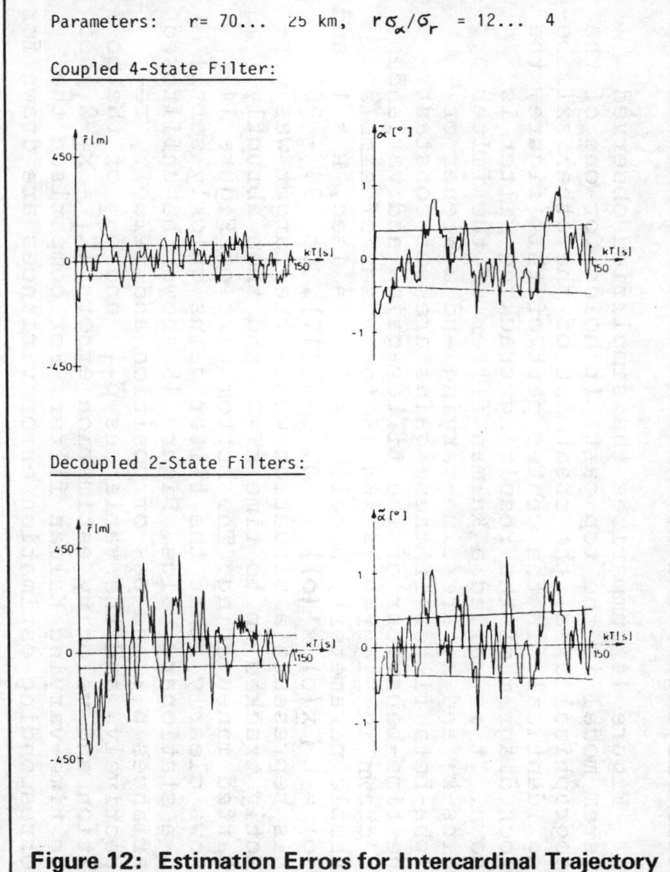

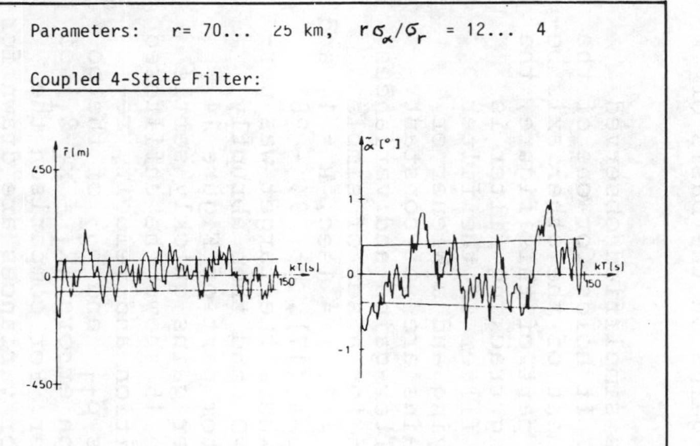
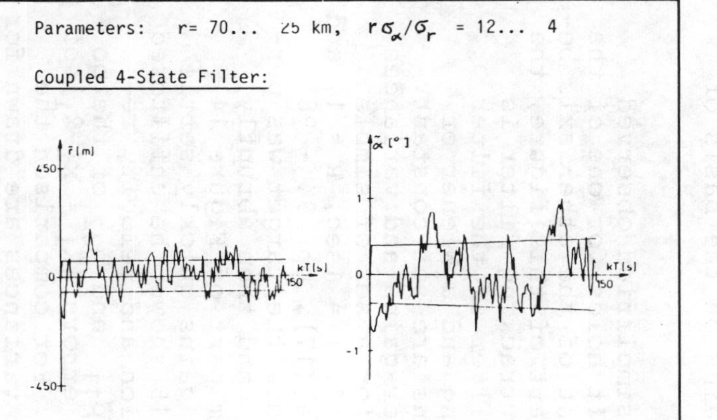

Figure 12: Estimation Errors for Intercardinal Trajectory

- <u>Assumptions:</u>
  - Maneuver Acceleration is white noise
  - No Doppler Measurements
  - Cross Coupling Terms in $\underline{R}$ neglected

- <u>Observed System Model (in 1 axis):</u>

$$\underline{x}(k+1) = \begin{bmatrix} 1 & T \\ 0 & 1 \end{bmatrix} \underline{x}(k) + \underline{w}(k),$$

$$y_1(k) = \begin{bmatrix} 1 & 0 \end{bmatrix} \underline{x}(k) + v_1(k)$$

$$\underline{Q}(k) = \begin{bmatrix} T^3/3 & T^2/2 \\ T^2/2 & T \end{bmatrix} \cdot q \qquad R(k) = \sigma^2_{v_1}$$

- <u>Kalman-Filter and α-β-Filter have same form:</u>

$$\underline{x}^*(k+1) = \begin{bmatrix} 1 & T \\ 0 & 1 \end{bmatrix} \underline{x}^*(k)$$

$$\hat{\underline{x}}(k) = \underline{x}^*(k) + \begin{bmatrix} \alpha \\ \beta/T \end{bmatrix} \left\{ y_1(k) - x_1^*(k) \right\}$$

- <u>Choice of α and β</u>
  - By Trial and Error
  - By Least Squares Method ($\underline{Q} = \underline{O}$)
  - Using Wiener Gains (Stationary Kalman)
  - Using Kalman Gains (Time-Varying, Riccati)

Figure 13: Simplified KF vs. α-β- Filter

this procedure takes into account only measurement errors but no plant noise, it is equivalent to setting $\underline{Q} = \underline{O}$ in the Kalman-Bucy filter approach. It is well known that in a stationary situation the filter gains then tend to zero as time proceeds. Statements in the sense that alpha-beta filters are unusable for maneuvering targets refer to this phenomenon, but it should be clear that setting $\underline{Q} = \underline{O}$ is not reasonable anyway.

An appropriate way to chose alpha and beta is to use the Wiener gains, which are computed from the stationary solution of the Riccati equation of the Kalman filter theory. For given constant maneuver intensities $\underline{Q}$ and measurement characteristics R the corresponding stationary filter gains can be precomputed off-line and tabulated (note that the carthesian measurement variance is range-dependent). It has been shown that in many practical systems, the constant-gain Wiener filter provides tracking accuracy equivalent to that of the Kalman filter at one third the computational cost (16).

If enough capacity is available in the tracking computer, the filter gains can be determined on-line. Then it is not much additional effort to directly compute the time-varying Kalman gains on the basis of (time-varying) $\underline{Q}$ and R.

Figure 14 summarizes the simplified observed system model in the top part. It holds for one of the geographical axes, the treatment of the other axis being identical. In the center part of this figure, the block diagram of the resulting tracking filter is shown. It is called a Kalman filter if the filter gains k_1 and k_2 are time-varying and a Wiener or alpha-beta filter if these gains are set constant. The time-behaviour of the filter gains and variances is drawn in Figures 14 and 15 for a set of simple example parameters, namely q = 1, T = 1sec, R = 1, and $\underline{P}(o) = E\{\underline{x}(o)\ \underline{x}'(o)\} = \underline{O}$, see (17), pp. 94 - 98. This represents a situation where the target was perfectly tracked up to time kT=0 and then abruptly started maneuvering. The bottom part of Figure 14 shows clearly, that the filter gains quickly settle to a stationary value. Figure 15 shows the unfiltered variances p_{11} and p_{22} of position and velocity respectively, and the variances $\tilde{p}_{11}$ and $\tilde{p}_{22}$ of the position and velocity estimation errors ($\tilde{x}_i = x_i - \hat{x}_i$) of the time-varying Kalman filter. For comparison the corresponding estimation error variances are drawn for

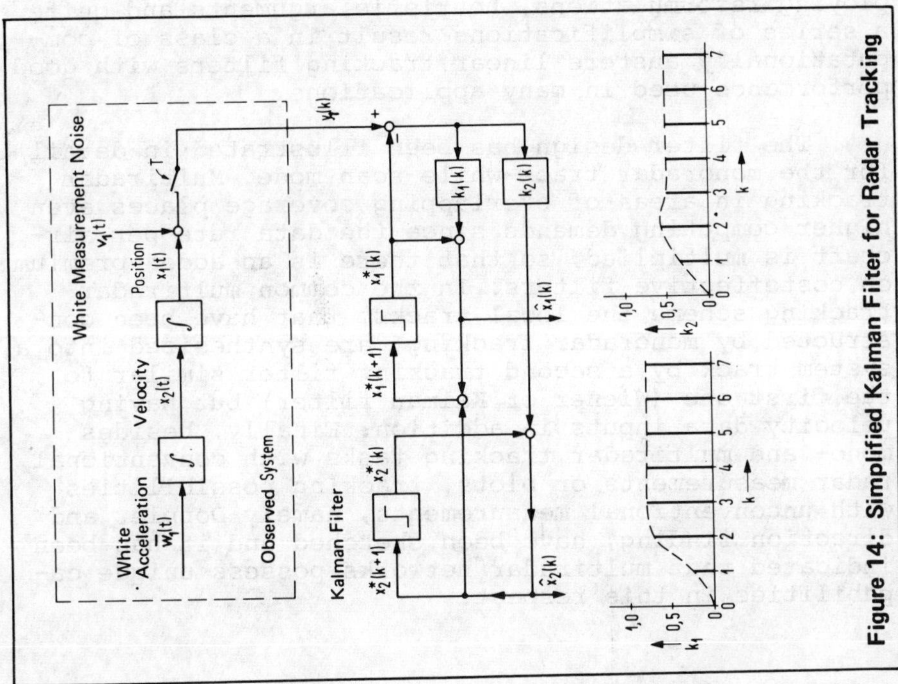

Figure 14: Simplified Kalman Filter for Radar Tracking

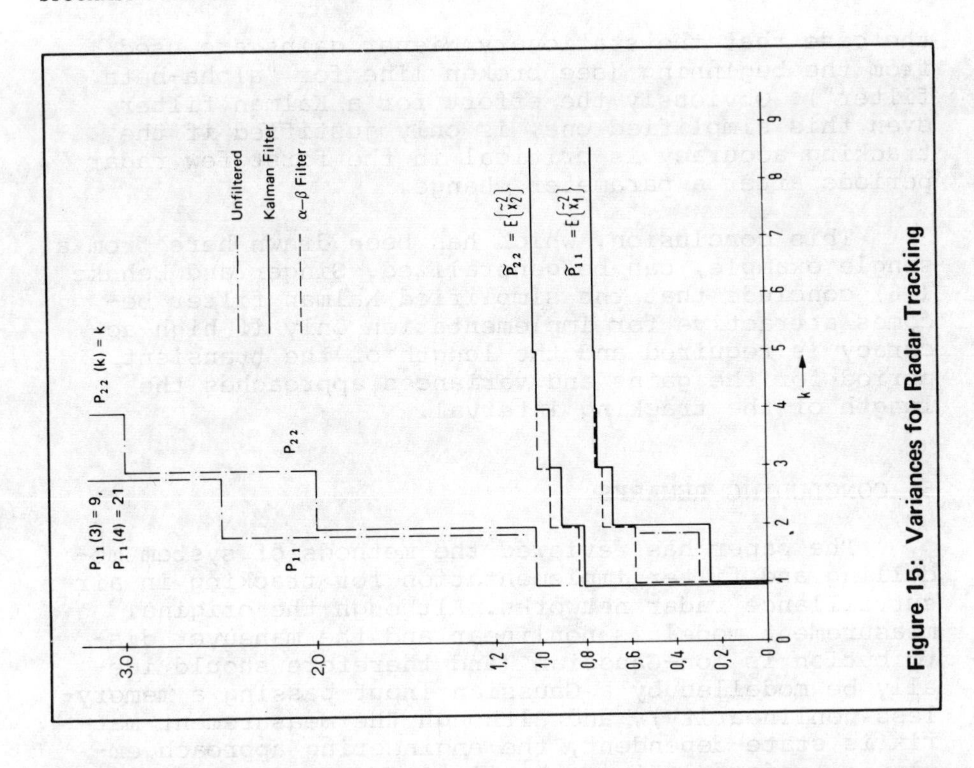

Figure 15: Variances for Radar Tracking

the case that the stationary Wiener gains are used from the beginning (see broken line for "alpha-beta filter"). Obviously the effort for a Kalman filter, even this simplified one, is only justified if the tracking accuracy is critical in the first few radar periods after a parameter change.

This conclusion, which has been drawn here from a single example, can be generalized. Singer and Behnke (16) conclude that the simplified Kalman filter becomes attractive for implementation only if high accuracy is required and the length of the transient period for the gains and variances approaches the length of the tracking interval.

5. CONCLUDING REMARKS

The paper has reviewed the methods of system modelling and filter implementation for tracking in air surveillance radar networks. Although the original measurement model is nonlinear and the maneuver distribution is non-Gaussian (and therefore should ideally be modelled by a Gaussian input passing a memoryless nonlinearity), and although the measurement matrix is state-dependent, the engineering approach employing reformulations, heuristic arguments and quite a series of simplifications result in a class of computationally austere linear tracking filters with good performance used in many applications.

The filter design has been illustrated in detail for the monoradar track-while-scan mode. Multiradar tracking in areas of overlapping coverage places even higher computing demands since the data rate per aircraft is multiplied, so that there is an added premium on costeffective filters. In the common multiradar tracking scheme the local tracks, that have been constructed by monoradar tracking, are synthesized into a system track by a second tracking filter similar to the first one (Wiener or Kalman filter) but having velocity data inputs in addition. Finally, besides mono- and multiradar tracking tasks with conventional radar measurements or plots, tracking possibilities with unconventional measurements, namely Doppler and direction finding, have been sketched and it has been indicated that multiradar networks possess unique capabilities in this respect.

REFERENCES

(1) Berry, J. P.: "Multisensor airspace surveillance, the results and conclusions of two experiments", Proc. Internat. Conference on Radar (SEE), Paris, Dec. 1978, pp. 93 - 99.

(2) Ebert, H.: "Techniques of automatic tracking of targets of surveillance and secondary radars (in German)", DGON-Vierteljahresmitteilungen, III/1977, pp. 57 - 76.

(3) Bonnefoy, J., Borsu, M., Maignan, G., and Storey, J. T.: "The development of a true multi-radar tracking system", Proc. Internat. Conference on Radar (SEE), Paris, Dec. 1978, pp. 109 - 117.

(4) Brammer, K., Herzmann, F., Kainzinger, A., and Knoppik, N.: "Algorithms for simultaneous automatic track initiation in multiple radar networks," AGARD CP 252, June 1979, pp. 6.1 - 6.17.

(5) Herzmann, F., and Sanders, H.: "Design and simulation of a CCC-system for surveillance purposes", AGARD CP 268, Jan. 1980, pp. 13.1 - 13.23.

(6) de Galard, J.: "Modernizing air traffic control in France", airport forum no. 1/1982, pp. 29 - 34.

(7) Farina, A., and Pardini, S.: "A new class of track-while-scan algorithms using the radial velocity measurements", Proc. Internat. Conference on Radar (SEE), Paris, Dec. 1978, pp. 118 - 128.

(8) van Keuk, G.: "Extended Kalman filter for triangulation in a variable multisensor system (in German)", Regelungstechnik, Heft 10, 1978, pp. 331 -336.

(9) Morgan, D.R.: "A target trajectory noise model for Kalman trackers", IEEE Trans., vol. AES-12, no. 3, May 1976, pp. 405 - 408.

(10) Singer, R.A.: "Estimating optimal tracking filter performance for manned maneuvering targets", IEEE Trans. vol. AES-6, no. 4, July 1970, pp. 473 - 483.

(11) Schumacher, W.: "Investigation of adaptive Kalman Filters for the solution of the radar tracking problem (in German)", Dissertation TU Berlin, FB 10, D 83, Berlin, 1979.

(12) McAuley, R.J., and Denlinger, E.J.: "A decision-directed adaptive tracker", IEEE Trans., vol. AES-9, no. 2, March 1973, pp. 229 - 236.

(13) van Keuk, G.: "Sequential estimation of kinematic flight parameters from radar data (in German)", Angewandte Informatik, Nov. 1973, pp. 471 - 480.

(14) Chan, Y. T., Plant, J. B., and Bottomley, I.R.T.: "A Kalman tracker with a simple input estimator", IEEE Trans., vol. AES-18, no. 2, March 1982, pp. 235 - 241.

(15) Friedland, B.: "Optimum steady-state position and velocity estimation using noisy sampled position data", IEEE Trans., vol. AES-9, no. 6, Nov. 1973, pp. 906 - 911.

(16) Singer, R.A., and Behnke, K.W.: "Real-time tracking filter evaluation and selection for tactical applications", IEEE Trans., vol. AES-7, no. 1, Jan. 1971, pp. 100 - 110.

(17) Brammer, K., and Siffling. G.: "Kalman-Bucy-Filter", R. Oldenbourg, München, 1975.

POPULATION EXTINCTION PROBABILITIES AND METHODS OF ESTIMATION FOR POPULATION STOCHASTIC DIFFERENTIAL EQUATION MODELS

Carlos A. Braumann

Universidade de Évora, 7000 Évora, Portugal

Abstract. The basic population growth models in a randomly fluctuating environment are the stochastic differential equations $d\ell nN/dt=r$ (Malthusian), $d\ell nN/dt=r(1-N/K)$ ($K > o$; logistic), and $d\ell nN/dt=r(\ell nK - \ell nN)$ ($\ell nK > o$; Gompertz), where $N = N(t)$ is the population size at time t and $r = r(t) = ro + \sigma\, \varepsilon(t)$ ($\sigma > o$) is a random process with $\varepsilon(t)$ "standard" white noise. A reference to "colored" noise is also made. In applications to real populations we need parameter estimates based on the usually available discrete observations of a single realization. This paper gives moment and ML estimators. It also gives estimates of the probability of the population dropping below an extinction threshold within a given time. These results can be applied to fisheries and environmental impact assessment.

1. INTRODUCTION

Let $N = N(t)$ be the population size at time t. Three basic deterministic models of population growth have been widely used in theory and applications. They are the *Malthusian model*

$$d\ell nN/dt = r, \tag{1.1}$$

the *logistic* (Pearl-Verhulst) *model*

$$d\ell nN/dt = r(1-N/K) \quad (K > o), \tag{1.2}$$

and the *Gompertz model*

$$d\ell nN/dt = r(\ell nK - \ell nN) \quad (\ell nK > o). \tag{1.3}$$

R. S. Bucy and J. M. F. Moura (eds.), Nonlinear Stochastic Problems, 553–559.

Here r is a growth parameter and K the carrying capacity of the environment.

When the environment is subjected to random fluctuations over time, we may model this by considering r to be a stochastic process of the form

$$r = ro + \sigma\,\varepsilon(t) \quad (\sigma > o), \tag{1.4}$$

where $\varepsilon(t)$ is a "standard" white noise, that is, is stationary Gaussian with zero mean and autocovariance function $\delta(\tau)$, δ being the Dirac delta function. The resulting stochastic differential equation (SDE) is interpreted as a Stratonovich equation. For the question of whether to use Ito or Stratonovich calculus, see Arnold (1974), Braumann (1979, 1980 b), Feldman and Roughgarden (1975), Guess and Gillespie (1977), Turelli (1977, 1978), Capocelli and Ricciardi (1979). The case where $\varepsilon(t)$ is not white noise but a second-order stationary Gaussian process is treated for example, in Soong (1973) and Braumann (1979, 1980 a,b).

The above stochastic models of population growth have been treated by several authors. To cite a few, we refer the reader to Capocelli and Ricciardi (1974), Feldman and Roughgarden (1975), Levins (1969), May (1973,1974), Nobile and Ricciardi (1979), and Tuckwell (1974).

To apply these models to real populations we need to estimate the parameters r,K, and σ from the observations. Unfortunately, the observations are usually made on a single realization N(t) of the Stochastic process. In fact, we can not replicate population growth processes, for we can not reasonably assume that the different populations envolved all have the same parameters. Since counting population size is expensive, time consuming, and difficult, we also can not continuously determine population size. We do it at regular time intervals δ. Therefore, the available information are the observed values of

$$N(0) = No,\ N(\delta) = N_1,\ N(2\delta) = N_2,\ \text{etc.}, \tag{1.5}$$

of a single realization. The parameter estimates are to be based on this information alone.

We will assume that the initial size is known, that is Prob $\{N(0) = No\} = 1$.

Another quantity that we would like to estimate, because of its importance for the comparison of fishing policies or the assessment of an environmental impact (dam, power plant, etc.) is the population extinction probability during the time T the fishing or the impact is supposed to last. By "extinction" we mean "population size dropping below a critical threshold value $N_c > 0$". Knowing the initial population size $N(o) = N_o > N_c$, let $\theta = \inf \{t > o: N(t) < N_c\}$ be the first passage time through the threshold. The required probability is $P = \text{Prob}\ \{\theta < T\}$.

In section 2 we give the solutions to the referred to SDE's and the formulae for P. Section 3 will study how to obtain moment estimators for r and σ and to perform statistical tests and prediction. Section 4 will deal with the maximum likelihood estimators of r, K, σ and P. In section 5 we will refer on how to approach the problem of estimation when the noise $\varepsilon(t)$ is not white and also how to solve the similar problem when K, instead of r, is a random process. Section 6 contains the concluding remarks.

2. SOLUTIONS OF THE SDE'S

Using Stratonovich Calculus, the change of variable

$$X = f(N) = \ell n N, \tag{2.1}$$

$$X = f(N) = \ell n\ (N/(1-N/K)), \tag{2.2}$$

$$X = f(N) = -\ell n\ (\ell n\ (K/N)), \tag{2.3}$$

respectively for models (1.1), (1.2), and (1.3), gives the SDE

$$dX/dt = r_o + \sigma\ \varepsilon(t). \tag{2.4}$$

Let $X_o = f(N_o)$. The solution of (2.4) is

$$X(t) = X_o + r_o\ t + \sigma\ w(t), \tag{2.5}$$

where w(t) is the Standard Wiener process, that is, a Gaussian process with zero mean and autocovariance function

$$h(s,t) = \min(s,t). \tag{2.6}$$

The solution of (2.4) and its asymptotic properties when $\varepsilon(t)$ is not white noise, can be seen in Braumann (1979, 1980 a).

The expression for the population extinction probability P is

$$P = \phi((X_c - X_o - r_o\ T)/\sigma\ T^{1/2}) + \phi((X_c - X_o + r_o\ T)/\sigma\ T^{1/2})\ (\exp(X_c - X_o))^{2r_o/\sigma^2} \tag{2.7}$$

where $X_c = f(N_c)$ and $\phi(z)$ is the cumulative d.f. of the standard normal distribution. This result can be seen in Capocelli and

Ricciardi (1974). Ricciardi (1981) extends this to the non-white noise case for $r = 0$ and Braumann (1981 b) gives an approximate expression for $r \neq 0$.

3. MOMENT ESTIMATORS

Let us assume first that $K(>N_o)$ is known. From the observations (1.5) we can obtain the values of $X_t = f(N_t)$ $(t = 0,1,2,\ldots)$. Since we can not obtain the derivatives of X, we use the finite time differnces

$$\Delta(t) = \{X_t - X_{t-1}\}/\delta = r_o + \sigma\{w(\delta t)-w(\delta(t-1))\}/\delta \quad (3.1)$$

for $t = 1,2,\ldots$. From (2.6), we see that $E\{\Delta(t)\} = r_o$ and

$$\mathrm{COV}\{\Delta(t),\ \Delta(t+\tau)\} = \sigma^2/\delta \text{ for } \tau = 0, \text{ and } = 0 \text{ for } \tau \neq 0. \quad (3.2)$$

Notice that the $\Delta(t)$'s are Gaussian i.i.d. and obtainable from the observations. The *sample time-average mean* of n observations

$$\overline{r}_\tau = \sum_{t=1}^{n} \Delta(t+\tau)/n = \{X_{n+\tau} - X_\tau\}/\delta n \quad (3.3)$$

is Gaussian with mean r_o and variance $\sigma^2/(\delta n)$ and converges to r_o with probability one when $n \to +\infty$. It is a *Gaussian unbiased estimator* of r_o and to compute it we only need two population censuses, at times $\delta\tau$ and $\delta(n+\tau)$. The *sample time-average autocovariances*

$$S(\tau) = \sum_{t=1}^{n} \{\Delta(t) - \overline{r}_o\}\{\Delta(t+\tau) - \overline{r}_\tau\}/(n-1) \quad (3.4)$$

converge with probability one and in the mean to (3.2) when $n\to+\infty$ and the sample variance $S(0)$ is an *unbiased estimator* of σ^2/δ. Since $(\overline{r}_o - r_o)\ /\ (S(o)/n)^{1/2}$ has a Student t-distribution with n-1 df, we can construct *confidence intervals* for r_o. Since $\delta(n-1)S(o)/\sigma^2$ has a chi-square distribution with n-1 df, we can also construct *confidence intervals* for σ^2. *Statistical tests* for comparing populations and analysis of variance can be preformed with the usual t and F statistics.

If we have population censuses at times $0,\delta,2\delta,\ldots,n\delta$, we can *predict* future population sizes using

$$X(\delta(n+\tau)) = X(\delta n) + r_o\delta\tau + \sigma\{w(\delta(n+\tau))-w(\delta n)\}; \quad (3.5)$$

an unbiased predictor of $X(\delta(n+\tau))$ is $X(\delta n) + \overline{r}_o\delta\tau$. The error envolved is Gaussian with zero mean and variance $\sigma^2\delta\tau(1+\tau/n)$ and $\{X(\delta n) + \overline{r}_o\delta\tau - X(\delta(n+\tau))\}/\{S(o)\ \delta^2\tau(1+\tau/n)\}^{1/2}$ is t-distributed with n-1 df, allowing us to construct confidence prediction intervals.

The problems of parameter estimation, testing, and prediction are therefore solved if we know K, which we need (except for model 1.1.) to obtain the X_t's from the N_t's we observe. One way to obtain K is to wait long enough, for we know that Prob$\{N(t)\uparrow K$ when $t \to \infty\} = 1$. There are other cumbersome ways (see Braumann, 1979) to estimate r and σ (but not K) approximately. This is as far as the moment estimators go now.

4. MAXIMUM LIKELIHOOD ESTIMATORS

Maximum likelihood (ML) allows one to obtain estimators $\hat{r}_o$, $\hat{\sigma}$, and $\hat{K}$ for r_o, σ, and K simultaneously. Of course, plugging in those estimators in expression (2.7) in place of r_o, σ, and K, we obtain the ML estimator $\hat{P}$ of P; notice that $\hat{K}$ is necessary to obtain X_c and X_o.

For known $N(0) = N_o$, and noticing that N(t) is a Markov process (see, for example, Arnold, 1974), the log-likelihood function of the observations $N_1,\ldots,N_n$ is

$$L(N_1,\ldots,N_n|r_o,\sigma,K) = \sum_{t=1}^{n} \ln h_t(N_t|N_{t-1}), \quad (4.1)$$

where $h_t(N_t|N_{t-1}) = h_t(N_t|N_{t-1};\ r_o,\sigma,\ K)$ is the density of $N_t{=}N(\delta t)$ conditioned on $N_{t-1} = N(\delta(t-1))$. Let $g_t(X_t|X_{t-1}){=}g_t(X_t|X_{t-1};r_o,\sigma,K)$ be the density of X_t conditioned on X_{t-1}. Since f is a monotonic function, we have

$$h_t(N_t|N_{t-1}) = g_t(X_t|X_{t-1})\ f'(N_t), \quad (4.2)$$

where $f'(N) = N^{-1}$ for model (1.1), $f'(N) = N^{-1}(1-N/K)^{-1}$ for model (1.2), and $f'(N) = N^{-1}\ (\ln(K/N))^{-1}$ for model (1.3). On the other hand, from (2.5) and (2.6), we have

$$g_t(X_t|X_{t-1}) = (\sigma^2\ 2\pi\delta)^{-1/2}\ \exp\{-(X_t-X_{t-1}-r_o\delta)^2/(2\sigma^2\delta)\}. \quad (4.3)$$

From (4.1), (4.2), and (4.3), we obtain

$$L=-n\ln\sigma - n\ \ln(2\pi\delta)/2-(2\sigma^2\delta)^{-1}\sum_{t=1}^{n}(X_t-X_{t-1}-r_o\delta)^2 + \sum_{t=1}^{n}\ln(f'(N_t)). \quad (4.4)$$

Setting $\partial L/\partial r_o = 0$, $\partial L/\partial\sigma = 0$, and $\partial L/\partial K = 0$ gives a system of three equations in the unknowns r_o,σ, and K. Solving it we obtain the ML estimators $\hat{r}_o,\hat{\sigma}$, and $\hat{K}$. The system is

$$\hat{r}_o = (\hat{X}_n - \hat{X}_o)/(n\delta) \quad (4.5)$$

$$\hat{\sigma}^2 = (n\delta)^{-1}\ \Sigma\ (\hat{X}_t - \hat{X}_{t-1} - \hat{r}_o\delta)^2 \quad (4.6)$$

and

$$\Sigma(\hat{X}_t-\hat{X}_{t-1}-(\hat{r}_o+\hat{\sigma}^2)\delta)\exp(\hat{X}_t) - \Sigma(\hat{X}_t - \hat{X}_{t-1} - \hat{r}_o\delta)\exp(\hat{X}_{t-1}) = 0, \quad (4.7)$$

where $\hat{X}_t$ is given (2.2) or (2.3), according to the model, by substituting $\hat{K}$ for K. For model (1.1), we only have the first two equations and we have $\hat{X}_t = X_t$; we conclude, in this case, that $\hat{r}_o = \bar{r}_o$ and $\hat{\sigma}^2/\delta = (n-1)S(o)/n$. For models (1.2) and (1.3), equations (4.5)-(4.7) can be solved only numerically by an iterative procedure.

For large enough n, the ML estimators $\hat{r}_o,\hat{\sigma}$, and $\hat{K}$ are approx. Gaussian with means r_o,σ^2, and K and variance-covariance matrix V, where $nV = R^{-1}$, R being the matrix $R = [\rho_{ij}]$, where $\rho_{ij} = -E\{\partial^2 L/\partial\theta_i\, \partial\theta_j\}$ $(i,j = 1,2,3)$ and $\theta_1 = r_o, \theta_2 = \sigma^2$, and $\theta_3=K$. The expression so obtained for V are cumbersome but, as n gets large, $\hat{r}_o$ and $\hat{\sigma}^2$ have about the same variance as the moment estimators. On the contrary, $\hat{K}$ tends to have a comparatively larger variance. From theses results one can arrive at an approximate expression for the variance of $\hat{P}$, the estimator of the population extinction probability, which turns out to be quite huge.

5. NON-WHITE NOISE AND NOISE IN K

If $\varepsilon(t)$ is not white noise, we can obtain good results if its autocovariance has a known expression with few parameters. Otherwise, we can still obtain moment estimators, in much the same way as we did in section 3, for r_o,σ, and the autocovariance of the noise (at discrete time steps), although things are not so clean. See Braumann (1979, 1980 a). ML estimators are feasible but too cumbersome, since N(t) is no longer a Markov process.

Braumann (1979, 1980 a) also gives moment estimators and is currently working out the ML estimators for the case where r is now fixed and K is a random process. The approach is however somewhat different.

6. CONCLUSION

For the most common stochastic population growth models, we have now available methods of estimation, testing, and prediction based on observations made at discrete time instants on a single realization of the stochastic process. This allows the models to be used in practical applications. One can even estimate population extinction probabilities, which can be used to determine how "bad" is a fishing policy or an environmental impact by means of their effects on those probabilities.

Since the above SDE models are also applied in other fields of Science and are not exclusive of Population Dynamics, we hope these results can be helpfull to other people.

There are still many loose ends to tie up in the problem of estimation and there are other SDE models quite more difficult that the ones commonly used in population growth studies. We feel this is a subject deserving future research.

REFERENCES

Arnold, L.: 1974, *Stochastic Differential Equations: Theory and Applications*. Wiley, New York, p. 221.

Braumann, C.A.: 1979, *Population Growth in Random Environment*. State University of N. Y. at Stony Brook, Ph.D.Thesis.

Braumann, C.A.: 1980 a, International Summer School on Statistical Distributions in Scientific Work, Trieste, Ms.Nr.93.

Braumann, C.A.: 1980 b, Math. Biosciences (in print).

Braumann, C.A.: 1981 a, Proceedings of the Simpósio de Estatística e Investigação Operacional, Fundão, pp. 74-101.

Braumann, C.A.: 1981 b, Contribution to a Report on the EPA Environmetal Risk Assessment Project (manuscript).

Capocelli, R.M. and Ricciardi, L.M.: 1974, Theoret. Popul. Biol. 5, pp. 28-41.

Capocelli, R.M. and Ricciardi, L.M.: 1979, J. of Cybernetics 9, pp. 297-312.

Guess, H.A. and Gillespie, J.H.: 1977, J. Appl. Probab. 14, pp. 58-74.

Levins, R.: 1969, Proc. Natl. Acad. Sci. USA 62, pp. 1061-1065.

May, R.M.: 1973, Amer. Natur. 107, pp. 621-650.

May, R.M.: 1974, *Stability and Complexity in Model Ecosystems*, Princeton University Press, New Jersey, 2nd. ed., p. 265.

Nobile, A.G. and Ricciardi, L.M.: 1979, Proc. INFO II, Patras (in press).

Ricciardi, L.M.: 1981, Contribution to a Report on the EPA Environmental Risk Assessment Project (manuscript).

Soong, T.T.: 1973, *Random Differential Equations in Science and Engineering*, Academic Press, New York, p. 327.

Turelli, M.: 1977, Theoret. Popul. Biol. 12, pp. 140-178.

Turelli, M.: 1978, Theoret. Popul. Biol. 13, pp. 244-267.

DYNAMIC SHIP POSITIONING CONTROL SYSTEMS DESIGN INCLUDING NONLINEAR THRUSTERS AND DYNAMICS

P.T.K. Fung *, Y.L. Chen ** and M.J. Grimble

Spar Aero Space Ltd., Ontario, Canada *
Qinghua University, Beijing, People's Republic of China **
University of Strathclyde, Glasgow, U.K.

ABSTRACT

The control systems for dynamically positioned oil-rig drill ships, support vessels and survey vessels include filter to remove the wave motion signal. This is necessary because the thrust devices are not intended and are not rated to suppress the wave induced motions of the vessel. A combined self-tuning filter and Kalman filter scheme is proposed for this role and an optimal state estimate feedback control scheme is described. The ship dynamic equations are nonlinear and significant nonlinearities are present in the thrust devices. The self-tuning filter has the advantage of maintaining optimum performance in different weather conditions and of compensating for the modelling errors introduced by the nonlinearities.

1. INTRODUCTION

Dynamic ship positioning control can be defined as the process of automatically controlling a ship or floating platforms position and heading above a preselected fixed position on the sea bed by using the vessels thrusters. The definition can be extended to include vessels which must follow a given track at a fixed speed. The advantages of dynamic positioning systems may be summarised as follows:

(a) The need for conventional anchors or moorings is eliminated.
(b) Suitable for drilling in deep water such as the North Sea.
(c) Delays due to the laying and retrieving of anchors are avoided.
(d) Obstructions on the sea bed, such as oil and gas pipes, cannot be damaged, as is possible when using a cluster of anchors.

R. S. Bucy and J. M. F. Moura (eds.), Nonlinear Stochastic Problems, 561–579.

The dynamic positioning (DP) accuracy demanded is typically 3% of water depth. However, cyclic variations in the thrust demand, referred to as thruster modulation, must be suppressed. The first order oscillatory component of the wave forces which causes these high frequency (HF) motions is very large and cannot be counteracted effectively. Any attempt to do so would lead to unnecessary wear of the thrusters and waste energy. The second-order wave forces, current forces and wind forces which cause the low frequency (LF) drift of the vessel must be offset. The ship will respond to wind gusts in the frequency range 0-0.25 rads/sec and thus the control system is required to reduce these LF motions but ignore the HF (0.3-1.6 rads/sec) wave motions. The use of a combined self-tuning/Kalman filter scheme for filtering the wave motions was first proposed by Fung and Grimble[1] but the system was assumed linear and only one axis of motion was considered.

Since the surge motions are only weakly coupled to the sway and yaw motions the discussion here will be concerned with the latter two motions. The main components of the DP system (position measurements, nonlinear thrusters and vessel dynamics) are described first. The LF Kalman motion estimator and the HF self-tuning filter are then considered. Finally, the ship simulation results are presented.

2. POSITION MEASUREMENT

There are many types of acoustic position measurement systems but the GEC/Marconi system was a single beacon on the sea bed with a multi-head transponder on a pod beneath the ship. The transponder electronics evaluates the position of the beacon with respect to the ship by phase measurements on the echoes received by three or more heads. The beacon includes batteries and is automatically triggered at a fixed pulse rate. A vertical reference unit is also required for the compensation of the roll and pitch of the vessel. Acoustic beacons normally have about a two month life. The signals can be upset by gas bubbles from divers or from the ocean bottom. However, vessels often use more than one position reference system including taut wire, rig mounted radio beacons and satellite fixes.

3. THRUSTERS

The thrust devices for positioning the vessel can take several forms but the ship model used in the following is based upon Wimpey Sealab which has retractable ac motor driven thrusters with variable pitch propellers. The vessel has two rotatable bow and two rotatable stern thrusters (capable of 360° rotation and each rated at 12.5 tonnes).

The nonlinear model for the thrusters is shown in Fig.1. For brevity the nonlinear equations will not be given here. The most severe nonlinearity occurs due to the propeller pitch versus thrust relationship $T = p^{1.76}$. It is usual to compensate for this nonlinearity using the input compensator of the form $s_o = s_i^{0.57}$. The main servo includes a bang-bang element with a small dead zone of 0.005 per unit. The factor of 454 on the input of the thruster model, is to change from the per-unit quantities. Similarly, the factor of 1/454 = 0.0022 is to change back into per-unit variables, (the thrusters are each rated at 12.5 tonnes ≡ 122.62 kN and the base power = 55630 kN).

The Kalman filter for the low frequency motions includes a linearised model for the thrusters. This was obtained by fitting the best second order linear model to the nonlinear thrusters using frequency response tests and Bode diagrams. The following results were obtained:

Peak of sine wave I/P	Time constants T_1	T_2
0.0002	0.3981	0.3055
0.0005	0.7244	0.4266
0.001	1.059	0.6918
0.002	1.585	0.861

The time constants used in the linear model were taken as $T_1 = 1.059$ and $T_2 = 0.6918$ seconds which corresponds with a mid range input signal ($|u| < 0.002$ per-unit). The linearised model for the thrusters can be expressed in state equation form as:

$$\dot{x}_5 = -2.3895\,x_5 - 1.3646\,x_6 + 1.3646\,u_1 \tag{1}$$

$$\dot{x}_6 = x_5 \qquad (x_6 \equiv \text{thrust } y'_{13}) \tag{2}$$

$$\dot{x}_7 = -2.3895\,x_7 - 1.3646\,x_8 + 1.3646\,u_2 \tag{3}$$

$$x_8 = x_7 \qquad (x_8 \equiv \text{thrust } y'_{14}) \tag{4}$$

4. THE VESSEL MOTIONS

Surge motion has only a minor effect upon the directional stability of ships. Sway motion mainly occurs due to the imbalance of wind and tidal forces acting upon the vessel. This motion is associated with the yawing motions. Yawing is induced by orbital motions of the water in the wave, differential static pressure on the hull (because of shape) and the gyroscopic couple due to the imposition of rolling motion on the pitching ship.

The major disturbance which must be corrected and which affects the positional accuracy is the wind force. The wind produces (along and perpendicular to its average direction) force components of magnitude varying with the instantaneous wind speed. The perpendicular components of force have zero mean. Tidal changes of speed and direction are generally slow compared with wind direction and wave forces, and thus current speed and direction can be considered constant over considerable periods of time.

4.1 Low Frequency Motions

The LF equations of motion for the vessel Wimpey Sealab have been derived from tank tests[6] and are nonlinear:

$$\dot{x}_1 = -2.4022|x_1|x_1 + 0.03696\ x_3|x_3 - 0.5435|x_1|x_3 + 0.535\,u_1 + 0.5435\,\xi_1 \quad (5)$$

$$\dot{x}_2 = x_1 \quad (6)$$

$$\dot{x}_2 = 2.5245|x_1|x_1 - 1.585|x_3|x_3 - 1.634\,u_2 + 9.785\,\xi_2 \quad (7)$$

$$\dot{x}_4 = x_3 \quad (8)$$

where x_1 = sway velocity, x_2 = sway position, x_3 = yaw velocity x_4 = yaw angle, u_1 = thruster 1 output, u_2 = thruster 2 output, ξ_1, ξ_2 = process noise. The ship simulation is based upon the above model.

The Kalman filter for the low frequency motions depends upon a linearised version of the above equations which, under zero current flow conditions, becomes:

$$\dot{x}_1 = -0.0546\,x_1 + 0.0016\,x_3 + 0.5435\,u_1 + 0.5435\,\xi_1 \quad (9)$$

$$\dot{x}_2 = x_1 \quad (10)$$

$$\dot{x}_3 = 0.0573\,x_1 - 0.0695\,x_3 - 1.634\,u_2 + 9.785\,\xi_2 \quad (11)$$

$$\dot{x}_4 = x_3 \quad (12)$$

The linearised low frequency motion equation and the thruster equations can be combined and represented in the stabilizable and detectable linear state equation form:

$$\underline{\dot{x}}_\ell = A_\ell \underline{x}_\ell + D_\ell \underline{\omega} + B_\ell \underline{u} \quad (13)$$

$$\underline{y}_\ell = C_\ell \underline{x}_\ell \quad (14)$$

$$\underline{z}_\ell = \underline{y}_\ell + \underline{v} \quad (15)$$

where the process and measurement noise statistics follow as:

$$E\{\underline{\omega}(t)\underline{\omega}^T(\tau)\} = Q\delta(t-\tau)$$

$$E\{\underline{v}(t)\underline{v}^T(\tau)\} = R\delta(t-\tau)$$

$$E\{\underline{v}(t)\underline{\omega}^T(\tau)\} = 0$$

and $E\{\underline{\omega}(t)\} = 0$, $E\{\underline{v}(t)\} = 0$. The system matrices for Wimpey Sealab become:

$$A_\ell = \begin{bmatrix} -0.0546 & 0 & 0.0016 & 0 & 0 & 0.5435 & 0 & 0 \\ 1.0 & 0 & 0 & 0 & 0 & 0 & 0 & 0 \\ 0.573 & 0 & -0.0695 & 0 & 0 & 0 & 0 & -1.634 \\ 0 & 0 & 1.0 & 0 & 0 & 0 & 0 & 0 \\ 0 & 0 & 0 & 0 & -2.3895 & -1.3646 & 0 & 0 \\ 0 & 0 & 0 & 0 & 1.0 & 0 & 0 & 0 \\ 0 & 0 & 0 & 0 & 0 & 0 & -2.3895 & -1.3646 \\ 0 & 0 & 0 & 0 & 0 & 0 & 1.0 & 0 \end{bmatrix}$$

$$B_\ell = \begin{bmatrix} 0 & 0 \\ 0 & 0 \\ 0 & 0 \\ 0 & 0 \\ 1.3646 & 0 \\ 0 & 0 \\ 0 & 1.3646 \\ 0 & 0 \end{bmatrix} \qquad D_\ell = \begin{bmatrix} 0.5435 & 0 \\ 0 & 0 \\ 0 & 9.785 \\ 0 & 0 \\ 0 & 0 \\ 0 & 0 \\ 0 & 0 \\ 0 & 0 \end{bmatrix}$$

$$C_\ell = \begin{bmatrix} 0 & 1 & 0 & 0 & 0 & 0 & 0 & 0 \\ 0 & 0 & 0 & 1 & 0 & 0 & 0 & 0 \end{bmatrix}$$

The covariance matrices for the process and measurement noises were defined as: $Q = \text{diag}\{4\times10^{-6}, 9\times10^{-8}\}$ $R = \text{diag}\{10^{-5}, 1.22\times10^{-5}\}$. In calculating the Kalman filter gain for the low frequency subsystem the measurement noise was inflated artificially to $R_f = \text{diag}\{4\times10^{-5}, 19.52\,10^{-5}\}$. The increase can be justified theoretically[3] since the use of a self-tuning filter results in a sub-optimal scheme, even in the linear system case and the low frequency Kalman filter therefore has an additional high frequency input which can be treated as measurement noise.

4.2 High Frequency Motions

The high frequency motions of the vessel are due to the first order wave forces. The worst case high frequency motion is determined by the sea spectrum alone. The high frequency motions were simulated using two fourth order colouring filters driven by white noise. In state space notation:

$$\dot{\underline{x}}_h = A_h\underline{x}_h + D_h\underline{\xi}$$
$$y_h = C_h\underline{x}_h$$

$$A_h = \begin{bmatrix} A_h^s & 0 \\ 0 & A_h^y \end{bmatrix} \text{ and } D_h = \begin{bmatrix} D_h^s & 0 \\ 0 & D_h^y \end{bmatrix}$$

and the submatrices for the sway and yaw directions have the same form:

$$A_h^s = \begin{bmatrix} 0 & 1 & 0 & 1 \\ 0 & 0 & 1 & 0 \\ 0 & 0 & 0 & 1 \\ -a_4^s & -a_3^s & -a_2^s & -a_1^s \end{bmatrix} \quad D_h^s = \begin{bmatrix} 0 \\ 0 \\ 0 \\ k^s \end{bmatrix}$$

$$C_h^s = \begin{bmatrix} 0 & 0 & 1 & 0 \end{bmatrix}$$

The parameters of the system matrices are calculated to minimise the integral squared error between the modelled and Pierson Moskowitz sea spectra[2].

Collecting results, the state equations for the linearised low and high frequency ship models can be combined and can be written in the form:

$$\begin{bmatrix} \dot{\underline{x}}_\ell \\ \dot{\underline{x}}_h \end{bmatrix} = \begin{bmatrix} A_\ell & 0 \\ 0 & A_h \end{bmatrix} \begin{bmatrix} \underline{x}_\ell \\ \underline{x}_h \end{bmatrix} + \begin{bmatrix} \underline{B}_\ell \\ 0 \end{bmatrix} \underline{u}_\ell + \begin{bmatrix} D_\ell & 0 \\ 0 & D_h \end{bmatrix} \begin{bmatrix} \underline{\omega} \\ \underline{\xi} \end{bmatrix} + \begin{bmatrix} E_\ell \\ 0 \end{bmatrix} \underline{n}_\ell \tag{18}$$

The position of the vessel is given by the sum of the low and high frequency motions $\underline{y} = \underline{y}_\ell + \underline{y}_h$. The position measurement $\underline{z}$ is therefore:

$$\begin{bmatrix} z_1 \\ z_2 \end{bmatrix} = \begin{bmatrix} C_\ell & C_h \end{bmatrix} \begin{bmatrix} \underline{x}_\ell \\ \underline{x}_\ell \end{bmatrix} + \underline{v} = \underline{y}_\ell + \underline{y}_h + \underline{v} \tag{19}$$

The input $\underline{n}_\ell$ represents the measurable component of the wind disturbance signal.

5. DISCRETE MODEL FOR THE LF AND HF SUBSYSTEMS

The optimal estimators are based upon the linearised ship and thruster equations in discrete form. The low frequency subsystem is represented (using the same symbols for the discrete system matrices) as:

$$\underline{x}_\ell(t+1) = A_\ell \underline{x}_\ell(t) + B_\ell \underline{u}(t) + D_\ell \underline{\omega}(t) \tag{20}$$

$$S_\ell : \underline{y}_\ell(t) = C_\ell \underline{x}_\ell(t) \tag{21}$$

$$\underline{z}_\ell(t) = \underline{y}_\ell(t) + \underline{v}(t) \tag{22}$$

$$E\{\underline{\omega}(t)\} = 0, \quad E\{\underline{\omega}(k)\underline{\omega}^T(m)\} = Q\delta_{km} \tag{23}$$

$$E\{\underline{v}(t)\} = 0, \quad E\{\underline{v}(k)\underline{v}^T(m)\} = R\delta_{km} \tag{24}$$

and δ_{km} is the Dirac delta function, $\underline{x}_\ell(t) \varepsilon R^n$, $\underline{u}(t) \varepsilon R^m$ $\omega(t) \varepsilon R^q$ and $\underline{y}_\ell(t) \varepsilon R^r$. The observed plant output includes the wave disturbance signal $\underline{y}_h(t)$ and is given as:

$$\underline{z}(t) = \underline{z}_\ell(t) + \underline{y}_h(t) \tag{25}$$

The high frequency disturbance can be represented as:

$$S_h: \quad A_h(z^{-1})\underline{y}_h(t) = C_h(z^{-1})\underline{\xi}(t) \tag{26}$$

and $\underline{y}_h(t) \in R^r$ and $\underline{\xi}(t) \in R^r$. Here $\underline{\xi}(t)$ represents an independent zero mean random vector which is uncorrelated with $\underline{\omega}(t)$ and $\underline{v}(t)$ and has a diagonal covariance matrix Σ_ξ. The polynomial matrices $A_h(z^{-1})$ and $C_h(z^{-1})$ are assumed to be square and of the form:

$$A_h(z^{-1}) = I_r + A_1 z^{-1} + A_2 z^{-2} + \ldots + A_{n_a} z^{-n_a} \tag{27}$$

$$C_h(z^{-1}) = C_1 z^{-1} + C_2 z^{-2} + \ldots + C_{n_c} z^{-n_c} \tag{28}$$

where z^{-1} is the backward shift operator. The matrix $A_h(z^{-1})$ is assumed to be regular (that is A_{n_a} is non-singular). The zeros of $\det(A_h(x))$ and $\det(C_h(x))$ are assumed to be strictly outside the unit circle. The order of the polynomial matrices are known but the coefficient matrices are treated as unknowns, since in practice the wave disturbance spectrum varies slowly with weather conditions. The disturbances in each observed channel are uncorrelated so that the matrices $\{A_i\}$ and $\{C_j\}$ have diagonal form.

6. LOW FREQUENCY MOTION ESTIMATOR

Assume for the moment that the coloured noise signal $\underline{y}_h$ can be measured, and thence $\underline{z}_\ell$ can be calculated. The plant states $\underline{x}_\ell$ can be estimated using a Kalman filter with input $\underline{z}_\ell$ assuming the ship equations and noise covariances are known. The worst case process noise is normally assumed and this is taken to have a constant variance. Under these conditions the Kalman gain matrix is constant and may therefore be computed off-line.

The Kalman filter algorithm becomes:

Predictor:

$$\hat{\underline{x}}_\ell(t|t-1) = A_\ell \hat{\underline{x}}_\ell(t-1|t-1) + B_\ell u(t-1) \tag{29}$$

$$\hat{\underline{y}}_\ell(t|t-1) = C_\ell \underline{x}_\ell(t|t-1) \tag{30}$$

$$P(t|t-1) = A_\ell P(t-1|t-1) A_\ell^T + D_\ell Q D_\ell^T \tag{31}$$

Corrector:

$$\hat{\underline{x}}_\ell(t|t) = \hat{\underline{x}}_\ell(t|t-1) + K_\ell(t)\underline{\varepsilon}_\ell(t) \tag{32}$$

$$\hat{\underline{y}}_\ell(t|t) = C_\ell \hat{\underline{x}}_\ell(t|t) \tag{33}$$

$$P(t|t) = P(t|t-1) - K(t) C_\ell P(t|t-1) \tag{34}$$

$$K(t) = P(t|t-1)C^T[CP(t|t-1)C^T + R]^{-1} \tag{35}$$

$$\underline{\varepsilon}_\ell(t) \triangleq \underline{z}(t) - \hat{\underline{y}}_\ell(t|t-1) - \underline{y}_h(t) \tag{36}$$

$$= \underline{z}_\ell(t) - \hat{\underline{y}}_\ell(t|t-1) \tag{37}$$

and $K(t)$ is the Kalman gain matrix, $P(.)$ is the error covariance matrix.

7. HIGH FREQUENCY MOTION ESTIMATOR

In this section the high frequency motion estimator is constructed based upon the model (26).

Define the new variable $\underline{m}_h(t)$ as:

$$\underline{m}_h(t) = \underline{z}(t) - \hat{\underline{y}}_\ell(t|t-1) \tag{38}$$

and from (36):

$$\underline{m}_h(t) = \underline{\varepsilon}_\ell(t) + \underline{y}_h(t) \tag{39}$$

The innovations signal $\underline{\varepsilon}_\ell$ is white noise and covariance matrix for $\underline{\varepsilon}_\ell$ is denoted by $\Sigma_{\varepsilon_\ell}$. The innovations signal model becomes:

$$A_h(z^{-1})\underline{m}_h(t) = D_h(z^{-1})\underline{\varepsilon}(t) \tag{40}$$

where $\{\underline{\varepsilon}(t)\}$ is an independent random sequence with covariance matrix Σ_ε.

The matrix polynomial $D_h(z^{-1})$ has the form:

$$D_h(z^{-1}) = I_r + D_1 z^{-1} + \ldots + D_{n_d} z^{-n_d} \tag{41}$$

where the zeros of $\det(D_1(x))$ lie strictly inside the unit circle. The parameters of $D_h(z^{-1})$ are determined by the following spectral factorisation:

$$D_h(z^{-1})\Sigma_\varepsilon D_h^T(z) = C_h(z^{-1})\Sigma_\varepsilon C_h^T(z) + A_h(z^{-1})\Sigma_{\varepsilon_\ell} A_h^T(z) \tag{42}$$

Note that $n_d = n_a$ and that by taking the limit as $z \to 0$:

$$D_{n_d}\Sigma_\varepsilon = A_{n_a}\Sigma_{\varepsilon_\ell} \tag{43}$$

Since $A_h(z^{-1})$ is regular the following identity holds:

$$A_{n_a}^{-1}D_{n_a} = \Sigma_{\varepsilon_\ell}\Sigma_\varepsilon^{-1} \tag{44}$$

The optimal estimate of $\underline{y}_h(t)$ can be calculated using:

$$\hat{\underline{y}}_h(t|t) = \underline{m}_h(t) - \Sigma_{\varepsilon_\ell}\Sigma_\varepsilon^{-1}\underline{\varepsilon}(t) \tag{45}$$

where

$$\underline{\varepsilon}(t) = \underline{m}_h(t) - \hat{\underline{y}}_h(t|t-1) \tag{46}$$

Using the identity in equation (44), $\hat{\underline{y}}_h(t|t)$ becomes:

$$\hat{\underline{y}}_h(t|t) = \underline{m}_h(t) - A_{n_a}^{-1}D_{n_d}\underline{\varepsilon}(t) \tag{47}$$

The wave frequency model changes with weather conditions and these variations are accounted for in (47) by on-line estimation of A_{n_a}, D_{n_d} and the innovations $\underline{\varepsilon}(t)$.

7.1 Error Term

The signal $\underline{y}_h$ is not measurable and must be replaced in the low frequency Kalman filter by $\hat{\underline{y}}_h(t|t)$. This approximation causes a difference in the innovation which is calculated:

$$\underline{\bar{\varepsilon}}(t) \triangleq \underline{z}(t) - \hat{\underline{y}}_\ell(t|t-1) - \hat{\underline{y}}_h(t|t) = \underline{\varepsilon}_\ell(t) + \underline{n}_h(t) \qquad (48)$$

where $\underline{n}_h(t) \triangleq \underline{y}_h(t) - \hat{\underline{y}}_h(t|t)$. The signal $\underline{n}_h(t)$ for the high frequency motion estimator has a zero mean value if the errors in calculating $\hat{\underline{y}}_h(t|t)$ are neglected. The signal $\underline{\varepsilon}$ can be computed using[3]:

$$\underline{\bar{\varepsilon}}(t) = A_{n_a}^{-1} D_{n_d} \underline{\varepsilon}(t) \qquad (49)$$

The signal $\underline{m}(t)$ is similarly changed by the above approximation and becomes[3]:

$$\begin{aligned} \underline{\bar{m}}_h(t) &= \underline{m}_h(t) - \tilde{\underline{y}}_\ell(t|t-1) \\ &= A_h(z^{-1})^{-1} D_h(z^{-1}) \underline{\varepsilon}(t) - \tilde{\underline{y}}_\ell(t|t-1) \end{aligned} \qquad (50)$$

The error signal $\tilde{\underline{y}}_\ell(t|t-1)$ can be shown [3] to be generated from the output of the low frequency subsystem, driven by the zero mean signal $\underline{n}_h$. The resulting position variations are relatively slow in comparison with the high frequency motions. Define $\underline{s}(t) = A_h(z^{-1})\tilde{\underline{y}}_\ell(t|t-1)$ then this can be treated as a constant and identified together with the high frequency model parameters. Equation (5) now becomes:

$$A_h(z^{-1})\underline{\bar{m}}_h(t) = D_h(z^{-1})\underline{\varepsilon}(t) - \underline{s}(t) \qquad (51)$$

This innovations signal model can be represented in the usual form for parameter estimation:

$$\underline{\bar{m}}_h(t) = \psi(t)\underline{\Theta} + \underline{\varepsilon}(t) \qquad (52)$$

and the algorithm due to Panuska [4] can be employed to estimate the unknown parameters.

In the ship positioning problem the high frequency disturbances can be assumed to be decoupled, so that $A_h(z^{-1})^{-1}D_h(z^{-1})$ is a diagonal matrix and the parameters for each channel can be estimated separately. For the ith channel:

$$\bar{m}_{h_i}(t) = \psi_i(t)\underline{\Theta}_i + \varepsilon_i(t) \qquad (53)$$

where

$$\psi_i(t) = [-\bar{m}_{h_i}(t-1), \ldots, -\bar{m}_{h_i}(t-n_a); \varepsilon_i(t-1), \ldots, \varepsilon_i(t-n_d); 1] \qquad (54)$$

$$\underline{\Theta}_i^T = [a_{i_1}, \ldots, a_{in_a}; d_{i_1}, \ldots, d_{in_d}; s_i] \qquad (55)$$

Past values of the innovations signal are approximated by:

$$\hat{\varepsilon}_i(t) = \bar{m}_{h_i}(t) - \hat{\psi}_i(t)\hat{\underline{\Theta}}_i \qquad (56)$$

where $\hat{\psi}_i(t)$ is given by (54) with $\varepsilon_i(t-j)$ replaced by $\hat{\varepsilon}_i(t-j)$ $j = 1, 2, \ldots, n_d$ and $\hat{\underline{\Theta}}_i$ represents the estimated parameter vector.

8. KALMAN AND SELF-TUNING FILTER ALGORITHMS

The Kalman and self-tuning filter algorithms are combined below to produce the desired low frequency motion estimator. The structure of the self-tuning/Kalman filtering scheme is shown in Fig.2.

Algorithm 8.1

1. Initialise $\underline{\theta}_i$, initial parameter covariance for each channel and assign the forgetting factor β. Initialise state estimates.
2. Generate the Kalman filter low frequency estimates $\underline{\hat{x}}_\ell(t|t-1)$.
3. Calculate $\bar{m}_{h_i}(t)$ using (38) and form $\hat{\psi}_i(t)$.
4. Parameter update:

$$\hat{\underline{\theta}}_i(t) = \hat{\underline{\theta}}_i(t-1) + K_i^p(t)(\bar{m}_{h_i}(t) - \hat{\psi}_i(t)\hat{\underline{\theta}}_i(t-1)) \tag{57}$$

5. Covariance and gain update:

$$P_i^p(t) = \{P_i^p(t-1) - K_i^p(t)(\beta + \psi_i(t)P_i^p(t-1)\psi_i^T(t))K_i^p(t)^T\}/\beta$$
$$K_i^p(t) = P_i^p(t-1)\psi_i(t)(\beta + \psi_i(t)P_i^p(t-1)\psi_i^T(t))^{-1} \tag{58}$$

   where $0.95 \lesssim \beta \lesssim 1$.
6. Innovations update:

$$\hat{\varepsilon}_i(t) = \bar{m}_{h_i}(t) - \hat{\psi}_i(t)\hat{\theta}_i(t) \tag{59}$$

7. Calculate $\bar{\varepsilon}_{\ell i}(t)$ for channel i using (49):

$$\bar{\varepsilon}_{\ell_i}(t) = \hat{a}_{n_a}^{-1}\,\hat{d}_{n_d}\,\hat{\varepsilon}_i(t) \tag{60}$$

8. If i< number of channels (r) go to step 3.
9. Generate the state $\underline{\hat{x}}_\ell(t|t)$.
10. Calculate the estimated steady state $\underline{y}_\ell(t|t-1)$ as:

$$\hat{\tilde{y}}_{\ell_i}(t|t-1) = \hat{s}_i(t)/A_{h_i}(1) \tag{61}$$

11. Correct the position estimates using $\hat{\tilde{y}}(t|t-1)$. Return to step 2.

9. OPTIMAL CONTROL GAIN

The optimal control gain depends upon the fixed low frequency subsystem and is independent of the variations of the wave model. The same gain is therefore employed for all weather conditions. The controller with input $\underline{z}$ and output $\underline{u}$ is found to minimise the performance criterion:

$$J = \lim_{T\to\infty} \frac{1}{2T} E\{\int_{-T}^{T} (\underline{x}_\ell - \underline{r}_\ell)^T Q_1 (\underline{x}_\ell - \underline{r}_\ell) + \underline{u}^T R_1 \underline{u}\, dt\} \tag{62}$$

where Q_1 and R_1 are positive definite weighting matrices, and the system is represented by the linearised equations (18). The optimal control becomes $\underline{u}_O(t) = -K_C\hat{\underline{x}}_\ell(t|t)$ where K_c is found from the Riccati equation in the usual way [5], [9], [10].

10. SIMULATION RESULTS

The performance of the optimal state estimate feedback scheme is illustrated in figures 3 to 6. The system has a constant current disturbance input of 0.002 pu to the sway subsystem at t = 50 seconds, (N.B. real time = 3.104 x simulated time). The disturbance is much larger than would occur in practice since it would involve the thrusters acting continuously at full load to counteract the effect. However, the results demonstrate that the system remains stable even in this extreme situation. The steady state error can be reduced to zero by the use of integral action (see figures 15 to 18).

The estimates of the LF motions (figures 4 and 6 are needed for control purposes and are good even in this non-linear situation. The parameter estimates of the HF subsystem (figure 7) converge rapidly and the cumulative loss functions ($J_s = \sum_{i=1}^{N} (y_\ell^s(t) - \hat{y}_\ell^s(t|t))^2$)for the low frequency estimator (figure 8) increase steadily after the parameters have settled down. The estimated thrusts, shown in figures 9 and 10, are as expected varying more than the thrusts generated from the modelled nonlinear thrusters. This also applies to the estimated and modelled velocities shown in figures 13 and 14. The estimates of the self-tuning filter parameter s(t) for sway and yaw (figures 11 and 12) show higher frequency variations than is obtained with linear ship models (recall that s(t) is related to the error term $\tilde{y}_\ell$).

11. CONCLUSIONS

The self-tuning/Kalman filtering scheme has several advantages over the more usual extended Kalman filtering DP systems[7]. The separation of the LF and HF estimation functions is convenient for both analysis and fault detection. The errors which are introduced due to the presence of the nonlinearities are compensated by the self-tuning identification of the term $\tilde{y}_\ell$. Existing constant gain Kalman filtering DP systems[8] can also be easily modified to include the self-tuning action.

REFERENCES

1. Fung, P.T.K. and Grimble, M.J., 'Self tuning control of ship positioning systems', IEE Workshop on Theory & Application of

Adaptive & Self-tuning Control, Oxford University, March 1981 Also published by Peter Peregrinus Ltd, edited by C.J. Harris and S.A. Billings, 1981.

2. Morgan, M.J., 'Dynamic positioning of offshore vessels', Divsion of the Petroleum Publishing Co., Tulsa, Oklahoma, USA 1978.
3. Fung, P.T.K. and Grimble, M.J., 'Dynamic ship positioning using a self-tuning Kalman filter', Research report, Univ of Strathclyde, No EE/8/March 1982.
4. Panuska, V., 'A new form of the extended Kalman filter for parameter estimation in linear systems with correlated noise', IEEE trans on Auto Contr, Vo. AC-25, No 2, pp 229-235, 1980.
5. Grimble, M.J., Patton, R.J. and Wise, D.A. 'The design of dynamic ship positioning control systems using stochastic optimal control theory', Optimal Control Applications & Methods, pp 167-202, 1980
6. Wise, D.A. and English, J.W., 'Tank and wind tunnel tests for a drill ship with dynamic position control', Offshore Technology Conf., Dallas, paper No OTC2345, 1975.
7. Balchen, J.G., Jenssen, N.A. and Saelid, S., 'Dynamic positioning using Kalman filtering and optimal control theory', Automation in Offshore Oil Field Operation, pp 183-188, 1976.
8. Grimble, M.J., Patton, R.G. and Wise, D.A. 'Use of Kalman filtering techniques in dynamic ship positioning systems', Proc IEE, Vol 127, Pt D, No 3,pp 93-102, May 1980.
9. Grimble, M.J. 'Design of optimal stochastic regulating systems including integral action', Proc IEE, Vol 126, No 9, pp 841-848, Sept 1979.
10. Grimble, M.J. 'Solution of the stochastic optimal control problem in the s-domain for systems with time delay', Proc IEE, Vol 126, No 7, July 1979.

ACKNOWLEDGEMENT

We are grateful for the help of Mr Peter Urry and Mr Dennis Wise of GEC Electrical Projects Ltd, Rugby and for the support of GEC and the United Kingdom Science & Engineering Research Council.

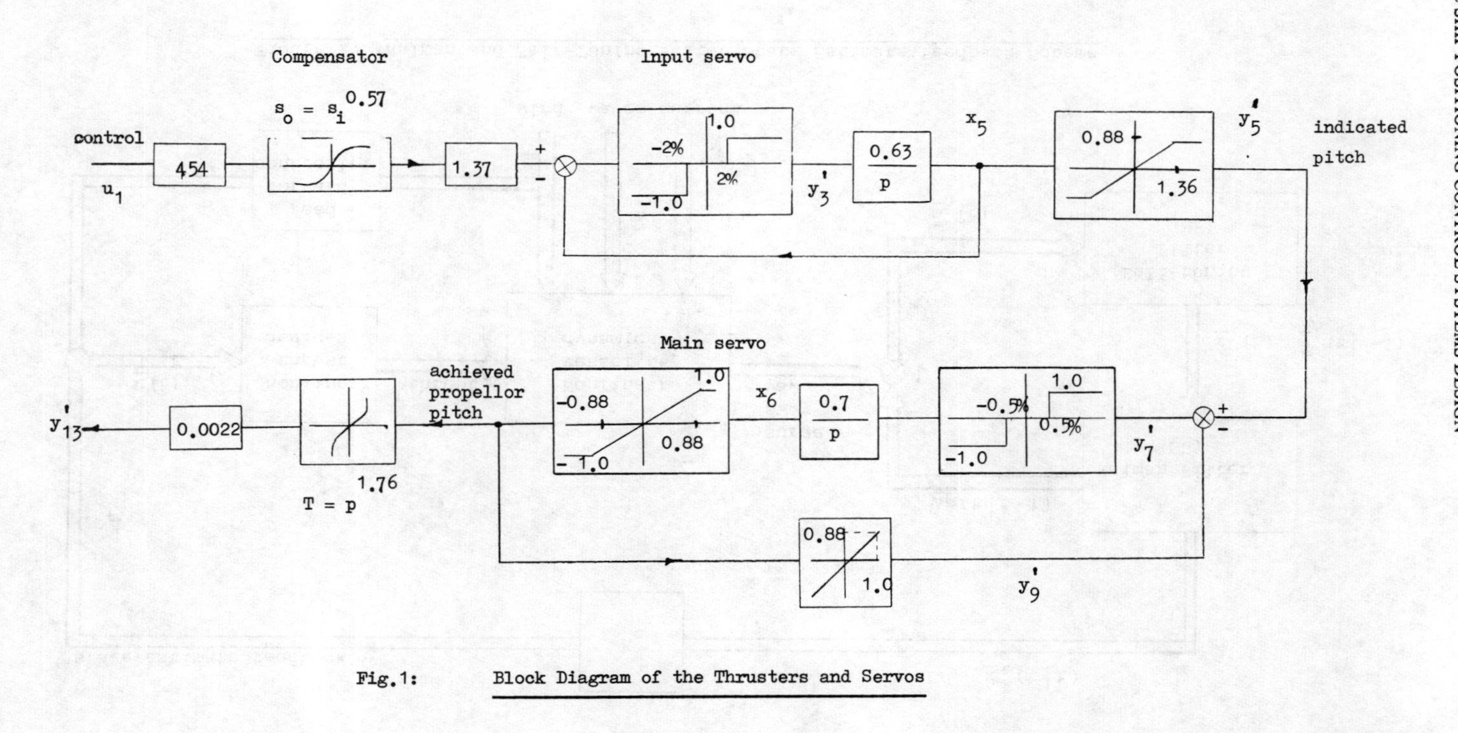

Fig.1: Block Diagram of the Thrusters and Servos

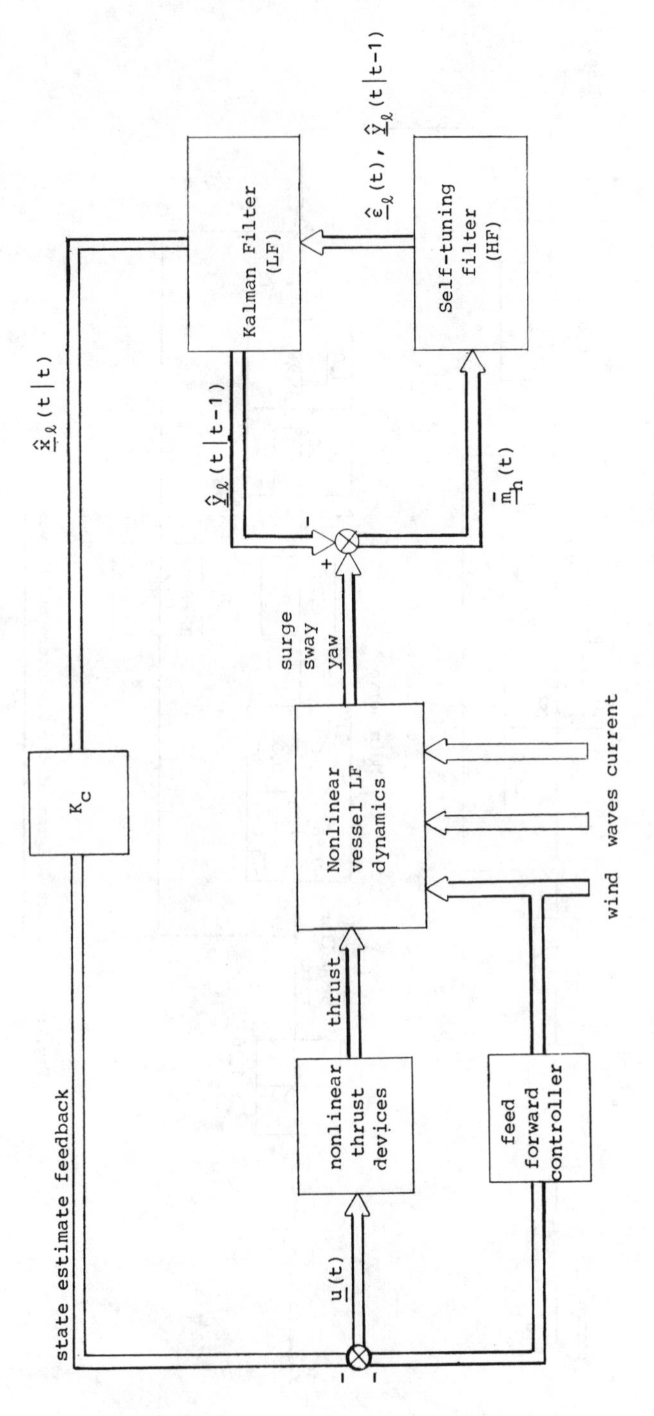

Figure 2: Kalman and Self-Tuning Filter State Estimate Feedback Scheme

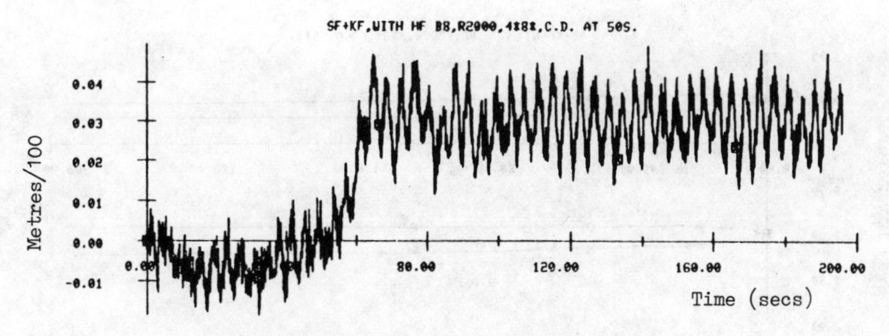

Fig.3 Total Observed Sway Motion (Beaufort 8)

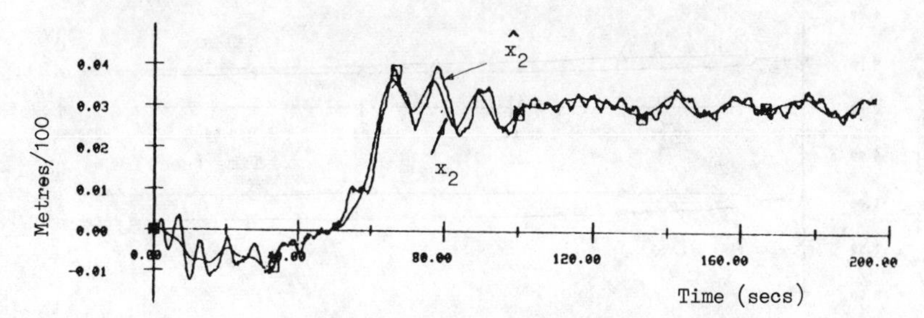

Fig.4 Modelled and Estimated L.F. Sway Motion (Beaufort 8)

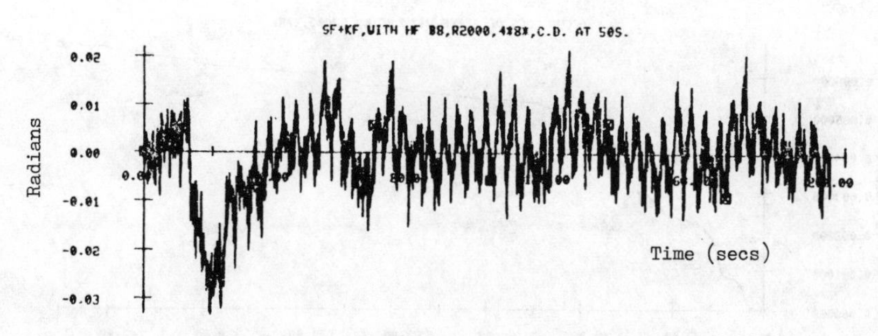

Fig.5 Total Observed Yaw Motion (Beaufort 8)

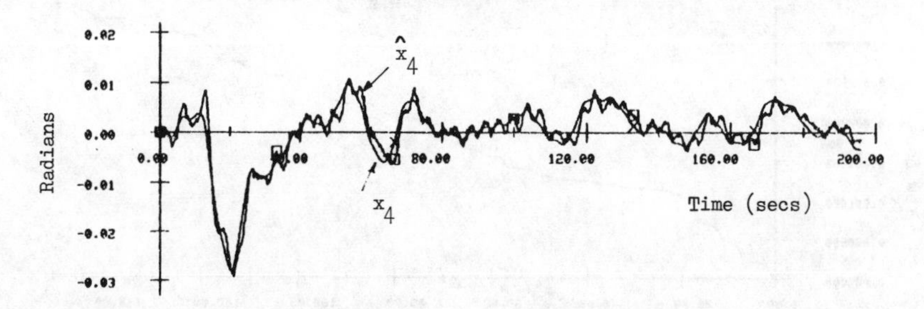

Fig.6 Modelled and Estimated L.F.Yaw Motion (Beaufort 8)

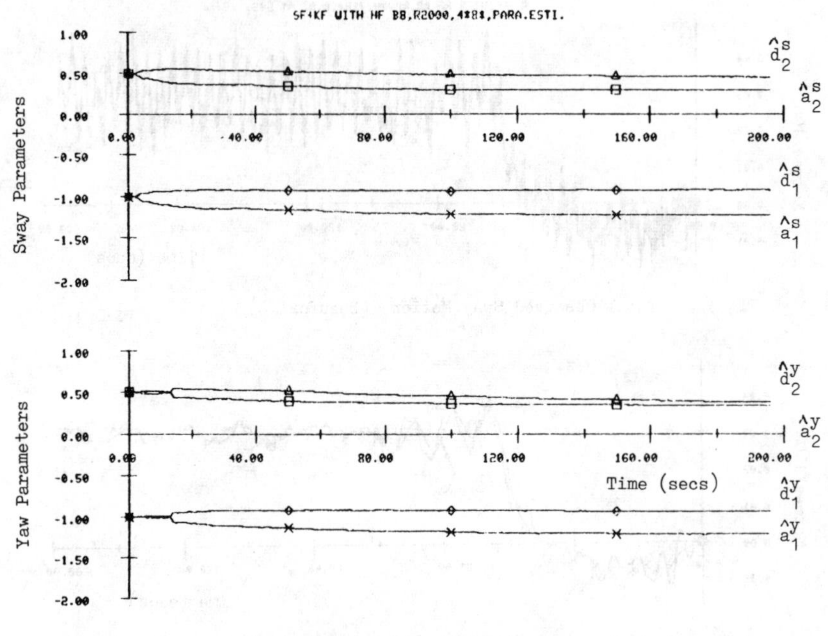

Fig.7 Sway and Yaw Estimated H.F. Parameters (Beaufort 8)

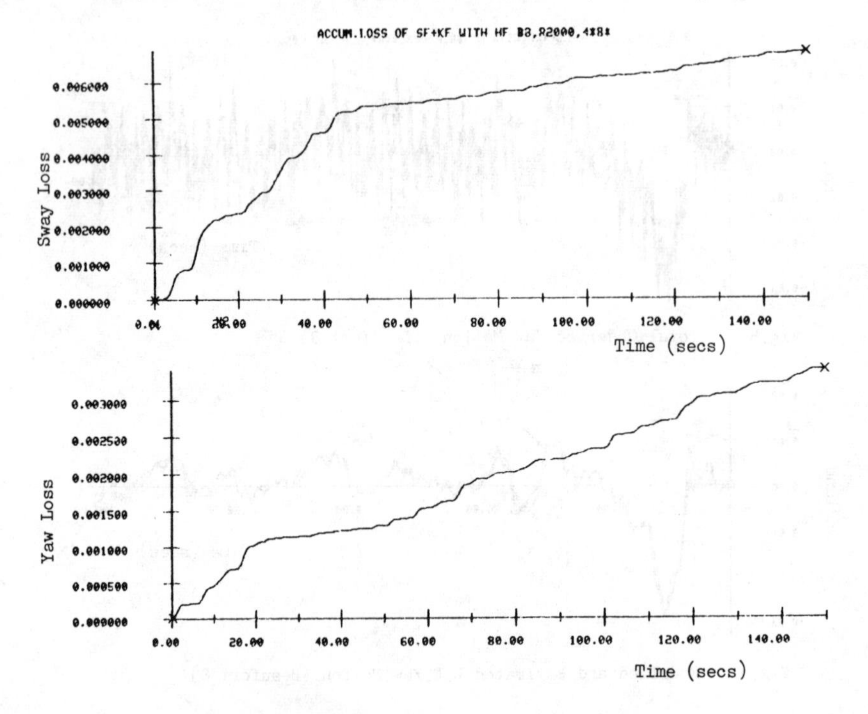

Fig.8 Sway and Yaw Cumulative Loss Functions (Beaufort 8)

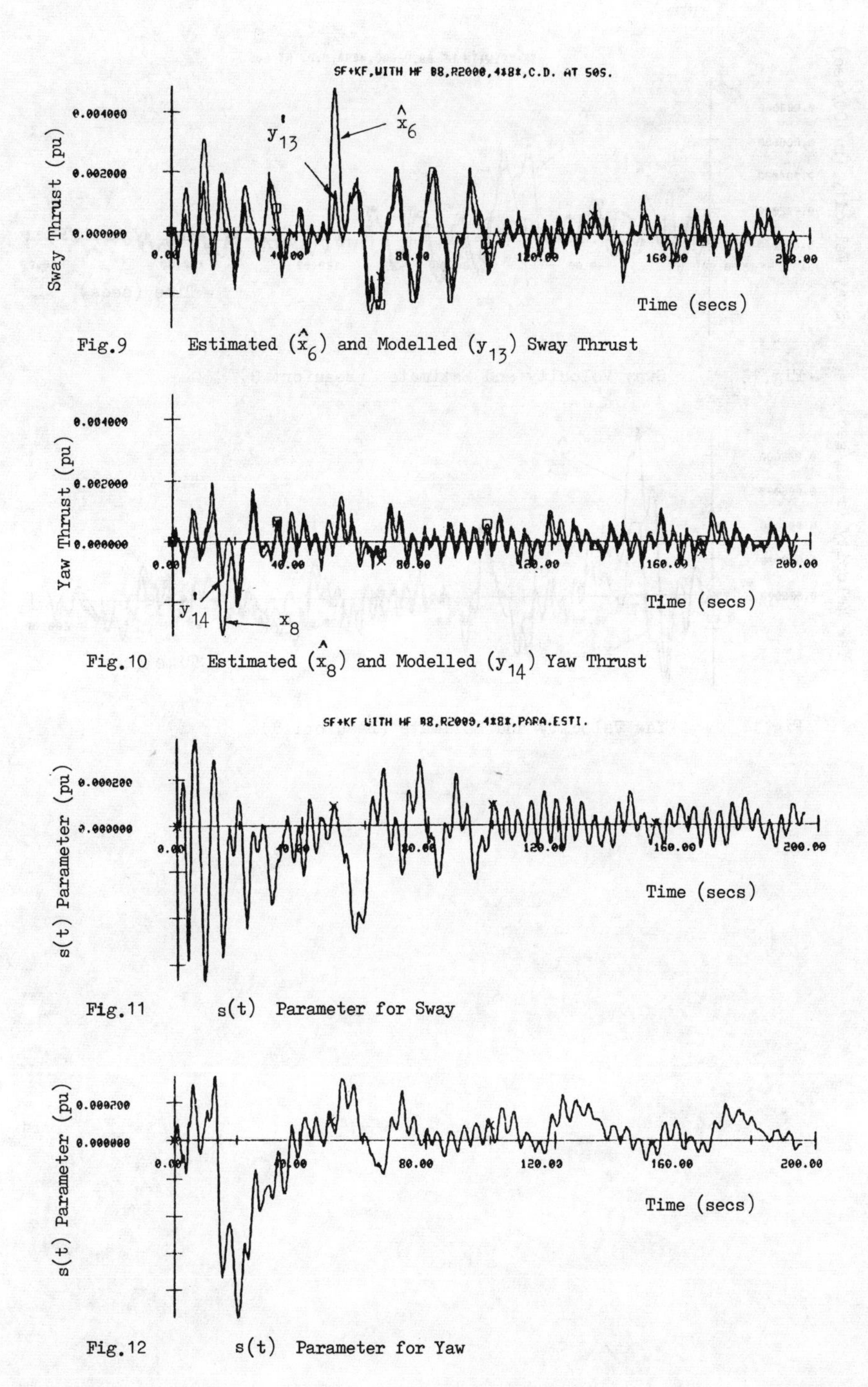

Fig.9 Estimated ($\hat{x}_6$) and Modelled (y_{13}) Sway Thrust

Fig.10 Estimated ($\hat{x}_8$) and Modelled (y_{14}) Yaw Thrust

Fig.11 s(t) Parameter for Sway

Fig.12 s(t) Parameter for Yaw

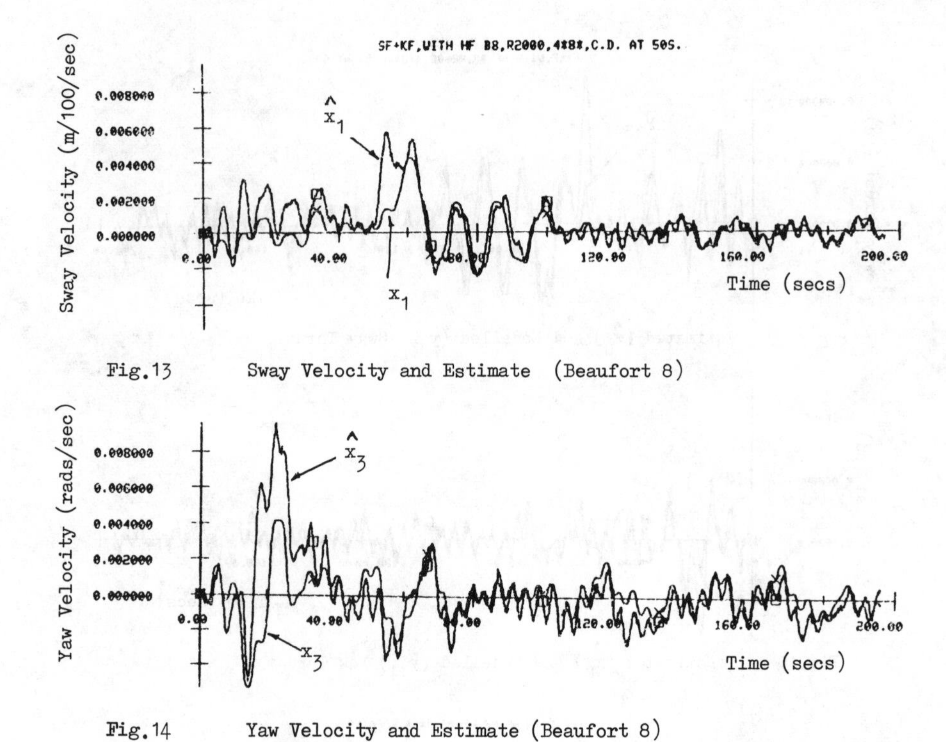

Fig.13 Sway Velocity and Estimate (Beaufort 8)

Fig.14 Yaw Velocity and Estimate (Beaufort 8)

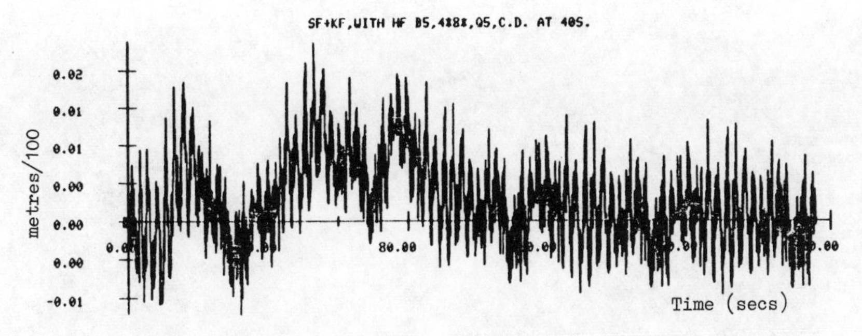

Fig.15: Total Observed Sway Motion (Integral Control, B8)

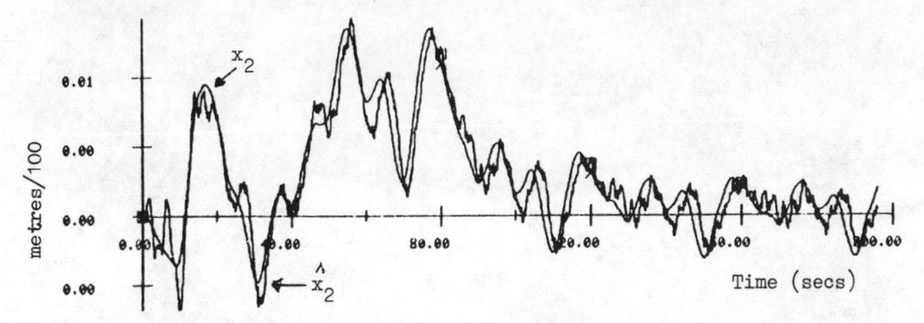

Fig.16: Modelled and Estimated LF Sway Motion (Int. Cont., B8)

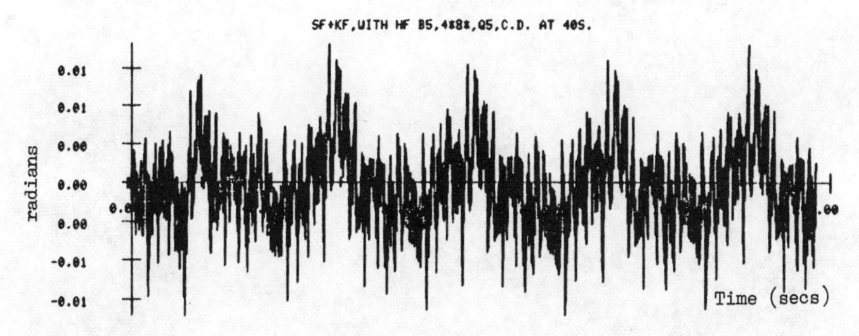

Fig. 17: Total Observed Yaw Motion (Integral Control, B8)

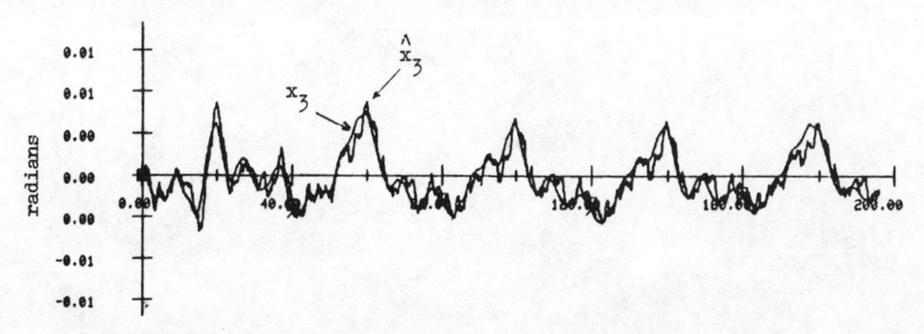

Fig.18: Modelled and Estimated LF Yaw Motion (Int. Cont., B8)

JOINT OPTIMIZATION OF TRANSMITTER AND RECEIVER FOR CYCLO-STATIONARY RANDOM SIGNAL PROCESSES.

Friedrich K. Graef

Institut für Angewandte Mathematik
der Universität Erlangen-Nürnberg

Abstract. Linear periodically varying filters are considered as modulators and demodulators for the transmission of a cyclo-stationary source signal over a stationary channel. A spectral representation of filters and signals of this type by operator-valued transfer functions and spectral densities is introduced and the joint optimization of transmitter and receiver is performed in terms of these spectral characteristics.

1. INTRODUCTION

This paper deals with the joint optimization of transmitter and receiver in a linear modulation system as illustrated in Fig. 1. A stochastic source signal x is to be transmitted over a channel introducing linear distortion and additive noise. Transmitter and receiver are assumed to be linear filters with impulse responses f,g, and optimization is to be performed with respect to some mean square error criterion.

For stationary source signals and time invariant filters this problem has been solved by SINAJ (8) in the time-continuous case and by BERGER, TUFTS (1) for pulse amplitude modulation of time discrete signals. ERICSON (2) extended the results - at the cost of a strict bandwidth restriction imposed on the source signal - to periodically varying filters.

In the present paper joint optimization of transmitter and receiver is carried out for periodically varying filters and cyclostationary source signals, i.e. stochastic processes whose first and second order moments fluctuate periodically with

R. S. Bucy and J. M. F. Moura (eds.), Nonlinear Stochastic Problems, 581–592.

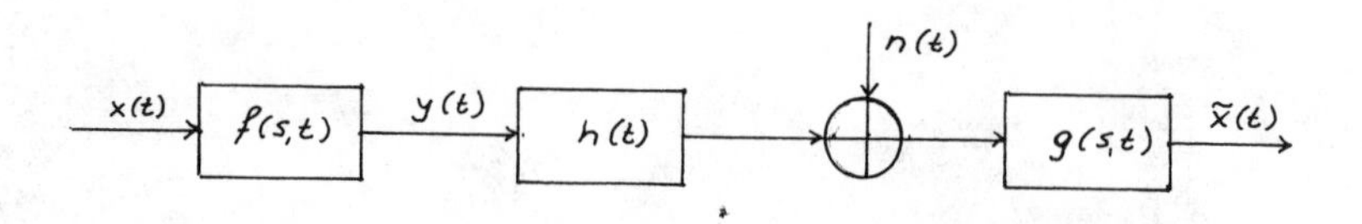

Fig. 1 The Transmission System

time - wide sense stationary processes being a specific case. Processes of this type have extensively been discussed by GARDNER, FRANKS in (3), where also receiver optimization was achieved by the aid of certain series representations.

Here a characterization in frequency domain of cyclostationary processes and periodically varying filters by means of operator-valued spectral densities and transfer functions is introduced. This allows us to apply the results obtained in (4), (5) where optimal filtering of infinite-dimensional stationary signals had been discussed. In Sec.s 3 and 4 spectral density and transfer operators are defined and characterized by the respective covariances and impulse responses. In Sec. 5 the time averaged mean square transmission error will be derived as a nonlinear functional of the spectral characteristics of the system, error minimization being carried out in Sec.s 6 through 8. Finally as a specific example wide sense stationary source signals with arbitrary power spectra are considered.

2. PRELIMINARIES

Trace-class operators (SCHATTEN (7)): Let $\mathcal{H}$ be some (real or complex) Hilbert space whose inner product will be denoted by $(.,.)$. A bounded linear operator $A:\mathcal{H}\rightarrow\mathcal{H}$ will be called of *trace-class*, if for arbitrary orthonormal bases (ONB's) $\{e_j\}$, $\{f_j\}$ of $\mathcal{H}$ we have $\sum_j|(Ae_j,f_j)|<\infty$. The set τc of all trace-class operators in $\mathcal{H}$ forms a Banach space with the property $BAC \in \tau c$ for $A \in \tau c$ and arbitrary linear bounded operators B,C in $\mathcal{H}$. For any $A \in \tau c$ the sum

$$\text{trace } A := \sum_j (Ae_j,e_j)$$

is independent of the specific ONB $\{e_j\}$ and defines a linear continuous functional on τc which has the property trace AB = trace BA provided that A is in τc and B is linear and bounded. For a square matrix A trace A is the sum of its diagonal elements.

Fourier-Operators: Throughout this paper T denotes a fixed positive real number, I the interval $(-\frac{T}{2},\frac{T}{2}]$, and $\tilde{I}$ the interval $(-\frac{1}{2T},\frac{1}{2T}]$. $L^2 = L^2(I)$ will be the Hilbert space of (equivalence classes of) Lebesgue square integrable complex valued functions on I with the usual inner product

$$(f,g) = \int_I f(t)\overline{g(t)}dt \quad ,$$

the bar indicating complex conjugation. l^2 denotes the Hilbert space of square summable sequences $z = (z_n)_{n\varepsilon Z}$ of complex numbers, where Z stands for the set of the integers. Interpreting these sequences as infinite dimensional column vectors we may characterize linear operators in l^2 by matrices (a_{ik}) where the indices i for rows and k for columns run through the integers.

For any $\lambda \varepsilon \tilde{I}$ the *frequency-translated Fourier operator* $\mathcal{F}(\lambda)\ L^2 \rightarrow l^2$ is defined by

$$[\mathcal{F}(\lambda)f]_n = \frac{1}{T}\int_I f(t)\ \exp(j2\pi(\lambda+\frac{n}{T})t)\ dt \quad , \tag{2.1}$$

$[\mathcal{F}(\lambda)f]_n$ denoting the n'th component of $\mathcal{F}(\lambda)f \ \varepsilon\ l^2$. $\mathcal{F}(\lambda)$ satisfies the equations

$$\mathcal{F}(\lambda)\,\mathcal{F}(\lambda)^* = \frac{1}{T}\mathbb{1}_{l^2} \text{ and } \mathcal{F}(\lambda)^*\,\mathcal{F}(\lambda) = \frac{1}{T}\mathbb{1}_{L^2} \ , \tag{2.2}$$

where $\mathbb{1}$. is the identity operator in the respective space and the star indicates the adjoint operator.

Diagonally periodic functions: We shall call a complex-valued function h of two real arguments s,t T-*diagonally periodic* (T-DP) if $h(s+T,t+T) = h(s,t)$ for all s,t. If h is locally square integrable then there exist almost uniquely defined functions h_k, $k \ \varepsilon\ Z$, such that h has the mean square convergent series representation

$$h(s,t) = \sum_k h_k(s-t)\cdot\exp(-j2\pi k\frac{t}{T}) \tag{2.3}$$

(see (3) and the references therein), which we shall call the F-series representation (FSR) of h.

3. THE SPECTRAL DENSITY OF A CYCLOSTATIONARY PROCESS

Let $T, I, \tilde{I}$, and L^2 be defined as before. A real or complex valued stochastic process $x(t)$, t real, is called T-*cyclostationary* (T-CS) if its mean function $m(t)$ is periodic and its covariance $C_x(s,t)$ is diagonally periodic both with period T. (Cf. (3), (2)). Since in this paper we are interested in second order properties of the process only we may w.r.o.g. assume that $m(t) \equiv 0$. x is said to be of *finite average power* if it is mean square Lebesgue measurable and if

$$\mathrm{avp}(x) := \frac{1}{T}\int_I \mathcal{E}|x(t)|^2 dt < \infty \qquad (3.1)$$

For such a T-CS we define a sequence $\mathcal{R}_x = (R_n)_{n \in Z}$ of bounded linear operators $R_n : L^2 \to L^2$ by

$$(R_n f, g) = \int_I \int_I f(s) C_x(s+nT, t) \overline{g(t)} ds dt \qquad (3.2)$$

for arbitrary $f, g \in L^2$. These operators have the following easily verified properties:

i) The sequence $\mathcal{R}_x$ is positive definite in the sense that for arbitrary integers $n_1, n_2, \ldots, n_k$ and functions $f_1, \ldots, f_k$ in L^2

$$\sum_i \sum_j (R_{n_i - n_j} f_i, f_j) \geq 0$$

holds, and

ii) the R_n are of trace-class with trace $R_0 = T\,\mathrm{avp}(x)$. These properties suggest to call $\mathcal{R}_x$ the *(operator) autocorrelation* of the process x.

In generalization of Bochner's theorem it can be shown that $\mathcal{R}_x$ admits a Fourier integral representation with respect to an operator valued spectral measure. See (6), (4). Here we will omit this step and introduce the notion of a spectral density by

Definition 3.1 A function $\lambda \to S_x(\lambda)$ defined on $\tilde{I}$ is called the *(operator) spectral density* of the T-CS x if for almost all $\lambda \in \tilde{I}$ the linear operator $S_x(\lambda)$ in L^2 is nonnegative definite and of trace-class and if for all integers n the equality

$$R_n = T \int_{\tilde{I}} \exp(j2\pi n\lambda T) \cdot S_x(\lambda) d\lambda \qquad (3.3)$$

holds, where the integral converges in the Banach space of trace-class operators.

We note that S_x is uniquely defined up to λ-sets of measure zero and completely specifies the second order properties of the process x.

Matrix representation of the spectral density. S_x may be characterized by means of the covariance as follows: Given the FSR

$$C_x(s,t) = \sum_k c_k(s-t)\exp(-j2\pi k\tfrac{t}{T}) \tag{3.4}$$

of the covariance C_x of the T-CS x, let

$$\tilde{c}_k(f) := \int_{-\infty}^{\infty} \exp(-j2\pi ft)\cdot c_k(t)dt$$

be the Fourier transforms of the functions c_k. We then obtain

Theorem 3.2 If the $\tilde{c}_k$ are absolutely integrable the spectral density S_x exists and is given by

$$S_x(\lambda) = T\mathcal{F}(\lambda)^*\tilde{S}_x(\lambda)\,\mathcal{F}(\lambda) \quad , \quad \lambda \in \tilde{I} \quad , \tag{3.5}$$

where $\tilde{S}_x(\lambda)$ is the (infinite dimensional) matrix with components $s_{ik}(\lambda) = \tilde{c}_{i-k}(\lambda + k/T)$, i.e. the n-th diagonal of $\tilde{S}_x(\lambda)$ is composed of the values $\tilde{c}_n(\lambda + k/T)$. Moreover the average power of the process x is given by

$$\text{avp}(x) = \int_{I} \text{trace } S_x(\lambda)d\lambda = \int_{\tilde{I}} \text{trace } \tilde{S}_x(\lambda)d\lambda \quad . \tag{3.6}$$

Proof rests, after inserting (3.4) into (3.2), on certain convergence arguments.

Notation: $\tilde{S}_x$ will be called the *matrix representation* of S_x.

Remark on stationary processes: If x is wide sense stationary then the c_k in (3.4) and consequently the $\tilde{c}_k$ are identically zero for $k \neq 0$, and $\tilde{c}_o$ is the usual spectral density (function) of this process. Thus it follows from the above theorem that wide sense stationary processes are characterized in this context by diagonal matrices $\tilde{S}_x(\lambda)$.

4. SPECTRAL REPRESENTATION OF LINEAR PERIODIC FILTERS

In time domain a linear periodic filter is characterized by an impulse response $h(s,t)$ which is diagonally periodic with some period $T > 0$, cf. (2), (3). This type of filter preserves cyclostationarity. In fact assume that x is T-CS with $\text{avp}(x) < \infty$, then, under suitable integrability conditions, as e.g.

$$\sum_k \left(\int_I \int_I |h(s+kT,t)|^2 ds dt \right)^{1/2} < \infty \quad , \tag{4.1}$$

the stochastic integrals

$$y(s) = \int_{-\infty}^{\infty} h(s,t)x(t)dt$$

are mean square convergent and y is T-CS with autocorrelation $\mathcal{R}_y = (\hat{R}_n)$ given by $\hat{R}_n = \sum_k \sum_l H_k^* R_{n-l+k} H_l$, where $\mathcal{R}_x = (R_n)$ is the autocorrelation of x and the H_k are the integral transforms in L^2 defined through

$$H_k f(t) = \int_I f(s)h(s+kT,t)ds \quad .$$

For each $\lambda \in \tilde{I}$ define the linear operator $H(\lambda)$ in L^2 by

$$H(\lambda) = \sum_k \exp(-j2\pi k\lambda T) \cdot H_k \quad , \tag{4.2}$$

which is bounded if (4.1) holds. If x possesses a spectral density S_x then it is seen from (3.3) that y has spectral density S_y given by

$$S_y(\lambda) = H(\lambda)^* S_x(\lambda) H(\lambda) \tag{4.3}$$

Thus $H(\lambda)$ completely characterizes the linear periodic filter as far as second order properties are concerned. Formula (4.3) suggests to call the function $\lambda \to H(\lambda)$ the *(operator) transfer function* of the filter. Analogous to theorem 3.2 we obtain a characterization of this transfer function via the FSR

$$h(s,t) = \sum_k h_k(s-t) \cdot \exp(-j2\pi k\tfrac{t}{T}) \tag{4.4}$$

of the T-DP impulse response h:

Theorem 4.1 If the functions h_k are absolutely integrable with Fourier transforms $\tilde{h}_k$ then the transfer function H is given by

$$H(\lambda) = T\,\mathcal{F}(\lambda)^*\tilde{H}(\lambda)\,\mathcal{F}(\lambda) \quad , \quad \lambda \in \tilde{I} , \tag{4.5}$$

where $\tilde{H}(\lambda)$ is the matrix with components $h_{ik}(\lambda) = \tilde{h}_{i-k}(\lambda + \frac{k}{T})$.

Proof: This may be seen by evaluating the components of $\mathcal{F}(\lambda)H(\lambda)f$, making use of (4.2) and (4.4).

We shall call $\tilde{H}$ the *matrix representation* of H.

Remark on time invariant filters: Similar to the case of wide sense stationary processes it is seen that time invariant filters in our context are characterized by diagonal matrix representations.

Remark on filters preserving cyclostationarity: The notion of a transfer function may be extended to the set of all linear-operator-valued functions $\lambda \to H(\lambda)$ such that (4.3) defines the spectral density of a T-cyclostationary process. This set may be shown to be a Hilbert space $\mathcal{L}_x^2$ characterizing the class of filters operating on x and preserving cyclostationarity. For the details we refer to (5). For the purposes of this paper only bounded $H(\lambda)$ are needed.

5. THE JOINT OPTIMIZATION PROBLEM

We now return to the transmission system described by Fig. 1. Let x be a T-cyclostationary process of finite average power with known spectral density S_x and f,g the impulse responses of linear periodic filters with period T whose transfer functions we denote by F,G respectively.

The linear distortion introduced by the channel is viewed as the result of a linear time invariant filter with impulse response h(t). We assume that the (usual) transfer function $\tilde{h}_o(f)$ of this filter is known and bounded. Let H denote the corresponding operator transfer function.

Assume further n(t) to be a stationary noise process which is uncorrelated to x and has power spectrum $\upsilon(f)$ satisfying $\upsilon(f) \geqq \sigma > 0$ for some positive constant σ. We define the (operator) spectral density S_N of n(t) to be the (unbounded) operator-valued function $\lambda \to S_N(\lambda)$ given by

$$S_N(\lambda) = T\,\mathcal{F}(\lambda)^* \tilde{S}_N(\lambda)\,\mathcal{F}(\lambda) \quad , \quad \lambda \in \tilde{I} \,,$$

where $\tilde{S}_N(\lambda)$ is the diagonal matrix with components $\upsilon(\lambda + \frac{k}{T})$.

In compliance with ERICSON (2) we use the time averaged mean square error as *performance criterion*. Thus

$$\text{avp}(\tilde{x}-x) = \frac{1}{T}\int_I \mathcal{E}\,|\tilde{x}(t) - x(t)|^2 dt$$

has to be minimized subject to a *power constraint* for the transmitted signal:

$$\text{avp}(y) = \frac{1}{T}\int_I \mathcal{E}\,|y(t)|^2 dt \le P \quad .$$

From (4.3) and (3.6) it follows that the power constraint may be written as

$$C(F) := \int_{\tilde{I}} \text{trace } F(\lambda)^* S_x(\lambda) F(\lambda) d\lambda \le P \,. \tag{5.1}$$

As to the spectral representation of the performance criterion, we have

Theorem 5.1 In frequency domain the time averaged mean square error is given by

$$\begin{aligned} \text{avp}(\tilde{x}-x) =: D^2(F,G) = \int_{\tilde{I}} & \text{trace}\{G(\lambda)^*[H(\lambda)^* F(\lambda)^* S_x(\lambda)\cdot \\ & \cdot F(\lambda)H(\lambda) + S_N(\lambda)]G(\lambda) - G(\lambda)^*[S_x(\lambda)F(\lambda)H(\lambda)]^* - \\ & - [S_x(\lambda)F(\lambda)H(\lambda)]G(\lambda) + S_x(\lambda)\}d\lambda \end{aligned} \tag{5.2}$$

Proof. For a T-CS x define the sequence $X = (X_n)_{n\varepsilon Z}$ of linear operators from L^2 into the space of second order random variables by the stochastic integrals $X_n f = \int_I f(t)x(t+nT)dt$, $f \varepsilon L^2$. Then x is a stationary sequence of operators of Schmidt-class, cf. (7), (4). Thus the general theory of filtering of stationary operator sequences as treated in (4), (5) is applicable, and (5.2) turns out to be a special case of the transmission error functional (11) of (5).

It goes without saying that for the case of a wide sense stationary process x the error $D^2(F,G)$ coincides with ERICSON's total distortion D.

6. RECEIVER OPTIMIZATION

In a first step we minimize the functional $D^2(F,G)$ with respect to the receiver transfer function G while F is being held fixed. This is straightforward, since the integrand in (5.2) is a quadratic functional of the argument $G(\lambda)$, and the result consists in the Wiener-type transfer function

$$G_o(\lambda) = [H(\lambda)^* F(\lambda)^* S_x(\lambda) F(\lambda) H(\lambda) + S_N(\lambda)]^{-1} \cdot \cdot H(\lambda)^* F(\lambda)^* S_x(\lambda) \quad . \tag{6.1}$$

Existence of the inverse of the bracketed operator is assured, since the latter is bounded away from zero.
Inserting (6.1) into (5.2) we obtain after some transformations $D^2(F,G) = \int_{\tilde{I}} \text{trace}\, S_x(\lambda) d\lambda - E^2(F)$, where

$$E^2(F) = \int_{\tilde{I}} \text{trace}\{F(\lambda)^* S_x^2(\lambda) F(\lambda) H(\lambda) [H(\lambda)^* F(\lambda)^* S_x(\lambda) \cdot \cdot F(\lambda) H(\lambda) + S_N(\lambda)]^{-1} H(\lambda)^*\} d\lambda \tag{6.2}$$

Thus to solve the joint optimization problem the functional $E^2(F)$ measuring *transmission efficiency* has to be maximized subject to the constraint (6.1).

7. EFFICIENCY MAXIMIZATION

For technical reasons the following assumption concerning the transfer function $\tilde{h}_o$ has to be made which will be removed later on in this section:

Assumption 7.1 For all frequencies f there is $0 < |\tilde{h}_o(f)| \leq \eta < \infty$.

Under this assumption the infinite diagonal matrix $\tilde{S}_c(\lambda)$ with diagonal elements

$$c_k(\lambda) = \upsilon(\lambda + \tfrac{k}{T})[\,|h_o(\lambda + \tfrac{k}{T})|^2\,]^{-1}$$

is well defined and for all integers k we have $0 < \sigma/\eta^2 \leq c_k(\lambda) < \infty$. Let now $\tilde{S}_x$ and $\tilde{F}$ designate the matrix representations (3.5) and (4.5) of S_x and F respectively. By use of (2.2) one obtains

$$\tilde{E}^2(\tilde{F}) = \int_{\tilde{I}} \text{trace}\{\tilde{F}(\lambda)^* \tilde{S}_x^2(\lambda) \tilde{F}(\lambda) [\tilde{F}(\lambda)^* \tilde{S}_x(\lambda) \tilde{F}(\lambda) + \tilde{S}_c(\lambda)]^{-1}\} d\lambda \tag{7.1}$$

as an expression for the transmission efficiency, and the power constraint reads

$$\tilde{C}(\tilde{F}) = \int_{\tilde{I}} \text{trace}\{\tilde{F}(\lambda)^* \tilde{S}_x(\lambda)\tilde{F}(\lambda)\}d\lambda \leq P. \tag{7.2}$$

Maximization of a functional of type (7.1) subject to (7.2) has been carried out in Sec. 6 of (5) by means of the Kuhn-Tucker theorem applied to the Fréchet-differentiable integrands. Therefore we may restrict ourselves here to citing the assumptions required and the results obtained.

Since for any λ the matrices $\tilde{S}_x(\lambda)$ are nonnegative definite and of trace-class, they possess a countable number of nonnegative eigenvalues $\sigma_j(\lambda)$, $j\varepsilon Z$, satisfying $\sum_j \sigma_j(\lambda) = \text{trace}\, \tilde{S}_x(\lambda) < \infty$. These properties imply the existence of an ordering $j_1^x(\lambda), j_2(\lambda)$, ... of the integers such that

$$\sigma_{j_1(\lambda)}(\lambda) \geq \sigma_{j_2(\lambda)}(\lambda) \geq \ldots \tag{7.3}$$

i.e., for any λ, the eigenvalues may be arranged in decreasing order. Let further denote $D_x(\lambda)$ the diagonal matrix composed of these eigenvalues and $U(\lambda)$ a unitary matrix of related 1^2-eigenvectors.
As concerns the matrices $\tilde{S}_c(\lambda)$, the following assumption (tacitly used in (2)) is to be made:

Assumption 7.2 For any λ there exists an ordering $k_1(\lambda)$, $k_2(\lambda)$, ... of the integers such that

$$c_{k_1(\lambda)}(\lambda) \leq c_{k_2(\lambda)}(\lambda) \leq \ldots \tag{7.4}$$

For real numbers α set $(\alpha)_+ := \max(\alpha, 0)$ and $(\alpha)_+^{1/2} := \sqrt{(\alpha)_+}$. One then obtains

Theorem 7.3 Under the above assumptions and under constraint (7.2) the functional (7.1) has the following maxima $\tilde{F}$:

$$\tilde{F}(\lambda) = U(\lambda)^* V_\lambda \tilde{F}_o(\lambda) W_\lambda \quad , \quad \lambda \,\varepsilon\, \tilde{I} \;.$$

Here V_λ and W_λ are unitary matrices commuting with $D_x(\lambda)$ and $\tilde{S}_c(\lambda)$ resp., and $\tilde{F}_o(\lambda)$ is a matrix with components $\tilde{f}_{ik}^x(\lambda)$ given by

$$\tilde{f}_{j_\nu(\lambda)k_\nu(\lambda)}(\lambda) = \left(\varkappa \sqrt{\frac{c_{k_\nu(\lambda)}(\lambda)}{\sigma_{j_\nu(\lambda)}(\lambda)}} - \frac{c_{k_\nu(\lambda)}(\lambda)}{\sigma_{j_\nu(\lambda)}(\lambda)} \right)_+^{1/2},$$

for $\nu = 1, 2, \ldots$, and $\tilde{f}_{jk}(\lambda) = 0$ otherwise.

The constant $\varkappa > 0$ has to be chosen such that

$$\tilde{C}(\tilde{F}) = \int_{\tilde{I}} \sum_{\nu} (\varkappa \sqrt{c_{k_\nu(\lambda)}(\lambda)\sigma_{j_\nu(\lambda)}(\lambda)} - c_{k_\nu(\lambda)}(\lambda)\sigma_{j_\nu(\lambda)}(\lambda))_+ d\lambda = P .$$

Remark on assumption 7.1: By setting the term $(\varkappa\sqrt{\alpha} - \alpha)_+ = 0$ for $\alpha = +\infty$ it is easily verified that the assumption may be dispensed with.

8. CONCLUSIONS AND FINAL REMARKS

Given the eigenvalues and eigenvectors of the matrix representation $\tilde{S}_x$ of the source signal the matrix representation $\tilde{F}$ of an optimal transmitter is obtained from theorem 7.3 and the corresponding optimal receiver after inserting $F(\lambda) = T\mathcal{F}(\lambda)^*\tilde{F}(\lambda)\mathcal{F}(\lambda)$ into (6.1). From the FSR (4.4) it is seen that the structure of optimal transmitters and receivers is that of a "bank" of linear time invariant filters (3) the transfer functions of which are given by the diagonals of the respective matrix representations.

A comment on the unitary matrices V_λ and W_λ is in place. Free choice of these matrices within the frame set by theorem 7.3 means, in the first place, that rows and columns of $\tilde{F}_o(\lambda)$ corresponding to eigenvalues $\sigma_j(\lambda)$ resp. diagonal elements $c_k(\lambda)$ of equal size may be transformed by unitary transformations without affecting optimality. Secondly, it means that all rows and columns may be multiplied by arbitrary complex numbers of modulus 1 expressing the fact that, due to the lack of causality restrictions, the phases of the optimal filters remain unspecified. Phase distortions introduced by the optimal transmitter are compensated by the corresponding receiver.

Determination of eigenvalues and eigenvectors of $\tilde{S}_x$ depends on the specific type of signal under consideration. The simplest case is that of a wide sense stationary source signal with spectral density function s(f):
As pointed out after theorem 3.2, the matrix $\tilde{S}_x(\lambda)$ then is a diagonal matrix, $U(\lambda)$ thus being the identity and the eigenvalues equal to the diagonal elements: $\sigma_j(\lambda) = s(\lambda + j/T)$ for all integers j.

REFERENCES

(1) Berger, T., Tufts, D.W.: 1967, IEEE Trans.Inf.Th. 13, pp. 196-208
(2) Ericson, T.H.E.: 1981, IEEE Trans.Inf.Th. 27, pp. 322-327
(3) Gardner, W.A., Franks, L.E.: 1975, IEEE Trans.Inf.Th. 21, pp. 4-14
(4) Graef, F.K.: *An Optimization Problem for Stationary Operator Sequences*, Ph.D. Dissertation, Erlangen 1976 (in German)
(5) Graef, F.K.: 1978, pp. 63-75, in *Measure Theory, Applications to Stochastic Analysis*, G. Kallianpur and D. Kölzow ed., Springer Lecture Notes in Mathematics 695
(6) Payen, R.: 1967, Ann.Inst. Poincaré III, pp. 323-396
(7) Schatten, R.: *Norm Ideals of Completely Continuous Operators*, Berlin 1970
(8) Sinaj, Ya.G.: 1959, Probl. Pered. Inf. 2, pp. 40-48 (in Russian)

FUNDAMENTAL PROPERTIES AND PERFORMANCE OF NONLINEAR ESTIMATORS FOR BEARINGS-ONLY TARGET TRACKING

S.C. NARDONE, A.G. LINDGREN[1], K.F. GONG

U.S. NAVAL UNDERWATER SYSTEMS CENTER[2],

NEWPORT, R.I. USA 02840

This paper considers the problem of estimating the position and velocity of an unaccelerated target from noise corrupted bearing measurements obtained by a single moving observation platform. The process is nonlinear and exhibits unusual observability characteristics that are geometry dependent. Although the general passive motion analysis problem is treated, attention is focused on the large-to-baseline geometries employing a symmetric observer maneuver strategy. For this limited but important class of geometries analytical results are derived for the Cramer-Rao bound (CRB). Using a maximum likelihood estimate to realize a solution, the results of a Monte-Carlo simulation are presented for the long range situation. Although the MLE nearly achieves the predicted CRB, it displays a gradual departure as the noise level and range-to-baseline ratio increases.

1.0 INTRODUCTION

Bearings-only target motion analysis (TMA), often stated in the context of the ocean environment, is a nonlinear state estimation problem. The relative target state, comprised of the Cartesian components of range and velocity, is $X = [r_x, r_y, v_x, v_y]$ Here the target velocity is constant while the observation platform or "ownship" is free to maneuver. The discrete-time state equation is

$$X(t_k) = \Phi(t_k, t_{k-1})X(t_{k-1}) + U(t_k), \tag{1}$$

where $\Phi(t_k, t_{k-1}) = \left[\begin{array}{c|c} I & (t_k - t_{k-1})\, I \\ \hline 0 & I \end{array}\right]$; $I = \begin{bmatrix} 1 & 0 \\ 0 & 1 \end{bmatrix}$, $U(t_k)$ accounts

R. S. Bucy and J. M. F. Moura (eds.), Nonlinear Stochastic Problems, 593–600.

for observer accelerations and t_k = time at the kth sample. The available measurements are noise corrupted bearings viewed from the observer's platform and given by

$$\beta_k = \theta_k + \nu_k , \quad (2)$$

where ν is zero-mean independent Gaussian noise with variance σ_ν^2, and θ is the noise-free bearing,

$$\theta_k = \tan^{-1}[r_x(t_k)/r_y(t_k)]. \quad (3)$$

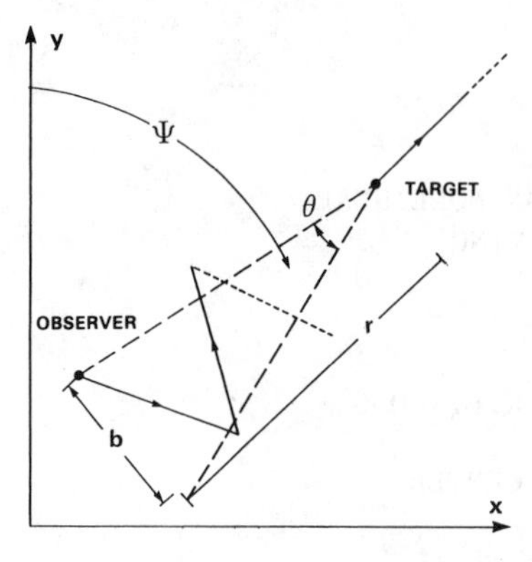

Fig. 1. Typical Geometry

The four-dimensional state equation (1) together with the nonlinear measurement of eqs.(2) and (3) define bearings-only TMA. The process is inherently nonlinear and some of the states are not observable through the measurements prior to an ownship acceleration [1-3]. This property is illustrated by noting that a family of constant velocity target tracks generate the same history when the observation platform velocity is constant. A typical encounter is shown in figure 1 where the observer's trajectory is composed of constant velocity segments, termed "legs". As a consequence of ownship's piecewise constant velocity motion, the TMA process is not completely observable for any single leg. Although two distinct legs permit a unique solution, the degree of convergence attained on a given leg is restriced [4] and several legs are typically required to achieve an acceptable estimate. Thus, in addition to measurement noise, the performance of any bearings-only TMA estimation technique is affected by the observer geometry. This required maneuver distinguishes single platform TMA from other localization techniques such as triangulation ranging.

2.0 MAXIMUM LIKELIHOOD ESTIMATE (MLE)

Given the history of measured bearings $\underline{\beta} = [\beta_1,\ldots,\beta_k]$ the likelihood function is

$$P_{\beta|x} = [(2\pi)^k \det W]^{-\frac{1}{2}} \exp\{-\tfrac{1}{2}[\underline{\beta}-\underline{\hat{\beta}}]^T W^{-1}[\underline{\beta}-\underline{\hat{\beta}}]\}, \quad (4)$$

where $\underline{\hat{\beta}} = \underline{\beta}(\hat{X})$ is the vector of estimated bearings and $W = \mathrm{diag}[\sigma_i^2]$. The maximum likelihood estimate (MLE) is the solution to the likelihood equation

$$\partial \ln P_{\beta|x}/\partial x = \underline{0}. \quad (5)$$

For the Gaussian distribution in eq.(4), the solution to eq.(5) is equivalent to the weighted least squares estimate. Performing the indicated operation in eq.(5) results in the gradient equation

$$A^T(\hat{\underline{\beta}},t_m)\hat{R}^{-1}W^{-1}[\underline{\beta}-\hat{\underline{\beta}}] = \underline{0}, \tag{6a}$$

where,

$$A(\hat{\beta},t_m) = \begin{bmatrix} \vdots & \vdots & \vdots & \vdots \\ \cos\hat{\beta}_i, & -\sin\hat{\beta}_i, & (i-m)\cos\hat{\beta}_i, & -(i-m)\sin\hat{\beta}_i \\ \vdots & \vdots & \vdots & \vdots \end{bmatrix}, \tag{6b}$$

$\hat{R} = \text{diag}[\hat{r}_i]$, $r = [r_x^2+r_y^2]^{1/2}$ and $m = t_m$ is the reference time of the state estimate. The Gauss-Newton iterative solution to this non-linear system of equations yields the MLE

$$\hat{X}^{\ell+1} = \hat{X}^{\ell}-s_{\ell}[\hat{A}^T\hat{R}^{-1}W^{-1}\hat{R}^{-1}\hat{A}]^{-1}\hat{A}^T\hat{R}^{-1}W^{-1}[\underline{\beta}-\hat{\underline{\beta}}], \tag{7}$$

where $\hat{A}$, $\hat{R}$, and $\hat{\underline{\beta}}$ are evaluated at X^{ℓ} and the step size, s_{ℓ}, is selected at each iteration to insure convergence. For this estimate, the information matrix is

$$\hat{\Omega}_{MLE} = [\hat{A}^T\hat{R}^{-1}W^{-1}\hat{R}^{-1}\hat{A}], \tag{8}$$

and the covariance estimate is

$$\hat{P}_{MLE} = \hat{\Omega}^{-1}_{MLE}. \tag{9}$$

If the true state were available to obtain the correct density in eq.(4), then the error resulting from the maximization of eq.(4) would be approximately

$$\tilde{X} = [A^T(\underline{\theta})R^{-1}W^{-1}R^{-1}A(\underline{\theta})]^{-1}A^T(\underline{\theta})R^{-1}W^{-1}\underline{V}, \tag{10}$$

where $\underline{V} = [\ldots,\nu_i,\ldots]$ is the measurement noise sequence and $\underline{\theta}$ is the vector of error-free bearing measurements. The expected value of $\tilde{X}$ is zero. The information matrix in this "ideal case" is

$$\Omega_{MLE} = [A^TR^{-1}W^{-1}R^{-1}A], \tag{11}$$

and the covariance matrix is

$$P_{MLE} = \Omega^{-1}_{MLE}. \tag{12}$$

The conditions eq.(11) and eq.(12) constitute the Cramer-Rao bound CRB [5,6] and define the minimum variance in the state estimation error that can be obtained by any unbiased estimator. The quadratic form involving eq.(11) is the hyperellipsoid of uncertainty in the state estimate and is defined by its eigenvectors and eigenvalues. Thus identification of the factors affecting the eigenvalues and eigenvectors of the information matrix becomes critical to an understanding of the bearings-only TMA process.

3.0 PROPERTIES OF BEARINGS-ONLY TMA

For large range-to-baseline geometries, the factors impacting the eigenvalues and eigenvectors of the ideal information matrix, and therefore the CRB, are examined. A more detailed development is presented in [7].

3.1 General Properties

The normalized information matrix, applicable to long range conditions (i.e., $R \approx rI$ and $W \approx \sigma^2 I$) is defined as $\Omega = r^2\sigma^2\Omega_{MLE}$, or

$$[A^T A] = \sum_{i=1}^{k} \left[\begin{array}{c|c} \Omega_i & (i-m)\Omega_i \\ \hline (i-m)\Omega_i & (i-m)^2\Omega_i \end{array}\right] = \left[\begin{array}{c|c} \Omega_{11} & \Omega_{12} \\ \hline \Omega_{12} & \Omega_{22} \end{array}\right], \tag{13a}$$

where

$$\Omega_i = \begin{bmatrix} \cos^2\theta_i & -\frac{1}{2}\sin 2\theta_i \\ -\frac{1}{2}\sin 2\theta_i & \sin^2\theta_i \end{bmatrix}. \tag{13b}$$

Note that:

i) The matrix is singular prior to an ownship acceleration. Following an appropriate maneuver Ω is positive definite and the process is observable.

ii) The trace $[\Omega(m)] = k_o + k_1$; where $k_n = \Sigma(i-m)^n$; and is therefore independent of any particular bearing history. This applies to the partitioned matrices as well.

iii) The determinant, which is inversely proportional to the volume of the uncertainty ellipse, is invariant when the state estimate is propagated to any other time. Thus, since the trace is variant, the apparent effect of time translations is a rotation and reshaping of the hyperellipsoid while perserving the total volume.

iv) Ω_{11} is the information matrix for the equivalent two-state localization problem that would result if the target velocity was known. It will be shown that this is a lower bound on the TMA solution.

3.2 Special Case Eigenvalues

Consider the effects of a specific maneuver strategy where ownship motion is symmetric as shown in figure 1. This situation is often approximated in long range tracking applications. By rotating the coordinate system through the average angle θ^*, the information matrix becomes a function of the angular difference $\underline{\Delta\theta} = [\ldots, \theta_i - \theta^*, \ldots]^T$. As a result of the symmetry in $\underline{\Delta\theta}$, Ω_{11} is diagonalized. Further, if the state estimate is referenced to the mid-interval (m=k/2), the form of Ω is simplified. When the number of observer legs (L) is even, Ω_{12} vanishes, and Ω becomes a block diagonal matrix. Here the two smallest eigenvalues, λ_1 and λ_2, are those of Ω_{11} and are equivalent to the localization problem. Assuming the small angle approximation applies, these eigenvalues are readily found and are listed in Table I.

$\lambda_1 = k\theta^2/12$	$\lambda_3 = (k^2/12)\lambda_1 p(L)$
$\lambda_2 = k-\lambda_1$	$\lambda_4 = (k^3/12)-\lambda_3$
$\theta \approx$ baseline/range $p(L) = 1+(4/5)L^{-2}-12L^{-4}$	

Table I: Special Case Eigenvalues

The major axis of the uncertainty ellipse aligns with the average bearing. The remaining two eigenvalues, λ_3 and λ_4, in Table I, are associated with the velocity components. Note that $p(L) \approx .45$ for L=2 and is approximately 1 for L>2. When the state estimate is obtained for current time (m=k) the two position eigenvalues are obtained from the approximate relationship

$$\{\lambda_1, \lambda_2\} \approx \text{e'val}\{\Omega_{11} - \Omega_{12}(k)\Omega_{22}^{-1}(k)\Omega_{12}(k)\}. \tag{14}$$

For L>4, λ_1 and λ_2 at current time are approximately ¼ of their value at the mid-interval. The effect of translating the state estimate from k/2 to k is to increase the variance of the position estimate by a factor of four. A similar result holds for an odd number of legs.

3.3 General Eigenvalue Bounds

For less restrictive geometries the eigenvalues can be bounded in a manner that provides insight into performance. Using Weyl's theorem and the inclusion principle, it has been shown [7] that the two largest eigenvalues, λ_3 and λ_4, are within a distance k of the eigenvalues of Ω_{22}. The two smallest eigenvalues, corresponding to the position estimates, are always less than or equal to the eigenvalues of Ω_{11} from the two-dimensional localization process. Finally the maximum value for the smallest eigenvalue, λ_1, attainable for any scenario, is k/2 at the mid-interval (m=k/2) and k/8 at current time (m=k).

3.4 Ideal Filter Performance

Filter performance, as defined by the statistics of the state estimation error was given previously by eq.(10-12). The ideal estimate is unbiased; and since

$$E\{\|X\|^2\} = \sigma^2 r^2 \Sigma \lambda_i^{-1}, \tag{15}$$

the variance depends largely on the growth of the smallest eigenvalue which cannot exceed the bounds given previously. For the special long range conditions employing the symmetric observer strategy, the variances at m=k/2 are

$$\sigma_\perp^2 = r^2\sigma_\nu^2/k \quad \text{and} \quad \sigma_\parallel^2 = r^2\sigma_\nu^2/(k\theta^2/12). \tag{16}$$

Here the coordinates are parallel ($\parallel$) and perpendicular ($\perp$) to θ^*. The range to cross-range resolution (i.e., depth to lateral resolution) is

$$\sigma_\parallel/\sigma_\perp = \sqrt{12}/\theta = (2\sqrt{3})\text{range/baseline}. \tag{17}$$

For large range-to-baseline ratios, the error in range dominates the total error and the variance of the position components decreases asymptotically with the number of measurements.

4.0 EXPERIMENTAL RESULTS

Monte-Carlo experiments were conducted for the symmetric long range geometry depicted in figure 1. The normalized range standard deviation (STD) vs effective noise ($\sigma_{eff} = r\sigma_\nu/b$) is shown in figure 2 and is in agreement with results predicted by eq.(16). In this experiment, range-to-baselines ratios of 10 to 40 were used while ownship executed two, three, and four leg geometries with 200 measurements per leg. The results for each σ_{eff} value were obtained by averaging the filter solutions for 120 trials. The dashed curve applies to the two, three, and four leg geometries at mid-interval while the solid line shows the results for 3 and 4 leg geometries at final time. (The two leg geometry was omitted because p(L) is not near its steady state value). Scatter plots of the target position solutions, resulting from 500 Monte-Carlo trials are shown in figures 3-5.

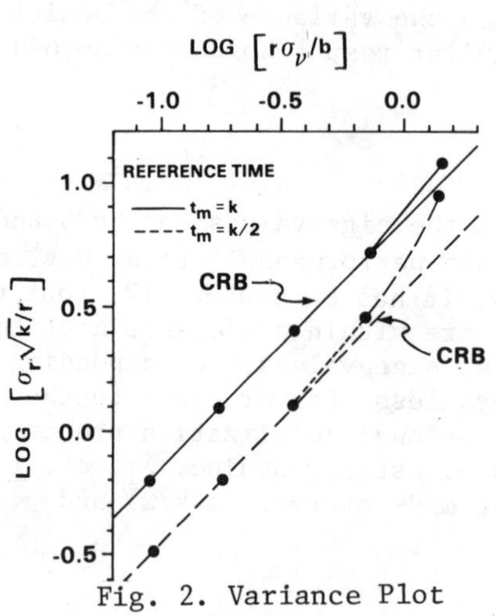

Fig. 2. Variance Plot

Using figure 3, where r/b = 40 and $\sigma_\nu = \sigma_o$, as a reference, σ_{eff} was varied by a factor of 2 for both figures 4 and 5. In figure 4, r/b = 20 and $\sigma_\nu = 4\sigma_o$, while for figure 5 r/b = 40 and $\sigma_\nu = 2\sigma_o$. The observed changes in the scatter plots with r/b and σ_ν are reasonably predicted by eq.(16). However, in figure 5 the

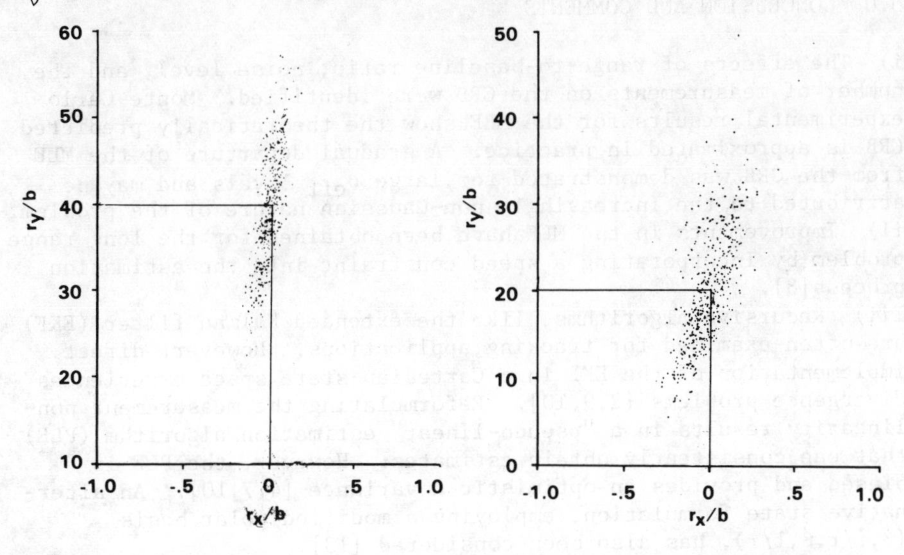

Fig. 3. Scatter Plot ($\sigma_\nu=\sigma_0$)   Fig. 4. Scatter Plot ($\sigma_\nu=4\sigma_0$)

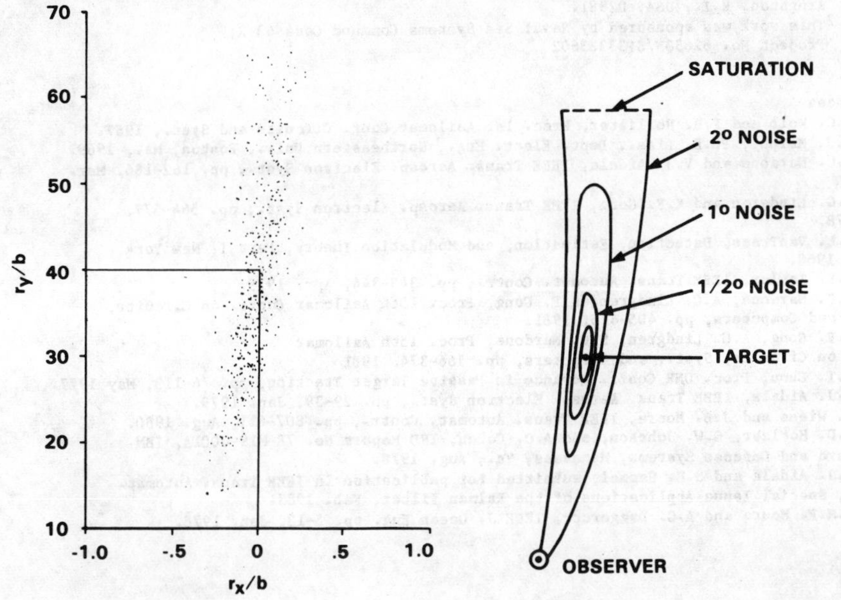

Fig. 5. Scatter Plot ($\sigma_\nu=2\sigma_0$)   Fig. 6. Approx. Distribution vs. σ

non-Gaussian nature is becoming more apparent. In fact, as σ_{eff} is increased the distribution of solutions transitions to the non-Gaussian shape depicted in figure 6.

5.0 CONCLUSION AND COMMENTS

i) The effects of range-to-baseline ratio, noise level, and the number of measurements on the CRB were identified. Monte-Carlo experimental results for the MLE show the theoretically predicted CRB is approximated in practice. A gradual departure of the MLE from the CRB was demonstrated for large σ_{eff} levels and may be attributed to the increasingly non-Gaussian nature of the problem.
ii) Improvements in the MLE have been obtained for the long range problem by incorporating a speed constraint into the estimation process [8].
iii) Recursive algorithms, like the extended Kalman filter (EKF) are often examined for tracking applications. However, direct implementation of the EKF in a Cartesian state space experiences divergence problems [2,9,10]. Reformulating the measurement non-linearity results in a "pseudo-linear" estimation algorithm (PLE) that can consistently obtain estimates. However, the PLE is biased and provides an optimistic covariance [4,7,10]. An alternative state formulation, employing a modified polar basis $(\dot{\beta}, \dot{r}/r, \beta, 1/r)$, has also been considered [13].

NOTES: [1]Department of Electrical Engineering, University of Rhode Island, Kingston, R.I., USA, 02881.
[2]This work was sponsored by Naval Sea Systems Command Code 63-R, Project No. 62633N/SF33323602

References

(1) R.C. Kolb and F.H. Hollister, Proc. 1st Asilomar Conf. Circuits and Syst., 1967.
(2) D.J. Murphy, Ph.D. Diss., Dept. Elect. Eng., Northeastern Univ., Boston, Ma., 1969.
(3) S.C. Nardone and V.J. Aidala, IEEE Trans. Aerosp. Electron Syst., pp. 162-166, Mar. 1981.
(4) A.G. Lindgren and K.F. Gong, IEEE Trans. Aerosp. Electron Syst., pp. 564-577, Jul.1978.
(5) H.L. VanTrees, Detection, Estimation, and Modulation Theory, Part I, New York, Wiley, 1968.
(6) J.H. Taylor, IEEE Trans. Automat. Contr., pp. 343-344, Apr. 1979.
(7) S.C. Nardone, A.G. Lindgren, K.F. Gong, Proc. 15th Asilomar Confr. on Circuits, Syst., and Computers, pp. 405-415, 1981.
(8) K.F. Gong, A.G. Lindgren, S.C. Nardone, Proc. 15th Asilomar Confr. on Circuits, Syst., and Computers, pp. 366-374, 1981
(9) S.I. Chou, Proc. ONR Confr. Advance in Passive Target Tracking, pp. 76-113, May 1977.
(10) V.J. Aidala, IEEE Trans. Aerosp. Electron Syst., pp. 29-39, Jan. 1979.
(11) H. Wiess and J.B. Moore, IEEE Trans. Automat. Contr., pp. 807-811, Aug. 1980.
(12) H.D. Hoelzer, G.W. Johnson, and A.O. Cohen, IRD Report No. 78-M19-0001A, IBM Shipboard and Defense Systems, Manassas, Va., Aug. 1978.
(13) V.J. Aidala and S.E. Hammel, submitted for publication in IEEE Trans. Automat. Contr., Special Issue Applications of the Kalman Filter, Feb. 1983.
(14) J.M.F. Moura and A.G. Baggeroer, IEEE J. Ocean Eng. pp. 5-13, Jan. 1978.

LIST OF PARTICIPANTS

INDEX OF SUBJECTS

INDEX OF AUTHORS

LIST OF PARTICIPANTS

ABOUTADJINE Driss Prof., LEESA Faculte des Sciences, B.P.1014 Rabat, MOROCO

ABU EL ATA Salwa Dr., ADERSA GERBIOS, 2 Av. du 1er Mai, 91120 Palaiseau, FRANCE

ALENGRIN G. Prof., Univ.of Nice, Lab.Signaux et Syst., 41 Boulevard Napoleon III, 06041 Nice Cedex FRANCE

ALLINGER Deborah F. Dr., MIT, Dept.of Mathematics 2-032 Cambridge, Mass. 02139, USA

ANDERSEN Kim Prof., Univ.of Aarhus, Dpt.Operations Res., Building 530 NY Munkegade, 8000 Aarhus C., DENMARK

BARAM Yoram Prof., Dept Electr.Commun., School of Eng., Tel Aviv Univ., Tel Aviv 69978, ISRAEL

BARKANA Atalay Prof., The State Acad.of Eng.and Archit. Yunos Emre Cad., Sezer Apt. No. 96/4, Eskisehir, TURKEY

BELO Carlos A.C.Eng., CAPS-Complexo I, Inst.Sup.Tecnico, Av. Rovisco Pais 1, P-1096 Lisbon Codex, PORTUGAL

BERROU Jean L. Mr., SACLANT ASW Res. Centre, Viale San Bartolomeo 400, I-19026 San Bartolomeo, (SP) La Spezia, ITALY

BORG Peter Mr., Inst. Energy Technology, P.O.Box 40, N-2007 Kjeller, NORWAY

BRAMMER Karl Dr.Ing., ESG-FEG, Vogel Weider Platz 9, D-8000 Munich 80, WEST GERMANY

BRATTENG Ove Prof., Institute of Math.and Physical Sciences, Univ. of Tromso, P.O.Box 953, N-9001 Tromso, NORWAY

BRAUMANN Carlos A. Prof., Universidade de Evora, Dept. de Matematica, 7000 Evora, PORTUGAL

BROWN Graeme Nicholas Mr., Optical Communications Group Dept. Electr. Eng. Science, Univ. of Essex, Colchester CO4 3SQ, UNITED KINGDOM

BUCY R. S. Prof., Dept. of Aerospace Eng., OHE-300 F, Univ. of Southern Calif., Los Angeles, CA 90007, USA

BUENO G. Ing.,Informatique Syst.Reseaux,523 Terrasse De l'Agora, 91034 Evry Cedex, FRANCE

BURG J. P. Dr., Time & Space Processing Inc., 3410 Central Expressway,Santa Clara, California 95051, USA

BYRNES Christopher I. Prof., Applied Mathem.,Division of Applied Science, Harvard Univ., Pierce Hall, Cambridge MA 02138, USA

CANABAL Jose R. Dr., Beethovenring 76, 6104 Seeheim - - Jungenheim, WEST GERMANY

CARVALHO Jorge L. M. Prof.,Faculdade Engenharia-Porto, Dept. Eng. Electrotecnica, Rua dos Bragas, 4099 Porto, PORTUGAL

DE LA RUBIA Francisco Javier Mr., Ciencias-UNED, Apdo. Correos 50487, Madrid, SPAIN

DEL GROSSO Gabriela Prof., Istit. Matem.'Guido Castel-Nuovo', Univ. di Roma, Citta Universitaria Pale Aldo Moro, 00100 Roma, ITALY

DIEULESAINT E. Prof.,Univ. Pierre et Marie Curie, Lab. D'Acoustoelectricite, 10 Rue Vauquelin, 75231 Paris Cedex, FRANCE

DI MASI Giovani Battista Prof.,LADSEB-CNR, Corso Stati Uniti 4, I-35100 Padova, ITALY

DUARTE A. Manuel de Oliveira Eng.,University of Essex Dept.of Electrical Eng.Science, Colchester CO4 3SQ, UNITED KINGDOM

ELTOFT Torbjorn Mr.,Inst.of Mathem.and Physical Scien. Univ. of Tromso,P.O.Box 953,N-9001 Tromso, NORWAY

ENCHEV Ognian B.Dr.,Centre of Mathematics, 7000 Rousse P.O.Box 325, BULGARIA

ESTANDIA A. Mr.,Imperial College,Dept. of Electr. Eng. (Control Sect.),Exhibition Road, London SW7 2BT, UNITED KINGDOM

FRAGOSO M. D. Mr., Imperial College, Dept. of Electr. Eng. (Control Sect.), Exhibition Road, London SW7 2BT, UNITED KINGDOM

FUHRER Claus Dipl. Math., Institut fur Dynamic der Flugsysteme, DFVLR,Oberpfaffenhofen,D-8031 Wessling/OBB, WEST GERMANY

GRAEF Friedrich K.Dr.,Univ.Erlangen-Nurnberg,Inst.fur Angewandte Mathematik, Martensstr 3, D-8520 Erlangen, WEST GERMANY

GRIMBLE Mike J. Prof., Dept. of Electrical Eng., Univ. of Strathclyde, 204 George Street, Glasgow G1 1XW, UNITED KINGDOM

HALLINGSTAD Oddvar Dr.Ing.,Norwegian Defence Research Est., NDRE, Postboks 25, N-2007 Kjeller, NORWAY

HAVER Sverre Dr.Ing., Norwegian Hydrodynamic Lab.,P.O. Box 4118 - Valentinlyst, N-7001 Trondheim, NORWAY

HAYES Annie R. Ms. Dr., Office of Naval Res,Statistics & Probability Program,800 N.Quincy St., Arlington, Virginia 22217, USA

HAYKIN Simon Prof., McMaster Univ., Communivations Research Lab., 1280 Main Street West, Hamilton Ontario L8S 4L7, CANADA

HELMES Kurt Dr.,Inst. fur Angewandte Mathematik, Univ. Bonn,Wegelerstrabe 6, 5300 Bonn, GERMANY

HIJAB Omar Prof., Case Western Reserve University, Cleveland, Ohio 44106, USA

INCERTIS Carro F. Dr., Centro IBM Investigation, IBM, P. de la Castellana 4, Madrid 1, SPAIN

ISTEFANOPULOS Yorgo Dr.,Bogazici Univ.,Dept.of Electr Eng., P.K. 2 Bebek,Istanbul, TURKEY

KLIEMANN Wolfgang Dr., Forschungsschwerpunlit 'Dynamische Systeme',Universitat Bremen, Postfach 330440,2800 Bremen 33, WEST GERMANY

KNUDSEN Tor Mr., SACLANT ASW Res. Centre, Viale San Bartolomeo 400,I-19026 San Bartolomeo (SP) La Spezia, ITALY

KOREZLIOGLU H. Prof.,ENST,46 Rue Barrault,75634 Paris Cedex 13, FRANCE

LEITAO José M.N.Eng,CAPS-Complexo I,Instituto Superior Tecnico, Av.Rovisco Pais, P-1096 Lisbon Codex, PORTUGAL

LEITHEAD W. Dr., Dept. of Mathematics and Comput., Paisley College of Technology, High St., Paisley, Renfrewshire, PA1 2BE Scotland, UNITED KINGDOM

LEMOS João M.L.Eng.,CAPS-Complexo I,Instituto Superior Tecnico, Av. Rovisco Pais, P-1096 Lisbon Codex, PORTUGAL

LINDQUIST Anders Prof.,Coll.of Arts and Sciences,Dept. of Mathematics,Univ.of Kentucky,Lexington, Kentucky 40506, USA

LLOYD Lewis John Dr.,SACLANT ASW Res.Centre, Viale San Bartolomeo 400,I-19026 San Bartolomeo(SP), La Spezia, ITALY

LO James T. Prof., Dept. of Mathematics, Univ. of Maryland Baltimore County,Catonsville, MD. 21228, USA

LOURENCO Miguel Eng., R.Prof.Vieira de Almeida 6-2D, 1600 Lisbon, PORTUGAL

LUNDE Even Borten Mr., SACLANT ASW Res. Centre, Viale San Bartolomeo 400,I-19026 San Bartolomeo (SP),La Spezia, ITALY

MACCONE Claudio Dr., Via Martorelli 43, 10155 Torino ITALY

MANFREDI Claudia, Univ. of Florence, ISIS, Via Di S. Marta 3, 50139 Florence, ITALY

MELO António F. P. Prof., Dept. Electronica e Telecom. Univ. de Aveiro, 3800 Aveiro, PORTUGAL

MITTER Sanjoy K. Prof.,MIT, Room 35-233, MIT Cambridge Mass. 02139, USA

MORO Antonio Prof., Univ. of Florence, Istituto Matem. "U.Dini",Viale Morgagni 67/A,50134 Firenze ITALY

MOSCA Edoardo Prof., ISIS Facolta di Engenharia, Univ. di Firenze, Via S. Marta 3, 50139 Firenze ITALY

MOURA Carlos Alberto Mendonça Eng.,Dpt.de Eng. Electr. Fac.Engenharia, Univ.Porto, Rua dos Bragas 4099 Porto Codex, PORTUGAL

MOURA José M.F.Prof., CAPS-Complexo I, Inst. Superior. Tecnico, Av.Rovisco Pais, P-1096 Lisbon Codex, PORTUGAL

NAESS Arvid Dr., Norwegian Hydrodynamic Laboratories, Pob 4118 Valentinlyst, N-7001 Trondheim, NORWAY

NARDONE Steven C. Eng., Bldg.1171 (1) Code 3512, U.S. Naval Underwater Syst. Center, Newport, RI. 02840, USA

NITHILA Markku T. Dr., Helsinki University of Techn., Control Eng. Laboratory, Otakaari 5A, SF--02150 Espoo 15, FINLAND

O'REILLY John J. Prof., Dept. of Electr. Eng. Science, University of Essex, Colchester CO4 3SQ, UNITED KINGDOM

ORTIGUEIRA Manuel Duarte Eng., CAPS-Complexo I, Instituto Superior Tecnico, Av. Rovisco Pais, P-1096 Lisbon Codex, PORTUGAL

PAGES Jaume Prof.,Facultat d'Informatica, Jordi Girona Salgado 31, Barcelona 34, SPAIN

PICCI Giorgio Prof., Dept. of Mathematics, Univ. of Kentucky, Lexington KY 40506 , USA

PRABHAKAR J. C. Prof.,California State Univ. Dept. of Electrical & Computer Eng., 18111 Nordhoff Street, Northridge, CA 91330, USA

PROTTER Philip Prof., Mathematics Dept., Purdue Univ., W. Lafayette, Indiana 47907, USA

REAL Jose Anguas Dr., Universidad de Sevilla, Facultat De Matematicas, c/Tarfia, Sevilla, SPAIN

RIBEIRO M. Isabel Eng., CAPS-Complexo I, Instituto Superior Tecnico, Av.Rovisco Pais,P-1096 Lisbon Codex, PORTUGAL
RUCKEBUSCH Guy Dr., Schlumberger - Doll Research, P.O. Box 307, Ridgefield CT 06877, USA
SARKAR Tapan Kumar Prof., Rochester Inst. of Techn., Dept. of Electrical Eng., Rochester, New York - 14623, USA
SHEPP Lawrence A. Prof.,Bell Laboratories Inc.,Murray Hill N. J. 07974, USA
SOUSA António Manuel M. Mr., IUTAD - Instituto Univ. Apt. 202, 5000 Vila Real, PORTUGAL
TALAY Denis Mr., Mathematiques Appliquees, Univ. de Provence, Place Victor Hugo, 13331 Marseille Cedex, FRANCE
TELES Maria Lucia Dr., Inst. de Fisica , Univ. S.Paulo Dept.Fisica Nuclear, Caixa Postal 20516, 01000 Sao Paulo SP, BRASIL
TRONSTAD Yngvar Cmmd.,Royal Norwegian Navy Mat.Command Bureau of Weapons, P.O.Box 3, 5078 Haakonsvern, NORWAY
USTUNEL A.S.Dr., CNET, 2 Bd. Auguste Blangui, Paris 75013, FRANCE
VAL João Bosco Ribeiro Mr.,Imperial College, Dept. Electr. Eng. (Control Sect.), London SW7 2AZ, UNITED KINGDOM
VELOSO M. Manuela Eng.,CAPS-Complexo I, Inst.Sup.Tecn. Av. Rovisco Pais, P-1096 Lisbon Codex PORTUGAL
WELLINGS Peter Mr., Imperial College Science Techn., Dept. of Electrical Eng.,Exhibition Road, London SW7 2AZ, UNITED KINGDOM
ZAPPA Giovani Dr., Istituto di Informatica e Sistemistica, Facolta di Ingegneria, Univ.Firenze, Via S.Marta 3, 50139 Firenze, ITALY

INDEX OF SUBJECTS

INDEX OF AUTHORS